WAVE PHENOMENA

WAVE PHENOMENA

DUDLEY H. TOWNE
Professor of Physics
Amherst College

Dover Publications, Inc.
New York

Published in Canada by General Publishing Company, Ltd., 30 Lesmill Road, Don Mills, Toronto, Ontario.
Published in the United Kingdom by Constable and Company, Ltd.

This Dover edition, first published in 1988, is an unabridged and corrected republication of the work first published by Addison-Wesley Publishing Company, Reading, MA, in 1967.

Manufactured in the United States of America
Dover Publications, Inc., 31 East 2nd Street, Mineola, N.Y. 11501

Library of Congress Cataloging-in-Publication Data

Towne, Dudley H.
Wave phenomena / Dudley H. Towne.
p. cm.
Reprint, with corrections. Originally published: Reading, Mass. : Addison-Wesley, 1967.
Includes bibliographical references and index.
ISBN 0-486-65818-X (pbk)
1. Wave mechanics. I. Title.
QC174.2.T66 1988
530.1′24—dc19

88-28206
CIP

Preface

The undergraduate curriculum in physics should be subjected to constant reexamination to determine whether it is giving the best foundation in what is considered to be the mainstream of present-day physics. Viewed in this light there are a number of topics in both optics and acoustics which are essentially technical in nature and are needed only by specialists in these fields. On the other hand, the intersection between these two classes of subject matter—the topics common to both optics and acoustics—constitutes a set of concepts and techniques which are central to advanced work in physics. For example, the study in the context of either optics or acoustics of the wave equation, wave propagation, boundary value problems, and normal modes is an excellent introduction to the mathematical techniques (and physical reasoning) of quantum mechanics. It is advantageous for the student to become gradually accustomed to the sophisticated language of "boundary conditions," "eigenfunctions," etc., in a physical context that is within his grasp.

The present text was written for a one-semester course (four credit-hours per week) required of all physics majors at Amherst College in the second semester of Junior year. However, more material is included in the text than can be covered in a one-semester course, so that variability in the selection of topics is possible. Complete coverage of all topics could easily occupy two full semesters. The course as taught at Amherst replaces separate elective courses in optics and acoustics. One advantage is the more efficient use of faculty time in offering one less course. It has also made it possible for the student to get essential material in both optics and acoustics at the expense of only one course instead of two. Such an economy is valuable in achieving the aim of providing adequate preparation for graduate work while still enabling the student to take advantage of outside courses in the liberal arts curriculum.

An essential prerequisite is the calculus up to the level of partial differentiation. It is also desirable to have had experience, through a prior course in mechanics, electricity and magnetism, or differential equations, with the representation of sinusoidal functions by means of the complex exponential. However, Appendix I can be used to introduce the properties of complex numbers which are required in the text. The discussion of electromagnetic waves is initiated by postulation of Maxwell's equations. A minimum of manipulation of these equations is required, and this is done in a way that can be followed by students who have not had a course in electricity and magnetism or vector calculus. Use of the vector calculus is not essential except in the final chapter on acoustic waves in three dimensions, thus permitting this material to be presented when advanced calculus is a corequisite.

The book stresses the mastery of certain fundamental mathematical techniques. A number of the problems are nontrivial and call for a full exercise of the kinds of physical and mathematical reasoning learned from a study of the text material. To permit this degree of participation on the part of the student, the first seven chapters deal only with one-dimensional waves (i.e. plane waves, normal incidence). Thus the calculations remain simple to perform at a time when the student is becoming acquainted with the use of complex numbers and the abstract approach to boundary value problems. This restriction to the one-dimensional case also permits the use of isomorphisms in the treatment of a wider range of physical situations than would be possible in higher dimensions. Appendix V describes a number of situations which give rise to equations of the same form as those describing transverse waves on a string and acoustic and electromagnetic plane waves, which are dealt with in the text.

Considerable emphasis is placed upon establishing a close connection between mathematical expressions and the associated physical ideas. The solutions for the acoustic plane wave, for example, are interpreted by verbal statements which are aimed at developing a clear picture of "what is going on in the medium." The standard approximations that must be made in developing certain equations are subjected to scrutiny to emphasize the limitations which they impose and the motivation for making the approximations. Thus, although the acoustic approximation ultimately washes out the distinction between the Eulerian and Lagrangian meanings of the variables, basic considerations such as mass conservation, energy flow, and Newton's second law cannot be made clear unless one or the other formulation is consistently adopted. This avoids sloppy physics and makes it possible to point specifically to the kind of mathematical difficulties which arise if subsequent approximations are not made.

Intermixed with the analytical chapters are others of a more descriptive type. The inclusion of this material represents a compromise to the total discarding of subjects which are secondary to the central purpose of the text.

It seems reasonable to suppose that there are many things a student should know about without taking the time to follow their development in detail. The treatment of such material that the student will find in other texts is in most cases too lengthy to be read for simple informational content. The abbreviated treatment presented here is not intended to be complete, but is given to enable the student to become aware of a variety of topics which branch out from the system of ideas being examined. Unless the instructor wishes to enlarge upon this material, the chapters in question can be given as assigned reading upon which little class time will be spent. A suggested way of having the student pursue some of this auxiliary material further is to assign a term paper on a topic of his choice. The specific chapters of relatively discursive nature are: Chapter 5 (Experimental Aspect of Acoustics); Chapter 6 (Sections 6–3 through 6–5 on the Electron Theory of Matter and Dispersion Theory); Chapter 10 (Additional Optical Properties of Matter). Other descriptive material which can be assigned as reading chiefly in connection with the laboratory work is to be found in: Chapter 8 (The Production and Detection of Linearly Polarized Light); Chapter 9 (The Production and Detection of Elliptically Polarized Light); Chapter 11 (Sections 11–8 through 11–15 on practical details in Young's experiment, reflection from thin films, and the Michelson interferometer).

Interference and diffraction effects are approached semi-empirically and are not derived as analytical consequences of the wave equation. It is pointed out that the Huygens-Fresnel principle was initially a postulate, but can now be shown to follow from the wave equation. An empirical summary of the behavior of retardation plates is given in the first part of Chapter 9 so that experiments with polarized light can be understood without a description of the wave surfaces within a crystal medium. This description is reserved for the latter part of the chapter and would probably be omitted in a one-semester course. A similar grading of subject matter is followed in other chapters. Specialized topics and detailed elaborations are placed at the ends of the chapters so that they are available for interested students or for use in a two-semester version of the course.

Entire chapters which are optional in a one-semester course are: Chapter 13 (Fresnel Diffraction); Chapter 16 (Waves in a Dispersive Medium); Chapter 17 (The Acoustic Wave Equation in Three Dimensions).

The first part of Chapter 15 contains important material dealing with the concept of the normal modes of a vibrating string. The approach is analytical and makes use of language and techniques characteristic of a broad class of eigenvalue problems, such as those of quantum mechanics. The significance of the analytical results is thoroughly comprehensible, since the concept of normal modes is usually presented by means of lecture demonstrations in introductory courses. When the student later confronts quantum mechanics he can then concentrate his attention on the novel aspects of physical interpretation without being distracted by what seem to be bizarre mathematical techniques. Chapter 15 can be taken up at any time after Chapter 4. It is given this placement in the text to allow for the development of material pertinent to the laboratory work. The spacing out of the relatively more abstract material also accords with the fact that at this point in his career a student is maturing rapidly in his ability to handle a more sophisticated mathematical approach. The sections after Section 15–11 may be considered optional in a one-semester course.

The writing of this text was inspired by the example of many authors whose works convey a depth of insight. Most notable among these are Arnold Sommerfeld, Lord Rayleigh, Sir Horace Lamb, Max Born, Robert W. Wood, Richard C. Tolman, and Mark W. Zemansky. The author is grateful to many students for helping to shape the ideas in this text and to Miss Alice Russell for typing voluminous class notes. Special acknowledgment is due to Mr. Jay Atlas for careful criticism and proofreading of the final manuscript.

Amherst, Mass.
January 13, 1967

D.H.T.

Contents

WAVE PHENOMENA

1 Transverse waves on a string

1-1 INTRODUCTION

Elementary definitions of the term *wave* usually consist of a series of examples of situations which are referred to as waves. In addition it may be stated that one feature common to all waves is the propagation of some kind of "disturbance" (elevation of the water surface in the case of water waves or pressure variations in the case of sound) with a velocity characteristic of the medium. A more satisfactory definition is not available without the use of mathematics, primarily the concept of partial differentiation, which goes beyond the level of most introductory texts. In fact, what is required for a physical situation to be referred to as a wave, is that its mathematical representation give rise to a partial differential equation of particular form, known as *the wave equation*. Many of the properties of waves can be learned most simply by experimental demonstration rather than through manipulation of the mathematics associated with the wave equation. However, it is only if the latter approach is used that one can be sure he is dealing with a property common to all wave situations and not special to the particular kind of wave exhibited. Furthermore, it is desirable that one's first acquaintance with the more abstract approach be at the level of *simple* phenomena so the language and techniques will be familiar when applied in a more advanced context.

As our first example of a wave we consider the motion which results when a string under tension is given a transverse displacement. We shall deal with an "ideal" string subject to certain simplifying assumptions concerning the nature of its motion. In principle, we ought to examine the motivation for and the limitations imposed by each of these assumptions. However, our major interest lies in showing in bold outline strokes how the wave equation originates and what properties may be deduced from it. We are not interested at this point in describing the most general possible motion of a real string.*

* For a more complete derivation see J. B. Keller, "Large Amplitude Motion of a String," *Am. J. Phys.* **27**, 584 (1959).

1-2 DERIVATION OF THE WAVE EQUATION

Consider a string of indefinite length which in the undisturbed condition coincides with the x-axis. Restrict the possible motions of the string to those cases in which the string at all times lies in a fixed plane. Choose a y-axis lying in this plane. Further assume that the displacements of each particle of the string are *wholly transverse*. That is to say, assume that the x-coordinate of a given particle of the string does not change with time and that the only displacements the particle undergoes are parallel to the y-axis. The motion of the string can then be specified by a function $y(x, t)$ which is a function of the two independent variables x and t. Thus, for example, the graph of y versus x determined by the equation $y = y(x, t_1)$ depicts the *shape* of the string at the fixed time t_1 (see Fig. 1-1), and the graph of y versus t determined by the equation $y = y(x_1, t)$ specifies the transverse displacement of the single particle located at x_1 as a function of the time. The latter graph is sometimes referred to as the *history* of the particle at x_1. Note that $(\partial y/\partial t)(x_1, t)$ and $(\partial^2 y/\partial t^2)(x_1, t)$ are the instantaneous velocity and acceleration, respectively, of this particle.* Assume that the string in the undisturbed configuration has a uniform linear mass density, σ, and is under a uniform tension T. Also assume that any changes in either of these quantities which occur during the motion of the string are sufficiently small so that they may be neglected.

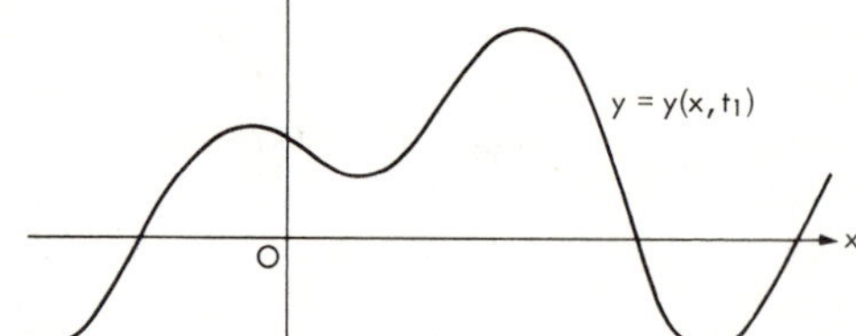

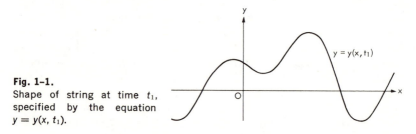

Fig. 1-1. Shape of string at time t_1, specified by the equation $y = y(x, t_1)$.

We are now in a position to apply Newton's second law to analyze the motion of that portion of the string which lies between x and $x' = x + \Delta x$ (Fig. 1-2). Although the tension is uniform in *magnitude* throughout the string, the *direction* along which it acts, being along the tangent to the string, is different at the two ends of the segment. Let $\theta = \theta(x, t)$ designate the inclination of the tangent at x and $\theta' = \theta(x', t)$ the inclination at x'. The

---

* When it is important to indicate that a specific substitution is being made for the arguments of a partial derivative, we shall use the notation $(\partial y/\partial x)(x_1, t_1)$ in formulas appearing in a line of text and

$$\frac{\partial y}{\partial x}(x_1, t_1)$$

in displayed formulas.

tension contributes a net transverse force on the given segment expressed by

$$T \sin \theta' - T \sin \theta = \Delta(T \sin \theta). \tag{1-1}$$

We assume that this is the only significant force having a transverse component. Thus friction is assumed negligible; gravity and forces due to the air surrounding the string are either absent altogether or, if present, much smaller in order of magnitude than the force given by Eq. (1-1).

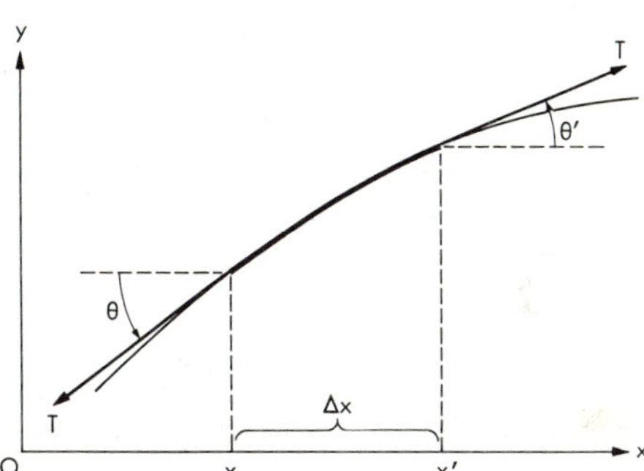

Fig. 1-2.
Isolated portion of a string.

The mass of the segment is $\sigma\,\Delta x$ and the acceleration of its center of mass may be represented by $(\partial^2 y/\partial t^2)(\bar{x}, t)$, where $\bar{x}$ is some value of x intermediate between x and x'. By Newton's second law we therefore obtain

$$\sigma\,\Delta x\, \frac{\partial^2 y}{\partial t^2}(\bar{x}, t) = \Delta(T \sin \theta). \tag{1-2}$$

We may now divide by Δx and take the limit as Δx approaches zero. The coordinate of the center of mass, $\bar{x}$, then approaches x and, since we assume that $y(x, t)$ and hence $(\partial^2 y/\partial t^2)(x, t)$ are continuous functions of x, the left-hand side becomes $\sigma(\partial^2 y/\partial t^2)(x, t)$. On the right-hand side

$$\lim_{\Delta x \to 0} \left\{ \frac{\Delta(T \sin \theta)}{\Delta x} \right\}$$

can be recognized as the definition of the partial derivative of the function $T \sin \theta$ with respect to x. The dynamics of the motion of the string thus requires that the functions $y(x, t)$ and $\theta(x, t)$ be connected by the relation

$$\sigma \frac{\partial^2 y}{\partial t^2} = \frac{\partial}{\partial x}(T \sin \theta). \tag{1-3}$$

To simplify this result further we now make the assumption that the motion of the string is such that the inclination angle θ is always small compared with one radian. In that case $\sin \theta \doteq \tan \theta$. But $\tan \theta$ is simply the slope of the string, which may also be designated as $\partial y/\partial x$. The right-hand side of (1-3) then becomes

$$\frac{\partial}{\partial x}\left(T \frac{\partial y}{\partial x}\right) = T \frac{\partial^2 y}{\partial x^2},$$

since the tension T is a constant. The resultant equation may then be written

$$\frac{1}{c^2}\frac{\partial^2 y}{\partial t^2} = \frac{\partial^2 y}{\partial x^2}, \tag{1-4}$$

where c^2 is used to designate the combination of constants T/σ. Thus, if Newton's second law is to be satisfied, the function $y(x, t)$ cannot be an arbitrary function of x and t, but is subject to a restriction imposed by Eq. (1–4). This condition is a linear second-order partial differential equation which has the form referred to as *the one-dimensional wave equation.* The equation is a valid representation of the physical conditions in the system only so long as the inclination of the string remains everywhere small. Since this implies that the slope of the string must also be small, we shall refer to this restriction as *the small slopes approximation.**

1-3 SOLUTION OF THE ONE-DIMENSIONAL WAVE EQUATION

Before tackling the problem of the most general type of function $y(x, t)$ which will satisfy Eq. (1–4), it is profitable to examine a special form which, as can easily be checked, is indeed a solution. Consider $f(u)$, an arbitrary function f of the single argument u. Assume that f is a twice-differentiable function of its argument, and designate the first and second derivatives by $f'(u)$ and $f''(u)$ respectively. From such a function we can form a function y of the two variables x and t by the prescription

$$y(x, t) = f(x - ct). \tag{1-5}$$

It is now a simple matter to prove that this function satisfies the conditions of Eq. (1–4). By the "chain rule" for differentiating a function-of-a-function form,† we have

$$\frac{\partial y}{\partial x} = f'(x - ct), \qquad \frac{\partial^2 y}{\partial x^2} = f''(x - ct),$$

$$\frac{\partial y}{\partial t} = -cf'(x - ct), \qquad \frac{\partial^2 y}{\partial t^2} = c^2 f''(x - ct).$$

Comparison of the two right-hand equations shows that the function defined by Eq. (1–5) satisfies Eq. (1–4) identically and allows f to be an arbitrary twice-differentiable function.

---

* Symbolically, this will be written $\partial y/\partial x \ll 1$. The notation $a \ll b$ means "$|a|$ is very much less than $|b|$."

† Thus, for example, with $y = f(u)$ and $u = x - ct$: $\partial y/\partial t = (df/du)(\partial u/\partial t) = -cf'(u)$.

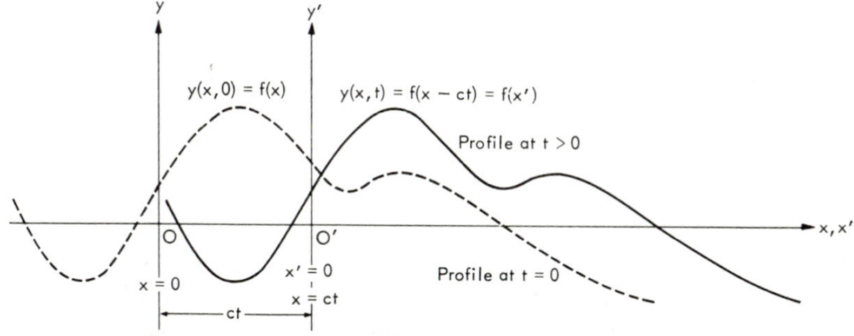

Fig. 1–3. Progressive waveform represented by $f(x - ct)$.

What is the meaning of a solution having the form of Eq. (1–5)? In asking this question we are not yet inquiring into the more difficult question, "How must the string be treated in order that a solution of this form will prevail?" Instead we first ask the simpler question: "Suppose the string *is* moving in a manner described by a solution of this form, is there anything special we can say about the appearance of the string?" At $t = 0$ the profile of the string is given by $y = y(x, 0) = f(x)$. The function f therefore specifies the initial shape of the string. But the graph of $y = f(x - ct)$ differs from the graph of $y = f(x)$ merely by a translation parallel to the x-axis. For, if we shift the origin on the x-axis through the change of variables $x' = x - ct$, the graph of $f(x - ct) = f(x')$ bears the same relation to the new origin located at $x' = 0$ (or $x = ct$) that the graph of $f(x)$ bears to the old origin at $x = 0$ (Fig. 1–3). This implies that the profile of the string at $t > 0$ is obtained from that at $t = 0$ by a translation in the positive x-direction by an amount ct. As time progresses, then, Eq. (1–5) represents a shape which moves without distortion in form in the positive x-direction with velocity c. We may refer to this as the propagation of a *waveform* with *wave propagation velocity c*.

In summary of Section 1–3, we have shown that given any twice-differentiable function $f(u)$, the function $y(x, t) = f(x - ct)$ is a formal solution to the one-dimensional wave equation (1–4), and represents a waveform propagating without change of shape in the positive x-direction with constant velocity c. We shall refer to such a situation as a $+$*wave*.

1–4 WAVE PROPAGATION VELOCITY ON A STRING

Section 1–3 contains statements which apply in general to any equation having the form of the one-dimensional wave equation. Since we have seen that the transverse motion of the particles of a string is governed by an equation of this form, we infer the possibility of propagation of a waveform on a string with propagation velocity $c = \sqrt{T/\sigma}$. Note carefully that the wave propagation velocity is not to be confused with the transverse velocity of the individual particles of the string. The latter is designated by $\partial y/\partial t$. In a +wave given by $y = f(x - ct)$ we have

$$\frac{\partial y}{\partial t} = -cf'(x - ct) = -c\left(\frac{\partial y}{\partial x}\right).$$

Since we required $\partial y/\partial x \ll 1$, we find that the small slopes approximation implies $\partial y/\partial t \ll c$. That is, the transverse particle velocity is necessarily small in the type of wave we are considering compared with the wave propagation velocity. The particles of the string are moving in a transverse direction, whereas the waveform propagates along the string. The "propagation" is one of form, but not of substance. (We shall see later, however, that there is a transport of energy and momentum which coincides with the propagation of the waveform.)

The predicted dependence of the wave propagation velocity on the tension and linear mass density ($c = \sqrt{T/\sigma}$) can be demonstrated in a number of ways:

a) Suspend a piece of rubber tubing between two fixed posts so that the velocity with which a small pulse moves back and forth can be observed. As the tubing is pulled tighter so as to increase T (and also somewhat decrease σ) the velocity increases.

b) A second piece of rubber tubing can be loaded with shot so as to increase the value of σ. When compared with an unloaded tube under approximately the same tension the heavier tube propagates pulses at a markedly slower velocity.

c) Load the second tube so that $\sigma_2 = 4\sigma_1$ and suspend both tubes under equal tensions between the same pair of fixed posts. The fundamental modes of vibration (with which it is assumed the reader has some familiarity from an elementary course) should now have the frequency ratio of 2:1. It can easily be seen that the lighter tube completes two of its fundamental vibrations while the heavier tube completes just one. Alternatively, the first overtone of the heavier tube is synchronous with the fundamental of the lighter tube.

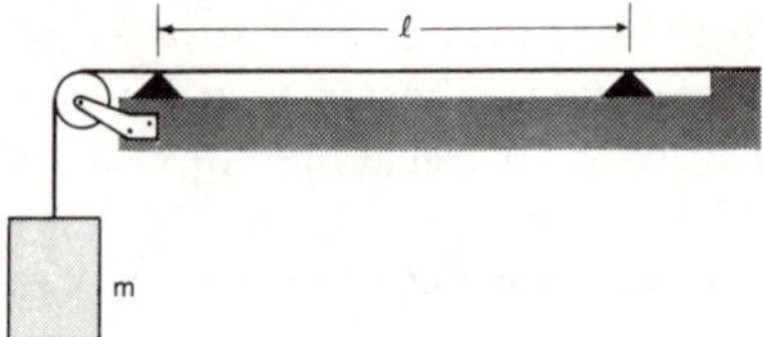

Fig. 1-4. Sonometer.

d) In a sonometer (Fig. 1-4), the tension of a piece of wire is measured by the suspended weight. Quantitative tests of the relation $c \propto \sqrt{T}$ can be made by tuning the wire to different frequencies by comparison with a set of tuning forks.

1-5 THE MOST GENERAL SOLUTION TO THE ONE-DIMENSIONAL WAVE EQUATION

It is easy to verify that another form of solution to Eq. (1–4) can be obtained from an arbitrary* function $g(u)$ by the prescription

$$y(x, t) = g(x + ct). \tag{1–6}$$

By similar reasoning to that employed in Section 1–3, this solution can be seen to represent a propagation with constant velocity c in the *negative* x-direction, of a wave of constant shape given initially by the graph of $y = g(x)$. Such a situation is referred to as a $-$*wave*. It is also clear that any function of the form

$$y(x, t) = f(x - ct) + g(x + ct), \tag{1–7}$$

where f and g are unrelated arbitrary functions, will surely satisfy Eq. (1–4). That is to say, Eq. (1–7) is a *sufficient* form for a solution to the wave equation. This follows from the *linearity* of Eq. (1–4). The sum of any two solutions yields a new function which will also satisfy the wave equation. To show this, let the wave equation be written symbolically as $\Box^2 y = 0$, where $\Box^2$ is the *D'Alembertian operator*,

$$\Box^2 \equiv \frac{\partial^2}{\partial x^2} - \frac{1}{c^2}\frac{\partial^2}{\partial t^2}.$$

The D'Alembertian operator distributes over a sum of functions. That is, $\Box^2(y_1 + y_2) = \Box^2 y_1 + \Box^2 y_2$, where y_1 and y_2 are any functions. Thus, in general, if both y_1 and y_2 are solutions of the wave equation, $\Box^2 y_1 = 0$ and $\Box^2 y_2 = 0$, and hence $\Box^2(y_1 + y_2) = 0$, or $y_1 + y_2$ is also a solution to the wave equation.

What we wish to show now is that Eq. (1–7) is a *necessary* form for a solution to the wave equation. That is to say, there are no solutions of the wave equation which are not of this form. The proof consists of changing variables in Eq. (1–4) to reduce it to a simpler form from which the general integral can be recognized even by those who have no previous experience with partial differential equations. Consider a change in the independent variables (x, t) to a set (u, v) defined by the equations $u = x - ct$, $v = x + ct$. The dependent variable which was previously given by the functional form $y(x, t)$ produces a new functional form $Y(u, v)$ of the variables u, v. The connection is given by the relation

$$y(x, t) = Y[u(x, t), v(x, t)]. \tag{1–8}$$

---

* Arbitrary except for the requirement that it be twice-differentiable with respect to its argument. This condition will be presupposed for both the functions f and g throughout the following.

Thus
$$\frac{\partial y}{\partial x} = \frac{\partial Y}{\partial u}\frac{\partial u}{\partial x} + \frac{\partial Y}{\partial v}\frac{\partial v}{\partial x} = \frac{\partial Y}{\partial u} + \frac{\partial Y}{\partial v}$$
and
$$\frac{\partial^2 y}{\partial x^2} = \left\{\frac{\partial}{\partial u}\left[\frac{\partial Y}{\partial u} + \frac{\partial Y}{\partial v}\right]\right\}\frac{\partial u}{\partial x} + \left\{\frac{\partial}{\partial v}\left[\frac{\partial Y}{\partial u} + \frac{\partial Y}{\partial v}\right]\right\}\frac{\partial v}{\partial x}$$
$$= \frac{\partial^2 Y}{\partial u^2} + 2\frac{\partial^2 Y}{\partial u\,\partial v} + \frac{\partial^2 Y}{\partial v^2}. \tag{1–9}$$

Likewise, since $\partial u/\partial t = -c$ and $\partial v/\partial t = +c$, we have
$$\frac{\partial y}{\partial t} = -c\frac{\partial Y}{\partial u} + c\frac{\partial Y}{\partial v},$$
and
$$\frac{\partial^2 y}{\partial t^2} = -c\frac{\partial}{\partial u}\left[-c\frac{\partial Y}{\partial u} + c\frac{\partial Y}{\partial v}\right] + c\frac{\partial}{\partial v}\left[-c\frac{\partial Y}{\partial u} + c\frac{\partial Y}{\partial v}\right]$$
$$= c^2\left\{\frac{\partial^2 Y}{\partial u^2} - 2\frac{\partial^2 Y}{\partial u\,\partial v} + \frac{\partial^2 Y}{\partial v^2}\right\}. \tag{1–10}$$

Direct substitution from (1–9) and (1–10) into (1–4) shows that the wave equation is equivalent to the equation
$$\frac{\partial^2 Y}{\partial u^2} - 2\frac{\partial^2 Y}{\partial u\,\partial v} + \frac{\partial^2 Y}{\partial v^2} = \frac{\partial^2 Y}{\partial u^2} + 2\frac{\partial^2 Y}{\partial u\,\partial v} + \frac{\partial^2 Y}{\partial v^2},$$
or
$$\frac{\partial^2 Y}{\partial u\,\partial v} = 0. \tag{1–11}$$

Now Eq. (1–11) may be written in the form
$$\frac{\partial}{\partial u}\left(\frac{\partial Y}{\partial v}\right) = 0,$$
which shows that $\partial Y/\partial v$ is not a function of u, but may be an arbitrary function of v, say
$$\frac{\partial Y}{\partial v} = h(v). \tag{1–12}$$

Equation (1–12) can be integrated directly providing that the "constant" of integration is considered to be an arbitrary function of u, say $f(u)$. Thus, $Y(u, v) = \int h(v)\,dv + f(u)$. However, since $h(v)$ is an arbitrary function of v, its indefinite integral is some arbitrary function of v which we shall designate as $g(v)$. The general solution to Eq. (1–11), which we have seen is equivalent to the original wave equation, is then $Y(u, v) = f(u) + g(v)$. Now expressing this in terms of x, t as prescribed by Eq. (1–8), we have
$$y(x, t) = f(x - ct) + g(x + ct). \tag{1–13}$$

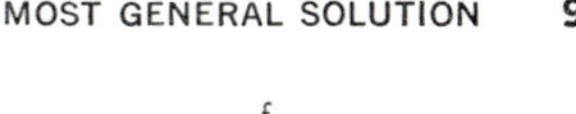

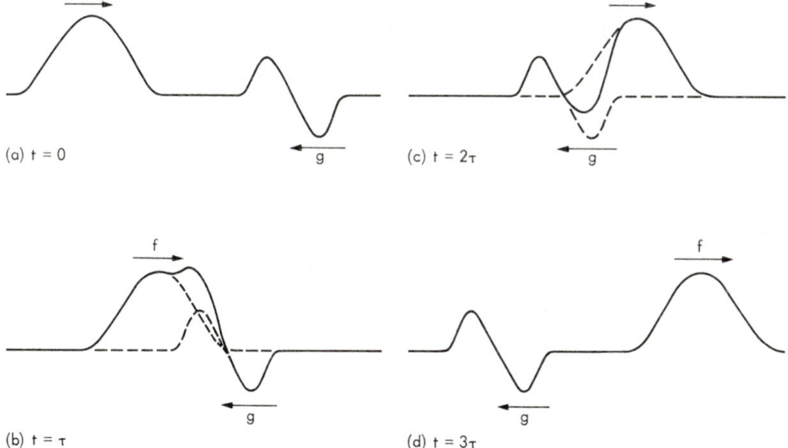

Fig. 1-5. Typical shapes produced by a superposition of + and − waves.

The most general form of solution to the one-dimensional wave equation therefore consists of a superposition of + and −waves of arbitrary form. If either f or g is zero, the wave is referred to as a *progressive* wave, i.e., either a pure − or a pure +wave respectively. If neither f nor g is zero, the string as a whole does not remain unchanged in shape. The successive forms of the string are to be obtained by locating the + and −waves in the appropriate position for the given time and then adding them together (see Fig. 1–5). If f and g are both waveforms of finite duration (i.e., which vanish identically outside of a given interval), the + and −waves at first progress without change in form. There is then a period of overlap during which the original forms are not apparent, and finally a separation of the two waveforms each of which is unchanged from its original shape. The fact that these two waveforms can "pass through" one another in this way and reappear without distortion is associated with the linearity of the wave equation. If, for example, we were dealing with a situation in which the small slopes approximation could not be made, the differential equation governing the motion would turn out to be nonlinear. We would find that progressive waves traveling in opposite directions are changed in shape after passing through one another, the modifications depending in a complicated way upon the original shapes of both.

Since we have succeeded in finding the general form of solution to the one-dimensional wave equation, a great deal of the subsequent discussion will proceed directly from Eq. (1–13). There will be little need to consult the wave equation itself, since all of the information contained therein is expressed by the general solution. We are not as fortunate in the case of the two- and three-dimensional wave equations, for which such a convenient form of general solution does not exist.

1-6 KINEMATICS ASSOCIATED WITH THE WAVEFORM

A certain amount of information about the instantaneous state of motion of a string can be obtained just from a knowledge of its shape at that instant. The wave equation shows that the instantaneous acceleration of a point on the string is proportional to $\partial^2 y/\partial x^2$. The latter can be associated with the *curvature* of the string at this point. The precise expression for the curvature is

$$\frac{\partial^2 y/\partial x^2}{[1 + (\partial y/\partial x)^2]^{3/2}}$$

but this reduces to $\partial^2 y/\partial x^2$ in the small slopes approximation, $\partial y/\partial x \ll 1$. Thus there is a positive acceleration wherever the curve is concave *upward* ($\partial^2 y/\partial x^2 > 0$) regardless of the signs of y and $\partial y/\partial t$. The instantaneous acceleration is zero at a point of inflection ($\partial^2 y/\partial x^2 = 0$). The forces acting at the two ends of a segment of the string about such a point are oppositely directed and produce a zero resultant.

The remaining associations are restricted to cases in which we know that we are dealing with a progressive wave. As a matter of convenience in notation, a function of the form $f(x - ct)$ which is associated with a progressive +wave will be designated by $y_+(x, t) = f(x - ct)$. Likewise for the function associated with a −wave: $y_-(x, t) = g(x + ct)$. Differentiation by the chain rule establishes the identities

$$\frac{\partial y_\pm}{\partial t} = \mp c\,\frac{\partial y_\pm}{\partial x}\cdot \tag{1-14}$$

Thus for either of the special cases of progressive waves, the transverse particle velocity is directly proportional to the slope of the string. For the +wave the transverse velocity is negative where the slope is positive and conversely. We can see from this how the string is "preparing itself" for the advance of the waveform. In Fig. 1-6 the instantaneous velocity at various points is rep-

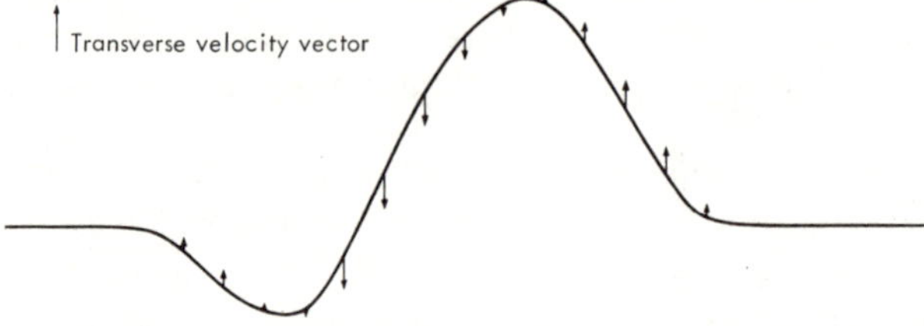

Fig. 1-6. Proportionality of transverse velocity to slope in a + wave.

resented by arrows. The tips of the arrows outline the position of the string a short interval of time later than the instant shown. We see from this diagram that the relation expressed by Eq. (1–14), which was derived formally from the mathematics, makes good sense physically, for it shows that the particles are moving up or down with appropriate velocities to allow the waveform to progress in the positive direction. The particles at a crest or a trough of course have zero instantaneous velocity, since they are reversing their direction of motion.

1–7 DESCRIPTION OF A SINUSOIDAL PROGRESSIVE WAVE

Contrary to the impression which may be created by the fact that waves of sinusoidal form are the most frequently cited examples of waves, it is to be noted that neither *wavelength* nor *periodicity* is an essential characteristic of a wave. An initial waveform of any desired shape determines an allowable solution to the wave equation. There are, however, several reasons that waves of sinusoidal form play an important role in the study of wave phenomena. In the first place, the sine function is one of the simplest analytic functions which is bounded on an infinite interval. Other simple waveforms can be constructed using sections of polynomials (see Example 2, Section 1–8), but the piecewise character of the definition makes the situation somewhat more complicated to handle. The sine function is also of importance because of its relation to the pure tone in sound and the spectral color in light. Furthermore, sinusoidal waves give rise to interference effects which are of considerable practical importance and they will be the subject of later chapters. In some contexts, involving the phenomenon of *dispersion*, the wave equation is not satisfied *except* for waves of sinusoidal form and then only when a wave propagation velocity appropriate to the given frequency is substituted. Finally, the sinusoidal wave achieves an overwhelming importance through Fourier's theorem (Chapter 15), which shows that a pulse of arbitrary form can be thought of as a superposition of a number of waves of sinusoidal form. It follows from this that many properties which can be demonstrated for a sinusoidal wave and are independent of its frequency will also be properties of a waveform of arbitrary shape.

An initial waveform which is sinusoidal is described by the function*

$$y(x, 0) = \mathcal{R}e\,\{\mathbf{y}e^{ikx}\}, \tag{1–15}$$

---

* See Appendix I for a discussion of the use of the complex exponential to represent sinusoidal functions. The applications made in Chapter 1 are simple enough so that it would be easier to use the cosine function directly rather than the complex exponential. (The student is advised to see for himself by using this alternative notation.) Since the complex exponential notation is ultimately the most advantageous, it is introduced at this point so the student can begin to become familiar with it.

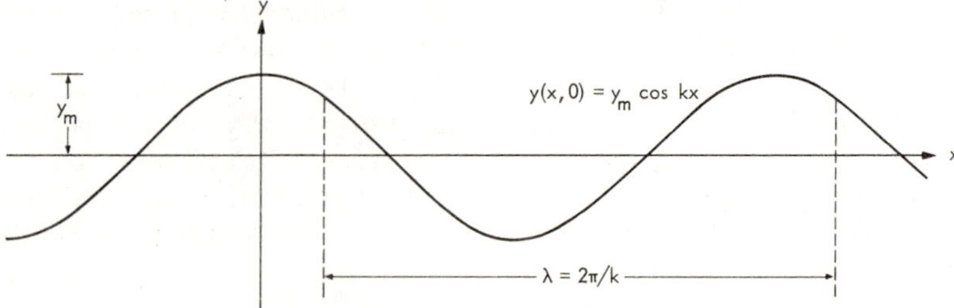

Fig. 1-7. Sinusoidal waveform, $\phi = 0$.

where k and $\mathbf{y}$ are given constants. If the *complex amplitude* $\mathbf{y}$ is written in the form $\mathbf{y} = y_m e^{i\phi}$, the *real amplitude* y_m determines the maximum displacement of the string from the x-axis and the phase ϕ determines the position of the curve with respect to translation parallel to the x-axis. (The case shown in Fig. 1-7 is for $\phi = 0$.) When the quantity (kx) increases by 2π, the function $y(x, 0)$ goes through one cycle of its values. The corresponding increment in x is, of course, the wavelength, λ. Thus $k(x + \lambda) = kx + 2\pi$, or $k\lambda = 2\pi$, $k = 2\pi/\lambda$. The parameter k is referred to as the *wave number* and may be thought of as the number of waves contained in a distance of 2π meters.

Let us endow this initial waveform with motion in the positive x-direction. This implies that we are taking $g = 0$ in Eq. (1-13) and $f(x) = y(x, 0)$. Then the future motion of the string is described completely by the function

$$y(x, t) = f(x - ct) = \mathcal{R}e\,\{\mathbf{y}e^{ik(x-ct)}\}. \tag{1-16}$$

Note that if we focus attention upon any particular particle of the string, say x_1, the motion of this particle as a function of the time is sinusoidal.* This can be seen by letting $\mathbf{y}e^{ikx_1} = y_m e^{i(kx_1+\phi)} = y_m e^{i\phi'}$, where $\phi' = \phi + kx_1$. Equation (1-16) then becomes $y(x_1, t) = y_m \cos[kct - \phi']$. This function is a periodic function of the time with a period T defined by the relation $kcT = 2\pi$. Thus the frequency $\nu = 1/T = kc/2\pi$ and the associated angular frequency $\omega = 2\pi\nu = kc$. The relation $\omega = kc$ is of course equivalent to the familiar $c = \lambda\nu$, but since it will be more convenient to use angular frequency and wave number as the fundamental parameters rather than frequency and wavelength, the relation will bear memorization in its new form. In this notation the sinusoidal +wave is therefore written

$$y(x, t) = \mathcal{R}e\,\{\mathbf{y}e^{i(kx-\omega t)}\}. \tag{1-17}$$

---

* This illustrates the general proposition that in any progressive wave the curve which describes the history of a single particle is of the same geometric form as the wave profile of the string.

It is pertinent to inquire what restriction is imposed upon our choice of parameters in the function by the requirement of small slopes. By differentiation,

$$\frac{\partial y}{\partial x} = \mathcal{R}e\,\{ik\mathbf{y}e^{i(kx-\omega t)}\}$$

and the requirement $\partial y/\partial x \ll 1$ implies that the amplitude of $\partial y/\partial x$ must be small ($\partial y/\partial x$ is itself a sinusoidal function), that is

$$|ik\mathbf{y}| = ky_m \ll 1 \qquad \text{or} \qquad y_m \ll \lambda/2\pi. \tag{1-18}$$

Thus for a sinusoidal wave to be properly described by the one-dimensional wave equation the amplitude must be quite small in comparison with the wavelength. In the context of sinusoidal waves the small slopes approximation is often referred to as the *small amplitudes* approximation.

We conclude the discussion of the sinusoidal +wave with the remark that since each point on the string is in sinusoidal motion, the function in Eq. (1–17) naturally satisfies the identity $\partial^2 y/\partial t^2 = -\omega^2 y$; that is to say, the acceleration of each particle is at all times proportional to its displacement from the equilibrium position. This invites comparison with an example such as the simple harmonic oscillation of a mass on the end of a spring. There is an important distinction between the two cases. In the case of the oscillator there is a physical mechanism which requires the restoring force to be proportional to the displacement regardless of the state of motion. In the case of the string carrying a sinusoidal +wave, the individual particle is not in fact "attached" to an equilibrium position. It just so happens that in this state of motion the neighboring elements of the string exert forces in such a way that the net force and the displacement remain proportional. The frequency is determined solely by the wavelength of the passing wave and may be chosen arbitrarily. The frequency of the simple harmonic motion of an oscillator does not have this arbitrariness.

*1-8 INITIAL CONDITIONS APPLIED TO THE CASE OF A STRING OF INFINITE LENGTH

The one-dimensional wave equation which describes the motion of the infinite set of particles on a string is a second-order *partial* differential equation which contains two arbitrary *functions* in its general solution. This is somewhat analogous to the fact that the equation describing the motion of a single mass point is a second-order *ordinary* differential equation whose general solution contains two arbitrary *constants*. In the latter case the initial position

* May be omitted without loss of continuity.

and velocity of the mass point determine unique values of the constants. In a similar fashion we shall show that a knowledge of the initial position of all the particles on the string (i.e. the initial wave profile) and their initial velocities (the velocity profile) enables us to uniquely determine the functions f and g in Eq. (1–13). Let us assume that the functions $y(x, 0)$ and $(\partial y/\partial t)(x, 0)$ are both given functions, say $\phi(x)$ and $\psi(x)$ respectively. From Eq. (1–13) we have $y(x, t) = f(x - ct) + g(x + ct)$ and $(\partial y/\partial t)(x, t) = -cf'(x - ct) + cg'(x + ct)$. Applying the given initial conditions, we therefore have

$$\phi(x) = f(x) + g(x), \tag{1–19}$$

$$\psi(x) = -cf'(x) + cg'(x). \tag{1–20}$$

Since $\psi(x)$ is a *given* function we may integrate Eq. (1–20). Let

$$\frac{1}{c}\int_0^x \psi(x)\,dx = \chi(x). \tag{1–21}$$

Then (1–20) becomes

$$\chi(x) = -f(x) + g(x) + C, \tag{1–22}$$

where C is a constant of integration which will soon be seen to be irrelevant. Now Eqs. (1–19) and (1–22) are a pair of equations in the desired functions f and g. Elimination between these equations leads to

$$\begin{aligned} f(x) &= \tfrac{1}{2}\{\phi(x) - \chi(x) + C\}, \\ g(x) &= \tfrac{1}{2}\{\phi(x) + \chi(x) - C\}. \end{aligned} \tag{1–23}$$

The unique solution to the wave equation subject to the given initial conditions is therefore

$$\begin{aligned} y(x, t) &= f(x - ct) + g(x + ct) \\ &= \tfrac{1}{2}[\phi(x + ct) + \phi(x - ct) + \chi(x + ct) - \chi(x - ct)]. \end{aligned} \tag{1–24}$$

The arbitrary constant C has cancelled out in this last step and has no relevance to the quantity which is directly observable, namely $y(x, t)$. The + and − components themselves are not directly observable, since all we can see at any time is the net shape of the string, obtained by their superposition. In any situation we can, without changing the physical situation, add an arbitrary constant to the function describing the +wave, providing we subtract the same value from the −wave. Since the value of C is not physically relevant, we may therefore choose C equal to zero and consider that Eq. (1–23) uniquely determines the functions f and g.

EXAMPLE 1. Suppose the string is given an initial form $y(x, 0) = \phi(x)$ by being held at rest in a template. At $t = 0$ the template is suddenly removed and the string is free to move. Determine the future motion of the string.

We might expect from the symmetry of this situation that the + and −waves will both be of the same form, each contributing one-half of the initial waveform. Turning to our formal solution, we have $\psi(x) = 0$ and hence $\chi(x) = 0$. Thus from Eq. (1–23) we have $f(x) = \frac{1}{2}\phi(x)$ and $g(x) = \frac{1}{2}\phi(x)$, which confirms our expectation.

EXAMPLE 2. A string which is initially straight is struck with a broad mallet of width a. The mallet imparts an initial velocity v_0 to the particles which lie between $x = 0$ and $x = a$. Determine the future motion of the string.

We are given $\phi(x) = 0$ and

$$\psi(x) = \begin{cases} 0, & x < 0, \\ v_0, & 0 \leq x \leq a, \\ 0, & x > a. \end{cases}$$

Integrating the expression for $\psi(x)$, we find

$$\chi(x) = \begin{cases} 0, & x < 0, \\ v_0 x/c, & 0 \leq x \leq a, \\ v_0 a/c, & x > a. \end{cases}$$

From Eq. (1–23) we therefore find for the shape of the +wave component:

$$f(x) = \begin{cases} 0, & x < 0, \\ -v_0 x/2c, & 0 \leq x \leq a, \\ -v_0 a/2c, & x > a. \end{cases}$$

The initial form of the −wave is given by $g(x) = -f(x)$. The shape of the string at various later times can be obtained by shifting the + and −waves an appropriate distance and adding graphically (Fig. 1–8). The ultimate effect of this action is seen to be a shifting upward of the entire string through a distance $v_0 a/2c$.

This example provides an interesting case in which to check the law of conservation of momentum. Initially a section of string of mass σa is moving upward with velocity v_0, giving a total momentum of $\sigma a v_0$. Choosing a time $t > a/2c$ (as in Fig. 1–8d), when the sloping edges of the + and −waves no longer overlap, the momentum may be calculated as follows: In the +wave a section of string of mass σa is moving upward with a velocity

$$\frac{\partial y_+}{\partial t} = -c\,\frac{\partial y_+}{\partial x} = -c\left(\frac{-v_0}{2c}\right) = \frac{v_0}{2},$$

giving a momentum contribution $\sigma a v_0/2$. An equal momentum is contributed by the −wave, the total adding up to the original value.

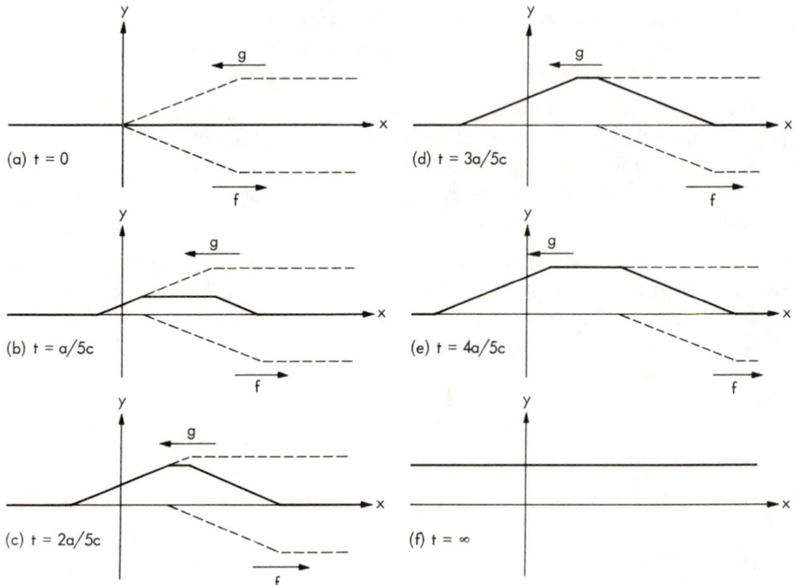

Fig. 1-8. Successive shapes of a string struck with a broad mallet.

The derivations in this section have assumed that the wave equation is satisfied at every point in the interval $-\infty < x < \infty$, and that the initial wave and velocity profiles, $\phi(x)$ and $\psi(x)$, are given for all values of x. Suppose, however, that we are dealing with a finite length of string suspended between two fixed ends at $x = 0$ and $x = l$. The preceding analysis breaks down because the initial conditions can give us information only about $\phi(x)$ and $\psi(x)$ for $0 \leq x \leq l$. Outside this range the functions are undefined. The solution given by Eq. (1-24) remains valid at a given point on the string up to the time when the arguments $(x - ct)$ and $(x + ct)$ exceed the domain of definition. Thus, for a point lying in the first half of the string, $0 < x < l/2$, the solution is valid so long as $x - ct > 0$ or $t < x/c$. At a point lying in $l/2 < x < l$ the solution is valid so long as $x + ct < l$, or $t < (l - x)/c$. These two times are the length of time which would be required for a pulse to reach the point in question starting from the nearest end of the string. Thus we can state that the motion of a given point on the string will be the same as it would be in the case of an infinite string having the same initial conditions in the range $0 \leq x \leq l$ up to the time which would be required for a pulse traveling with velocity c to reach the given point from one of the ends. A more complete discussion of the unique determination of the motion of a noninfinite string will be deferred until Chapter 15.

PROBLEMS*

1–1. Which of the following are solutions to the one-dimensional wave equation for transverse waves on a string?

a) $x^2 - 2cxt + c^2t^2$ b) $10(x^2 - c^2t^2)$ c) $\sigma x^2 + Tt^2$
d) $\sqrt[3]{\sin[(x - ct)^3]}$ e) $2x - 3ct$ f) $10(\sin x)(\cos ct)$

1–2. If $h(u)$ is an arbitrary twice-differentiable function of u, show by direct calculation that $y(x, t) = h(t + x/c)$ satisfies the one-dimensional wave equation. What relation does this solution have to Eq. (1–13)?

1–3. The note produced by a sonometer wire (Fig. 1–4) is in tune with a tuning fork of frequency 128 sec^{-1} when $l = 1$ m and $m = 7.76$ kg.
a) Determine the linear mass density of the wire.
b) If the midpoint of the wire is executing simple harmonic motion given by the expression $y = A \sin \omega t$, where $A = 0.2$ cm, calculate the maximum velocity of this point. Compare this value with the velocity of wave propagation on the string.

1–4. Is it possible to put a steel wire under sufficient tension so that the velocity of transverse waves on the wire will be the same as the velocity of sound waves in air? (Compute the implied stress and compare with the tensile strength of steel.)

1–5. When a rubber band is stretched between two fingers, the pitch of the "twang" does not seem to vary significantly as the distance between the fingers is varied. Explain.

1–6. Show that the dimensions of the ratio T/σ are the same as velocity squared.

1–7. Let $v(x, t)$ designate the transverse velocity at time t of the point x on a string. Show that this function satisfies the same partial differential equation as $y(x, t)$.

1–8. Given $y(x, t) = \mathcal{R}e\,\{\mathbf{A}e^{i(\omega t - kx)}\}$.
a) Verify by direct substitution that this function is a solution of the one-dimensional wave equation. What kind of wave is this?
b) Verify that this function conforms with the appropriate relation asserted in Eq. (1–14).

1–9. Examine the correctness of the following assertion: "Equation (1–14) shows that for both + and −waves the magnitude of the transverse velocity is c times the magnitude of the slope. Since the general solution is a superposition of + and −waves, the same property must hold for it."

1–10. A +wave on an infinite string has the initial waveform $y(x, 0) = e^{-x^2}$. Sketch the initial velocity profile.

1–11. A string of infinite length is initially straight and is given a "triangular" initial velocity profile:

$$\psi(x) = \begin{cases} 0, & x < -a, \\ v_0(1 + x/a), & -a \le x \le 0, \\ v_0(1 - x/a), & 0 \le x \le a, \\ 0, & x > a. \end{cases}$$

Sketch the shape of the string at a few representative times.

---

* An asterisk preceding the problem number indicates a more difficult problem.

1–12. The velocity relative to a fixed observer of the propagation of a waveform on a string can be slowed down by moving the string in a direction opposite to the direction of propagation of the wave. In particular, suppose that the pulleys in Fig. 1–9 are rotated so that the linear speed of the string is the same as c, the wave propagation velocity. The observer will then see a waveform which maintains a fixed position. The particles of the string are following curved paths and therefore have centripetal acceleration. The required centripetal force on a section of the string of length dx is $(\sigma\, dx)c^2/R$, where R is the radius of curvature of the string at this location. The net force acting on this section of the string is the vector difference between the tension forces acting at the two ends of the segment. (Note that if T is constant in magnitude, the direction of this net force is perpendicular to the tangent at the midpoint of the segment and is therefore directed toward the center of curvature.)

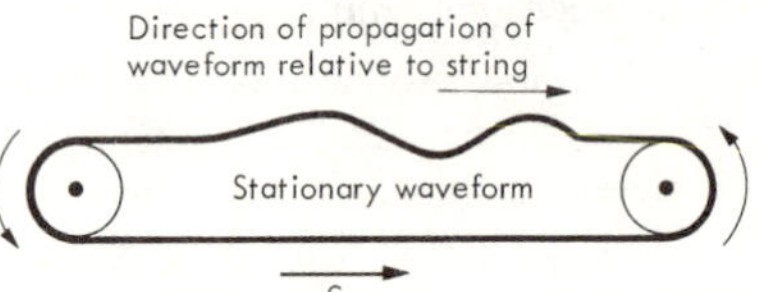

Figure 1–9

Show that when the small slopes approximation is made and the two appropriate expressions for the centripetal force are equated, the previously known relation $c^2 = T/\sigma$ results. (*Note:* This method, known as *reduction to a steady wave*, is sometimes used in elementary texts as a derivation of the expression for the wave velocity. Since this deals with a steady-state condition, the use of partial derivatives is avoided. This method shows that a wave can be propagated with velocity $\sqrt{T/\sigma}$ in one direction without distortion of form, but it is not capable of analyzing a superposition of waves traveling in opposite directions.)

1–13. The tension in a section of rope which is hanging vertically is clearly not a constant, but is a linear function of position. Another example of a string under nonuniform tension is the case of a whirling string (free at one end, attached at the other end to a hub rotating with constant angular velocity). Show that, in general, if $T = T(x)$, other approximations being made as before, the equation satisfied by the displacement of the string is

$$\frac{\partial}{\partial x}\left(T\,\frac{\partial y}{\partial x}\right) = \sigma\,\frac{\partial^2 y}{\partial t^2}\cdot$$

(This equation is not of the form that is usually called "the wave equation," but is a generalization thereof. The relatively simple techniques by means of which we obtained the general solution of Eq. (1–4) are not applicable here. For most forms of $T(x)$ it is necessary to resort to numerical calculation or to simplifying approximations in order to obtain a solution.)

***1–14.** The integration constants in the problem of the motion of a single mass point can sometimes be determined from a knowledge of the positions of the body at two different times. Investigate the corresponding situation for a string. That is, if $y(x, 0) = \phi(x)$ and $y(x, t_1) = \xi(x)$, what can you determine about $y(x, t)$?

2 The acoustic plane wave

2-1 DEFINITION OF THE VARIABLES

For the development of the concept of acoustic waves we shall deal with the mathematical fiction of an *ideal fluid medium*. The discrete, molecular structure of matter is ignored and the substance is treated as being *continuous*, that is, infinitely subdivisible. When we speak of the motion of a "particle" of the fluid we are referring in fact to the average motion of the molecules which are contained within an element of volume which is "macroscopically small" but "microscopically large." By the former we mean that the linear dimensions of the volume element are small compared with any relevant macroscopic quantities having the dimension of length, such as the wavelength of a sinusoidal wave. The volume element is "microscopically large" if it contains a large number of molecules of the substance and if its dimensions are large compared with the molecular mean free path (the average distance traveled by a molecule between collisions). For a given substance there will be conditions under which it is not possible to pick an element of volume which satisfies both these restrictions. In a gas at low pressure, for example, the mean free path may be comparable to the wavelength of a given sinusoidal disturbance. Under this condition our equations developed for the ideal fluid will not be an adequate representation.

Since the motion of a particle of the fluid represents the *average* motion of a large number of molecules of the substance, we are of course not describing the random thermal motions that the individual molecules undergo. It requires the techniques of advanced statistical mechanics to arrive at a description of the phenomenon of sound starting from a microscopic specification of the motion of the molecules of the system.*

---

* J. H. Jeans, *The Dynamical Theory of Gases*, 1st ed. Cambridge: Cambridge University Press, 1904, Chapter XVI. (This material has been omitted from the more recent editions.)

In order to define what is meant by an ideal fluid, imagine an element of area ΔA drawn at an arbitrary location within the medium. The portions of the substance on the two sides of this surface can be considered to be exerting forces on one another. If these forces are *normal* to the surface (no tangential stress), and if the magnitude of the force, $|\Delta F|$, is independent of the orientation of the surface (the stress is isotropic), then the situation can be characterized by the single parameter P, the *pressure* at a given point defined by the relation

$$P \equiv \lim_{\Delta A \to 0} (|\Delta F|/\Delta A).$$

A medium satisfying these conditions is referred to as an *ideal fluid.*

Consider a fluid medium which is not in equilibrium: that is, suppose that there are pressure gradients in the medium which are not balanced by forces of some other nature. (A counterexample is the atmosphere in equilibrium under gravity, where there exists a vertical pressure gradient which balances the gravitational field.) We expect under these conditions that motion of the fluid will result and a disturbance which was originally confined to one portion of the fluid will be transmitted to other portions. The study of the general motion of a fluid constitutes the field of *hydrodynamics.* As we shall see presently, when a certain restricted class of fluid motions is considered, the associated spreading disturbance corresponds exactly with what we recognize as the propagation of sound. Acoustics may thus be looked upon as a branch of hydrodynamics, which in turn is a part of the general subject of mechanics. In this chapter we will develop a description for the special case of plane waves.

In the undisturbed state let the pressure P_0 and the density ρ_0 be uniform throughout the medium. This assumption implies that in cases where external force fields are present we must restrict ourselves to consideration of a portion of the medium throughout which variations associated with these fields are negligible. Thus we suppose either that our fluid is in gravity-free space, or, if it is in a gravitational field, that the extent of the fluid is sufficiently limited so that the spatial variations of P_0 and ρ_0 can be neglected. Let us consider disturbances of the medium for which:

a) all particle displacements are parallel to a particular direction which we shall designate as the x-axis;

b) the values of any physical quantities are the same at any two points lying in the same plane perpendicular to the x-axis.

After we have related this situation to the wave equation, (b) will be interpreted to mean that we have restricted our discussion to *plane waves;* (a) will imply the *longitudinal* character of these waves.

Consider those particles of the fluid which would be assigned coordinate x in the undisturbed state, that is, all those particles originally in a given plane perpendicular to the x-axis. Let the displacement of these particles from their normal location be represented at time t by the function $\xi(x, t)$. The coordinate of these particles at time t is thus $x + \xi(x, t)$. It is important to note that we are using x as a parameter to *label* a set of particles according to their location in the equilibrium state. The function $\xi(x, t)$ does not tell us the displacement of those particles which *at time t* have coordinate x, but rather the displacement of those particles which have coordinate x *in the undisturbed condition*. Likewise we will let $P(x, t)$ and $\rho(x, t)$ be the pressure and density being experienced at time t by those particles *originally* at x. We are here choosing what is known as the *Lagrangian* formulation. An alternative formulation is the *Eulerian* (see Appendix III), in which $P(x, t)$ refers to the pressure at time t being experienced by the particles which are *then* at x.

2-2 DERIVATION OF THE WAVE EQUATION

We now seek to determine relations among the three functions P, ρ, and ξ which will enable us to eliminate and find a differential equation involving only one of these functions.

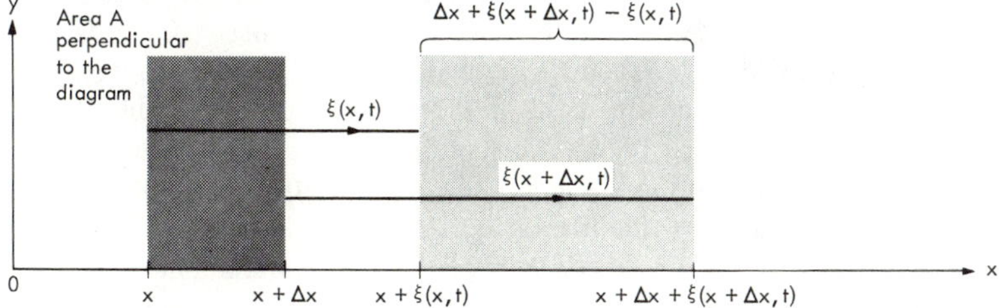

Fig. 2-1. Location of particles in undisturbed and disturbed configurations.

1) It should be clear that the function $\xi(x, t)$ contains complete information about the locations of all the particles of the fluid at time t, and thus enables us to calculate the density throughout the fluid. Consider a volume of the fluid contained in a cylinder having cross-sectional area A perpendicular to the x-axis, which is bounded in the undisturbed state by the planes x and $x + \Delta x$. The mass contained within this cylinder is $\rho_0 A\,\Delta x$. Under the disturbance the particles will have moved to new locations, as indicated in Fig. 2-1. The new volume occupied by these particles is $A[\Delta x + \xi(x + \Delta x, t) - \xi(x, t)]$. The *average* density throughout this volume is obtained by dividing the total mass (which is the same as it was in the undisturbed state) by the volume:

$$\bar{\rho} = \rho_0\left[1 + \frac{\xi(x + \Delta x, t) - \xi(x, t)}{\Delta x}\right]^{-1}. \tag{2-1}$$

If we now let Δx approach zero, $\bar{\rho}$, which is a function of x, Δx, and t, approaches $\rho(x, t)$. On the right-hand side of Eq. (2-1) we use the fact that

$$\lim_{\Delta x \to 0} \left[\frac{\xi(x + \Delta x, t) - \xi(x, t)}{\Delta x} \right]$$

is the defining expression for $\partial \xi / \partial x$. Thus we obtain

$$\rho(x, t) = \rho_0 \left[1 + \frac{\partial \xi}{\partial x}(x, t) \right]^{-1}. \quad \textit{Equation expressing the conservation of mass} \tag{2-2}$$

No approximations have been made in deriving this relation. Of course, the ideal fluid is an approximate representation of a real medium, but within the context of the ideal fluid this is an exact relation for the restricted type of motion being considered.

In deriving Eq. (2-2) we have used a notation which expresses all quantities accurately both before and after the limit is taken as Δx approaches zero. It is standard practice in physics not to be so meticulous in the steps which precede the limit process. For example, in labeling Fig. 2-1, we would use the differential approximation immediately to replace $\xi(x + \Delta x, t)$ by its approximate value $\xi(x, t) + (\Delta x)(\partial \xi / \partial x)(x, t)$. The new volume is then $A(\Delta x)[1 + (\partial \xi / \partial x)(x, t)]$. When this is divided into the mass, $\rho_0 A \, \Delta x$, the factors of $A \, \Delta x$ cancel and Eq. (2-2) is the direct result. The abbreviated proof does not distinguish between instantaneous and average values, nor does it make *explicit* mention of the fact that a limit must be taken. To have a proper understanding of this form of proof it must be realized that these ideas are tacitly implied. It *appears* as if the replacement made for $\xi(x + \Delta x, t)$, which is not a strict identity, were introducing an approximation (the differential approximation) which might limit the generality of the results and contradict the claim made earlier that Eq. (2-2) is an exact relation for the ideal fluid. This misunderstanding can be cleared up by the tacit implication that the mean value theorem of the differential calculus is being employed. Instead of "approximating," we may use the exact relation

$$\xi(x + \Delta x, t) = \xi(x, t) + (\Delta x) \frac{\partial \xi}{\partial x}(\bar{x}, t), \tag{2-3}$$

where $\bar{x}$ is some value between x and $x + \Delta x$. After the limit is taken, $\bar{x}$ becomes replaced by x. The abbreviated method of proof is thus adequately justified by these tacit assumptions and will be employed throughout the text without further comment. The reader should note that the outlawed notion of an "infinitesimal" has not appeared in this discussion. Equations (2-1) and (2-3), for example, are valid for any value of Δx, no matter how large, and do not require that we think of Δx as being "vanishingly small."

It is instructive to look at some results which follow from Eq. (2–2) and note that they make good sense physically. For example, if $\partial\xi/\partial x = 0$ for some value of x and t, this means that the particles in plane x have undergone displacements which are "to first order" the same as the displacements of particles in "neighboring" planes. We would then expect the density surrounding these particles to be the same as the density in the undisturbed state, which is what Eq. (2–2) does tell us. Likewise, $\partial\xi/\partial x > 0$ implies that the particles at $x + \Delta x$ have undergone larger displacements than the particles at x. From this we anticipate that the density will have a value less than normal. This result is also confirmed by Eq. (2–2).

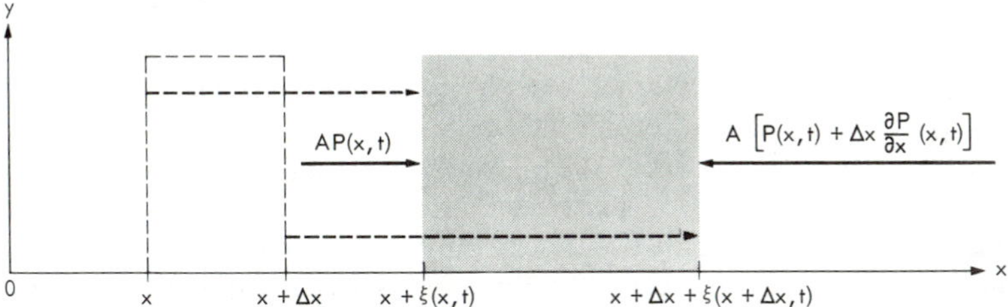

Fig. 2–2. Forces acting in the disturbed state on particles having parameters between x and $x + \Delta x$.

2) For a second relation among P, ρ, and ξ, consider the application of Newton's second law to our volume element in the disturbed state (Fig. 2–2). Recall that we defined $P(x, t)$ to represent the pressure at the plane of particles which was *originally* at x. It is implicit in the notation that $P(x, t)$ is the pressure at whatever displaced location these particles occupy. From the figure, it can be seen that the net force in the positive x-direction is given by the expression $-A\,\Delta x(\partial P/\partial x)$. The mass of the volume element is $\rho_0 A\,\Delta x$ and the acceleration is $\partial^2\xi(x, t)/\partial t^2$. As previously mentioned, the center of mass would be associated with some value of x between x and $x + \Delta x$, but this distinction is not made since we have in mind taking the limit as Δx approaches zero. Dividing the expression for the net force by the mass and equating this to the acceleration, we obtain

$$-\frac{\partial P}{\partial x}(x, t) = \rho_0 \frac{\partial^2\xi}{\partial t^2}(x, t). \quad \textit{Equation expressing Newton's second law} \tag{2–4}$$

3) We now have two equations involving the three quantities P, ρ, and ξ. For a third relation we require a statement of the way in which the density of the medium changes with changing pressure. We refer to this as the *equation of state* of the medium. In the general case this is an empirical relation. Let us assume that it is expressed in the form

$$P = P(\rho). \quad \textit{Equation of state} \tag{2–5}$$

We are now in a position to eliminate P and ρ in favor of ξ: From Eq. (2–5), $\partial P/\partial x = P'(\rho)(\partial\rho/\partial x)$, but from Eq. (2–2),

$$\frac{\partial\rho}{\partial x} = -\rho_0\left[1 + \frac{\partial\xi}{\partial x}\right]^{-2}\frac{\partial^2\xi}{\partial x^2}.$$

Substituting these expressions into Eq. (2–4), we obtain

$$\frac{\partial^2\xi}{\partial t^2} = P'\left[\rho_0\left(1 + \frac{\partial\xi}{\partial x}\right)^{-1}\right]\left(1 + \frac{\partial\xi}{\partial x}\right)^{-2}\frac{\partial^2\xi}{\partial x^2}. \tag{2–6}$$

This is an equation in which x and t are independent variables and the only dependent variable is ξ. It is assumed, of course, that the functional connection $P(\rho)$ is given.

Equation (2–6) resembles the one-dimensional wave equation; in fact, it would be precisely of that form if the coefficient of $\partial^2\xi/\partial x^2$ were a constant. An appropriately chosen function $P(\rho)$ would make this coefficient identically constant (see Problem 2–15), but since no real substance has an equation of state of the required form, the coefficient must in general be looked upon as a function of $\partial\xi/\partial x$. This means that Eq. (2–6) is a *nonlinear* equation in the function $\xi(x, t)$. However, if the condition $\partial\xi/\partial x \ll 1$ is fulfilled throughout the disturbance, the variability of this coefficient will be small, and its value may be replaced by the constant $P'(\rho_0)$. For the restricted class of motions for which $\partial\xi/\partial x \ll 1$, known as *acoustic* disturbances (the approximation being known as the *acoustic approximation*), the function $\xi(x, t)$ is therefore determined by the one-dimensional wave equation

$$\frac{\partial^2\xi}{\partial t^2} = c^2\frac{\partial^2\xi}{\partial x^2}, \tag{2–7}$$

where the constant $P'(\rho_0)$ is designated by c^2.

2–3 THE VELOCITY OF SOUND

Up to now we have been describing the dynamics of some kind of motion of the fluid, but have had no assurance that this has any connection with the propagation of sound. Of course, we recognize immediately that Eq. (2–7) has a solution of the form $f(x - ct)$, which implies the propagation of "something" with the velocity $c = \sqrt{P'(\rho_0)}$. Calculation of this number will therefore give an important clue toward identifying the phenomenon to which our equations refer. The equation of state for an ideal gas is known to us, so we shall use this as a first example. If we consider a volume V containing a unit mass of the gas, then $\rho = 1/V$ is the density, and $n = 1/M$ is the number of moles, where M is the molecular weight (or average molecular weight for a mixture of gases). The equation of state $PV = nRT$ may then be written equivalently as

$$P = RT\rho/M, \tag{2–8}$$

where R is the universal gas constant and T is the absolute temperature of the gas. Using this for the function $P(\rho)$, we have $P'(\rho) = RT/M = P/\rho$. For air at standard temperature and pressure, $\rho_0 = 1.2933\ \text{kg/m}^3$ and $1\ \text{atm} = 1.0135 \times 10^5\ \text{N/m}^2$. From these data the calculated value of the propagation velocity is $c = \sqrt{P_0/\rho_0} = 279.95$ m/sec. Now the measured velocity of sound in air at 0°C and 1 atm pressure is 331.00 m/sec. There is a 16% discrepancy in these values, all of which are known to the number of significant figures quoted. Certainly, unless there is some fallacy in our reasoning subsequent to Eq. (2–7), it would not be very satisfactory to conclude that Eq. (2–7) describes the propagation of sound.

A calculation of the type we have just performed was first made by Newton in 1687. At that time the available data were not known to the same accuracy as those above, and Newton was willing to conclude that the calculated velocity was in agreement with the measured velocity of sound. It was not until 1807 that Laplace pointed out an oversight in the method we have used to calculate $P'(\rho)$ from Eq. (2–8). Are we justified in assuming that T is not a function of ρ? As the disturbance propagates through the medium, portions of the gas are either compressed or expanded from the equilibrium state. Will these compressions and expansions take place *isothermally?* Consider the short length of time available for these processes. It does not seem likely that a portion of the gas which is being compressed at this rapid rate will be able to transfer heat energy to the neighboring elements in such a way as to remain at a constant temperature.* Laplace considered it more plausible to suppose that these successive compressions and expansions take place *adiabatically*, that is, without heat flow between neighboring volume elements of the gas. If we assume that the process is adiabatic, Eq. (2–8) is still an appropriate relation to describe the equilibrium state:

$$P_0 = RT_0\,\rho_0/M. \tag{2–9}$$

Equation (2–8) is also a correct relation among the variables in the disturbed state, providing we understand that T varies with ρ (see Problem 2–2). To obtain P directly as a function of ρ under the conditions of adiabatic expansion, we make use of the adiabatic law, $PV^\gamma = P_0V_0^{\ \gamma}$. Expressing this in terms of the density we have the adiabatic equation of state

$$P_a(\rho) = P_0(\rho/\rho_0)^\gamma, \tag{2–10}$$

where γ is the ratio of the specific heat at constant pressure to the specific heat at constant volume. This is a quantity which is found to be constant over a wide range of the variables.

---

* This depends, of course, on the heat conductivity of the substance. For a discussion of the relation between this quantity and the parameters associated with the wave see Mark W. Zemansky, *Heat and Thermodynamics*, 4th ed. New York: McGraw-Hill, 1957, p. 132.

A simple model in classical statistical mechanics predicts 5/3, 7/5 and 8/6 as the respective values of γ for monatomic, diatomic and triatomic molecules. The experimental values listed in Table IV–1 (Appendix IV, Page 444) are in good agreement with the predicted values. Although the classical theory accounts for the approximate values of the specific heats, it is unable to predict the precise values, nor is it able to account for the fact that γ does vary with large changes in temperature. The theory of specific heats is one of the weak points in classical physics which remained unsatisfactory until the advent of quantum mechanics.

With Eq. (2–10) taken as the equation of state, we obtain

$$P_a'(\rho) = \gamma P_0 \frac{\rho^{\gamma-1}}{\rho_0^{\gamma}}, \tag{2–11}$$

and hence

$$c^2 = P_a'(\rho_0) = \frac{\gamma P_0}{\rho_0} = \gamma \frac{RT_0}{M}. \tag{2–12}$$

The values of c calculated from Eq. (2–12) are in agreement with the measured value of the velocity of sound in air. We therefore conclude that ordinary sound is a kind of disturbance described by the variables we have introduced, and that it fulfills the condition of the acoustic approximation $\partial\xi/\partial x \ll 1$. We should be wary of taking tabulated values of γ for various gases and inserting them in Eq. (2–12) to see how close the agreement is, for we must inquire how the values of γ were measured. In principle, γ is to be calculated from the ratio of the two specific heats of the gas, which can be obtained by thermodynamic measurement. In practice, however, Eq. (2–12) is looked upon as one of the most precise means of determining γ. Our confidence in the identification of Eq. (2–7) as the wave equation for the propagation of sound is sufficient so that the measured value of the velocity of sound is commonly used to find γ.

Consider now a more general fluid medium. We wish to relate $c = \sqrt{P_a'(\rho_0)}$ to convenient parameters which describe the behavior of the medium under adiabatic compression. A given change in the pressure ΔP produces a *strain*, measured by the dimensionless ratio $\Delta V/V$. The stress-strain ratio $-\Delta P/(\Delta V/V)$ is defined as the *bulk modulus of elasticity*, symbolized by $\mathcal{B}$. (The minus sign is chosen since $\Delta P > 0$ implies $\Delta V < 0$.) More accurately, since the limit as ΔV approaches zero is implied, and since it must be specified that the process is adiabatic, the *adiabatic bulk modulus* $\mathcal{B}_a$ is defined as $\mathcal{B}_a = -VP_a'(V)$, where $P_a'(V)$ is the slope of an *adiabatic* curve in the P-V plane. When the variables are changed to $\rho = m/V$, where m is the mass contained in volume V, this becomes $\mathcal{B}_a = \rho P_a'(\rho)$, where $P_a(\rho)$ is the function which gives the pressure as a function of density in an adiabatic variation.

The desired expression for the velocity of sound is thus

$$c = \sqrt{P_a'(\rho_0)} = \sqrt{\mathcal{B}_a/\rho_0}\,. \tag{2–13}$$

It is understood that the adiabatic bulk modulus of the medium is evaluated under the conditions of the undisturbed state. A laboratory experiment to determine the bulk modulus is usually made under *isothermal* conditions, which means that the most commonly tabulated quantity is the isothermal bulk modulus, $\mathcal{B}_{\text{isoth}}$. It can be shown from thermodynamics that the ratio of the adiabatic to the isothermal bulk modulus is the same as the ratio of the specific heats at constant pressure and constant volume: $\mathcal{B}_a = \gamma\mathcal{B}_{\text{isoth}}$.

The quantity

$$K \equiv \frac{1}{\mathcal{B}} = -\frac{1}{V}\frac{\Delta V}{\Delta P},$$

which represents the fractional change in volume produced by a unit change in pressure, is called the *compressibility* of the medium. Some tables list values of either the isothermal or adiabatic compressibility, rather than the bulk modulus. Since the velocity varies inversely with the square root of K, we associate highly compressible media with low propagation velocities. An "incompressible medium" ($K = 0$, or $\mathcal{B} = \infty$) would propagate sound with "infinite velocity."

For a given medium we must consult tables to determine the bulk modulus, since this is not derivable from any theoretical relation as it is in the case of ideal gases. Thus, for water at 20°C: $\gamma = 1.004$, $\mathcal{B}_{\text{isoth}} = 2.18 \times 10^{19}\ \text{N/m}^2$ and $\rho = 0.998 \times 10^3\ \text{kg/m}^3$. From these values the calculated velocity of sound is 1.48×10^3 m/sec, which agrees with the experimental value. The value of γ for most fluids is very close to unity, though there are notable exceptions, such as turpentine, for which $\gamma = 1.27$.

A great deal of information is contained in Eq. (2–12) concerning the dependence of the velocity of sound in a gas upon such factors as barometric pressure, temperature, and the nature of the gas. Thus, for a given gas at a given temperature we find that the velocity of sound is independent of the base pressure P_0. Although the form $c = \sqrt{\gamma P_0/\rho_0}$ makes it look as if c may depend upon P_0, c is in fact independent of P_0, since P_0 and ρ_0 are proportional to each other at fixed temperature. This feature is most readily seen from the form $c = \sqrt{\gamma R T_0/M}$.

The *temperature* dependence of the sound velocity is given by $c \propto \sqrt{T_0}$. In the vicinity of 0°C this expression can be expanded to yield a linear function of t, the temperature in degrees centigrade, by writing $T_0 = T' + t$, where $T' = 273.2$. This yields

$$c = c'\left[1 + \frac{t}{T'}\right]^{1/2} \doteq c' + \left(\frac{c'}{2T'}\right)t,$$

where c' is the velocity at 0°C. For air the coefficient $c'/2T'$ has the approximate value 0.6 m/sec-°C, which means that the velocity of sound increases by about 0.6 m/sec for each degree rise in temperature.

If we compare two different gases at the same temperature, we find $c \propto \sqrt{\gamma/M}$. Thus sound travels fastest in gases which have low molecular weight, such as hydrogen and helium.

A simple demonstration of the dependence of c on the type of gas is afforded by attaching a tin flute to a source of illuminating gas. The wavelength of the sound within the tube is determined by the length of the tube, and remains fixed. Hence the frequency of the emitted note will change in direct proportion to c. A rough estimate of the molecular weight of the gas relative to that of air can be obtained by the simple process of judging the musical interval of the change of pitch which occurs when the gas enters the tube. A more striking demonstration is provided by listening to the voice quality of a subject who attempts to speak normally after temporarily inhaling some helium gas. The frequency of vibration of the vocal chords is determined by muscular tension and is presumably not altered by the presence of the helium. However, the increase in the velocity of sound raises the resonant frequency of the various resonant cavities of the body (the lungs and the sinuses) which reinforce the output of the vocal chords. The fundamental and lower harmonics are de-emphasized and the higher harmonics are strengthened, resulting in a dramatic change in timbre.

2-4 SIMPLIFIED FORM OF THE EQUATIONS FOR ACOUSTIC WAVES

In the context of the acoustic approximation it is convenient to introduce two new variables, and to consider the simplified form which the fundamental equations take in this restricted case. The difference between normal pressure and the pressure in the disturbance is referred to as the *excess acoustic pressure*, p:

$$p \equiv P - P_0. \tag{2–14}$$

The *relative* change in density, which is a dimensionless measure of the extent to which the fluid is compressed, is referred to as the *condensation variable*, s:

$$s = (\rho - \rho_0)/\rho_0. \tag{2–15}$$

Both p and s can take on positive and negative values. Note that "negative pressure" in this sense merely means $P < P_0$. Regions where $s > 0$ are referred to as *condensations*, and regions where $s < 0$, as *rarefactions*. The acoustic approximation is seen to be equivalent to the assumption that the first term in a Taylor series expansion of $P(\rho)$ will be adequate. That is, $P(\rho) \doteq P_0 + P'(\rho_0)(\rho - \rho_0)$. Since $P'(\rho_0) = c^2 = \mathcal{B}_a/\rho_0$, the last equation becomes expressed in terms of the new variables as $p = \mathcal{B}_a s$. Furthermore, with $\partial\xi/\partial x \ll 1$, the right-hand side of Eq. (2–2) can be expanded to give $\rho = \rho_0(1 + \partial\xi/\partial x)^{-1} \doteq \rho_0(1 - \partial\xi/\partial x)$. Expressing this in terms of the condensation variable yields $s = -\partial\xi/\partial x$. In summary, the fundamental equations specialized to the acoustic approximation are:

$$s = -\partial\xi/\partial x. \qquad \textit{Conservation of mass} \qquad (2\text{–}16)$$

$$p = \mathcal{B}_a s. \qquad \textit{Adiabatic equation of state} \qquad (2\text{–}17)$$

$$-\partial p/\partial x = \rho_0(\partial^2\xi/\partial t^2). \qquad \textit{Newton's second law} \qquad (2\text{–}18)$$

Naturally, if we were interested only in the small amplitude case it would have been simplest to specialize to these equations right from the start. In doing this, however, we would not appreciate why the restrictions made are *necessary* in order to obtain the linear wave equation. Also, we would not have been able to anticipate, as will be done below, what kind of modifications in the nature of the results are to be expected if $\partial\xi/\partial x$ is not small.

Elimination among the three fundamental equations will show that any of the variables p, s, and ξ satisfies the one-dimensional wave equation with $c = \sqrt{\mathcal{B}_a/\rho_0}$. This implies immediately that we may write

$$p = p_+ + p_-, \qquad (2\text{–}19)$$

$$s = s_+ + s_-, \qquad (2\text{–}20)$$

and

$$\xi = \xi_+ + \xi_-, \qquad (2\text{–}21)$$

where p_+, s_+, and ξ_+ are functions of the combination $(x - ct)$, and p_-, s_-, and ξ_- are functions of the combination $(x + ct)$. We may look upon ξ_+ and ξ_- as functions which are chosen arbitrarily. Once this choice has been made, however, the remaining functions p_+, s_+, p_-, and s_- are determined. For example, substitution from Eq. (2–20) into (2–17) yields $p = \mathcal{B}_a s = \mathcal{B}_a s_+ + \mathcal{B}_a s_-$. Thus

$$p_+ + p_- = \mathcal{B}_a s_+ + \mathcal{B}_a s_-,$$

or

$$(p_+ - \mathcal{B}_a s_+) = -(p_- - \mathcal{B}_a s_-).$$

Now since the left-hand side is to be a function of $u = x - ct$ only, and the right-hand side a function of $v = x + ct$ only, and since u and v can vary independently of each other, the only possibility is that both sides remain constant:

$$p_+ = \mathcal{B}_a s_+ + C, \qquad (2\text{–}22)$$

$$p_- = \mathcal{B}_a s_- - C. \qquad (2\text{–}23)$$

By arguments similar to those employed in Section 1–8 it can be seen that the constant C is of no physical significance and can be set equal to zero without loss of generality. Thus Eqs. (2–22) and (2–23) show that Eq. (2–17) is a relation which holds *separately* for both the + and − components of the wave. In a like manner it can be shown that Eqs. (2–16) and (2–18) apply to the individual components as well as to their superposition.

The function $(\partial\xi/\partial t)(x, t)$, which we shall designate as $\dot{\xi}(x, t)$, represents the distribution of instantaneous longitudinal velocities of the particles of the medium. The two most convenient parameters for describing an acoustic field are $\dot{\xi}(x, t)$ and $p(x, t)$. A useful connection between these two variables can be obtained as follows: Since $\xi_+ = f(x - ct)$, we have

$$\dot{\xi}_+ = -cf'(x - ct) = -c\,\frac{\partial \xi_+}{\partial x} = cs_+.$$

Comparison with Eq. (2–22) shows that

$$p_+ = \mathcal{B}_a s_+ = (\mathcal{B}_a/c)\dot{\xi}_+ = Z\dot{\xi}_+, \tag{2–24}$$

where $Z = \mathcal{B}_a/c$ is a parameter referred to as the *acoustic wave impedance of the medium.* (As motivation for the term *impedance* note that the equation $p_+ = Z\dot{\xi}_+$ resembles the equation $V = ZI$ for a circuit element.) From $c = \sqrt{\mathcal{B}_a/\rho_0}$ the bulk modulus can be expressed as

$$\mathcal{B}_a = \rho_0 c^2, \tag{2–25}$$

and the impedance as

$$Z = \rho_0 c. \tag{2–26}$$

It can be correspondingly shown that the −wave components satisfy the relation

$$p_- = -Z\dot{\xi}_-. \tag{2–27}$$

Adding Eqs. (2–24) and (2–27) we obtain a prescription for obtaining the pressure function in terms of given information about the velocity function:

$$p = p_+ + p_- = Z(\dot{\xi}_+ - \dot{\xi}_-). \tag{2–28}$$

The acoustic approximation, $\partial\xi/\partial x \ll 1$, is reflected in restrictions on the magnitudes of the other variables as follows:

From Eq. (2–16):

$$s \ll 1. \tag{2–29}$$

Thus the relative changes in density of the medium must be small if the disturbance is to be described by the equations of the acoustic approximation.

From Eq. (2–17):

$$p \ll \mathcal{B}_a. \tag{2–30}$$

In an ideal gas ($\mathcal{B}_a = \gamma P_0$) this means that the pressure variations must be small compared with atmospheric pressure. In a liquid such as water ($\mathcal{B}_a \sim 2 \times 10^4$ atm) the restriction is less severe. Another restriction is imposed, however, by the simplified form we have used for the equation of state. If the total pressure $P = P_0 + p$ falls below the vapor pressure of the liquid, a more complicated equation is required. In such a region bubbles containing vapor may be formed, this phenomenon being known as *cavitation*. Demonstration of this effect is provided by the passage of large-amplitude ultrasonic waves through a tank of water. The planes of minimum pressure are rendered visible through light scattered by the cavitation bubbles.

From the relation $\dot{\xi}_\pm = \mp c(\partial\xi_\pm/\partial x)$ we observe that the acoustic approximation implies the additional restriction for progressive waves:

$$\dot{\xi} \ll c. \tag{2–31}$$

The longitudinal velocity of a fluid particle is therefore required to be small compared with the wave propagation velocity.

On the other hand, kinetic theory shows that the mean molecular velocity of thermal motion is the same order of magnitude as the wave propagation velocity. For example, the root-mean-square velocity, $\sqrt{\overline{v^2}} = \sqrt{3RT/M}$, differs from the wave propagation velocity $c = \sqrt{\gamma RT/M}$ by a factor of $\sqrt{3/\gamma}$. Thus when the conditions of the acoustic approximation prevail, the longitudinal velocity associated with the wave motion is only a slight addition to the thermal motion of the individual molecules.

2-5 DETAILED DESCRIPTION OF A PROGRESSIVE SINUSOIDAL WAVE

Consider the +wave specified by

$$\xi = \mathcal{R}e \left\{\boldsymbol{\xi} e^{i(kx-\omega t)}\right\}. \tag{2–32}$$

The graph of ξ versus x at a given time t is of sinusoidal form. The interpretation of this graph is somewhat more abstract than was the case for transverse waves on a string. In the case of the sound wave, no object or set of particles actually takes on the physical shape of a sine wave. The displacement $\xi(x, t)$ is the displacement of the entire set of particles in the plane labeled x, and is parallel to the x-axis. Where ξ is positive, the displacement is in the positive x-direction; where it is negative, the displacement is in the negative x-direction.

Differentiating Eq. (2–32), we obtain

$$s = -\frac{\partial \xi}{\partial x} = \mathcal{R}e \left\{-ik\boldsymbol{\xi} e^{i(kx-\omega t)}\right\}. \tag{2–33}$$

We can summarize this result by stating the relation between the complex amplitudes of s and ξ:

$$\mathbf{s} = -ik\boldsymbol{\xi}. \tag{2–34}$$

Similarly, we find that

$$\mathbf{p} = \mathcal{B}_a\mathbf{s} = -ik\mathcal{B}_a\boldsymbol{\xi} = -i\omega Z\boldsymbol{\xi} \tag{2–35}$$

and

$$\dot{\boldsymbol{\xi}} = -i\omega\boldsymbol{\xi}. \tag{2–36}$$

By comparison of Eqs. (2–35) and (2–36), we see that the relation between the complex amplitudes of p and $\dot{\xi}$ is

$$\mathbf{p} = Z\dot{\boldsymbol{\xi}}. \tag{2–37}$$

This relation is also obvious from Eq. (2–24), since we are here dealing with a +wave.

Equation (2–37) shows that the pressure and particle velocity functions are *in phase* with each other. On the other hand, Eq. (2–35) shows that the pressure and particle displacement functions are $\pi/2$ out of phase with each other. This at first sight seems paradoxical. Suppose we select $\boldsymbol{\xi} = \xi_m$ and thus write $\xi(x, t) = \xi_m \cos (kx - \omega t)$. Each particle is executing simple harmonic motion, and we might expect that at the extremes of the motion, where the *acceleration* is a maximum, the *pressure* would also be a maximum. However, in this case,

$$p(x, t) = -\mathcal{B}_a \frac{\partial \xi}{\partial x} = k\mathcal{B}_a\xi_m \sin (kx - \omega t). \tag{2–38}$$

From this we see that when the particle displacement is a maximum $(kx - \omega t = n\pi)$, the pressure is zero. What induces the particles to turn around and move in the opposite direction if they are not subject to extremes of pressure at their maximum displacement? The source of the paradox lies in confusion between pressure and pressure *gradient*. The net force that acts upon a volume element depends upon a *difference* in pressure at opposite sides of the bounding surface, and hence upon the existence of a pressure gradient. By differentiation of Eq. (2–38) we obtain

$$\frac{\partial p}{\partial x} = k^2\mathcal{B}_a\xi_m \cos (kx - \omega t) = k^2\mathcal{B}_a\xi,$$

which shows correctly that the pressure gradient is zero at the equilibrium positions and is a maximum where the displacement is a maximum.

In Fig. 2–3 arrows are used to indicate the values of ξ and $\dot{\xi}$ at various locations at some particular time t for a sinusoidal wave traveling in the positive direction along the x-axis. A study of the various features of this figure make it apparent that the analytical results make good sense in terms of the mechanisms involved. Inspection of the figure at the lettered positions shows the following:

a) An extreme value of ξ implies $\partial\xi/\partial x = 0$. This we may interpret as meaning that "neighboring particles" have moved equal distances from their equilibrium positions. We therefore expect normal density and $p = 0$.

b) Particles have moved in from both sides and converged on this location; thus $p > 0$.

c) Contrariwise to (b).

d) Particles are in the process of moving away on both sides of this location, and hence we expect the pressure to decrease as t increases. This is what occurs as the wave progresses toward the right.

e) "Neighboring particles" here are moving along together ($\partial\dot{\xi}/\partial x = 0$); hence the change in p during a small time Δt will be an "infinitesimal of higher order." This corresponds to the fact that the pressure is at its maximum value here and therefore $\partial p/\partial t = 0$.

f) Contrariwise to (d).

g) The same as (e), except that the pressure is at a minimum. One aspect of the situation at (g) poses a problem, however. Since $\dot{\xi} < 0$, it looks as if the rarefaction should be moving toward the *left*, which contradicts the fact that the entire waveform is progressing toward the right. However, recall that the wave we are describing must satisfy the condition $\dot{\xi} \ll c$ (Eq. 2–31). By the time the particles at this point can move a significant distance at the velocity $-\dot{\xi}_m$, the transfer of the waveform from one set of particles to the next shifts the locus of the rarefaction a much greater distance in the positive direction.

To obtain an idea of the order of magnitude of the quantities dealt with in our theoretical description, consider a sound wave in air at standard temperature and pressure having a frequency near the middle of the audible

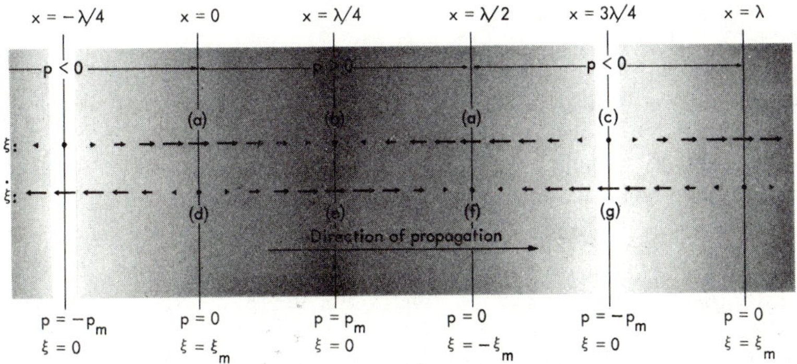

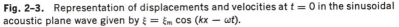

Fig. 2-3. Representation of displacements and velocities at $t = 0$ in the sinusoidal acoustic plane wave given by $\xi = \xi_m \cos(kx - \omega t)$.

frequency range (e.g. two octaves above middle C, $\nu \sim 10^3\ \text{sec}^{-1}$). The values of s_m, the amplitude of the condensation variable, range from $s_m \sim 10^{-10}$ for a sound which can barely be heard, to $s_m \sim 10^{-4}$ for a sound which is so intense that it is painful to the ear. These values show that the acoustic approximation is adequate for all sound waves of tolerable intensity. Ordinary conversation involves values of s_m near the logarithmic midpoint of the stated range, namely $s_m \sim 10^{-7}$. From Eq. (2–35)

$$p_m = \mathcal{B}_a s_m = \gamma P_0 s_m,$$

and the pressure amplitude of ordinary conversation is $p_m \sim 1.4 \times 10^{-2}\ \text{N/m}^2$. The ear is thus engaged constantly in detecting pressure variations which are the order of one ten-millionth of one atmosphere. The velocity amplitude is $\dot{\xi}_m = cs_m \sim 3.31 \times 10^{-5}$ m/sec. For a frequency $\nu = 10^3\ \text{sec}^{-1}$ the displacement amplitude is $\xi_m = \dot{\xi}_m/\omega \sim 5.3 \times 10^{-9}$ m, a mere 53 angstroms (Å). The acoustic approximation for sinusoidal waves requires from Eq. (2–34) that $k\xi_m \ll 1$ or $\xi_m \ll \lambda$. This is surely satisfied in the present example, for a frequency $\nu = 10^3\ \text{sec}^{-1}$ corresponds to a wavelength of 0.331 m.

2-6 NONLINEAR WAVE PROPAGATION IN AN IDEAL GAS

Mathematical expediency caused us to inquire first into the simplified form obtained for the basic dynamical equation (2–6) when $\partial\xi/\partial x$ is small enough in comparison with unity to be neglected. We return now to a brief discussion of the unapproximated equation specialized to the medium of an ideal gas. Substituting from Eq. (2–11) into Eq. (2–6) yields the nonlinear wave equation for an ideal gas:

$$\frac{\partial^2 \xi}{\partial t^2} = c_0^2 \left(1 + \frac{\partial \xi}{\partial x}\right)^{-1-\gamma} \frac{\partial^2 \xi}{\partial x^2}, \tag{2–39}$$

where $c_0^2 = \gamma P_0/\rho_0$.

The general solution of this equation cannot be expressed conveniently as we were able to do with the linear wave equation.* A qualitative notion of the nature of the implied propagation can be obtained by considering Eq. (2–39) to be of the same form as the linear wave equation except that the propagation velocity is now "amplitude dependent." That is, Eq. (2–39) can be written in the form

$$\frac{\partial^2 \xi}{\partial t^2} = c_e^2 \frac{\partial^2 \xi}{\partial x^2}, \quad \text{where} \quad c_e = c_0 \left(1 + \frac{\partial \xi}{\partial x}\right)^{-(1+\gamma)/2}. \tag{2–40}$$

---

* For a discussion of formal solutions to the nonlinear wave equation see H. Lamb, *Hydrodynamics*, New York: Dover, 1945, p. 481.

The effective velocity, c_e, is greater than c_0 in extreme condensations (i.e. where $\rho > \rho_0$ and hence $\partial\xi/\partial x < 0$), and less than c_0 in extreme rarefactions. Thus with a pulse of given initial shape (Fig. 2–4), the portions where the pressure is large tend to overtake the rest of the waveform and concentrate at the leading edge, thereby creating a *shock front* which travels at a velocity greater than c_0. The total width of the pulse increases since the small amplitude portions in the "tail" propagate with normal velocity. There is thus distortion in form in several respects. Linearity of superposition does not hold for solutions of Eq. (2–39); hence we can no longer conclude that pulses traveling in opposite directions will pass through one another without permanent effects of the interaction. We can anticipate, for example, that weak pulses which meet a shock front at an oblique angle will give rise to a reflected pulse similar to that produced at any surface across which the density is discontinuous.

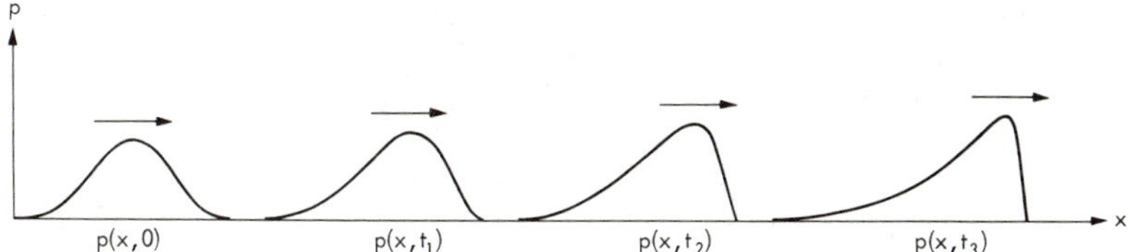

Fig. 2–4. Development of a shock front in a wave not satisfying the acoustic approximation.

Since there is no waveform for which $\partial\xi/\partial x$ is identically zero, all waves must partake to a certain extent of the properties of the nonlinear wave. A wave for which $\partial\xi/\partial x$ is small may propagate for some distance before a distortion in form becomes noticeable. Most observations of sound waves of ordinary intensity are made within this region. However, the effects of the nonlinearity are cumulative, and the wave becomes progressively more distorted as it propagates over long distances.* The question as to whether one should use Eq. (2–7) or (2–39) depends not only on the initial values of $\partial\xi/\partial x$ but also on how long a distance one is going to follow the progression of the wave. It is shown in Problem 2–16 that for sinusoidal waves the relevant distance beyond which distortion becomes important is inversely proportional to the initial pressure amplitude and is a constant number of wavelengths for waves of different frequency.

---

* The context here is that of plane waves. The same consideration does not apply to a spherical wave diverging from a point source, since in this case the pressure decreases inversely with distance from the center and the conditions of the acoustic approximation are met more ideally at large distances.

PROBLEMS

2–1. From the definition of the adiabatic bulk modulus show that $\mathcal{B}_a = \gamma P_0$ for an ideal gas.

2–2. If an ideal gas is compressed adiabatically from a state in which the density is ρ_0 and the temperature is T_0 to a state in which the density is ρ and the temperature is T, find T as a function of ρ.

2–3. When a tin flute is attached to a source of illuminating gas it is found that the pitch is approximately a musical third lower than the corresponding pitch in air. Estimate the molecular weight of the gas.

2–4. If the gas filling an organ pipe is changed from air to helium, find the change in pitch of the fundamental vibration. Express the result as a frequency ratio and give the nearest musical interval to which this corresponds.

2–5. An organ pipe is tuned to a pitch of 440 sec^{-1} when the temperature is 25°C. What will the pitch be at a temperature of 0°C?

2–6. Show from the basic equations of the acoustic approximation that p, s, ξ, and $\dot{\xi}$ all satisfy the one-dimensional wave equation.

2–7. Derive Eq. (2–4) in an accurate notation which does not make use of the differential approximation.

2–8. Given an acoustic plane wave traveling in the positive direction, whose wave profile at $t = 0$ is specified by the graph in Fig. 2–5. Draw sketches of the initial profiles for s, p, and $\dot{\xi}$, labeling appropriate magnitudes.

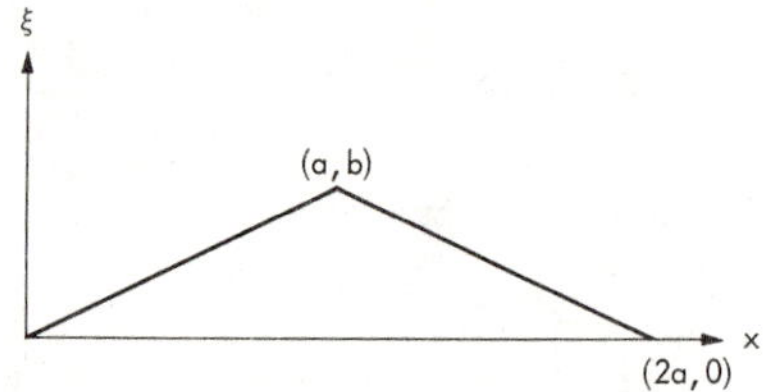

Figure 2–5

2–9. The pressure amplitude of a fairly intense sound wave in air is the order of 2N/m^2. Calculate the associated values of ξ_m, $\dot{\xi}_m$, and s_m for a sinusoidal wave of frequency 1000 sec^{-1}. How much above room temperature is the temperature at the center of a condensation? *Note:* When a calculation depends upon the small difference between two large numbers it is unwise to use a slide rule. Use of the binomial expansion (or equivalently, the differential approximation) will yield more accurate results.

2–10. Consider a progressive sinusoidal plane wave of given frequency ω and displacement amplitude ξ_m traveling in air, and a wave having the same values of ω and ξ_m traveling in water. How do the pressure amplitudes compare? If the value of s_m for the wave in water is sufficiently small to satisfy the acoustic approximation, is it necessarily so for the wave in air?

2–11. The velocity distribution in an acoustic plane wave is given by the function $\dot{\xi}(x, t) = \dot{\xi}_m \sin kx \cos \omega t$. Determine the corresponding pressure distribution.

2–12. Prepare a sketch similar to Fig. 2–3 showing displacements and instantaneous velocities at time $t = 0$ for the solution to the acoustic wave equation given by $\xi(x, t) = \xi_m \cos kx \cos \omega t$. Discuss the relation between particle displacement and pressure. Predict from the sketch what conditions will be a short time later, and compare your predictions with results obtained analytically.

2–13. When a sound wave is passing through air the temperature will be a function of x and t. Let $\tau \equiv T - T_0$. Show that for an acoustic wave the one-dimensional wave equation is satisfied by $\tau(x, t)$. What restriction is placed on the value of τ by the acoustic approximation?

2–14. Through the relation $p = -\mathfrak{B}_a(\partial\xi/\partial x)$ for acoustic waves, it is evident that the maximum, minimum, and inflectional points and the curvature of the graph of ξ versus x will have significance in the graph of p versus x. Suppose that $\xi(x, 0) = Ae^{-ax^2}$. Sketch the graphs of $\xi(x, 0)$ and $p(x, 0)$. Discuss the correspondence between their features in physical terms wherever possible.

2–15. From Eq. (2–6) we see that if the equation of state of a substance were such that $P'(\rho)(\rho/\rho_0)^2$ were to be a constant (designate this by c_0^2), the general equation would no longer be nonlinear. This implies that a wave of arbitrarily large amplitude would propagate without change in form with velocity c_0. Show that the equation of state would have to be $P(\rho) = P_0 + \rho_0 c_0^2(1 - \rho_0/\rho)$. (Since no known substance has an equation of state of this form, it can be concluded that there are no conditions under which a large-amplitude wave can propagate without distortion.)

2–16. *Distortion of a sinusoidal wave due to nonlinearity.* Consider a wave which is initially sinusoidal, having a condensation amplitude s_m. Show that if s_m is small but not entirely negligible, the expression for the effective velocity can be expanded to yield $c_{\max} = c_0[1 + (1 + \gamma)s_m/2]$ for a crest, and $c_{\min} = c_0[1 - (1 + \gamma)s_m/2]$ for a trough. Each crest will tend to overtake the trough ahead of it. If we agree to call the wave distorted when the distance from crest to trough has been reduced to $\frac{9}{10}$ its original value, show that the distance the wave must travel before it has this much distortion is given by

$$d = \frac{1}{20(\gamma + 1)s_m}\,\lambda.$$

Calculate numerical values from the data of Problem 2–9.

***2–17.** *The nonlinear equations satisfied by $\rho(x, t)$ and $P(x, t)$.* Show that without any approximations, $\rho(x, t)$ satisfies the equation

$$\frac{\partial^2}{\partial t^2}\left(\frac{1}{\rho}\right) = \frac{\partial}{\partial x}\left[c^2\frac{\partial}{\partial x}\left(\frac{1}{\rho}\right)\right],$$

where $c^2(\rho) = P'(\rho)(\rho/\rho_0)^2$, and likewise

$$\frac{\partial}{\partial t}\left[\frac{1}{c^2}\frac{\partial P}{\partial t}\right] = \frac{\partial^2 P}{\partial x^2},$$

where c^2 is the quantity $P'(\rho)(\rho/\rho_0)^2$ expressed as a function of P.

***2–18.** *Equilibrium density distribution in a vertical column of gas.* The following problem is most commonly approached from the Eulerian point of view. It is, however, instructive to formulate the problem in Lagrangian form and to compare the solution with the well-known results of the other method. Consider a tube of gas extending vertically from $x = 0$ to $x = l$. The gas is initially in equilibrium in the absence of a gravitational field. This state is used as the reference state for assigning the Lagrangian parameter x to the particles which initially lie in a horizontal plane of coordinate x. At $t = 0$ a uniform gravitational field parallel to the negative x-axis is switched on. Let this field be associated with an acceleration which increases slowly to a final value g which then remains constant. If the gas is kept at constant temperature show that the basic equations are

$$\rho_0(\ddot{\xi} + g) = -\frac{\partial P}{\partial x}, \qquad \rho_0 = \rho\left(1 + \frac{\partial \xi}{\partial x}\right), \qquad P = \frac{P_0 \rho}{\rho_0}.$$

Show that these equations possess a time-independent solution which describes the new equilibrium state of the gas. [The conditions $\xi(0) = \xi(l) = 0$ will determine the constants of integration.] The density distribution in this state should, of course, turn out to be the familiar distribution associated with an isothermal atmosphere. The latter is usually expressed in Eulerian form. Transform your result for $\rho(x)$ to express the density as a function of the space variable $X = x + \xi(x, t)$.

***2–19.** Consider an acoustic disturbance possessing spherical symmetry in which the only displacement of the particles is in a radial direction. Let this be symbolized in Lagrangian notation by $\xi(r, t)$. Set up the equations governing the motion of a displaced element of volume. Make suitable approximations and obtain an equation in p as the only dependent variable. By suitable rearrangement this equation can be put in the form of the regular one-dimensional wave equation. Obtain from this the general solution to the original problem.

2–20. *Determination of an acoustic wave from given initial conditions.* For plane waves satisfying the linearized one-dimensional wave equation in a uniform fluid medium of infinite extent, assume that the function $\dot{\xi}(x, 0)$, which specifies the initial velocities of all the particles, and the initial pressure distribution, $p(x, 0)$, are given. Find the general expression for $p(x, t)$.

Apply your general result to the special cases:

a) particles initially at rest,

$$p(x, 0) = \begin{cases} p_0, & x < 0, \\ 0, & x \geq 0. \end{cases}$$

Sketch the graph of $p(x, t)$ at some later time. (Such a situation would result if a long tube were divided in the middle by a membrane, the pressure on the two sides of the membrane being P_0 and P_0', and the membrane were ruptured at $t = 0$. For the linearized wave equation to apply, we must have $p_0 = P_0 - P_0' \ll P_0$.)

b) Given that the initial pressure and velocity distributions satisfy the condition characteristic of a +wave, $p(x, 0) = \rho_0 c \dot{\xi}(x, 0)$, show that your general result reduces to the pure +wave $p(x, t) = p(x - ct, 0)$.

3 Boundary value problems

3-1 REFLECTION AT A FIXED END OF TRANSVERSE WAVES ON A STRING

The unique determination of the function $y(x, t)$ specifying the transverse motion of a string can be looked upon as a process of finding physical requirements which progressively narrow the set of permissible functions until there is only a single function which meets all the conditions imposed. Thus it was illustrated in Section 1–8 that the solution is uniquely determined for an *infinite* string by the requirements that $y(x, t)$ satisfy the wave equation and that $y(x, t)$ and $\dot{y}(x, t)$ reduce at $t = 0$ to given functions. In this chapter we consider a type of condition which must be imposed when the region over which the wave equation is satisfied does not extend from $-\infty$ to $+\infty$. These conditions, known as *boundary conditions*, restrict the set of functions which are solutions of the wave equation to a specified subset which can be characterized by means of a general form. The unique solution can then be determined by the additional requirement that certain initial conditions be satisfied. A variety of boundary conditions will be illustrated through individual examples.

As the first example consider a semi-infinite string extending from $x = -\infty$ to $x = 0$. We assume that the wave equation must be satisfied *within* the interval $-\infty < x < 0$. The solution is thereby restricted to the general form

$$y(x, t) = f(t - x/c) + g(t + x/c). \tag{3–1}$$

We have chosen the arguments $(t \mp x/c)$ rather than $(x \mp ct)$ to simplify algebraic manipulations throughout this chapter (cf. Problem 1–2). Let us suppose that the string is rigidly clamped at the boundary $x = 0$. This imposes the requirement that the transverse displacement be identically zero at all times at this location. The boundary condition to be applied is thus

$$y(0, t) = 0. \tag{3–2}$$

Together with Eq. (3–1), this implies $0 = f(t) + g(t)$, or

$$g(t) = -f(t). \tag{3-3}$$

By application of the boundary condition, the function $y(x, t)$ has been restricted to the form

$$y(x, t) = f(t - x/c) - f(t + x/c), \tag{3-4}$$

where the choice of $f(t)$ still remains arbitrary. As a check, note that the condition $y(0, t) = 0$ is surely satisfied by Eq. (3–4).

Having learned this much about $y(x, t)$, we are now in a position to determine its unique form by specifying a set of initial conditions (Problem 3–2). There is another method of defining the problem which is actually more common in the present context than is the specification of initial conditions. In this method the function $f(t)$ is associated with a +wave which approaches the boundary. We may consider that the form of this *incoming* or *incident* wave is given. The problem then consists of determining the function $g(t)$ associated with the −wave, which in this context is called the *reflected* wave. When the problem is looked at in this way, we see that the solution has already been obtained in Eq. (3–3). The reflected wave is specified by $g(t + x/c) = -f(t + x/c)$. The shape of the reflected wave at a fixed time t_1 is given by the graph of $-f(t_1 + x/c)$ for $-\infty < x \leq 0$. This graph is simply related to the incident waveform by a series of geometrical transformations illustrated in Fig. 3–1. The reflection in the x-axis shows that the reflected pulse is inverted in form relative to the incident. The reflection in the y-axis corresponds to the common-sense requirement that the "leading edge" of the reflected pulse be on the left, whereas in the incident pulse it is

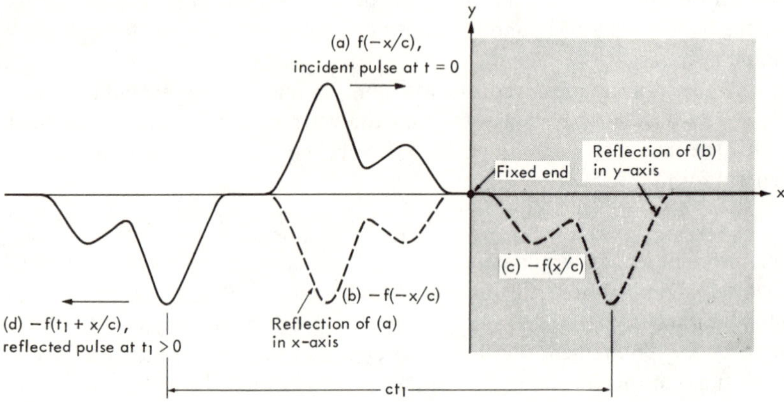

Fig. 3–1. Reflection of an arbitrary pulse at the fixed end of a string.

on the right. Since both the incident and the reflected waves have been determined, the net shape of the string at any desired time (when overlap occurs) can be obtained by graphical addition.

The net shape of the string produced by the superposition of a *sinusoidal* incident wave and its reflection is worthy of special description. If we choose $f(t) = y_m \cos \omega t$, Eq. (3–4) becomes

$$y(x, t) = y_m \cos(\omega t - kx) - y_m \cos(\omega t + kx).$$

Upon expanding the cosines, we find that this reduces to

$$y(x, t) = 2y_m \sin kx \sin \omega t. \tag{3–5}$$

The shape of the string is plotted at a series of representative times in Fig. 3–2.

Various features of the motion can be read from Eq. (3–5):

a) The motion of any fixed point on the string is a sinusoidal function of the time with period $T = 2\pi/\omega$.

b) Since $y(x, nT/2) = 0$ $(n = 0, 1, 2, \ldots)$, there are moments when the string is perfectly straight over its entire length.

c) At any time $t \neq nT/2$ the instantaneous shape of the string is sinusoidal. The amplitude varies as a function of the time. There are certain positions, referred to as *nodes*, at which there is never any displacement of the string. The nodes occur at $kx = -n\pi$, or $x = -n\lambda/2$, where the factor $\sin kx$ causes

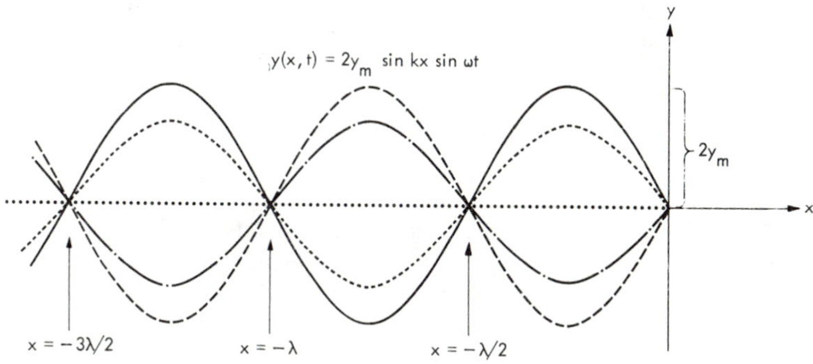

Fig. 3–2. A standing wave produced by the reflection of a sinusoidal wave at a fixed end.

$y(x, t)$ to vanish, regardless of the value of t. The fixed end is a node, and the spacing between successive nodes is a half-wavelength. Unlike the progressive wave, the sinusoidal waveform does not undergo translation parallel to the x-axis. This type of motion is referred to as a *standing wave*.

d) The amplitude of the simple harmonic motion of an individual particle depends upon its location. The particles halfway between the nodes have the largest amplitude for their motion. These halfway positions are referred to as *antinodes*.

3-2 REFLECTION OF ACOUSTIC WAVES AT A RIGID SURFACE

Suppose that a given waveform $\xi_{\text{inc}}(x, t) = f(t - x/c)$ is normally incident from $x < 0$ on a rigid wall at $x = 0$. We assume that the ideal rigid wall remains perfectly stationary despite the forces which act upon it due to the acoustic wave. We further assume that a vacuum cannot be created between the fluid and the wall, that is, the particles directly adjacent to the wall remain in that position. The appropriate boundary condition is therefore

$$\xi(0, t) = 0. \tag{3-6}$$

Since $\xi(x, t)$ must satisfy the wave equation within $x < 0$, it has the general form

$$\xi(x, t) = f(t - x/c) + g(t + x/c). \tag{3-7}$$

Application of the boundary condition therefore shows that $g(t) = -f(t)$, and the form of the reflected wave is related to that of the incident wave exactly as it was in the case of the string. For the sinusoidal incident wave $f(t) = \xi_m \cos \omega t$, the total disturbance (the sum of incident and reflected waves) is

$$\xi(x, t) = 2\xi_m \sin kx \sin \omega t. \tag{3-8}$$

In the resulting system of acoustic standing waves the rigid wall is a node for ξ and additional nodes are spaced at half-wavelength intervals.

The corresponding pressure field can be deduced from the relation $p = -\mathcal{B}_a(\partial\xi/\partial x)$:

$$p = -2\xi_m k \mathcal{B}_a \cos kx \sin \omega t. \tag{3-9}$$

A plot of pressure as a function of position yields a system of standing waves for which the nodes are located where $\cos kx = 0$, or $x = -(\lambda/4 + n\lambda/2)$, $n = 0, 1, 2, \ldots$ That is to say, the spacing between successive pressure nodes is a half-wavelength, the first node occurring at a distance of a quarter-wavelength from the rigid wall. At the wall the amplitude of the pressure variation is a maximum. Additional antinodes of pressure are located at $x = -n\lambda/2$.

Acoustic standing waves produced by reflection can be demonstrated by setting up a tuning fork at a reasonable distance from a flat surface such as a blackboard. The field in the vicinity of the surface can then be explored with a microphone. The positions at which nodes will be found depend on whether the microphone responds to pressure variations on the diaphragm, or whether a pressure gradient is required (Section 5–4). Using a pressure-sensitive microphone, for example, distinct minima can be found at $\lambda/4$ and $3\lambda/4$ from the reflecting surface.

Since the normal modes of a string or the harmonic vibrations of an organ pipe are often pointed to as examples of standing waves, the erroneous impression may be created that standing waves are a *resonance* phenomenon requiring the presence of two "ends" to the string or pipe. The above analysis shows that the standing wave results from the overlap of an incoming sinusoidal wave of any desired frequency and its reflection at a single boundary. Even if the incoming wave is a wavetrain consisting of a limited number of cycles of sinusoidal form, a standing wave exists in the region of overlap of the incident and reflected waves.

3–3 WAVES PRODUCED BY THE SPECIFIED MOTION OF A BOUNDARY SURFACE

Consider an acoustic problem in which a piston is set at one end of a long tube containing a fluid medium of semi-infinite extent (Fig. 3–3). Suppose that for $t < 0$ the piston is stationary and the fluid is in the undisturbed state. For $t \geq 0$, let the displacement of the piston from its initial position be given by the function $B(t)$. Since the wave equation must be satisfied in $x > 0$, the function $\xi(x, t)$, which specifies the particle displacements of the fluid, must have the general form

$$\xi(x, t) = f(t - x/c) + g(t + x/c). \tag{3–10}$$

We assume that there are no reflecting surfaces in $x > 0$ and no other possible sources of waves which would approach the piston. The statement of the problem therefore implies that no $-$wave is present, and that the function $g = 0$. The outgoing wave, f, is determined by the boundary condition which

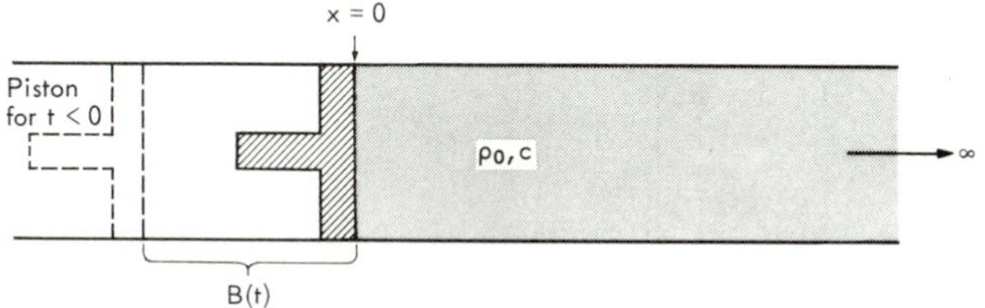

Fig. 3–3. Specified motion of a piston bounding a semi-infinite fluid.

must be met at the surface of the piston. We take this condition to be that the particles adjacent to the piston surface (those particles permanently labeled with the parameter $x = 0$) have the same motion as the piston itself. In symbols,

$$\xi(0, t) = B(t). \tag{3-11}$$

So long as the piston continues to move in the positive direction this condition is inescapable. If $\dot{B} < 0$, we must suppose that the velocity of the piston is small enough that cavitation does not develop. When we substitute into Eq. (3–11), the unknown function $f(t)$ is seen to be identical with the given $B(t)$:

$$B(t) = \xi(0, t) = f(t). \tag{3-12}$$

The entire motion of the fluid is therefore determined:

$$\xi(x, t) = f(t - x/c) = B(t - x/c). \tag{3-13}$$

This result is simple enough, and might have been anticipated without such a formal development; that is:

We assumed that only a +wave is generated. But in any +wave the disturbance which occurs at x is the same as that which occurred at $x = 0$ at a time x/c earlier.

It is wise, however, to become acquainted with the formalism in trivial cases so that the techniques will be familiar when applied in more complex circumstances.

EXAMPLE 1. Suppose that the piston moves with constant velocity v for $t \geq 0$:

$$B(t) = \begin{cases} 0, & t < 0, \\ vt, & t \geq 0. \end{cases}$$

Then

$$\xi(x, t) = \begin{cases} 0, & x > ct, \\ v(t - x/c), & x \leq ct. \end{cases}$$

By differentiation we obtain

$$\dot{\xi}(x, t) = \begin{cases} 0, & x > ct, \\ v, & x < ct. \end{cases}$$

From the relation $p = Z\dot{\xi}$ for a +wave,

$$p(x, t) = \begin{cases} 0, & x > ct, \\ Zv, & x < ct, \end{cases}$$

and

$$s(x, t) = p/\mathcal{B}_a = \begin{cases} 0, & x > ct, \\ v/c, & x < ct. \end{cases}$$

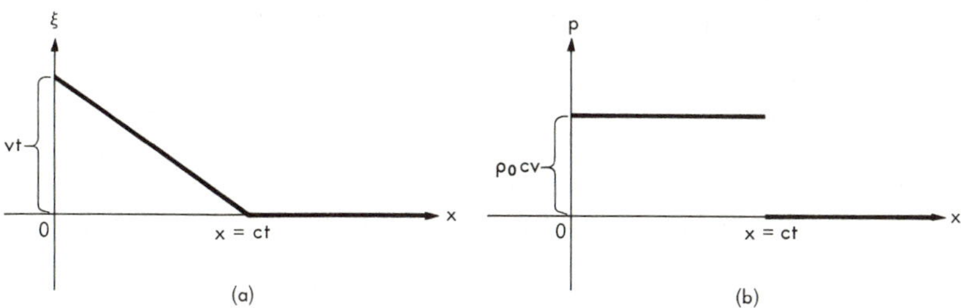

Fig. 3-4. Conditions resulting from uniform motion of boundary piston: (a) particle displacement, (b) excess pressure.

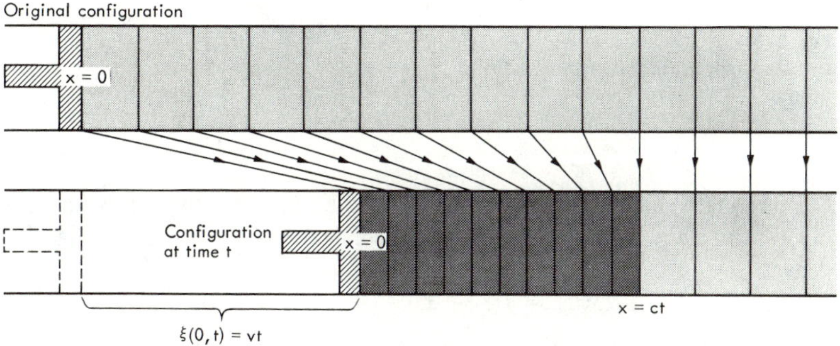

Fig. 3-5. Interpretation of Fig. 3-4(a) showing original and final locations of equally spaced planes of particles.

We thus find that at a particular location the medium remains undisturbed for the length of time required by a signal traveling with velocity c to reach there from the origin. After this time, the fluid particle starts moving with constant velocity v and continues to do so indefinitely. The pressure and density behind the advancing wavefront are uniform. Since we must have $s \ll 1$ for the acoustic approximation to be valid, the velocity with which the piston moves must be very much smaller than the velocity of sound. The graphs of ξ and p with a choice of $v > 0$ are shown in Fig. 3-4. The displacements of the particles at various locations are indicated in Fig. 3-5.

EXAMPLE 2. Consider

$$B(t) = \begin{cases} 0, & t < 0, \\ A \sin \omega t, & t \geq 0. \end{cases}$$

Then

$$\xi(x, t) = \begin{cases} 0, & x > ct, \\ A \sin (\omega t - kx), & x \leq ct. \end{cases}$$

In the region $0 \leq x \leq ct$, this is a sinusoidal wave of the type discussed previously.

3-4 REFLECTION AND TRANSMISSION AT AN INTERFACE

Let the plane $x = 0$ designate an interface between two semi-infinite media of acoustic parameters ρ_1, c_1 in $x < 0$, and ρ_2, c_2 in $x > 0$ (Fig. 3–6). If a plane wave of given form is incident from $x < 0$ upon this boundary surface, we inquire what kind of disturbance is created in medium 2 (transmitted wave) and what additional disturbance arises in medium 1 (reflected wave).

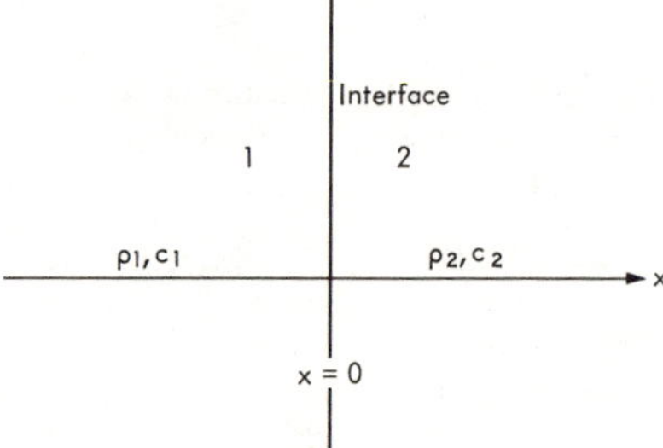

Fig. 3–6. Interface separating two semi-infinite media.

Let the total velocity field in 1 be designated by $\dot{\xi}_1(x, t)$. Since this function must satisfy the wave equation with propagation velocity c_1, it has the general form

$$\dot{\xi}_1(x, t) = f(t - x/c_1) + g(t + x/c_1). \tag{3–14}$$

Similarly, the velocity field in 2 has the general form

$$\dot{\xi}_2(x, t) = F(t - x/c_2) + G(t + x/c_2). \tag{3–15}$$

From the statement of the problem we assume that $G = 0$, for if not, this would imply the presence of a source or an additional reflecting surface which could produce a $-$wave traveling toward the boundary surface from $x > 0$. We assume that the incident waveform f is given, and that the *reflected* wave g and the *transmitted* wave F are the unknowns in the problem. Two boundary conditions are available for their determination:

i) The function $\xi_1(0, t)$ designates the displacement of fluid particles of medium 1 just adjacent to the interface, and $\xi_2(0, t)$, the corresponding displacement of the particles of medium 2. In the absence of cavitation these two displacements must be equal at all times:

$$\xi_1(0, t) = \xi_2(0, t). \tag{3–16}$$

Since we are working directly with the velocity functions in Eqs. (3–14) and (3–15) (a matter of algebraic convenience), we differentiate Eq. (3–16) with respect to t and take as the first boundary condition

$$\dot{\xi}_1(0, t) = \dot{\xi}_2(0, t). \tag{3–17}$$

This equation is a statement of the continuity of the function $\dot{\xi}(x, t)$ across the interface.

ii) Consider a small pillbox having one face in either medium (Fig. 3-7). The mass contained within this volume element approaches zero as Δx approaches zero, provided we assume that the "interface" is merely a mathematical surface separating the two continuous media. The net force acting on this volume element must therefore approach zero as Δx approaches zero. From this we conclude that the pressure must be a continuous function across the interface:

$$p_1(0, t) = p_2(0, t). \tag{3-18}$$

If an actual physical membrane having a finite mass per unit area separates the two media, this boundary condition will be altered.

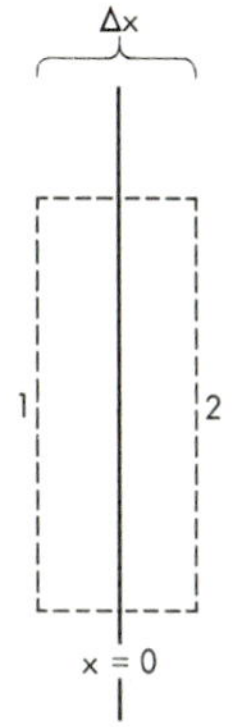

Fig. 3-7.
Imaginary pillbox surface constructed about the interface. (Consider unit area perpendicular to the plane of the figure.)

To apply the second boundary condition, we make use of Eq. (2-28):

$$p_1(x, t) = Z_1[f(t - x/c_1) - g(t + x/c_1)],$$
$$p_2(x, t) = Z_2F(t - x/c_2),$$

and therefore, at $x = 0$,

$$f(t) - g(t) = Z_{21}F(t), \tag{3-19}$$

where $Z_{21} = Z_2/Z_1$ is the relative impedance of the two media. The first boundary condition leads to

$$f(t) + g(t) = F(t). \tag{3-20}$$

The last two equations are a pair of linear equations in the unknown functions g and F. By elimination we obtain

$$g(t) = R_{\dot{\xi}}f(t), \qquad F(t) = T_{\dot{\xi}}f(t), \tag{3-21}$$

where $R_{\dot{\xi}} = (1 - Z_{21})/(1 + Z_{21})$, and $T_{\dot{\xi}} = 1 + R_{\dot{\xi}} = 2/(1 + Z_{21})$. The constants $R_{\dot{\xi}}$ and $T_{\dot{\xi}}$ are referred to as the *reflection* and *transmission coefficients* for the particle velocity function. The explicit solutions for the reflected and transmitted waves are therefore

$$\dot{\xi}_{\text{refl}}(x, t) = R_{\dot{\xi}}f(t + x/c_1), \qquad \dot{\xi}_{\text{trans}}(x, t) = T_{\dot{\xi}}f(t - x/c_2). \tag{3-22}$$

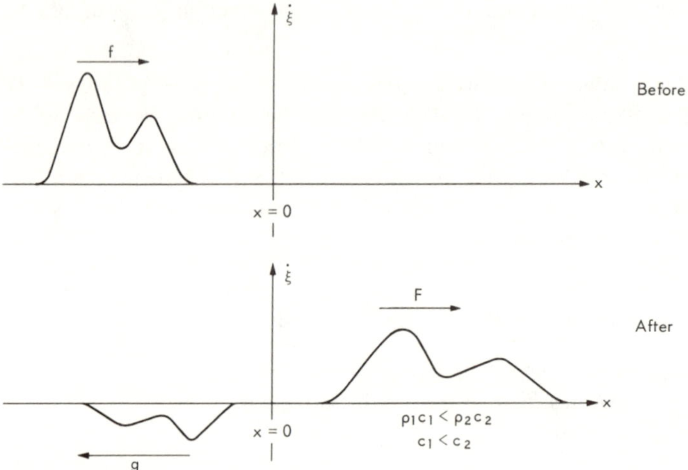

Fig. 3–8. Reflection and transmission of an arbitrary pulse incident from the low impedance side of an interface.

This completes the formal solution of the problem. The remainder of this section is concerned with interpretation of the results.

The form of the reflected wave is basically the same as that of the incident wave except for multiplication by the factor $R_{\dot{\xi}}$, which changes the scale of the ordinate. Since the defining equation for $R_{\dot{\xi}}$ shows that $|R_{\dot{\xi}}| < 1$, the reflected wave is always *reduced* in scale relative to the incident. The sign of $R_{\dot{\xi}}$ can be either negative or positive, depending on whether $Z_1 < Z_2$ or $Z_1 > Z_2$. Thus when the wave is reflected from the *low* impedance side the particle velocity function is inverted. That is to say, a crest in the graph of $\dot{\xi}$ is reflected as a trough (Fig. 3–8). Inversion in $\dot{\xi}$ does not take place when the wave is reflected from the *high* impedance side. An interesting case arises when the impedances of the two media are equal, for then $R_{\dot{\xi}} = 0$, and no reflection is produced at the interface. The values of ρ and c separately can be quite different for the two media, but, providing that $\rho_1 c_1 = \rho_2 c_2$, the boundary is "transparent" to an incident wave. An application of this principle is made in the design of a rubber encasement to protect the piston face of a sonar transducer from the corrosive action of sea-water. A type of rubber (known as "rho-c rubber") has been developed whose impedance matches that of the water, thus eliminating undesired reflections at the rubber-water interface.

The form of the transmitted wave is obtained from that of the incident by two scale transformations. First, the ordinate is multiplied by the factor $T_{\dot{\xi}}$. From the definition of $T_{\dot{\xi}}$ it can be seen that $T_{\dot{\xi}}$ is never negative, and the transmitted wave is therefore never inverted relative to the incident. The second scale factor applies to the abscissa. Since the coefficient of x in the argument of the function f is $1/c_2$ for the transmitted wave and $1/c_1$ for the incident wave, the abscissa is affected by a scale factor c_2/c_1. If $c_2 > c_1$, the transmitted waveform is stretched relative to the incident in a direction parallel to the x-axis, whereas if $c_2 < c_1$, the transmitted wave is compressed into a narrower range of values of x. The interpretation of this result is clear: the time it takes an entire pulse to pass a given point is the same in either medium, and hence the spatial widths of the pulses must be in the same ratio as the velocities.

The pressure functions in the incident, reflected, and transmitted waves follow from the relation $p_\pm = \pm Z\xi_\pm$:

$$\begin{aligned} p_{\text{inc}}(x, t) &= Z_1 f(t - x/c_1), \\ p_{\text{refl}}(x, t) &= -Z_1 R_\xi f(t + x/c_1), \\ p_{\text{trans}}(x, t) &= Z_2 T_\xi f(t - x/c_2). \end{aligned} \tag{3-23}$$

We define the reflection and transmission coefficients for pressure as the ratio at $x = 0$ of the pressure in the respective waves to that in the incident wave:

$$R_p \equiv \frac{p_{\text{refl}}(0, t)}{p_{\text{inc}}(0, t)} = -R_\xi = \frac{1 - Z_{12}}{1 + Z_{12}}, \tag{3-24}$$

$$T_p \equiv \frac{p_{\text{trans}}(0, t)}{p_{\text{inc}}(0, t)} = Z_{21} T_\xi = \frac{2}{1 + Z_{12}} = 1 + R_p. \tag{3-25}$$

It is interesting to note that we can obtain the pressure coefficients from the velocity coefficients by interchanging the role of the two media, i.e., by the replacement $Z_{21} \to Z_{12}$. The fact that $T_p > 0$ shows that the transmitted pressure wave is never inverted with respect to the incident. Likewise, the inequality $|R_p| < 1$ shows that the scale of the reflected pressure wave is reduced compared with that of the incident wave. The relation $R_p = -R_\xi$ points out the important fact that inversion of p and inversion of ξ are mutually exclusive. When $Z_1 < Z_2$, ξ is inverted, but p is not. That is, a condensation is reflected as a condensation. Conversely, when $Z_1 > Z_2$, ξ is not inverted, but a condensation is reflected as a rarefaction. Note that if it were not for this complementarity of sign change, the relation $p = Z\xi$ for the incident wave would not be properly converted into $p = -Z\xi$ for the reflected wave.

3-5 REFLECTION OF A SINUSOIDAL WAVE; PARTIAL STANDING WAVE

As an application of the last section let $f(t) = \cos \omega t$. This means that we are choosing an incident wave which is sinusoidal and of unit velocity amplitude. From Eq. (3-22), the reflected wave is then

$$\xi_{\text{refl}}(x, t) = R_\xi \cos \omega(t + x/c_1) = R_\xi \cos(\omega t + k_1 x), \tag{3-26}$$

where $k_1 = \omega/c_1$ is the wave number associated in medium 1 with a wave of frequency ω. The quantity R_ξ emerges in this context as the *amplitude* of the reflected wave. If R_ξ is positive, the reflected wave is in phase with the incident; if R_ξ is negative, we speak of a *phase reversal.* The transmitted wave is also obtained from Eq. (3–22):

$$\xi_{\text{trans}}(x, t) = T_\xi f(t - x/c_2) = T_\xi \cos \omega(t - x/c_2) = T_\xi \cos (\omega t - k_2 x),$$

where $k_2 = \omega/c_2$. The amplitude of the transmitted wave (relative to unit amplitude for the incident) is thus T_ξ. From the argument of the cosine function we observe that our general method of solution has proved that the transmitted wave has the same frequency as the incident. The scale change on the abscissa referred to in the last section has, in the context of a sinusoidal wave, the simple interpretation that the wavelengths in the two media are in the same proportion as the propagation velocities.

From the fact that $c = \lambda\nu$ and $c_1 \neq c_2$, we know that the frequencies or the wavelengths, or possibly both must differ in the two media. Direct application of the boundary conditions has shown rigorously that the frequency does not change. Although it was not necessary to introduce this as an *ad hoc* assumption, there is a simple argument which leads to this conclusion: each crest arriving at the boundary from the incident wave originates a crest in the transmitted wave; therefore if ν crests arrive in each second, ν crests must leave. In future problems dealing with sinusoidal waves we shall consider it permissible to assume at the outset that the frequency is the same throughout.

The net disturbance in medium 1 resulting from the superposition of the incident and reflected waves is a phenomenon referred to as a *partial* standing wave. Thus, if the incident wave has unit pressure amplitude, the net pressure in medium 1 is

$$p_1(x, t) = \cos (\omega t - k_1 x) + R_p \cos (\omega t + k_1 x). \tag{3–27}$$

Considered as a function of either x or t this is the sum of two sinusoidal functions, and hence we may assert (Appendix I) that the shape of the pressure curve is sinusoidal at all times, and that the pressure history of any particle is sinusoidal. It is left as an exercise (Problem AI–3) for the reader to show that p_1 may be written in the form

$$p_1(x, t) = A(x) \cos [\omega t - \psi(x)],$$

where $A(x) = \sqrt{1 + 2R_p \cos 2k_1 x + R_p^2}$, and $\psi(x)$ is a phase angle which does not concern us here. In contrast to the full standing wave, the zeros of p_1 and its extreme values do not occur at fixed locations. If we assume $Z_1 < Z_2$ and hence $R_p > 0$, we find by inspection that $A(x)$ ranges from a maximum value $(1 + R_p)$ at the locations where $\cos 2k_1 x = +1$

to a minimum value $(1 - R_p)$ where $\cos 2k_1x = -1$. The former positions, which are analogous to the pressure antinodes in a full standing wave, occur at $x = -n\lambda_1/2$, $n = 0, 1, 2, \ldots$ The positions analogous to the nodes occur at $x = -n\lambda_1/2 - \lambda_1/4$. The *standing wave ratio* $(1 - R_p)/(1 + R_p)$ is a measure of the contrast available when one explores the field with a pressure-sensitive microphone. By substitution from Eq. (3–24), we find that the standing wave ratio reduces to $(1 - R_p)/(1 + R_p) = Z_{12}$.

3–6 EXTREME MISMATCH OF IMPEDANCES; RIGID AND FREE SURFACES

The formulas of Section 3–4 reduce to two interesting special cases when the impedance of the second medium is assumed to be either infinite or zero in comparison with the first:

i) $Z_{21} = \infty$. Then from Eqs. (3–21), (3–24), and (3–25), $R_\xi = -1$, $T_\xi = 0$, $R_p = +1$, and $T_p = 2$. The field in medium 1 is

$$\xi_1(x, t) = f(t - x/c_1) - f(t + x/c_1),$$

which we recognize to be identical with the solution obtained in Section 3–2 for reflection from an ideal rigid surface. The field obtained in medium 2 under these extreme conditions is not of direct interest because of the fictitious nature of infinite impedance. On the other hand, consider the transmission of a wave from air to water. The velocity of sound in water is about five times that in air, and the density of water is about 800 times that of air. The impedance ratio Z_{21} (2 being water, 1 air) is therefore on the order of 4000, which, though not infinite, is nevertheless very large in comparison with unity. The appropriate thing to do is to express the reflection and transmission coefficients as functions of the small quantity Z_{12}, and to expand by the binomial theorem. Thus,

$$\begin{aligned} R_p &= (1 - Z_{12})/(1 + Z_{12}) \doteq (1 - Z_{12})^2 \doteq 1 - 2Z_{12}, \\ R_\xi &= -R_p \doteq -1 + 2Z_{12}, \\ T_p &= 1 + R_p \doteq 2 - 2Z_{12}, \\ T_\xi &= 1 + R_\xi \doteq 2Z_{12}. \end{aligned} \tag{3–28}$$

The reflection coefficients show that the field in air is adequately close to that described by ideal reflection at a rigid surface. The significant new feature which we learn from this expansion is that the velocity amplitude (assuming an incident sinusoidal wave of unit velocity amplitude) is not zero in medium 2, but is the small quantity $2Z_{12}$.

ii) $Z_{21} = 0$. For this case, $R_\xi = +1$, $T_\xi = 2$, $R_p = -1$, and $T_p = 0$.

The field in medium 1 has velocity function

$$\dot{\xi}_1(x, t) = f(t - x/c_1) + f(t + x/c_1) \tag{3-29}$$

and pressure function

$$p_1(x, t) = Z_1[f(t - x/c_1) - f(t + x/c_1)]. \tag{3-30}$$

The latter satisfies the identity $p_1(0, t) = 0$. Thus, whenever a wave is incident upon a medium of relatively negligible impedance, the excess acoustic pressure must be zero at all times at the boundary surface. Under these conditions the surface is referred to as a "free surface." Equation (3-29) shows that the extent of the motion of the particles at the boundary is just twice what it would have been if the incident wave alone had been present without the reflected wave: $\dot{\xi}_1(0, t) = 2f(t) = 2\dot{\xi}_{\text{inc}}(0, t)$. In the case of a sinusoidal wave, the *partial* standing wave of the last section reduces to the special case of a *full* standing wave, since the minimum pressure amplitude $(1 + R_p)$ is in this case zero, and the corresponding locations are true nodes. The free surface itself is a pressure node and a particle velocity and displacement antinode.

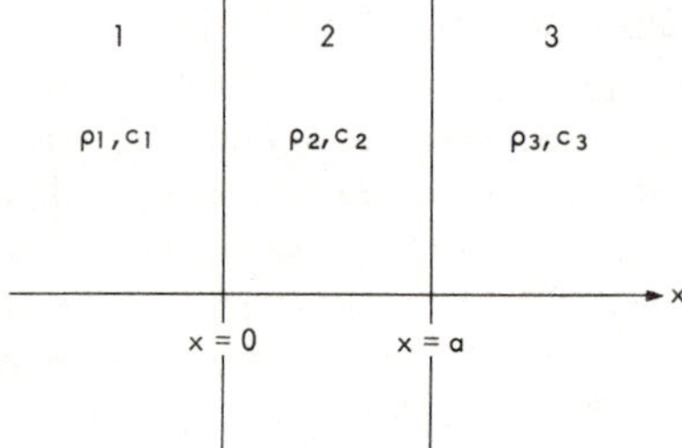

Fig. 3-9. Layer of thickness *a* separating two semi-infinite media.

3-7 REFLECTION OF A SINUSOIDAL WAVE FROM A PAIR OF INTERFACES

Consider three media of acoustic parameters ρ_i, c_i, $i = 1, 2, 3$, where medium 1 extends throughout $x < 0$, medium 2 throughout $0 \leq x \leq a$, and medium 3 throughout $x > a$ (Fig. 3-9). A given sinusoidal wave is normally incident from $x < 0$ on the interface at $x = 0$. The general problem consists in determining the reflected wave in medium 1, the transmitted wave in medium 3, and the kind of disturbance which occurs in medium 2. The present discussion will concentrate on the reflected wave, the others being brought in only when necessary. The most straightforward approach is to treat this as a boundary value problem with conditions to be met at the surfaces $x = 0$ and $x = a$. If the incident wave is of the form

$$\dot{\xi}_{\text{inc}} = \mathcal{R}e \left\{e^{i(\omega t - k_1 x)}\right\},$$

then we may safely assume* from our experience in Section 3–5 that all the other relevant wave functions will be sinusoidal with the same frequency ω. For example, the reflected wave can be taken to be of the form $\xi_{\text{refl}} = \mathcal{R}e\,\{\mathbf{R}e^{i(\omega t + k_1 x)}\}$, where the *complex* reflection coefficient $\mathbf{R}$ is an unknown to be determined. The author feels at this point that the presentation of further details on this problem would deprive the reader of an excellent opportunity to enjoy an experience which is perhaps more typical of what a working physicist does than are most "exercises" that the reader has encountered up to now. Some additional physical thinking is required to set up the problem, and its execution involves maximal use of mathematical techniques already at the reader's disposal. From this point on a series of results will be stated and then left for the reader to verify. The worked problem in Section 3–9 can be used as a model for techniques which are convenient for problems of this type.

1)
$$\mathbf{R} = \frac{R_{12} + R_{23}e^{-2ik_2 a}}{1 + R_{12}R_{23}e^{-2ik_2 a}}, \tag{3–31}$$

where R_{ij} is the ordinary reflection coefficient for ξ at the i-j interface: $R_{ij} = (1 - Z_{ji})/(1 + Z_{ji})$.

2) This result reduces correctly to the expected value in each of the following special cases: $Z_2 = Z_3$, $Z_1 = Z_2$, $a = 0$.

3) Under certain conditions it is possible for the reflection coefficient $\mathbf{R}$ to be zero. This implies that even though different impedances are involved, the layer is completely transparent to the incident wave. The possible conditions under which there will be no reflected wave are:

a) The impedances of 1 and 3 are equal and the layer is an integral number of half-wavelengths thick. (The wavelength in question is that of the wave in medium 2.)

b) The impedance of 2 is the geometric mean of the impedances of the outer media and the layer is an odd number of quarter-wavelengths thick.

c) The impedances of all three media are equal (essentially a trivial case).

4) In any of the cases for which $\mathbf{R} = 0$ the reflections at the separate interfaces are "equally strong," that is $|R_{12}| = |R_{23}|$.

We note in passing that because of the presence of k_2 in Eq. (3–31), the reflection coefficient is dependent upon the frequency of the incident wave. Although the significance of this dependence cannot be fully appreciated until Fourier analysis has been studied, it turns out to be associated with the fact that an incident pulse of *arbitrary* shape is deformed upon reflection. It is for this reason that the statement of the problem was given directly in

* If the assumption should be unwarranted we would learn about it through our inability to meet all the boundary conditions.

terms of a sinusoidal wave, rather than by means of more general functional notation. When the latter approach is used, the solution proceeds easily up to the application of the boundary conditions, but the functional equations which result require more advanced techniques for their solution.

The problem discussed in this section has a number of important applications in other situations governed by the wave equation. The design of a nonreflecting film for coating lenses and other optical surfaces is based upon this principle, and will be given as a problem in a later chapter (Problem 7–5). In quantum mechanics the reflection and transmission of an electron beam by a thin film is a direct application of the results of this section, including the surprising fact that the film is transparent to the beam when the required conditions are met. Simplified models of neutron-proton scattering and of the radioactive decay of a nucleus through emission of an alpha particle also make use of the same mathematical analysis.

3-8 REFLECTION OF A SINUSOIDAL WAVE AT A PAIR OF INTERFACES, ALTERNATE METHOD

The last section recommended treating the problem of reflection at a pair of interfaces as a boundary value problem as-a-whole. A less elegant method, which may nevertheless yield some new insights into the phenomenon, determines the reflected wave by successive application of the solution to the problem of reflection at a single interface. Consider an incident pulse of short duration, $\xi_{\text{inc}} = f(t - x/c_1)$, where $f(t)$ vanishes outside the interval $0 \leq t \leq \tau$. If $\tau \ll a/c_2$, we may certainly treat the interaction of the pulse with the first boundary as a single-interface problem, ignoring temporarily the presence of the interface at $x = a$. A reflected wave g_1 and a transmitted wave F_1 are produced in this interaction, both of these being given by Eq. (3–22). The transmitted wave proceeds across the layer and encounters the interface at $x = a$ (Fig. 3–10). The interaction at this interface produces a transmitted wave, $\mathfrak{F}_1$, and a reflected wave, G_1, which proceeds back across the layer. When it reaches $x = 0$, G_1 gives rise to a transmitted wave, g_2, which proceeds without further mishap toward $x = -\infty$, and a reflected wave, F_2, which is sent back across the layer. Continuing in this manner, the net disturbance in $x < 0$ is seen to consist of the primary reflection g_1 plus a series of "echoes" $g_2, g_3, \ldots$ (Fig. 3–11). Each of the functions g_i can be found by using the single-interface formulas. The reflected wave is given by the sum

$$\xi_{\text{refl}}(x, t) = \sum_{i=1}^{\infty} g_i. \tag{3–32}$$

If the incident wave is an infinite sinusoid, it is not possible to separate this process into a succession of clearly defined stages. However, it should be

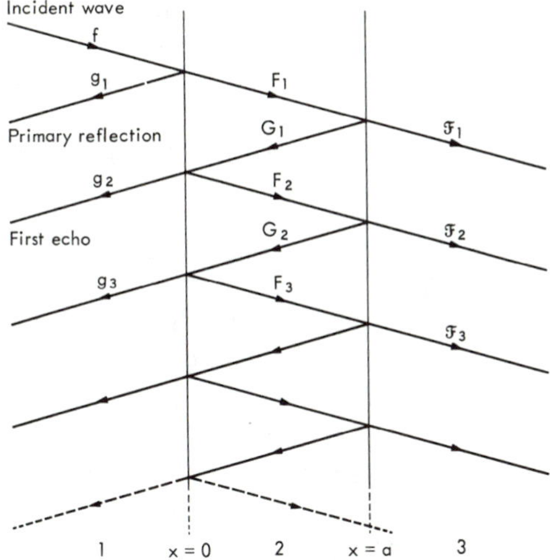

Fig. 3-10. Schematic diagram of multiple reflections between a pair of interfaces. (The rays represent plane waves which are parallel to the interfaces. They are drawn obliquely to prevent overlap.)

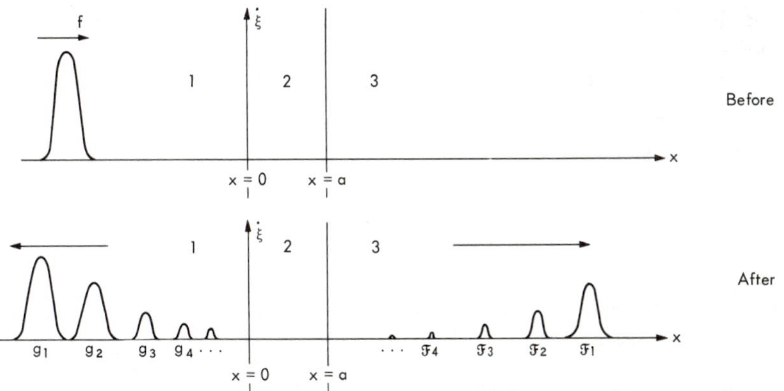

Fig. 3-11. Succession of pulses reflected and transmitted by a layer.

clear that a set of functions constructed along these lines will satisfy the boundary conditions at both surfaces, and hence constitutes a solution to the problem. The reader is invited to write out the explicit form of the functions g_1, F_1, G_1, etc., and to show that the resulting solution is identical with that given in the last section. The infinite series which results from Eq. (3–32) may at first sight seem unfamiliar, but will be recognized as a series whose sum is quite well known if use is made of the identity $e^{in\theta} = (e^{i\theta})^n$.

For the case of a sinusoidal wave, a simple comparison of the relative phases of g_1 and g_2 (the primary reflection and the first echo) will contribute to an understanding of the thickness requirements for nonreflection. If the phase factor in the exponential representation of the incident wave f is taken as unity at $x = 0$, then that of g_1 is ± 1 depending upon the sign of the reflection coefficient R_{12}. The phase factor of F_1 is $+1$ at $x = 0$, and e^{-ik_2a} at $x = a$. The phase factor of G_1 at $x = a$ is $\pm e^{-ik_2a}$, depending on the sign of R_{23}. Now, since G_1 is a $-$wave, its phase is changed by the factor e^{-ik_2a} in traveling back across the layer, and is thus $\pm e^{-2ik_2a}$ at $x = 0$. No change of phase occurs upon transmission at $x = 0$, so the phase factor of g_2 is $\pm e^{-2ik_2a}$. Consider now the two cases:

a) Z_2 *intermediate* in value between Z_1 and Z_3. The signs of R_{12} and R_{23} are equal, i.e. both -1 for $Z_1 < Z_2 < Z_3$, or $+1$ for $Z_1 > Z_2 > Z_3$. The relative phase factor between g_1 and g_2 is then e^{-2ik_2a}. If now $e^{-2ik_2a} = -1$ (which requires $a = (2n + 1)\lambda_2/4$), these two waves are in a destructive relation with each other, and will *tend* to cancel each other. Of course, the amplitude of g_2 is not the same as that of g_1, so these two alone do not produce complete destructive interference. The explicit solution shows that the remaining echoes are in phase with g_2, so that they are all contributing to cancellation with g_1. The complete solution further shows that if $Z_2 = \sqrt{Z_1Z_3}$, the amplitudes are correct to yield complete cancellation.

b) If Z_2 is not between Z_1 and Z_3, then R_{12} and R_{23} have opposite sign. The relative phase factor between g_1 and g_2 is $-e^{-2ik_2a}$, and the maximum tendency to destructive interference occurs when $a = n(\lambda_2/2)$. Simple phase considerations therefore lead us to expect that these will be the appropriate widths if the layer is to be nonreflecting. The complete solution further shows that complete cancellation is achieved only if $Z_1 = Z_3$.

*3-9 REFLECTION AND TRANSMISSION OF A SINUSOIDAL WAVE BY A THIN PLATE

Let the plane $x = 0$ coincide with a thin plate (membrane, pane of glass, etc.) having mass μ per unit area. The region $x < 0$ is filled with a fluid medium of acoustic parameters ρ_1, c_1, and the region $x > 0$, with a fluid of parameters ρ_2, c_2. A given sinusoidal wave is normally incident upon the plate from $x < 0$. We wish to determine the reflected and transmitted waves. The complete disturbance in medium 1 has the general form

$$\dot{\xi}_1(x, t) = \dot{\xi}_{\text{inc}}(t - x/c_1) + \dot{\xi}_{\text{refl}}(t + x/c_1),$$

and in medium 2,

$$\dot{\xi}_2(x, t) = \dot{\xi}_{\text{trans}}(t - x/c_2).$$

If the displacement of the plate as a function of the time is the unknown function $B(t)$, the assumption of no cavitation gives the two equations

$$\begin{aligned} \dot{\xi}_1(0, t) &= \dot{B}(t), \\ \dot{\xi}_2(0, t) &= \dot{B}(t). \end{aligned} \tag{3-33}$$

An additional condition is obtained when we apply Newton's second law to the dynamics of motion of a section of the plate of unit area. The mass is μ, the acceleration is $\ddot{B}(t)$, and the net force is

$$P_1(0, t) - P_2(0, t) = p_1(0, t) - p_2(0, t).$$

(We assume that the equilibrium pressure P_0 is the same on both sides of the plate.) Therefore the boundary condition is

$$p_1(0, t) - p_2(0, t) = \mu\ddot{B}(t). \tag{3-34}$$

Let the incident wave have unit velocity amplitude:

$$\dot{\xi}_{\text{inc}} = \mathcal{R}e\,\{e^{i(\omega t - k_1 x)}\}.$$

We assume the reflected and transmitted waves to be of the form*

$$\dot{\xi}_{\text{refl}} = \mathcal{R}e\,\{\mathbf{R}e^{i(\omega t + k_1 x)}\}, \tag{3-35}$$

$$\dot{\xi}_{\text{trans}} = \mathcal{R}e\,\{\mathbf{T}e^{i(\omega t - k_2 x)}\}. \tag{3-36}$$

From Eq. (3–33), the condition $\dot{\xi}_1(0, t) = \dot{\xi}_2(0, t)$ gives the equation

$$\mathcal{R}e\,\{(1 + \mathbf{R})e^{i\omega t}\} = \mathcal{R}e\,\{\mathbf{T}e^{i\omega t}\}. \tag{3-37}$$

Although one might be tempted to "cancel" the $\mathcal{R}e$ and $e^{i\omega t}$ from both sides of this equation, this is in fact not a legitimate operation with complex numbers.† However, we are helped by the fact that Eq. (3–37) must hold at all times t. Collecting terms on the left hand side, we have

$$\mathcal{R}e\,\{(1 + \mathbf{R} - \mathbf{T})e^{i\omega t}\} = 0. \tag{3-38}$$

We now proceed to prove a lemma:

If $\mathcal{R}e\,\{\mathbf{A}e^{i\omega t}\} = 0$ for all t, then $\mathbf{A} = 0$.

---

* See footnote, p. 53.

† That $\mathcal{R}e\,(\mathbf{u}\cdot\mathbf{v}) = \mathcal{R}e\,(\mathbf{u}\cdot\mathbf{w})$ in general implies neither $\mathbf{v} = \mathbf{w}$ nor $\mathcal{R}e\,\mathbf{v} = \mathcal{R}e\,\mathbf{w}$ can be seen from a counterexample, say $\mathbf{u} = 1 + 2i$, $\mathbf{v} = i$, and $\mathbf{w} = -2$. Thus $\mathcal{R}e\,(\mathbf{u}\cdot\mathbf{v}) = \mathcal{R}e\,(i - 2) = -2$ and $\mathcal{R}e\,(\mathbf{u}\cdot\mathbf{w}) = \mathcal{R}e\,(-2 - 4i) = -2$.

Proof. Let $\mathbf{A} = a + bi$. Then the choice of $t = 0$ shows $\mathcal{R}e\,\mathbf{A} = a = 0$, and the choice of $t = \pi/2\omega$ shows $\mathcal{R}e\,(\mathbf{A}\cdot i) = -b = 0$. Hence $\mathbf{A} = 0 + 0i = 0$.

Thus we can be sure that the equation

$$\mathbf{T} = 1 + \mathbf{R} \tag{3–39}$$

is not only a sufficient but also a necessary condition for Eq. (3–38).

From Eqs. (3–33) and (3–36) we may write

$$\mu\ddot{B}(t) = \mu\ddot{\xi}_2(0, t) = \mathcal{R}e\,\{i\mu\omega\mathbf{T}e^{i\omega t}\}.$$

The well-known relation between pressure and particle velocity functions enables us to write

$$p_1(0, t) - p_2(0, t) = Z_1[\dot{\xi}_{\text{inc}}(0, t) - \dot{\xi}_{\text{refl}}(0, t)] - Z_2\dot{\xi}_{\text{trans}}(0, t).$$

Thus the dynamical boundary condition (Eq. 3–34) can be written

$$\mathcal{R}e\,\{[Z_1(1 - \mathbf{R}) - Z_2\mathbf{T}]e^{i\omega t}\} = \mathcal{R}e\,\{i\mu\omega\mathbf{T}e^{i\omega t}\}.$$

Invoking the recent lemma, we then have

$$Z_1(1 - \mathbf{R}) - Z_2\mathbf{T} = i\mu\omega\mathbf{T}. \tag{3–40}$$

Equations (3–39) and (3–40) are a pair of linear equations in the unknowns $\mathbf{T}$ and $\mathbf{R}$. By elimination, we obtain

$$\mathbf{R} = \frac{R_{12} - ir}{1 + ir} \quad \text{and} \quad \mathbf{T} = 1 + \mathbf{R} = \frac{T_{12}}{1 + ir}, \tag{3–41}$$

where

$$R_{12} = \frac{1 - Z_{21}}{1 + Z_{21}}, \qquad T_{12} = 1 + R_{12}, \qquad \text{and} \qquad r = \frac{\mu\omega}{Z_1 + Z_2}.$$

To study these results consider first the following special cases:

1) $r \ll R_{12}$ (and, since $|R_{12}| < 1$, this means also that $r \ll 1$). Both $\mathbf{R}$ and $\mathbf{T}$ then reduce to the same value they would have if the plate were not present. This condition prevails if the mass per unit area of the plate is "small," i.e., small compared with the quantity $|Z_1 - Z_2|/\omega$. Another way of stating it is that this condition prevails at "sufficiently low frequencies," i.e., $\omega \ll |Z_1 - Z_2|/\mu$.

2) $r \gg 1$. Then $\mathbf{R} \doteq -1$ and $\mathbf{T} \doteq 0$. Thus, if the plate is sufficiently heavy, it behaves like a rigid boundary. Equivalently, any plate acts like a rigid boundary to waves of sufficiently high frequency.

3) $Z_1 \ll Z_2$. Then $R_{12} \doteq -1$ and $R \doteq (-1 - ir)/(1 + ir) = -1$. As might be expected, therefore, when the second medium is of relatively high impedance, the interface behaves like a rigid boundary regardless of the mass of the plate.

4) $Z_1 \gg Z_2$. Then $R_{12} \doteq +1$ and $\mathbf{R} \doteq (1 - ir)/(1 + ir)$. If we write $1 + ir = \sqrt{1 + r^2}e^{i\phi}$ and $1 - ir = \sqrt{1 + r^2}e^{-i\phi}$, where $\tan\phi = r$, then $\mathbf{R} \doteq e^{-2i\phi}$. This means that the amplitude of the reflected wave is the same as that of the incident, but there has been a change in phase. The interface acts like a free surface only if $r \ll 1$. But if $r \gg 1$, then $\phi \doteq \pi/2$, $\mathbf{R} \doteq e^{-\pi i} = -1$, and the interface acts like a rigid boundary.

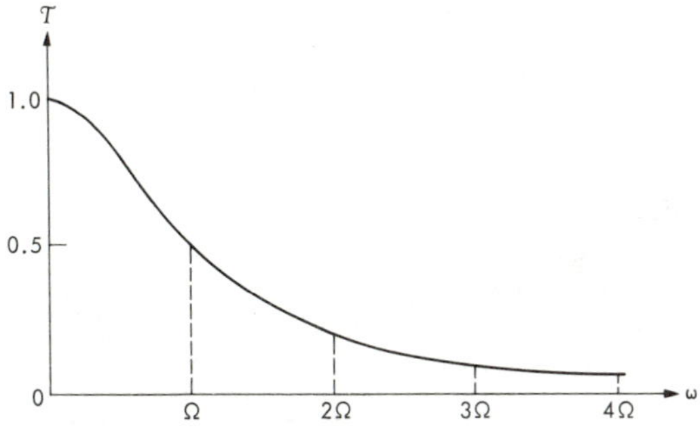

Fig. 3-12. Transmissivity of a plate as a function of frequency.

For the further study of Eq. (3-41), we anticipate the result which will be shown in the next chapter that the intensity of a wave within a given medium is proportional to the square of the amplitude. The *transmissivity of the plate*, defined by $\mathcal{T} = |\mathbf{T}|^2/(T_{12})^2$, represents the ratio of the intensity which the transmitted wave has when the plate is present to the intensity which would be obtained with the plate removed. In particular, if the two outside media are identical, then $T_{12} = 1$, and we can interpret $\mathcal{T}$ directly as the intensity of the transmitted wave relative to unit intensity of the incident. From Eq. (3-41),

$$\mathcal{T} = 1/(1 + r^2) = 1/[1 + (\omega/\Omega)^2],$$

where $\Omega = (Z_1 + Z_2)/\mu$. The graph of $\mathcal{T}$ versus ω is given in Fig. 3-12. The plate acts like an acoustic filter which cuts out frequencies much above the *characteristic frequency* Ω. Thus for $\omega > 3\Omega$, $\mathcal{T} < 0.1$; in other words less than 10% of the energy is passed. The characteristic frequency Ω runs the gamut of frequencies in the audible range when various common objects such as a sheet of paper, a windowpane or a wood panel are considered (see Problem 3-16).

PROBLEMS

3–1. Compare the graph of $f(x)$ with the graphs of the following functions (a and b are positive constants):

a) $-f(-x)$ b) $f(ax)$ c) $bf(ax)$ d) $f(b - ax)$.

3–2. Show that the initial conditions $y(x, 0) = \phi(x)$, and $\dot{y}(x, 0) = \psi(x)$, where ϕ and ψ are defined only for negative values of their arguments, completely determine the future motion of the semi-infinite string with fixed end.

3–3. An incoming pulse on a semi-infinite string with fixed end has the initial triangular shape

$$y_{\text{inc}}(x, 0) = \begin{cases} 0, & x < -2b, \\ 2a + ax/b, & -2b \leq x \leq -b, \\ -ax/b, & -b < x \leq 0. \end{cases}$$

Sketch the shape of the string at $t = b/2c$, $t = 3b/4c$, and $t = b/c$.

3–4. A piston is located at $x = 0$ adjacent to a fluid extending throughout $x > 0$. The piston is initially at rest, and there is no disturbance in the fluid. At $t = 0$ the piston starts to move with constant negative acceleration: $B(t) = -at^2/2$. Describe the condition of the fluid at some later time.

***3–5.** A piston is located at $x = 0$, adjacent to a fluid extending throughout $x > 0$. At $t = 0$ there is an incoming wave having initial shape $\xi_{\text{inc}}(x, 0) = f(x)$. In addition, the piston is set in motion with a specified displacement $B(t)$. Find a formal solution for $\xi(x, t)$.

3–6. A layer of water overlies another liquid of unknown acoustic impedance. Assume both media to be of semi-infinite extent. A generator of ideal plane waves (parallel to the interface) is placed in the water. If a pressure condensation is reflected as a rarefaction of magnitude half that of the incident pulse, what is the unknown impedance?

3–7. The boundary conditions at a simple interface are equivalent to the assertion that the functions $\xi(x, t)$ and $p(x, t)$ are continuous across the boundary surface. Which of the functions $\partial p/\partial x$, $\dot{\xi}$, $\ddot{\xi}$, s, $\partial p/\partial t$ are likewise continuous?

3–8. Sketch the graph of p_1 versus x for the partial standing wave of Eq. (3–27) at $t = 0, T/4, T/2$ $(T = 2\pi/\omega)$.

3–9. Start with the boundary condition of a free surface, $p(0, t) = 0$, and apply this directly (in a manner similar to the method of Section 3–2) to deduce $p(x, t)$ and $\dot{\xi}(x, t)$.

3–10. Two pulses are simultaneously incident from opposite sides on a plane boundary separating two semi-infinite media of parameters ρ_1, c_1 and ρ_2, c_2. Is it possible to arrange the relation between these pulses in such a way that there will be no "reflected wave" in one of the two media? (That is, in such a way that the solution does not contain any function representing a wave traveling away from the boundary surface in one of the media.)

3–11. *The principle of impedance matching.* It is desired to transmit a pressure pulse between two media of different impedance. At first sight it would seem that the introduction of an *intervening* layer would reduce the efficiency of transmission, because there would then be a loss due to reflection at *two* surfaces instead of one. Show, however, that providing the impedance of the intervening layer is intermediate between the impedances of the outer media, the primary transmitted pulse (corresponding to $\mathfrak{F}_1$ in Section 3–8) will be stronger than that obtained by direct transmission. Find the impedance of the intervening medium which will give maximum transmission. (*Suggestion:* Remember the convenience of logarithmic differentiation.) (The results of this problem suggest that nearly 100% transmission might be obtained by the introduction of a large number of layers whose impedances vary in small steps between those of the outer two media. In the limiting case, the transition layer would be a region within which the impedance is a continuous function which matches the impedances at the two ends. More advanced analysis shows that 100% transmission is in fact obtained, provided that the transition is sufficiently gradual. For a sinusoidal wave the requirement is that the relative change in impedance over one wavelength be small: $\lambda dZ/dx \ll Z$.)

3–12. The bottom of a deep body of water is covered with a uniform layer of sediment of thickness a. Apply the results of Section 3–7 to determine the character of the reflection of a sinusoidal plane wave which originates in the water and is normally incident on the bottom layer. (Assume that the sediment has the properties of an ideal fluid, and that the interface between the sediment and the underlying rock acts like a rigid boundary.)

***3–13.** *Generalization of the condition for nonreflection.* Find the general criterion for zero reflection at a pair of interfaces of an incident pulse of arbitrary form. Interpret the results and compare the specialized case of a sinusoidal wave with the results of Section 3–7. [*Hint:* With

$$\dot{\xi}_1(x, t) = \dot{\xi}_{\text{inc}} = f(t - x/c_1),$$
$$\dot{\xi}_2(x, t) = F(t - x/c_2) + G(t + x/c_2),$$

and

$$\dot{\xi}_3(x, t) = \mathfrak{F}(t - x/c_3),$$

there are three unknown functions to be solved for in terms of the given function f. Since there are *four* conditions imposed by the continuity of p and $\dot{\xi}$ at $x = 0$ and $x = a$, the problem is "overdetermined" and results in a condition on the nature of f.]

***3–14.** *Reflection of a nonsinusoidal wave at a pair of interfaces.* Show that for a general pulse $\dot{\xi}_{\text{inc}} = f(t - x/c_1)$ the reflected wave will be given by $\dot{\xi}_{\text{refl}}(x, t) = g(t + x/c_1)$, where

$$g(t) = R_{12}f(t) + R_{23}(1 - R_{12}^2) \sum_{n=1}^{\infty} (-R_{12}R_{23})^{n-1} f(t - n\tau),$$

$$R_{ij} = (1 - Z_{ji})/(1 + Z_{ji}), \qquad \tau = 2a/c_2.$$

Compare the specialized case of a sinusoidal wave with previous results.

***3–15.** If the wave transmitted by a pair of interfaces (Section 3–7) is written

$$\xi_{\text{trans}} = \mathcal{R}e\,\{\mathbf{T}e^{i(\omega t - k_3 x)}\},$$

solve the equations of the boundary value problem to obtain **T**. Show that the identity $1 = |\mathbf{R}|^2 + Z_{31}|\mathbf{T}|^2$ is satisfied. (Considerations based on the material of the next chapter will enable the reader to recognize this identity as a statement of the law of conservation of energy.)

3–16. In the problem of the transmission of a sinusoidal wave by a thin plate assume that the medium on both sides of the plate is air at standard temperature and pressure.

a) Determine the characteristic frequency for a sheet of paper, assuming that an $8'' \times 10''$ sheet has a mass of 3.4 gm.

b) Show that, in general, the characteristic wavelength is $\Lambda_1 = \pi a(\rho/\rho_1)$, where a is the thickness of the plate, ρ its density, and ρ_1 the density of the air.

c) Find the thickness of a pane of glass which has a characteristic frequency corresponding to middle C.

3–17. Show that the reflection coefficient obtained in Section 3–9 for the reflection of a sinusoidal wave by a "thin plate" can be obtained from the solution for reflection at a pair of interfaces by taking the limit as $a \to 0$ and $\rho_2 \to \infty$, with $\rho_2 a = \mu$. How small must a be for the layer to be considered "thin" in this context?

***3–18.** *Transmission of an arbitrary pulse by a plate.* Show that the boundary conditions of Section 3–9 for an incident pulse of the form $\xi_{\text{inc}}(x, t) = f(t - x/c_1)$ are met by a transmitted pulse given by

$$\xi_2(x, t) = h(t - x/c_2),$$

where

$$h(t) = (2Z_1/\mu) \int_{-\infty}^{t} e^{-(Z_1+Z_2)(t-t')/\mu} f(t')\, dt'.$$

Find the motion of the plate, if

$$f(t) = \begin{cases} 0, & t < 0, \\ 1, & t \geq 0, \end{cases}$$

and determine $p_1(0, t)$ and $p_2(0, t)$ for this case.

4 Energy in a sound wave; isomorphisms

4-1 ENERGY DENSITY AND ENERGY FLUX FOR A PLANE SOUND WAVE

Consider a volume element of unit cross-sectional area containing those particles of the fluid whose parameters lie between x and $x + \Delta x$. The mass is $\rho_0 \, \Delta x$ and the instantaneous velocity is $\dot{\xi}$; hence the kinetic energy of these particles is $\frac{1}{2}(\rho_0 \, \Delta x)\dot{\xi}^2$. We divide by the *original* volume Δx to obtain the *instantaneous kinetic energy density*

$$W_{\text{kin}} = \tfrac{1}{2}\rho_0 \dot{\xi}^2. \tag{4-1}$$

While working in Lagrangian notation, we shall find it convenient to refer densities to the original state of the system. If desired, any such quantity can be converted, through multiplication by the factor ρ/ρ_0, to represent the amount per unit of volume in the disturbed configuration. If the acoustic approximation is used in the calculation of $\dot{\xi}$, then the quantities $\rho\dot{\xi}^2/2$ and $\rho_0\dot{\xi}^2/2$ are interchangeable, since they differ by a quantity of the same order as the inaccuracy introduced by using the linearized equations to calculate $\dot{\xi}$. Strict notation will be followed for the present, however, in order to show that the law of conservation of energy is an exact relation.

Consider the portion of the fluid that initially occupied the volume $V_0 = \Delta x$, when it is in a disturbed condition occupying the volume $V = (\rho_0/\rho) \, \Delta x$. The amount of work which would be done *on* the given portion of the fluid in an adiabatic compression (or expansion) from volume V_0 to V is $-\int_{V_0}^{V} P_a(V) \, dV$, where $P = P_a(V)$ is the equation of the adiabatic curve connecting the states (P_0, V_0) and (P, V). This amount of work may be thought of as stored in the form of potential energy relative to the undisturbed state. The integral can be expressed in terms of ρ as the independent variable. Setting $dV = -\rho_0 \, \Delta x(d\rho/\rho^2)$, and dividing by Δx, we obtain the potential energy per unit (original) volume:

$$W_{\text{pot}} = \rho_0 \int_{\rho_0}^{\rho} P_a(\rho)(d\rho/\rho^2), \tag{4-2}$$

where $P_a(\rho)$ expresses the pressure as a function of ρ in adiabatic changes of state from the undisturbed configuration. The *total* energy (per unit original volume) is the sum of the kinetic and potential energies:

$$W = W_{\text{kin}} + W_{\text{pot}}. \tag{4-3}$$

During motion of the fluid, the work which is done on (or by) a given volume element is attributable to the forces which are exerted by adjacent portions of the fluid. Consider, for example, a unit area on the plane of particles labeled x_0. The fluid in $x < x_0$ exerts the force $P(x_0, t)$ in the positive direction across this plane. During time Δt, the displacement of this plane is

$$\Delta\xi = \xi(x_0, t + \Delta t) - \xi(x_0, t) \doteq \dot{\xi}(x_0, t)\,\Delta t,$$

and the work done is

$$P\,\Delta\xi = P\dot{\xi}\,\Delta t.$$

Dividing by Δt, we obtain the instantaneous rate at which work is done per unit area on the fluid in $x > x_0$ by the fluid in $x < x_0$. This quantity is referred to as the *instantaneous intensity*, I, at the point in question:

$$I(x_0, t) = P(x_0, t)\dot{\xi}(x_0, t). \tag{4-4}$$

The units are power per unit area, or watts per square meter in the mks system.

4-2 THE LAW OF CONSERVATION OF ENERGY

If the correct assignment has been made for each of the energy quantities in the preceding section, we should be able to show that a net flow of energy into a volume results in an appropriate increase in the energy density. Consider the volume of unit cross-sectional area containing the particles whose parameters lie between any chosen values x_1, x_2, with $x_1 < x_2$. The quantity $I(x_1, t) - I(x_2, t)$ expresses the net flow of energy into this volume per unit of time. The total energy contained in the volume at any given time is $\int_{x_1}^{x_2} W(x, t)\,dx$. We expect the rate of increase of this quantity to be the same as the rate of energy flow into the volume:

$$\frac{d}{dt}\int_{x_1}^{x_2} W(x, t)\,dx = I(x_1, t) - I(x_2, t). \tag{4-5}$$

That this is in fact an identity when the assignments for I and W are made according to the last section may be verified by direct substitution:

$$\frac{d}{dt}\int_{x_1}^{x_2} W(x, t)\, dx = \int_{x_1}^{x_2} \frac{\partial W}{\partial t}\, dx = \int_{x_1}^{x_2} \frac{\partial}{\partial t}\left\{\frac{\rho_0 \dot{\xi}^2}{2} + \rho_0 \int_{\rho_0}^{\rho} \frac{P d\rho}{\rho^2}\right\} dx$$

$$= \int_{x_1}^{x_2} \left\{\rho_0 \dot{\xi}\ddot{\xi} + \rho_0 \frac{P}{\rho^2}\frac{\partial \rho}{\partial t}\right\} dx.$$

Now substitute $\rho_0 \ddot{\xi} = -\partial P/\partial x$ from the general dynamical equation (2–4), and integrate the first term by parts:

$$\frac{d}{dt}\int_{x_1}^{x_2} W(x, t)\, dx = -P\dot{\xi}\big|_{x_1}^{x_2} + \int_{x_1}^{x_2} \left\{P \frac{\partial^2 \xi}{\partial x\, \partial t} + \rho_0 \frac{P}{\rho^2}\frac{\partial \rho}{\partial t}\right\} dx.$$

From Eq. (4–4), the first term on the right-hand side is just $I(x_1, t) - I(x_2, t)$. We can show that the second term vanishes by differentiating the conservation-of-mass equation (2–2):

$$\frac{\partial \rho}{\partial t} = \frac{\partial}{\partial t}\left\{\rho_0 \left(1 + \frac{\partial \xi}{\partial x}\right)^{-1}\right\} = -\rho_0 \left(1 + \frac{\partial \xi}{\partial x}\right)^{-2} \frac{\partial^2 \xi}{\partial x\, \partial t} = -\frac{\rho^2}{\rho_0}\frac{\partial^2 \xi}{\partial x\, \partial t}.$$

Equation (4–5) is thus substantiated.

Equation (4–5), which expresses the law of conservation of energy in integrated form, can be written equivalently in differential form through the following considerations. The right-hand side, which represents the net flux of energy per unit time into the volume, can be replaced* by the identical quantity $-\int_{x_1}^{x_2} (\partial I/\partial x)\, dx$. The entire equation then becomes

$$\int_{x_1}^{x_2} \left\{\frac{\partial W}{\partial t} + \frac{\partial I}{\partial x}\right\} dx = 0.$$

As this expression vanishes for *arbitrary* choice of x_1 and x_2, the quantity in brackets must vanish identically:†

$$\frac{\partial W}{\partial t} + \frac{\partial I}{\partial x} = 0. \qquad \textit{Law of conservation of energy in differential form} \qquad (4\text{–}6)$$

---

* The reader will undoubtedly encounter a similar transformation in the study of electricity and magnetism or vector analysis. The present step, which is trivially simple in the one-dimensional case, corresponds to the application of the divergence theorem (Green's theorem) in a three-dimensional context.

† Given $\int_{x_1}^{x_2} \phi(x)\, dx = 0$ for arbitrary x_1, x_2. Suppose that for some x_0, $\phi(x_0) > 0$. Then if $\phi(x)$ is continuous, $\phi(x)$ must be greater than zero in some neighborhood of x_0. Choosing x_1 and x_2 within this neighborhood and applying the mean-value theorem, we have $\int_{x_1}^{x_2} \phi(x)dx = \phi(\bar{x})(x_2 - x_1) \neq 0$, since $\phi(\bar{x}) > 0$. There is thus a contradiction unless $\phi(x) \leq 0$ everywhere. Treat the case $\phi(x_0) < 0$ similarly.

Direct substitution of the expressions defining I and W shows that this equation is satisfied. The law of conservation of energy is thus not imposing any *new* conditions which must be satisfied by the functions describing the system, but is deducible from the equations which have previously been written.

4-3 SEPARABILITY OF ENERGY INTO + AND − COMPONENTS

The context of this section is that of waves which satisfy the acoustic approximation. For this case the general solution is the superposition of + and −waves, and from Eq. (2–28) we can write

$$\dot\xi = \dot\xi_+ + \dot\xi_-, \qquad p = Z(\dot\xi_+ - \dot\xi_-). \tag{4–7}$$

The instantaneous intensity is therefore

$$I = P\dot\xi = P_0\dot\xi + p\dot\xi = P_0(\dot\xi_+ + \dot\xi_-) + Z(\dot\xi_+^2 - \dot\xi_-^2).$$

The significant feature of this result is that the intensity separates into two terms, one of which is associated solely with the +wave, and the other, solely with the −wave. Each of these terms is precisely what would be obtained if only that component wave were present. Thus, $I = I_+ + I_-$, where

$$I_+ = P_0\dot\xi_+ + Z\dot\xi_+^2 = (P_0 + p_+)\dot\xi_+$$

and

$$I_- = P_0\dot\xi_- - Z\dot\xi_-^2 = (P_0 + p_-)\dot\xi_-.$$

Hence we may think of the two waves as contributing independently to the energy flux.

A similar examination of the kinetic energy density shows that this quantity does *not* so separate:

$$W_{\text{kin}} = \rho_0\dot\xi^2/2 = (\rho_0/2)(\dot\xi_+^2 + 2\dot\xi_+\dot\xi_- + \dot\xi_-^2). \tag{4–8}$$

The term $\rho_0\dot\xi_+\dot\xi_-$ depends jointly on the + and −waves.

We turn next to a consideration of the potential energy density, for which we require a simplified expression applicable under the acoustic approximation. The substitution $P = P_0 + p$ enables us to write

$$\begin{aligned} W_{\text{pot}} &= P_0\rho_0 \int_{\rho_0}^{\rho} d\rho/\rho^2 + \rho_0 \int_{\rho_0}^{\rho} p\, d\rho/\rho^2 \\ &= W'_{\text{pot}} + w_{\text{pot}}. \end{aligned} \tag{4–9}$$

The first term can be integrated directly. By use of Eq. (2–2), we obtain

$$W'_{\text{pot}} = P_0\rho_0(1/\rho_0 - 1/\rho) = -P_0(\partial\xi/\partial x).$$

For the second term in Eq. (4–9), we substitute $p = \mathcal{B}_a s$ and also $\rho = \rho_0(1 + s)$:

$$w_{\text{pot}} = \rho_0 \int_{\rho_0}^{\rho} p(d\rho/\rho^2) = \mathcal{B}_a \int_0^s s\, ds/(1 + s)^2.$$

Although this can be integrated directly, it is pointless to do so. The substitution for p is only correct to first power in s; consequently the entire integrand should be approximated to this order:

$$w_{\text{pot}} = \mathcal{B}_a \int_0^s s\, ds = \mathcal{B}_a s^2/2 = p^2/2\mathcal{B}_a.$$

The potential energy density for waves satisfying the acoustic approximation therefore takes the form

$$W_{\text{pot}} = -P_0(\partial\xi/\partial x) + p^2/2\mathcal{B}_a. \tag{4–10}$$

Substituting the solution of general form (Eq. 4–7), and using the relation $Z^2/\mathcal{B}_a = \rho_0$, we find

$$W_{\text{pot}} = -P_0\left(\frac{\partial\xi_+}{\partial x} + \frac{\partial\xi_-}{\partial x}\right) + \frac{\rho_0}{2}\,(\dot{\xi}_+^2 - 2\dot{\xi}_+\dot{\xi}_- + \dot{\xi}_-^2). \tag{4–11}$$

This quantity also fails to separate into terms which are independently ascribable to the + and − waves separately. On comparing Eq. (4–11) with (4–8), we see, however, that the "offending terms" will cancel if we take the *sum* of kinetic and potential energy densities. Thus separability *does* apply to the total energy density, and we may write $W = W_+ + W_-$, where

$$W_\pm = -P_0(\partial\xi_\pm/\partial x) + \rho_0\dot{\xi}_\pm^2.$$

It is easy to see that W_+ and W_- are the total energy densities which would be assigned if the corresponding component wave were the only one present.

Algebraic substitution will also establish the following identity:

$$I_+(x, t) = cW_+(x, t). \tag{4–12}$$

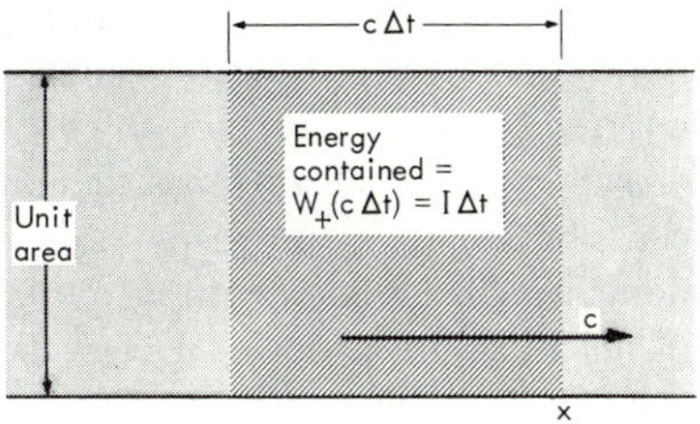

Fig. 4–1.
In a progressive wave the energy $I\,\Delta t$ which crosses a given surface in time Δt is contained in a cylinder of height $c\,\Delta t$.

The interpretation of this relation is interesting, for it shows that none of the energy stored in the field of a progressive wave is "dormant." We may think of all the energy as having forward motion with velocity c. If this is the case, then the energy $I_+ \Delta t$, which crosses a given surface in time Δt, is the energy which was contained in a volume of unit cross-sectional area and of length $c\,\Delta t$ (see Fig. 4–1). The amount of such energy is $W_+c\,\Delta t$. Equating these two expressions leads to Eq. (4–12) and confirms the interpretation. A similar interpretation applies to the identity for −waves: $I_- = -cW_-$.

4-4 CONVECTIVE AND RADIATIVE ENERGY TERMS

Let the intensity of a general acoustic wave be written in the form $I = P\dot{\xi} = P_0\dot{\xi} + p\dot{\xi} = I' + i$. The relative importance of the two terms in this sum depends upon the situation. The acoustic approximation requires that the *instantaneous* value of the first term be large compared with that of the second. In many applications, however, we are more interested in the total energy flux over a long time interval than we are in instantaneous values. To facilitate a comparison of the two terms, consider a situation in which +waves are generated by the motion of a piston. The value of I adjacent to the piston surface gives the instantaneous power delivered by the piston per unit cross-sectional area. The first term has the same sign as $\dot{\xi}$, the velocity of the piston. So long as $\dot{\xi}$ remains positive, the piston continues to deliver energy to the fluid. If $\dot{\xi}$ becomes negative however, the fluid does work (through this first term) on the piston, and the net energy delivered by the piston decreases. The first term therefore represents a fluctuating kind of energy transfer which the piston can take back by reversing its direction of motion. The net energy delivered by the piston up to time t associated with this term is

$$P_0 \int_0^t \dot{\xi}\,dt = P_0\xi,$$

the value of ξ being taken to be zero at $t = 0$. The net energy delivered thus depends only on the position of the piston. If the piston returns at any time to its original position, the net energy delivered is zero. A motion in which $\dot{\xi}$ maintains the same sign and the magnitude of ξ is an increasing function is of *convective* type in which the fluid is permanently shifted from its initial position. The term $I' = P_0\dot{\xi}$ is important in this type of motion and will be referred to as the *convective* intensity. In a *nonconvective* motion, for which ξ remains small at all times, it may be desirable to ignore the convective contribution to the intensity. Sinusoidal motion is an example of such a nonconvective motion. Since ξ returns to its original value at the completion of an integral number of cycles, the term I' makes no contribution to the total energy delivered over a complete cycle.

On the other hand, the second term, $i = p\dot{\xi}$, is always positive (or zero) in a +wave, since $p = Z\dot{\xi}$ and $p\dot{\xi} = Z\dot{\xi}^2 \geq 0$. This term represents energy which is permanently given up by the piston. The quantity $i = p\dot{\xi}$ will be referred to as the *radiative intensity*.

A similar consideration applies to the two terms $W'_{\text{pot}} = -P_0(\partial\xi/\partial x)$ and $w_{\text{pot}} = p^2/2\mathcal{B}_a$, which contribute to the potential energy density. The net potential energy over a finite volume extending from x_1 to x_2 is given by $\int_{x_1}^{x_2} W_{\text{pot}}\, dx$. The contribution from W'_{pot} integrates to yield

$$P_0[\xi(x_2, t) - \xi(x_1, t)],$$

and consequently depends only on the displacements at the end points. In a pulse of limited extent, x_1 and x_2 can be chosen so that both of these displacements vanish. Likewise, the contribution from W'_{pot} vanishes in a sinusoidal wave if x_1 and x_2 are chosen an integral number of wavelengths apart. By analogy with I', W'_{pot} will be referred to as the *convective* potential energy density. The quantity $w_{\text{pot}} = p^2/2\mathcal{B}_a$ will be referred to as the *radiative* potential energy density. Its value is positive in rarefactions as well as in condensations. (The *net* potential energy density is negative in a rarefaction. It is only in contexts where the convective potential energy is not of interest that we may think of the potential energy as positive in both rarefactions and condensations.)

Neglect of the convective term W'_{pot} is consistent with the neglect of I' in the energy relations derived in earlier sections. For example, if we consider the total radiative energy per unit volume, $w = W_{\text{kin}} + w_{\text{pot}}$, it can be seen by inspection of the earlier proofs that the radiative energies obey a separate conservation law: $\partial w/\partial t + \partial i/\partial x = 0$, the separability into + and −components, $i = i_+ + i_-$, $w = w_+ + w_-$, and the transport relations $i_\pm = \pm c w_\pm$.

4–5 RELATIVE RADIATIVE INTENSITIES IN REFLECTION AND TRANSMISSION AT A SINGLE INTERFACE

In the context of Section 3–4, it is convenient to compare the relative radiative intensities of incident, reflected, and transmitted waves by means of the reflection and transmission coefficients

$$\begin{aligned} R_i &= i_{\text{refl}}(0, t)/i_{\text{inc}}(0, t), \\ T_i &= i_{\text{trans}}(0, t)/i_{\text{inc}}(0, t). \end{aligned} \tag{4–13}$$

Substituting $i = p\dot{\xi}$, we obtain

$$R_i = \frac{p_{\text{refl}}(0, t)\dot{\xi}_{\text{refl}}(0, t)}{p_{\text{inc}}(0, t)\dot{\xi}_{\text{inc}}(0, t)} = R_p R_{\dot{\xi}} = -R_{\dot{\xi}}^2, \tag{4–14}$$

and

$$T_i = T_pT_{\dot\xi} = (1 + R_p)(1 + R_{\dot\xi}) = 1 - R_{\dot\xi}^2 = 1 + R_i. \tag{4-15}$$

Recall that intensity is an algebraic quantity whose sign indicates the direction of energy flow. The fact that $R_i < 0$ merely indicates that the energy flows in the incident and reflected waves are in opposite directions.

EXAMPLE 1. Consider the case of extreme impedance mismatch, $Z_{12} \ll 1$, covered by Eq. (3-28). To first order in Z_{12} we find that for an incident wave of unit intensity, the reflected wave has intensity

$$R_i = R_pR_{\dot\xi} \doteq (1 - 2Z_{12})(-1 + 2Z_{12}) \doteq -1 + 4Z_{12}, \tag{4-16}$$

and the transmitted wave has intensity

$$T_i = T_pT_{\dot\xi} \doteq (2 - 2Z_{12})(2Z_{12}) \doteq 4Z_{12}. \tag{4-17}$$

Since $R_i < 0$, we can write the relation $T_i = 1 + R_i$ in either of the equivalent forms:

$$|R_i| + |T_i| = 1,$$

or

$$|i_{\text{refl}}(0, t)| + |i_{\text{trans}}(0, t)| = |i_{\text{inc}}(0, t)|.$$

The relation may therefore be interpreted as a statement of the law of conservation of energy, for it shows that the energy which is brought into the interface in each second by the incident wave is divided between the reflected and transmitted waves. It is also clear from the form $|R_i| + |T_i| = 1$ that $|T_i| \leq 1$ and $|R_i| \leq 1$, which assures that neither the reflected nor the transmitted waves can have a greater intensity than the incident.

4-6 INTENSITY RELATIONS FOR PROGRESSIVE SINUSOIDAL WAVES

For a general +wave the relation $p = Z\dot\xi$ enables us to write the instantaneous radiative intensity in any of the forms

$$i = p\dot\xi = Z\dot\xi^2 = p^2/Z. \tag{4-18}$$

In a sinusoidal wave we may take $\dot\xi = \dot\xi_m \cos(\omega t - \phi)$, where ϕ is a constant with respect to t and depends upon the phase of the wave at the particular location being evaluated. Thus the instantaneous intensity is

$$i = Z\dot\xi_m^2 \cos^2(\omega t - \phi).$$

The time dependence of this quantity shows that radiative energy passes by a particular point of observation in a series of pulses rather than as a steady flow. The average value* of the cosine squared over a complete cycle is $\frac{1}{2}$; consequently,

$$\bar{\imath} = Z\dot{\xi}_m^2/2 = p_m\dot{\xi}_m/2 = p_m^2/2Z. \tag{4–19}$$

If we define root-mean-square values of p and $\dot{\xi}$, $p_{\text{rms}} = p_m/\sqrt{2}$ and $\dot{\xi}_{\text{rms}} = \dot{\xi}_m/\sqrt{2}$, we can also write

$$\bar{\imath} = p_{\text{rms}}\dot{\xi}_{\text{rms}} = Z\dot{\xi}_{\text{rms}}^2 = p_{\text{rms}}^2/Z. \tag{4–20}$$

We can summarize Eq. (4–19) in a loose way by stating that the average intensity of a sinusoidal wave is proportional to the square of the amplitude (p_m or $\dot{\xi}_m$). This statement is sufficient so long as comparisons are to be made in a single medium or in two media of the same impedance. Confusion may result if the impedance factor is forgotten. For example, in the transmission from air to water (Section 3–6), $T_p \doteq 2$, which implies that the pressure amplitude in the transmitted wave is double that of the incident. This does not imply that the intensity of the transmitted wave is four times that of the incident, however. When the impedance ratio is properly taken into account, the intensity of the transmitted wave is seen to be $4Z_1/Z_2 = 4Z_{12}$ times the intensity of the incident, which is in agreement with the result obtained in the last section.

The mathematics of this section is clearly the same as that involved in the calculation of the average power dissipated in a resistive element of an ac-circuit. The role of the power dissipated is taken by i, Z corresponds to the resistance, $\dot{\xi}$ to the current, and p to the potential difference. The analogy is purely algebraic and may be used as a mnemonic. The physical situations are quite different. For example, i represents the power *transported* across a unit area; there is no *dissipation* as in the case of the resistor. Not much is gained by trying to carry the analogy further.

In any progressive wave the instantaneous kinetic and radiative potential energy densities are equal:

$$w_{\text{kin}} = \rho_0\dot{\xi}^2/2 = (\rho_0/2)(\pm p/Z)^2 = p^2/2\mathcal{B}_a = w_{\text{pot}}.$$

The two densities are not, as one might have expected, complementary (one being zero where the other is a maximum), but because of the relation $p = \pm Z\dot{\xi}$, their maxima are coincident in location.

---

* $\cos^2(\omega t - \phi) = \frac{1}{2}[1 + \cos 2(\omega t - \phi)]$, and the average value of $\cos 2(\omega t - \phi)$ over a cycle (or half-cycle) is zero. The relation is also easily remembered by the fact that the graphs of $\cos^2(\omega t - \phi)$ and $\sin^2(\omega t - \phi)$ are identical except for a shift through a quarter of a cycle. Hence their average values are equal. Therefore

$$\overline{\cos^2(\omega t - \phi)} + \overline{\sin^2(\omega t - \phi)} = 1 = \overline{2\cos^2(\omega t - \phi)}.$$

*4–7 INTENSITY IN A SUPERPOSITION OF + AND − SINUSOIDAL WAVES

At a fixed location, the pressure and particle velocity functions can be written

$$p(x, t) = \mathcal{R}e\,\{\mathbf{p}e^{i\omega t}\} \qquad \text{and} \qquad \dot{\xi}(x, t) = \mathcal{R}e\,\{\dot{\boldsymbol{\xi}}e^{i\omega t}\},$$

where the complex amplitudes $\mathbf{p}$ and $\dot{\boldsymbol{\xi}}$ are functions of x. The instantaneous radiative intensity is $i = p\dot{\xi}$, which may *not* be written in the form $\mathcal{R}e\,\{\mathbf{p}\dot{\boldsymbol{\xi}}e^{i\omega t}\}$. We can calculate the intensity conveniently by introducing the *complex impedance* $\mathbf{Z}(x)$ defined as the ratio of the complex pressure and velocity amplitudes:

$$\mathbf{Z} \equiv \mathbf{p}/\dot{\boldsymbol{\xi}} = |\mathbf{Z}|e^{i\phi}. \tag{4–21}$$

Then, if $\dot{\boldsymbol{\xi}} = \dot{\xi}_m e^{i\theta}$, the explicit form of the instantaneous intensity is

$$i = [|\mathbf{Z}|\dot{\xi}_m \cos(\omega t + \theta + \phi)][\dot{\xi}_m \cos(\omega t + \theta)].$$

Taking average values over a complete cycle† yields

$$\bar{\imath} = \tfrac{1}{2}|\mathbf{Z}|\dot{\xi}_m^2 \cos\phi = \tfrac{1}{2}p_m\dot{\xi}_m \cos\phi = p_{\text{rms}}\dot{\xi}_{\text{rms}} \cos\phi.$$

The last two expressions for the average intensity resemble the formulas obtained in the case of progressive +waves, except for the addition of the "power factor" $\cos\phi$, which depends upon the relative phase of p and $\dot{\xi}$.

EXAMPLE 2. The formula for the average intensity in a general superposition reduces properly to either of the special cases of progressive waves. For a +wave:

$$\mathbf{Z} = Z, \quad \phi = 0, \quad \text{and} \quad \bar{\imath} = Z\dot{\xi}_m^2/2.$$

For a −wave:

$$\mathbf{Z} = -Z = Ze^{i\pi}, \quad \phi = \pi, \quad \text{and} \quad \bar{\imath} = -Z\dot{\xi}_m^2/2.$$

EXAMPLE 3. In the standing wave system of Section 3–2,

$$\dot{\xi} = 2\xi_m\omega \sin kx \cos\omega t = \mathcal{R}e\,\{2\xi_m\omega \sin kx\, e^{i\omega t}\},$$

and hence the complex amplitude of the velocity function is $\dot{\boldsymbol{\xi}} = 2\xi_m\omega \sin kx$. Likewise, we find from Eq. (3–9) that $\mathbf{p} = 2i\xi_m k\mathcal{B}_a \cos kx$. Thus the complex impedance is $\mathbf{Z} = \mathbf{p}/\dot{\boldsymbol{\xi}} = iZ\cot kx$. Since the phase of $\mathbf{Z}$ is $\phi = \pi/2$, we have

$$\bar{\imath} = p_{\text{rms}}\dot{\xi}_{\text{rms}} \cos\phi = 0.$$

The "purely reactive" character of the complex impedance is associated with the

† $\overline{\cos(\omega t + \theta)\cos(\omega t + \theta + \phi)}$

$$= \overline{\cos^2(\omega t + \theta)}\cos\phi - \overline{\cos(\omega t + \theta)\sin(\omega t + \theta)}\sin\phi$$

$$= \tfrac{1}{2}\cos\phi - \tfrac{1}{2}\sin\phi\,\overline{\sin 2(\omega t + \theta)} = \tfrac{1}{2}\cos\phi.$$

absence of a net energy transport. The instantaneous radiative intensity is

$$i = p\dot{\xi} = -\xi_m^2 \omega k \mathcal{B}_a \sin 2kx \sin 2\omega t.$$

This vanishes at the positions that are nodes for either p or $\dot{\xi}$. The picture we thus obtain for the standing wave is that of energy "shimmying" back and forth within isolated cells between the nodes.

4–8 INTERFERENCE BETWEEN SUPERPOSED WAVES

Let the function pairs $p_1(x, t)$, $\xi_1(x, t)$ and $p_2(x, t)$, $\xi_2(x, t)$ both be solutions to the one-dimensional acoustic equations. We have observed that the linearity of these equations implies that $p = p_1 + p_2$, $\xi = \xi_1 + \xi_2$ is another pair of functions that is also a solution. The radiative intensity corresponding to this new solution is

$$\begin{aligned} i = p\dot{\xi} &= (p_1 + p_2)(\dot{\xi}_1 + \dot{\xi}_2) = p_1\dot{\xi}_1 + p_2\dot{\xi}_2 + p_1\dot{\xi}_2 + p_2\dot{\xi}_1 \\ &= i_1 + i_2 + [p_1\dot{\xi}_2 + p_2\dot{\xi}_1]. \end{aligned} \tag{4–22}$$

Since the quantity in square brackets does not in general vanish, we observe that the intensity associated with the superposition of two solutions is not, in general, the sum of the associated intensities. To calculate the intensity, it is necessary first to obtain the explicit sum of the two solutions. The fact that intensity does not obey a simple superposition principle is referred to as the phenomenon of *interference*. The cross term [in square brackets in Eq. (4–22)] is referred to as the *interference term*. Wherever this is positive (or negative) the interference is referred to as *constructive* (or *destructive*). Linearity of superposition applies to linear combinations of the "field variables" p, ξ, $\dot{\xi}$, s, etc., but not to energy quantities such as i and w, which are quadratic in the field variables.

We have already observed that interference does not need to be taken into account in the superposition of a +wave with a −wave. We can investigate this more generally by assuming that $(p_1, \dot{\xi}_1)$ is a +wave, and inquiring what conditions are imposed on $(p_2, \dot{\xi}_2)$ in order for the interference term to vanish. With $p_1 = Z\dot{\xi}_1$, we have for the interference term $p_1\dot{\xi}_2 + p_2\dot{\xi}_1 = \dot{\xi}_1(Z\dot{\xi}_2 + p_2)$. For this to vanish, $p_2 = -Z\dot{\xi}_2$, which requires the second wave to be a −wave. Therefore the *only* case in which it is safe to add intensities directly is when the two waves are progressive waves traveling in opposite directions.

The superposition of two +waves results in $i = Z(\dot{\xi}_1 + \dot{\xi}_2)^2$. The intensity will be identically zero everywhere if and only if $\dot{\xi}_2(x, t) = -\dot{\xi}_1(x, t)$. In this case we have *complete destructive interference*, and the solution obtained by the superposition is the trivial one, $p = 0$, $\xi = 0$. If the two waves are sinusoidal, say

$$\dot{\xi}_1 = A\cos(\omega t - kx) \qquad \text{and} \qquad \dot{\xi}_2 = B\cos(\omega t - kx + \phi),$$

the interference term is

$$2ZAB \cos(\omega t - kx) \cos(\omega t - kx + \phi),$$

which averages over a complete cycle to $ZAB \cos \phi$. Thus when two superimposed +waves are in *quadrature* ($\phi = \pi/2$), interference does not need to be taken into account in a calculation of *average* intensity. More generally

$$\bar{\imath} = \bar{\imath}_1 + \bar{\imath}_2 + ZAB \cos \phi = (Z/2)[A^2 + B^2 + 2AB \cos \phi].$$

The maximum average intensity is obtained if the two waves are in phase, and, if the amplitudes are equal, it is four times that of either wave. The minimum average intensity (associated with the maximum tendency toward destructive interference) occurs with $\phi = \pi$. The destructive interference is complete only if the amplitudes are equal.

4-9 MEASUREMENT OF INTENSITY IN DECIBELS

The intensities* of audible sounds in the middle frequency range ($\nu \sim 10^3 \text{ sec}^{-1}$) vary from the *threshold of audibility* ($i \sim 10^{-12}$ W/m^2) to the *threshold of pain* ($i \sim 1$ W/m^2). In order to encompass such a wide range of variation it is convenient to employ a logarithmic scale. A reference intensity i_0 is chosen, and the *intensity level in decibels* (abbreviated db) of a wave of intensity i is defined as

$$\Delta \equiv 10 \log_{10} (i/i_0). \tag{4–23}$$

For sound waves in air, i_0 is usually chosen to be 10^{-12} W/m^2. A negative value of Δ then indicates a wave which is too faint to be heard. A list is given in Table 4–1 of the approximate intensity levels for a few common situations.

Given two sounds of intensities i_1 and i_2, the difference in their intensity levels $\Delta_{21} \equiv \Delta_2 - \Delta_1$ has the same form as Eq. (4–23) with i_1 in place of i_0:

$$\Delta_{21} = 10\{\log_{10} (i_2/i_0) - \log_{10} (i_1/i_0)\} = 10 \log_{10} (i_2/i_1). \tag{4–24}$$

It is clear that intensity differences in decibels add algebraically. Thus, if a second wave is 2 db louder than a first, and a third is 4 db louder than the second, the third is 6 db louder than the first. This statement is based on the relation

$$\Delta_{32} + \Delta_{21} = 10 \log_{10} (i_3/i_2) + 10 \log_{10} (i_2/i_1) = 10 \log_{10} (i_3/i_1) = \Delta_{31}.$$

---

* The context is that of sinusoidal waves. Average and not instantaneous intensity will be implied without further designation.

Table 4–1. Intensity levels for a few common situations*

Typical sound environment	Intensity level, decibels
Threshold of pain	120–130
Riveting machine 30–40 ft away	100
Subway with train passing	90
Average city street	70
Average restaurant	60
Average conversation 3 ft away	60–70
Outdoor minimum in city	30–40
Quiet office	30–40
Outdoor minimum in country	10
Threshold of audibility	0

* Adapted from Colby, *Sound Waves and Acoustics*, Henry Holt and Co., 1938. By permission.

The decibel is a convenient unit for describing the intensities of audible sounds, because two sounds which differ in intensity by a fraction of a decibel are indistinguishable to the ear. It is therefore usually sufficient to round off all calculations to the nearest decibel. The psychophysical experiment which demonstrates this feature of intensity discrimination can be described in the following terms. Given a sound of fixed intensity i_1, and one of variable intensity $i_2 = i_1 + \Delta i$, the investigation determines how small the value of Δi can be made while still allowing a subject to discriminate between the two sounds. "Discrimination" means that the subject is able to guess which sound has the greater intensity without making more than a specified percentage of wrong guesses. Experiment shows that the required minimum increment Δi is roughly proportional to i_1. The law may be stated as $\Delta i = \mathcal{W} i$, where $\mathcal{W}$ is a constant independent of intensity. The law breaks down near the thresholds, but $\Delta i/i$ remains fairly constant throughout a middle range of intensities. In the basic form of this experiment the two sounds to be compared are of the same frequency. The value of $\mathcal{W}$ does depend somewhat upon this choice of frequency. The law $\Delta i \propto i$ is an example of a general psychophysical law, known as the *Weber-Fechner* law, which applies to discrimination experiments in a wide variety of contexts. (For example, to discriminate between masses m and $m + \Delta m$ by the way it feels to lift them, the minimum Δm required is proportional to m.) The minimum increment is referred to as a *just noticeable difference* (jnd). The corresponding intensity difference in decibels is

$$\Delta_{21} = 10 \log_{10} \left(\frac{i_1 + \Delta i}{i_1} \right) = 10 \log_{10} (1 + \mathcal{W}).$$

The experimental value of $\mathcal{W}$ is about $\frac{1}{4}$, which yields an intensity difference of $10 \log_{10} 1.25$, or approximately 1 db. The intensity difference in decibels between any two sounds is thus a rough measure of the number of steps of jnd between them.

In reflection-transmission problems, Δ_{refl}, the decibel loss upon reflection, and Δ_{trans}, the decibel loss on transmission, are obtained from the formulas

$$\Delta_{\text{refl}} = 10 \log_{10} |i_{\text{inc}}/i_{\text{refl}}| = 10 \log_{10} (1/|R_i|)$$

and

$$\Delta_{\text{trans}} = 10 \log_{10} (1/T_i).$$

EXAMPLE 4. On transmission from air to water, we find from Eq. (4-17) that $\Delta_{\text{trans}} = 10 \log_{10} (Z_{21}/4)$, which gives a value of 30 db with $Z_{21} = 4000$. On the other hand,

$$\Delta_{\text{refl}} = 10 \log_{10} \left[\frac{1}{1 - 4Z_{12}}\right] \doteq 10 \log_{10} (1 + 4Z_{12})$$

$$= 10 (\log_{10} e) \log_e (1 + 4Z_{12}) \doteq 40Z_{12} \log_{10} e = 0.0043 \text{ db}.$$

Thus the intensity of the reflected wave differs imperceptibly from that of the incident.

Since $\bar{i} = p_{\text{rms}}^2/Z$, the intensity level of an acoustic wave is related to the rms pressure through the formula $\Delta = 20 \log_{10} (p_{\text{rms}}/p_0)$, where p_0 is the rms pressure corresponding to the reference intensity. For work in underwater sound, p_0 is customarily assigned the value $p_0 = 0.1 \text{ N/m}^2$. Using $Z = 1.5 \times 10^6 \text{ N/m}^2$ for seawater, we obtain the value $i_0 = 6.7 \times 10^{-9} \text{ W/m}^2$, which is the order of magnitude of general background noise under the sea.

4-10 ENERGY DEFINITIONS FOR TRANSVERSE WAVES ON A STRING

a) Power transported past a given point. The portion of the string to the left of any given point x_0 ("left" meaning $x < x_0$) is exerting a force on the portion to the right. Since we are assuming no longitudinal displacements, there is no work associated with the longitudinal component of this force. The component which is of interest is $F_y = -T \sin \theta$ (see Fig. 1-2). When we make the small slopes approximation, this becomes

$$F_y = -T(\partial y/\partial x). \tag{4-25}$$

The transverse velocity is $\dot{y} = \partial y/\partial t$, and consequently the instantaneous power delivered from left to right at x_0 is

$$\mathcal{P} = F_y \dot{y}. \tag{4-26}$$

b) Kinetic energy per unit length. A length Δx of the string has mass $\sigma \, \Delta x$ and kinetic energy $(\sigma \Delta x)\dot{y}^2/2$. Consequently, the kinetic energy per unit length is

$$\epsilon_{\text{kin}} = \sigma \dot{y}^2/2. \tag{4-27}$$

c) Potential energy per unit length. A section of the string in the disturbed configuration possesses potential energy by virtue of the fact that its length is altered. To stretch a section of original length Δx to the disturbed length $\Delta s = \sqrt{1 + (\partial y/\partial x)^2}\,\Delta x$ requires work to be done against the tension force T, which we have assumed to remain nearly constant. The work is

$$T[\Delta s - \Delta x] = T[\sqrt{1 + (\partial y/\partial x)^2} - 1]\,\Delta x \doteq (T/2)(\partial y/\partial x)^2\,\Delta x.$$

Consequently, the potential energy per unit length is

$$\epsilon_{\text{pot}} = (T/2)(\partial y/\partial x)^2. \tag{4–28}$$

4-11 ENERGY RELATIONS FOR TRANSVERSE WAVES ON A STRING; ISOMORPHISMS

It is natural to inquire next whether a law of conservation of energy can be deduced from the preceding definitions and the basic dynamical equations for transverse waves on a string. We are also interested in knowing whether the power flow past a given point can be separated into terms associated with the + and − waves, whether special relations exist between power and energy density in progressive waves, etc. It soon becomes clear after starting an investigation of these questions that the desired relations can be obtained by proofs which parallel exactly the mathematics in the corresponding derivations for acoustic waves. An inspection of the earlier sections of this chapter will show that the left-hand column of Table 4–2 contains the only information which was actually required concerning the variables p, ξ, i, etc. Other relations employed, such as the wave equation and the form of its general solution, are deducible by purely mathematical operations from those listed in the table. Furthermore, the deductions such as the law of conservation

Table 4–2. Isomorphism between basic equations for acoustic waves and transverse waves on a string

Acoustic waves	Transverse waves on a string
$p = -\mathcal{B}_a \dfrac{\partial \xi}{\partial x}$	$F_y = -T\dfrac{\partial y}{\partial x}$
$-\dfrac{\partial p}{\partial x} = \rho_0 \ddot{\xi}$	$-\dfrac{\partial F_y}{\partial x} = \sigma \dfrac{\partial^2 y}{\partial t^2}$
$i = p\dot{\xi}$	$\mathcal{P} = F_y \dot{y}$
$W_{\text{kin}} = \frac{1}{2}\rho_0 \dot{\xi}^2$	$\epsilon_{\text{kin}} = \frac{1}{2}\sigma \dot{y}^2$
$w_{\text{pot}} = \frac{1}{2}\mathcal{B}_a \left(\dfrac{\partial \xi}{\partial x}\right)^2$	$\epsilon_{\text{pot}} = \frac{1}{2}T\left(\dfrac{\partial y}{\partial x}\right)^2$

Table 4-3. Correspondence of variables in isomorphism between acoustic waves and transverse waves on a string

Acoustic waves	Transverse waves on a string
p	F_y
ξ	y
$\mathcal{B}_a$	T
ρ_0	σ
i	$\mathcal{P}$
W_{kin}	ϵ_{kin}
w_{pot}	ϵ_{pot}
$c = \sqrt{\mathcal{B}_a/\rho_0}$	$c = \sqrt{T/\sigma}$

of energy and the separability of intensity and energy density were obtained by purely mathematical operations on these equations with no further concern about their physical interpretation.

The equations in the right-hand column summarize the relations we have assumed for transverse waves on a string. [The second one comes from Eq. (1-3).] These equations can be obtained from those in the left-hand column by a consistent symbol-by-symbol translation according to the correspondence of variables listed in Table 4-3. We can therefore be sure that any deduction from the first set of equations can be translated into a corresponding deduction from the second set. The two sets of equations are said to be *isomorphic* to each other. The demonstration of the existence of an isomorphism is a compact way of proving numerous relations applicable to the transverse wave on a string. For example, the equation $p_+ = Z\dot{\xi}_+$ enables us to write $F_y^+ = Z\dot{y}_+$, where the wave impedance of the string is defined as $Z = \sigma c$. We thus learn that the transverse component of the tension is proportional to the instantaneous velocity at every point along a string carrying a +wave. Establishment of an isomorphism is certainly not the only way to prove this relation. A direct proof can be constructed using previous experience with the acoustic equations as a guide.

Since there are many different physical situations* which give rise to equations isomorphic to those in Table 4-2, our study of the single example of acoustic waves has in fact established a number of general properties which can be asserted for many other types of wave. One is tempted to say "for all waves," but this depends on how narrowly the term "wave" is defined. As is shown by the fact that the nonlinear equation for sound and the Schrödinger equation of quantum mechanics are called wave equations, the term has taken on a broad significance and is applied in situations that are not completely isomorphic to the acoustic wave.

---

* Appendix V contains a brief summary of a number of examples.

Table 4–4. Alternate correspondence of variables for establishing an isomorphism between the equations for acoustic waves and transverse waves on a string

Acoustic waves	Transverse waves on a string
p	$F_y/\sqrt{T}$
ξ	y
$\mathcal{B}_a$	$\sqrt{T}$
ρ_0	$\sigma/\sqrt{T}$
i	$\mathcal{P}/\sqrt{T}$
W_{kin}	$\epsilon_{\text{kin}}/\sqrt{T}$
w_{pot}	$\epsilon_{\text{pot}}/\sqrt{T}$
$c = \sqrt{\mathcal{B}_a/\rho_0}$	$c = \sqrt{T/\sigma}$

There are, of course, some aspects of the acoustic and transverse waves on a string which are not covered by the isomorphism. For example, there is no special class of strings which correspond to ideal gases; hence the relation between acoustic wave propagation velocity and temperature is irrelevant for transverse waves on a string. Similarly, the complete acoustic wave intensity contains the convective term $P_0\dot{\xi}$ which does not have any analog in the case of the string.

Since the isomorphism is concerned only with the mathematical structure of a set of basic equations there is no implication that corresponding quantities in the two systems are physically related in any way. For example, the dimensions of p and F_y are different, as are those of W_{kin} and ϵ_{kin}, etc. The way of setting up a correspondence of variables is not unique. Table 4–4 gives an alternative (albeit less convenient) prescription which translates all the acoustic equations into correct equations for the wave on a string. It is also sometimes possible to set up a self-isomorphism, that is, a translation between a given set of equations and the same set rearranged in different order. For example, if the first acoustic equation in Table 4–2 is replaced by its time derivative $\partial p/\partial t = -\mathcal{B}_a(\partial\dot{\xi}/\partial x)$, and w_{pot} is written as $w_{\text{pot}} = p^2/2\mathcal{B}_a$, a self-isomorphism results through the correspondence $p \leftrightarrow \dot{\xi}$, $\mathcal{B}_a \leftrightarrow 1/\rho_0$, $i \leftrightarrow i$, and $W_{\text{kin}} \leftrightarrow w_{\text{pot}}$. The diversity of possible ways of setting up a translation scheme to establish an isomorphism indicates that there is no ulterior connection between the variables which are matched in a given assignment.

4–12 BOUNDARY VALUE PROBLEMS FOR TRANSVERSE WAVES ON A STRING

Every solution which has been obtained to a boundary value problem for acoustic waves can be translated by Table 4–3 into the solution to some problem for transverse waves on a string. A similar translation can be effected for the other wave contexts listed in Appendix V. The following is a brief summary of the problems whose solutions are at our disposal.

a) Reflection at a fixed end. Since the boundary condition $y(0, t) = 0$ corresponds to $\xi(0, t) = 0$, the problem is completely isomorphic to the acoustic problem of reflection at a rigid surface (Section 3–2). Thus, as an alternative to the direct method of Section 3–1, the solution for the string can be obtained by translation of variables from the solution to the acoustic problem.

b) Reflection at a "free end." Suppose that a string extending over $x < 0$ is attached at $x = 0$ to a slip ring of mass m which is free to slide along a post perpendicular to the x-axis. The net transverse force acting on the ring is the force F_y due to the string. From Newton's second law, we require

$$F_y(0, t) = m(\partial^2 y/\partial t^2)(0, t). \tag{4–29}$$

If the ring has negligible mass, we have a *free-end* situation, and the boundary condition is

$$F_y(0, t) = 0. \tag{4–30}$$

Since $F_y(0, t) = -T(\partial y/\partial x)(0, t)$, this condition is equivalent to asserting that the slope of the string is zero at all times at the free end. Since this equation is isomorphic to the condition $p(0, t) = 0$ for a free surface, the results of this case can be applied. Of particular interest is the fact that a crest is reflected as a crest and not as a trough as in the case of a fixed end. For sinusoidal waves the free end is an antinode for y. The device of the slip ring is necessary if the string is to be maintained under tension. One of the conditions in the derivation of the wave equation is that the tension is uniform and sufficiently large in magnitude that changes in tension which occur during the motion can be neglected. Although the motion of a rope which is hanging vertically under gravity will resemble this case, the situations are not identical (see Problem 1–13).

c) Specified motion of an end support. Let a semi-infinite string be attached to an end support which is given a specified transverse motion. The problem is isomorphic to that solved in Section 3–3. In particular, Fig. 3–4(a) shows the shape which results from lifting one end of a string at constant velocity.

d) Reflection and transmission at a junction between two strings. Let a string under tension T_1 of mass per unit length σ_1 be attached at $x = 0$ to a string under tension T_2 of mass per unit length σ_2. If $T_1 \neq T_2$, the device of a massless slip ring will again be required. A simple attachment of two strings without the slip ring can be treated as the special case $T_1 = T_2$. The boundary conditions are $\dot{y}_1(0, t) = \dot{y}_2(0, t)$ and, because the slip ring is massless, $F_{y1}(0, t) = F_{y2}(0, t)$. The problem is isomorphic to the reflection-transmission problem at a single interface (Section 3–4). The character of the reflected and transmitted waves depends only on the relative impedances $Z = \sigma c$ of the two strings. No reflection occurs if $Z_1 = Z_2$; a partial standing wave results if $Z_1 \neq Z_2$. A fixed end can be conceived of as an attachment with $Z_2 \gg Z_1$ (i.e., the strings can be taken to have equal tensions, with the second string much more massive than the first).

e) Mass point attached to the middle of a string. Consider an infinite string with a point mass m attached at $x = 0$. The boundary conditions are

$$\dot{y}_1(0, t) = \dot{y}_2(0, t) \quad \text{and} \quad F_{y1}(0, t) - F_{y2}(0, t) = m\ddot{y}(0, t).$$

With the correspondence $\mu \leftrightarrow m$ the problem is isomorphic to that of reflection and transmission by a thin plate treated in Section 3–9, the latter being specialized to the case $Z_1 = Z_2$. An arbitrary pulse is changed in form upon being reflected at the mass point. Sinusoidal waves of short wavelength are completely reflected, whereas those of long wavelength are 100% transmitted. The characteristic wavelength which defines the transition between these regimes is $\Lambda = \pi m/\sigma$.

PROBLEMS

4–1. Find the exact expression for the potential energy density W_{pot} of an *ideal gas.* Show that this reduces to the correct form when the acoustic approximation is made. (*Note:* Since the desired quantity contains second-order terms in s, a more accurate expression than $s = -\partial\xi/\partial x$ is required to show the equivalence.)

4–2. Set up expressions for the momentum contained in a volume element of a general (nonacoustic) plane wave disturbance, and the impulse delivered to this volume. Show that the basic equations guarantee the appropriate relation between these quantities.

4–3. A plane acoustic wave is normally incident upon an interface between two media. It is found that the peak pressure in the transmitted wave is one-half that in the incident wave. What fraction of the energy is transmitted into the second medium? Will a condensation be reflected as a condensation or rarefaction? What is the impedance ratio of the two media?

4–4. Calculate the rms values of p, ξ, $\dot{\xi}$, and s in air at standard temperature and pressure for a sinusoidal wave of frequency $\nu = 1000\ \text{sec}^{-1}$ and average intensity $\bar{I} = 10^{-12}\ \text{W/m}^2$.

4–5. A piston of fixed displacement amplitude and frequency is radiating plane waves into a region filled with an ideal gas. In each of the following cases describe the variation in the power output of the piston as the indicated changes are made:

a) The temperature is held constant but the region is pumped out so that the base pressure is lowered.

b) The pressure remains constant but the temperature is lowered.

c) The region is originally filled with nitrogen which is then replaced by hydrogen at the same temperature and pressure.

4–6. Assume that the displacement amplitude of a vibrating piston is independent of the medium in which it is operating. Compare the power outputs of the piston in water and air. If the piston is under water and parallel to a water-air surface, compare the intensity of the wave transmitted into the air with the intensity of the wave obtained when the piston is operating directly in air.

4–7. Consider a sinusoidal wave normally incident on a layer or combination of layers extending throughout $0 \leq x \leq a$. The generalized structure of the transmission-reflection problem is as follows: Medium 1 occupies the region $x < 0$, and medium 2 the region $x > a$ (in special cases $a = 0$). An incident wave of the form $\xi_{\text{inc}} = \mathcal{R}e\,\{e^{i(\omega t - k_1 x)}\}$ gives rise to a reflected wave in $x < 0$ given by $\xi_{\text{refl}} = \mathcal{R}e\,\{\mathbf{R}_\xi e^{i(\omega t + k_1 x)}\}$ and a transmitted wave in $x > a$ given by $\xi_{\text{trans}} = \mathcal{R}e\,\{\mathbf{T}_\xi e^{i(\omega t - k_2 x)}\}$. Show that if there are no energy sources or sinks in $0 \leq x \leq a$, the law of conservation of energy requires that $|R_\xi|^2 + Z_{21}|T_\xi|^2 = 1$. (It must of course be assumed that whatever is present in $0 \leq x \leq a$ preserves the plane-wave character of the problem.) Show that the coefficients obtained for the single-interface and thin-plate problems satisfy this condition (cf. also Problem 3–15).

***4–8.** Consider the explicit superposition of + and −waves given by

$$\xi = \mathcal{R}e\,\{[\mathbf{A}e^{-ikx} + \mathbf{B}e^{ikx}]e^{i\omega t}\}.$$

Since intensity separates into + and −components, we can immediately write down the average intensity of this wave as $\bar{I} = (Z/2)\,\{|\mathbf{A}|^2 - |\mathbf{B}|^2\}$. In Section 4–7 the average intensity is alternatively expressed in terms of the complex impedance. Show by explicit calculation that these two results are equivalent.

4–9. A room having a volume of 1000 m^3 is filled with a sound wave of intensity level 60 db. Estimate the total energy present. At what intensity level would a total energy of 1 cal be achieved?

4–10. At what intensity level will the effective velocity of a sound wave exceed the acoustic wave velocity by 1%?

4–11. A plane sonar wave is normally incident on the plane underwater surface of an iceberg. What is the decibel level of the reflected wave relative to that of the incident wave? (Assume that ice has the acoustic properties of a fluid medium of impedance $Z = 2.9 \times 10^6$ mks.)

4–12. *Transmission characteristics of a wood panel.* Consider an oak panel 1 cm thick (the density of oak is 720 kg/m^3) with air on both sides. Show that the characteristic frequency (in the sense of Section 3–9) is at the low end of the audible range. Sketch a graph of Δ_{trans}, the decibel level of the transmitted wave relative to that of the incident, as a function of frequency over the audible range.

4–13. Express the law of conservation of energy for transverse waves on a string in either integral or differential form, and verify by direct substitution from the basic equations that the relation is satisfied.

4–14. The fundamental vibration of a violin string is a standing wave having nodes at the two fixed ends, and is described by the function $y(x, t) = A \sin kx \cos \omega t$, where $k = \pi/l$, l being the length of the string. Find the total instantaneous kinetic and potential energies (integrated over the length of the string) and show that their sum is constant. Find the numerical value of the total energy for the data of Problem 1–3.

4–15. A uniform string under a tension of 15 N has a displacement given by the following functions. In each case calculate the average power associated with the wave.

a) $y_1(x, t) = 1.73 \cos(60t - 12x)$
b) $y_2(x, t) = 3.00 \cos(60t - 12x + \pi/3)$
c) $y_3(x, t) = 3.00 \cos(60t + 12x + \pi/3)$
d) $y_4 = y_1 + y_2$ e) $y_5 = y_1 + y_3$ f) $y_6 = y_2 + y_3$
(t is in seconds; x is in meters; y is in centimeters.)

4–16. Consider an infinite string of linear mass density $\sigma = 1.14 \times 10^{-3}$ kg/m under a tension $T = 7.6 \times 10^4$ N. A point mass m is located at $x = 0$, and a sinusoidal wave of frequency $\nu = 128\ \text{sec}^{-1}$ is incident from $x < 0$. What value of m is required to produce a 50% reflection of energy?

5 Experimental aspects of acoustics

5-1 INTRODUCTION

Although many acoustic effects can be demonstrated qualitatively with relative ease, the *measurement* of acoustic variables in terms of specific units is often a problem of some difficulty. The present chapter aims to sketch by means of descriptive material some idea of the general nature of experimental acoustics. The majority of acoustical measuring instruments today are electronic. Since these instruments are usually based upon fairly elementary wave properties, their design is mostly a problem in electronics and will not be treated here in any detail. This subject, along with several others in this chapter, is mentioned only briefly in order to advertise it as a suitable topic for additional study.*

5-2 DIVISIONS OF THE ACOUSTIC SPECTRUM

The acoustic spectrum is customarily divided into the following three regions:

Infrasonic. This region comprises frequencies below 30 $\sec^{-1}$ with corresponding wavelengths in air greater than 10 m. Although human observers may "feel" such low frequency vibrations, they do not perceive them as sound. Infrasonic frequencies are of considerable importance in the seismic waves generated by earthquakes and man-made explosions.

---

* Suggested reference texts: Leo L. Beranek, *Acoustics*. New York: McGraw-Hill, 1954. Ludwig Bergmann, *Ultrasonics*. H. Stafford Hatfield, translator. New York: Wiley, 1938. E. G. Richardson, *Technical Aspects of Sound*. Amsterdam: Elsevier Publishing Co., 1953. A. B. Wood, *A Textbook of Sound*. London: G. Bell and Sons, 1949.

Sonic. This region comprises the audible frequencies ranging approximately from 30 sec^{-1} to $2 \times 10^4 \text{ sec}^{-1}$ with wavelengths in air ranging from 10 m to 1.6 cm. These wavelengths are comparable to the dimensions of common objects. This fact is unfortunate from one point of view, since it means that audible sound, with which we have so much experience, does not belong to either of the extremes of "short" or "long" wavelengths for which a simplified theoretical treatment is possible.

Ultrasonic. This region comprises frequencies above $2 \times 10^4 \text{ sec}^{-1}$ and wavelengths in air shorter than 1.6 cm. Frequencies as high as 10^9 sec^{-1} have been generated experimentally. Because of their short wavelength, ultrasonic waves tend to form well-defined beams. Many of their properties are adequately described in terms of the concept of rays which propagate in straight lines and which are subject to simple laws of reflection and refraction. For example, the focusing of a plane wave upon reflection in a concave surface can be easily demonstrated with ultrasonic waves. Since diffraction is a small effect with ultrasonic waves, as it is also with light, numerous ultrasonic experiments with gratings, interferometers, etc., are closely analogous to corresponding experiments in optics. A partial listing of the practical applications of ultrasonic waves includes: underwater signaling and depth sounding, emulsification of liquids, coagulation of aerosols, effects on biological organisms, detection of flaws in solids, catalysis of chemical reactions.

5-3 THE PROBLEM OF REFLECTION AND SCATTERING

In most circumstances only a negligible loss of energy occurs as a sound wave propagates through air. The major dissipation of energy occurs when the wave encounters obstacles in its path or when it reaches the solid boundaries of the region. Ordinary building materials such as brick or plaster reflect more than 95% of the energy of an incident wave. Even drapery materials, which we normally think of as absorbent, reflect about 50% of the incident energy. As this is a reduction in level of only 3 db, a wave of moderate intensity is re-reflected a number of times before it falls below the level of detection. These reflections must be avoided if an experimental measurement is to be made to check a theoretical calculation. The so-called *anechoic chamber*, whose walls consist of wedge-shaped cavities of absorbent material, makes it possible for us to perform experiments with essentially the same results as would be obtained if they had been performed in an infinite space with no boundaries. The sound which reaches the walls is attenuated as it reflects back and forth within the wedges.

The introduction of a measuring instrument into a field of sound may alter the acoustic situation in a manner that makes it difficult to relate the reading of the instrument to conditions which would prevail at the given location if the instrument were not there. Consider a microphone, for example. To waves of short wavelength the diaphragm of the microphone "looks like" a rigid surface. The wave reflected at this surface therefore results in a doubling of the pressure amplitude. For longer wavelengths, however, a different situation prevails. The presence of the microphone gives rise to a "scattered wave" which is more complicated than the simple reflected wave. The nature of this wave must be calculated so that we can apply the appropriate correction to the reading of the microphone. Alternatively, the microphone can be calibrated by experimental procedures. However, since the correction factors are dependent on wavelength, the calibration has to be carried out separately for different wavelengths.

5-4 TRANSDUCERS

A device which converts an input signal in one form of energy into an output signal in another form of energy is in general called a *transducer*. If the conversion can be effected in either direction, the transducer is said to be *reversible*. In particular, an *electroacoustic* transducer translates voltage variations in an electrical circuit into the pressure variations of an acoustic wave or conversely. Loudspeakers and underwater sound transmitters are examples of the former, and microphones or underwater sound receivers are examples of the latter. The ease with which electrical signals can be amplified, controlled, and measured results in the widespread use of electroacoustic transducers as sources and receivers in acoustic experiments.

When used as receivers, most transducers make use of a diaphragm which is set in motion by the acoustic wave. Two special cases may be distinguished:

1) *Only one face of the diaphragm exposed* (Fig. 5-1). The diaphragm responds to the pressure variations on this face and the device is said to be *pressure sensitive*.

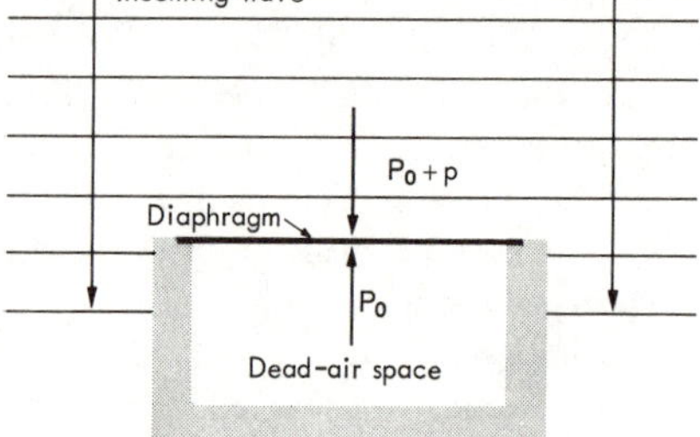

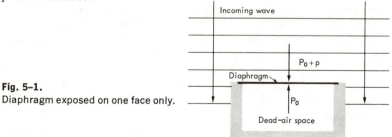

Fig. 5-1.
Diaphragm exposed on one face only.

2) *Both faces exposed* (Fig. 5-2). The diaphragm responds to the pressure difference between the two faces. When the wavelength is sufficiently long in comparison with the dimensions of the diaphragm that the incident wave

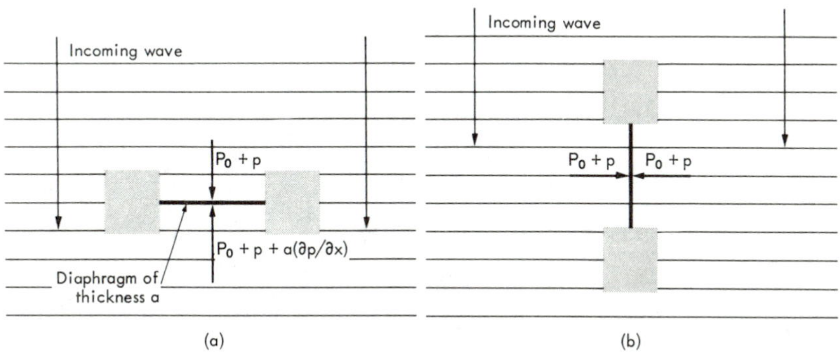

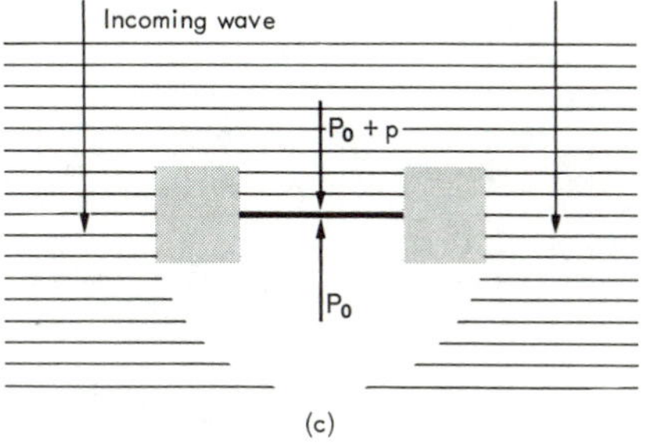

Fig. 5-2. Diaphragm with both faces exposed: (a) pressure-gradient sensitive (orientation for maximum response), (b) pressure-gradient sensitive (orientation for zero response), (c) pressure sensitive at high frequencies.

diffracts completely around the diaphragm, the motion of the latter depends on the pressure gradient of the incident wave, and the device is said to be *pressure-gradient sensitive*. As such a device is rotated, it will give a maximum response when the faces are perpendicular to the direction of propagation of the wave, and no response when the faces are parallel to this direction. In the presence of many sound sources the receiver therefore has *directionality*; that is, it responds most sensitively to sources which lie in a direction perpendicular to the faces.* At frequencies sufficiently high that a sound shadow is cast behind the diaphragm (Fig. 5–2c), the pressure gradient of the original wave is irrelevant and the functioning is again pressure sensitive. In contrast to the diaphragm with only one face exposed, however, this last case clearly has directional selectivity.

Numerous physical principles have been used as the basis for construction of electroacoustic transducers. To illustrate this diversity a few examples are indicated schematically in Fig. 5–3.

---

* The pattern of directional sensitivity of a reversible transducer functioning as a receiver is identical with the radiation pattern for the same device functioning as a transmitter. The pattern for the present case is that of an *acoustic dipole* (Section 11–2). In Fig. 11–4 the axis $\theta = \pi/2$ corresponds to the normal to the diaphragm.

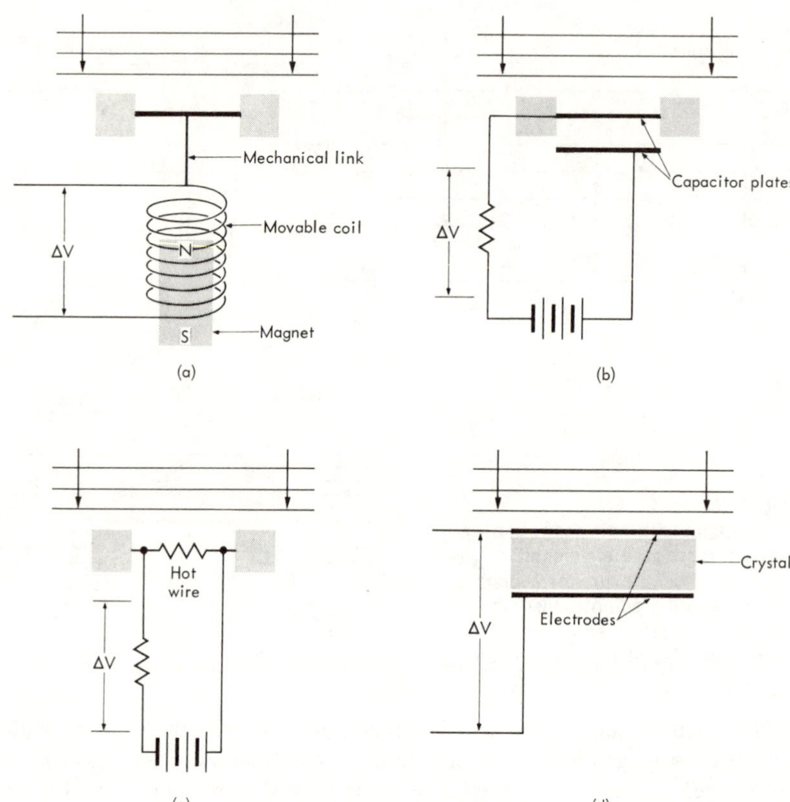

Fig. 5–3. Schematic illustration of principles used in the design of electroacoustic transducers: (a) electrodynamic, (b) electrostatic, (c) thermal, and (d) piezoelectric.

a) Electrodynamic. The diaphragm is linked mechanically to a coil in the field of a permanent magnet. Motion of the coil results in an induced potential difference across the coil. In this manner the system functions as a microphone which converts an incident acoustic disturbance into an electric signal. Conversely, an applied potential difference results in a current through the coil. The latter experiences a force in the magnetic field and is set into motion. The motion is communicated to the diaphragm, which now functions as a loudspeaker, and results in the generation of an acoustic wave.

b) Electrostatic. The diaphragm is employed as one plate of a parallel-plate capacitor. Motion of the diaphragm results in a change in capacitance which is detected by appropriate circuit arrangements. Conversely, an applied voltage variation in the circuit results in a variable force on the diaphragm and the generation of acoustic waves.

c) Thermal. As was noted in Chapter 2, the adiabatic character of the compressions of a sound wave results in temperature variations in the air. Since electrical resistance is a function of temperature, the resistance of a wire which is exposed to a passing sound wave will undergo small variations. These resistance changes result in readjustments in an electrical circuit of which the wire is a part. A suitably designed circuit therefore enables us to read the temperature variations of the sound wave as voltage variations. This principle is also reversible: A varying current in the wire produces temperature changes in the air which set up acoustic waves.

d) Piezoelectric. Certain crystals (e.g. Rochelle salt, quartz, tourmaline) develop a potential difference between opposite faces when they are deformed under mechanical stress. This phenomenon is known as the *piezoelectric effect*. To employ this effect in the design of an acoustic detector, we fasten plane electrodes to a pair of faces of a crystal slab. The pressure variations of an incident sound wave result in an electrical potential difference which is detected by an appropriate circuit connected to the electrodes. The effect is reversible. When a potential difference is applied across the electrodes the crystal is deformed, and the motion of one of its faces can be used to generate sound waves. If the driving frequency coincides with one of the frequencies of the normal modes of vibration of the crystal, a resonance response results and a relatively high power output is achieved. Conversely, the crystal has a relatively high sensitivity as a receiver of sinusoidal signals whose frequencies coincide with the natural vibration frequencies of the crystal. For the usual crystal these frequencies are in the ultrasonic range. Piezoelectric transducers are commonly used as reversible hydrophones for echo ranging and for continuous wave underwater signaling between ships. It is also possible to use the high-frequency ultrasonic wave as a carrier of audio signals through frequency or amplitude modulation.

Although the above principles are reversible, it does not necessarily follow that a given device will function equally efficiently as a transmitter and receiver. Thus, for example, the design requirements for a moving-coil type electrodynamic microphone are different from those for an electrodynamic loudspeaker. The variable-capacitor microphone is constructed differently from the electrostatic speaker, and the hot-wire microphone differently from the corresponding sound generator, the thermophone. The audio-frequency crystal microphone is commonly used as a receiver, whereas the audio-frequency crystal loudspeaker is used only rarely.

5-5 ABSOLUTE MEASUREMENT OF ACOUSTIC VARIABLES

In a great many acoustical experiments the instruments employed serve merely as uncalibrated *detectors* whose response measures the *relative* magnitudes of some acoustic variable. For example, the ear can be used to locate the nodes in an interference pattern, or the output of a pressure-sensitive microphone can be displayed on an oscilloscope to yield relative values of the pressure as a function of position. It is a problem of considerably greater difficulty to perform an *absolute* measurement, that is to say, a measurement which will yield the magnitudes of an acoustic variable in terms of the standard units of mass, length, and time. By far the most important present-day method of absolute measurement is the so-called *reciprocity technique* of calibration. By invoking a general property known as the *electroacoustical reciprocity theorem*, one can calibrate a microphone solely in terms of measured electrical quantities. For a description of this technique the reader is referred to the texts cited below.*

In the early development of experimental acoustics one of the few instruments available for absolute measurement was the *Rayleigh disk*. A brief description is included here, not only because of its historical importance, but also because it demonstrates a physical principle which is interesting in its own right. Consider an incompressible fluid of density ρ which is streaming in a steady parallel flow with velocity v. If an obstacle is introduced into the stream, the flow lines of the fluid in the vicinity of the obstacle will be altered in a manner which can be determined by solution of a boundary value problem in general hydrodynamics. (It is assumed that v is sufficiently small that *turbulence* does not develop.) The pattern of flow lines is indicated in Fig. 5-4 for an obstacle consisting of a thin ellipsoidal disk *ED* having a circular cross section of radius a in the plane perpendicular to the plane of the figure. Of primary importance is the symmetry of the pattern; the same pattern obtains if the directions of all the flow lines are reversed. The flow

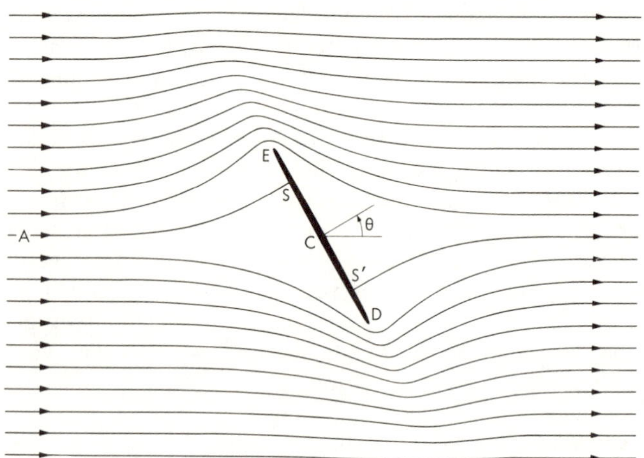

Fig. 5-4. General features of the pattern of fluid flow past a thin ellipsoidal disk.

---

* Leo L. Beranek, *op. cit.*, p. 377. L. E. Kinsler and A. R. Frey, *Fundamentals of Acoustics*. New York: Wiley, 1950, p. 353.

line AS divides the fluid into those particles which pass over the top of the disk and those which pass under it. The points S and S' are *points of stagnation* at which the velocity of the fluid is zero.

From Bernoulli's principle we observe that the fluid will exert a torque on the disk. The fluid pressure is greater at S and S' than it is at the corresponding points on the opposite sides of the disk, for the fluid has a greater velocity at the latter points. We therefore expect a clockwise torque to act on the disk when it is in the position indicated in the figure. The hydrodynamic calculation shows that the torque τ, taken about an axis through C, the center of the disk, is given by the expression

$$\tau = -\tfrac{4}{3}\rho a^3 v^2 \sin 2\theta. \tag{5–1}$$

In the absence of any other torques, the stable equilibrium position of the disk occurs at $\theta = 0$, i.e., when the disk is aligned *perpendicular* to the direction of the undisturbed flow lines.

One can provide a simple demonstration of this effect by dropping a small card from a horizontal position. As the card falls, it rocks gently back and forth around this equilibrium orientation. (It should be obvious that although the *fluid* is not in motion in this case, the velocity v in the above discussion can be taken generally as the *relative* velocity between the fluid and the obstacle.)

Suppose now that the fluid motion is associated with the rapid oscillations of a sinusoidal sound wave. The fact that Eq. (5–1) depends on the square of the velocity shows that the torque acts in the same direction throughout all portions of the cycle. This is associated with the previously mentioned symmetry of the flow pattern. We therefore expect a passing sound wave to exert a torque on the disk which will tend to turn it toward the equilibrium position $\theta = 0$.

One can exhibit this effect by suspending a rectangular card by means of a loose thread and placing it in the opening of the resonating base of a demonstration tuning fork. When the tuning fork is struck, the card turns so as to block the opening.

The Rayleigh disk method of absolute measurement consists in suspending a small disk from a delicate torsion fiber and placing it in the field of a steady sound wave. The torsion head is then rotated so that the disk achieves equilibrium at the angle $\theta = \pi/4$. (The hydrodynamic torque, which is balanced by the torque due to the fiber, will then be at a maximum.) Equation (5–1) is used to determine the value of v^2 averaged over many cycles of the sound wave. This is therefore an absolute measurement of the quantity designated as $\dot{\xi}^2_{\text{rms}}$. Numerous oversimplifications have gone into the calculation of Eq. (5–1); more elaborate formulas have been obtained for use in precise work.*

---

* Leo L. Beranek, *Acoustic Measurements*. New York: Wiley, 1949, pp. 148–158.

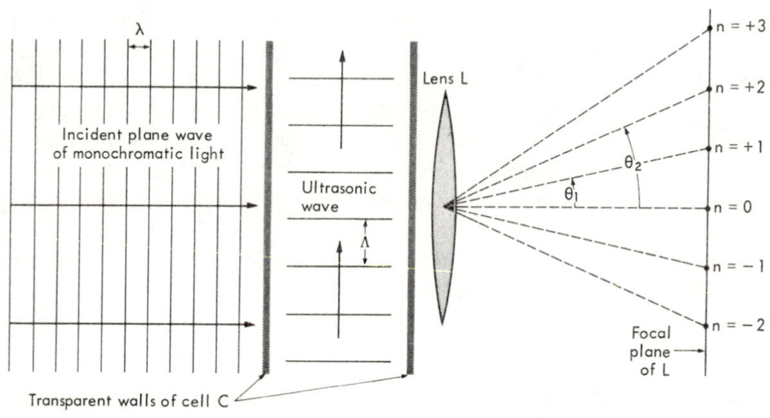

Fig. 5-5. Diffraction of light by an ultrasonic wave.

*5-6 DIFFRACTION OF LIGHT BY AN ULTRASONIC WAVE

A highly interesting series of experiments results from the discovery by Debye and Sears† that the density variations in a sinusoidal ultrasonic wave in a liquid can be used to diffract light in a manner analogous to the diffraction of light by the parallel rulings of a transmission grating. Thus in Fig. 5-5, a plane wave of monochromatic light of wavelength λ propagates across the liquid cell C in a direction transverse to the direction of propagation of a sinusoidal ultrasonic wave of wavelength Λ. In the focal plane of lens L a series of diffraction images appears whose angular positions θ_n are governed by the equation $n\lambda = \Lambda \sin \theta_n$. This relation is the same as the regular grating equation corresponding to a grating spacing Λ. It is interesting to note that the motion of the sound wave down the cell does not result in a washing-out of the pattern as might naively be expected when one thinks of the fact that the density at any particular location fluctuates rapidly about the average value. It is important to realize instead that the distance through which the ultrasonic wave moves during the passage of a light wave across the cell is small. Hence the acoustic wave is effectively stationary as far as the optical phenomenon is concerned. Note further that the location of the diffraction images produced by a grating is determined solely by their relation to the axis of the lens and is not affected by a shift of the grating parallel to itself. Thus the diffraction images are stationary and do not shift along the screen as the sound wave progresses. There is, however, an interesting effect associated with the motion of the sound wave. Crudely speaking, the different diffraction orders are produced by reflection from the set of planes of maxi-

† P. Debye and F. W. Sears, *Proc. Nat. Acad. Sci., Wash.* **18,** 410, 1932.

mum density in the fluid. Since these surfaces are in motion, it may be expected that there will be a Doppler shift in the frequency of the diffracted light. However, because the frequency shift is only slight, considerable ingenuity must be exercised to detect this effect experimentally.* The intensity of the various orders in the diffraction spectrum depends in a complex way on the intensity of the ultrasonic wave. As the latter is increased, the first-order diffraction image, for example, fluctuates in intensity, dropping to zero at regular intervals. This effect has been explained in a simple and elegant manner by Raman and Nath.†

PROBLEMS

5–1. a) Consider a room whose walls absorb a fraction α of the energy in an incident wave. Show that a sound of intensity level 60 db must undergo a number of reflections $N = -6/\log_{10}(1 - \alpha)$ before it reaches the threshold of audibility. Calculate N for $\alpha = 0.05$ and $\alpha = 0.5$.

b) Assume as a crude model that the sound wave must travel an average distance b between reflections, where b is comparable to the average dimension of the room. If T_r, the *reverberation time*, is defined as the length of time required for a sound of intensity level 60 db to become inaudible, show that

$$T_r = \frac{-6b}{c \log_{10}(1 - \alpha)}.$$

Calculate the reverberation time for $b = 10$ m, and the two values $\alpha = 0.05$ and $\alpha = 0.5$.

***5–2.** Consider a sinusoidal plane wave of sound traveling in the x-direction described by a pressure function $p(x, t)$. Let a thin rectangular diaphragm of thickness b and cross-sectional area A be inserted in the sound field with its normal making an angle θ with the x-axis. Assume that all dimensions of the diaphragm are small compared with the wavelength and that the acoustic pressure near the diaphragm is the same at any point as it would be in the absence of the diaphragm. Show that the instantaneous force on the diaphragm due to the sound wave is given by $Ab(\partial p/\partial x) \cos \theta$. What implication does this have for the directionality of a microphone whose diaphragm fulfills these conditions? (*Suggestion:* With $p = \Re e\{p_m e^{i(kx-\omega t)}\}$, the expression for the total force on each side of the diaphragm can be integrated explicitly. The results can then be approximated under the given conditions.)

5–3. Investigate the Doppler shift expected in the light emitted from a source moving with velocity $u \ll c$ as a function of the angular position of the point of

---

* An excellent description of the effects associated with the diffraction of light by ultrasonic waves is contained in Bergmann, *op. cit.*, pp. 63–89.

† C. V. Raman and N. S. Nagendra Nath, *The Diffraction of Light by High Frequency Sound Waves*, *Proc. Ind. Acad. Sci.*, **A2,** 406 (Part I) and 413 (Part II), 1935.

observation. Assuming that due to the motion of the sound wave, the nth diffraction image in Fig. 5–5 has undergone a Doppler shift associated with the angle θ_n, show that the frequency shift is $\delta\nu = nF$, where F is the frequency of the sound wave.

5–4. It can be seen from Eq. (5–1) that the torque on a Rayleigh disk vanishes when the disk is parallel to the flow lines as well as when it is perpendicular thereto. Show that the latter is a position of stable equilibrium, whereas the former is a position of unstable equilibrium.

6 The electromagnetic plane wave

6–1 MAXWELL'S EQUATIONS

At some point in his education the serious student will encounter a thorough presentation of electricity and magnetism which leads to the compact summarization of the basic laws in a set of partial differential equations known as *Maxwell's equations*. The present text assumes that the background of Maxwell's equations will ultimately be covered in another course of studies. Since it cannot be assumed that the reader has as yet completed such a course, we shall take an approach which does not require prior knowledge of electromagnetic theory. Maxwell's equations will be adopted as a set of *postulates* concerning the behavior of electric and magnetic fields. Indeed, this is close to the point of view taken by many theoretical physicists. The previous experimental work of Coulomb, Gauss, Ampère, Faraday, and others can be looked upon as part of a process of inductive reasoning which has finally led to a set of postulates now accepted as the starting point for deductive reasoning. The equations themselves define the necessary properties of the quantities to which they refer.

We shall confine our discussion to the properties of electric and magnetic fields within a region that does not contain free charges or currents. The region may consist of free space, or may contain a number of bodies composed of nonconducting dielectric substances. The complete description of an electromagnetic field requires the specification of four vectors **E**, **D**, **B**, and **H** at each point in space. If we further limit ourselves to *linear isotropic* media, the vector **D** is at all times parallel to **E** and proportional in magnitude to it, and a similar relation holds between **B** and **H**. In fact, from our postulational point of view, a linear isotropic medium is defined as one for which the relations

$$\mathbf{D} = \epsilon \mathbf{E} \qquad \text{and} \qquad \mathbf{B} = \mu \mathbf{H} \tag{6–1}$$

are satisfied, where ϵ and μ are parameters of the medium in question. They are referred to respectively as the *permittivity* and *permeability* of the medium. The requirement of linearity rules ferromagnetic media out of the context of our discussion, and the requirement of isotropy rules out crystalline substances. Since Eq. (6–1) enables us to dispense with **D** and **B** in succeeding equations, we shall discuss the names and units of **E** and **H** only. The system of units employed is the rationalized mks system. The vectors **E** and **H** are referred to respectively as the *electric and magnetic field intensity vectors.* The mks unit for **E** is the volt per meter, and for **H** it is the ampere per meter. Since **D** and **E** are defined experimentally by independent operational procedures, the permittivity ϵ is an empirical constant which has the dimensions of farads per meter in the mks system. The permittivity of free space is found to be

$$\epsilon_0 = 8.854 \times 10^{-12}\ \mathrm{F/m}. \tag{6–2}$$

The permittivity of a material medium relative to that of a vacuum is a pure number referred to as the *dielectric constant* of the medium, $K \equiv \epsilon/\epsilon_0$.

The permeability, μ, has the dimensions of henries per meter. By virtue of the way in which quantities are defined operationally in the mks system, the permeability of free space has the exact value

$$\mu_0 = 4\pi \times 10^{-7}\ \mathrm{H/m}. \tag{6–3}$$

The relations in Eq. (6–1) are known as the *constitutive relations* of the material medium and, as we shall see, play a role similar to that which the equation of state does in the derivation of the acoustic wave equation.

The Maxwell equations, limited to the situations described above, assert that the fields **E** and **H** must satisfy the conditions

*Maxwell's equations**

$$\operatorname{div} \mathbf{E} = 0, \tag{6–4}$$

$$\operatorname{div} \mathbf{H} = 0, \tag{6–5}$$

$$\operatorname{curl} \mathbf{E} = -\mu(\partial \mathbf{H}/\partial t), \tag{6–6}$$

$$\operatorname{curl} \mathbf{H} = \epsilon(\partial \mathbf{E}/\partial t). \tag{6–7}$$

We further postulate the following associations of energy quantities† with an electromagnetic field.

---

* Readers not yet acquainted with the vector operators div $= \nabla\cdot$ and curl $= \nabla\times$ may look upon these equations simply as abbreviations for certain longhand expressions in a cartesian coordinate system. These expressions are written out in full in Appendix VI.

† There are some debatable aspects to these energy assignments. See M. Mason and W. Weaver, *The Electromagnetic Field.* New York: Dover, pp. 266–272, and also Max Born and Emil Wolf, *Principles of Optics.* New York: Pergamon, 1959, p. 9.

Electrical energy density. At a point in a dielectric of permittivity ϵ where the electric field strength is **E**, the electrical energy per unit volume is

$$W_{\text{el}} = \epsilon E^2/2. \tag{6–8}$$

Magnetic energy density. At a point in a dielectric of permeability μ where the magnetic field strength is **H**, the magnetic energy per unit volume is

$$W_{\text{mag}} = \mu H^2/2. \tag{6–9}$$

Total energy density. The total energy stored in the electromagnetic field is the sum of its electric and magnetic parts:

$$W_{\text{tot}} = W_{\text{el}} + W_{\text{mag}}. \tag{6–10}$$

Energy flux. At a point where the electric and magnetic field intensities are **E** and **H** respectively, the rate of flow of electromagnetic energy is specified by the vector

$$\mathbf{S} = \mathbf{E} \times \mathbf{H}, \tag{6–11}$$

called the *Poynting vector.* The direction of energy flow is parallel to **S** and the energy which passes through an element of area $d\mathbf{A}$ per unit of time is given by $\mathbf{S} \cdot d\mathbf{A}$. Thus $|\mathbf{S}|$ is the power per unit area crossing a surface whose normal is parallel to **S**. The units of **S** are W/m^2 (watts per square meter).

To complete the list of postulates we state the boundary conditions which must be met at an interface between two dielectric media. If there are no surface distributions of free charge or surface currents at the interface, the requirement is that the tangential components of **E** and **H** and the normal components of **B** and **D** all be continuous across the boundary surface. If the two media are designated by subscripts 1 and 2, and if **n** is a unit vector perpendicular to the interface at any point, these boundary conditions can be written symbolically:

$$\begin{aligned} \mathbf{n} \times \mathbf{E}_1 &= \mathbf{n} \times \mathbf{E}_2, & \mathbf{n} \times \mathbf{H}_1 &= \mathbf{n} \times \mathbf{H}_2, \\ \epsilon_1(\mathbf{n} \cdot \mathbf{E}_1) &= \epsilon_2(\mathbf{n} \cdot \mathbf{E}_2), & \mu_1(\mathbf{n} \cdot \mathbf{H}_1) &= \mu_2(\mathbf{n} \cdot \mathbf{H}_2). \end{aligned} \tag{6–12}$$

6–2 A SOLUTION TO MAXWELL'S EQUATIONS FOR A SPECIAL SITUATION

At the time they were originally devised, Maxwell's equations represented a summary of a number of the "ordinary" experimental laws of electricity and magnetism, such as Faraday's law of electromagnetic induction and Ampère's law which determines the magnetic field produced by a given distribution of currents. It should also be added that Maxwell found it desirable on the basis of purely theoretical grounds to correct Ampère's law by the addition of a term whose effect would be small in the prior circumstances in which measurements were made. This term [which is, in fact, the right-hand side of Eq. (6–7)] will be quite important in the new deductions that follow.

For those who do have the preparation, the following review is given of the theoretical reasons for adding the extra term. In circuital form, the original version of Ampère's law is

$$\oint_L \mathbf{H}\cdot d\mathbf{r} = \iint_{A_L} \mathbf{j}\cdot d\mathbf{A},$$

where $\mathbf{j}$ is the current density vector, L is any closed loop, and A_L is an area having L as perimeter. Now unless the right-hand side is independent of how this area is chosen, the law has not been given an unambiguous formulation. Consider two surfaces A_L and A'_L having L as perimeter. If the flux of $\mathbf{j}$ is the same through these two surfaces, then the flux of $\mathbf{j}$ out of the closed surface $A_L + A'_L$ must be zero. But, since L is an arbitrary loop, the requirement can then be stated as

$$\oiint \mathbf{j}\cdot d\mathbf{A} = 0,$$

where the surface integral is taken over an arbitrary closed surface. A necessary condition for the latter is that div $\mathbf{j} = 0$. This, however, is not generally the case, since the law of conservation of charge states div $\mathbf{j} + \partial\rho/\partial t = 0$, where ρ is the charge density. Now from another of the Maxwell equations (when $\rho \neq 0$), $\partial\rho/\partial t = \text{div}\,(\partial\mathbf{D}/\partial t)$, and hence div $(\mathbf{j} + \partial\mathbf{D}/\partial t) = 0$. Thus the ambiguity of Ampère's law can be removed if $\mathbf{j}$ is replaced in the original statement by $\mathbf{j} + \partial\mathbf{D}/\partial t$. The added term $\partial\mathbf{D}/\partial t$ is referred to as the *displacement current density*.

It is important to realize that none of the reasoning or experimentation that went into the formulation of Maxwell's equations was connected with the field of optics. The parameters ϵ and μ are obtained from standard methods of electric and magnetic measurement. At the time the equations were originally written there was no strong suspicion that they would have anything to do with light. It was therefore a remarkable achievement on Maxwell's part to be able to show that among the diverse possible solutions to this set of equations (all electrostatic situations are, for example, comprised among the solutions to these equations) there exists a solution of a new type which describes the propagation of an electromagnetic disturbance at a velocity which coincides with the velocity of light.

To arrive at this solution, let us arbitrarily inquire whether or not Eqs. (6–4) through (6–7) permit a solution in which all the components of $\mathbf{E}$ and $\mathbf{H}$ except for E_y and H_z vanish identically everywhere. With this assumption, Eqs. (6–4) through (6–7) become

$$\partial E_y/\partial y = 0, \tag{6–4$'$}$$

$$\partial H_z/\partial z = 0, \tag{6–5$'$}$$

$$-(\partial E_y/\partial z)\mathbf{i} + (\partial E_y/\partial x)\mathbf{k} = -\mu(\partial H_z/\partial t)\mathbf{k}, \tag{6–6$'$}$$

$$(\partial H_z/\partial y)\mathbf{i} - (\partial H_z/\partial x)\mathbf{j} = \epsilon(\partial E_y/\partial t)\mathbf{j}. \tag{6–7$'$}$$

Equations (6–4′) and (6–5′) and the x-components of (6–6′) and (6–7′) will be satisfied only if E_y and H_z are functions of at most x and t. The equations which remain to be satisfied are the z-component of (6–6′) and the y-component of (6–7′):

$$(\partial E_y/\partial x) = -\mu(\partial H_z/\partial t), \tag{6–6″}$$

$$-(\partial H_z/\partial x) = \epsilon(\partial E_y/\partial t). \tag{6–7″}$$

We can eliminate H_z from these two equations by differentiating the first with respect to x and the second with respect to t. The result shows that $E_y(x, t)$ must satisfy the equation

$$\frac{\partial^2 E_y}{\partial x^2} = \mu\epsilon\,\frac{\partial^2 E_y}{\partial t^2}. \tag{6–13}$$

This has the form of the one-dimensional wave equation corresponding to a propagation velocity

$$c = 1/\sqrt{\mu\epsilon}. \tag{6–14}$$

The basic equations of electricity and magnetism have therefore shown that a solution of the restricted type we asked for is possible, and that this solution implies the propagation of a "disturbance" with a specified velocity along the x-axis. Substituting the values μ_0 and ϵ_0 for free space, the result turns out to be precisely the measured value of the velocity of light in free space, $c_0 = 3.00 \times 10^8$ m/sec. This astonishing coincidence of numbers which appear in two areas of physics previously thought to be unrelated is sufficient to suggest that light should be looked upon as an electromagnetic phenomenon, the complete description of which is contained within Maxwell's equations. This identification is further substantiated by the great number of detailed predictions which can be made about the behavior of light by analytical deduction from Maxwell's equations. Optics thus becomes subsumed as a subfield of electromagnetic theory in the same sense that acoustics is a subfield of mechanics.

6–3 IMPLICATIONS OF THE ELECTROMAGNETIC THEORY OF LIGHT

Before proceeding further with a discussion of the electromagnetic wave equation, we shall present a brief outline of some important areas of physics which originate from Maxwell's discovery (in the years around 1861) of the electromagnetic character of light.

i) The interaction of matter and radiation. It is possible to construct a fairly concrete model to predict the influence of a light wave on a material substance. If the composition of matter is visualized in terms of charged particles, the variable electric fields of an incident light wave are expected to exert forces on the charges and set them into oscillation. It is in turn a consequence of Maxwell's equations that an oscillating charge will act as the source of a radiated electromagnetic wave. The superposition of this "secondary radiation" and the field of the incident wave determines the net character of the wave which propagates through the material. This is the basic idea underlying the so-called *electron theory of matter* which was developed primarily by H. A. Lorentz.* Among its striking successes is its ability to explain the reduction in the propagation velocity of light in a material medium compared with the velocity of light in free space. The theory also provides an explanation of a magneto-optical effect which had been discovered experimentally by Faraday in 1845, and predicts numerous other influences which external electric and magnetic fields have on the optical properties of a material medium (Chapter 10).

ii) Extension of the electromagnetic spectrum. It was known prior to Maxwell's theory that visible light was some kind of wave whose wavelengths in a vacuum range from 4000 to 7000 Å. It was also known that ultraviolet and infrared radiations are associated with waves of the same character but with shorter and longer wavelengths respectively. The electron theory of matter presumes that the emission from a glowing solid or a gaseous discharge tube is due to the oscillation at corresponding frequencies of charge within the emitting substance. The theory suggests that electromagnetic waves with frequencies lying outside the above range can be produced if some means can be found for setting charge in oscillation at the required frequencies. Oscillations at a frequency lower than that of infrared radiation are provided by the spark discharge of an induction coil. In a classical series of experiments, Heinrich Hertz† showed in 1887 that such waves could be produced and detected. He demonstrated that these waves are reflected, refracted, and focused according to the laws of geometrical optics. A standing wave pattern was obtained by reflection of the waves from the surface of a plane conductor. The wavelength was determined from this measurement, and this, in conjunction with the calculated frequency of the oscillations in the induction coil circuit, enabled him to calculate the velocity. The result agreed with the velocity of light.

The success of Hertz's investigations stimulated the search for means of producing electromagnetic radiations of longer wavelength, such as those

* H. A. Lorentz, *The Theory of Electrons and its Application to the Phenomena of Light and Radiant Heat*. New York: Dover Publications, 1952.

† Translations of Hertz's research papers are contained in H. Hertz (D. E. Jones, Translator), *Electric Waves*. New York: Dover Publications, 1962. The introduction by Hertz makes fascinating reading, for it shows the extreme caution with which he interprets his results before arriving at the conclusion that these experiments may be looked upon as the first direct test of Maxwell's theory.

Table 6-1. List of wavelengths typical of different regions of the electromagnetic spectrum

Type	λ, meters	ν, sec^{-1}	Description
γ-ray	1.240×10^{-12}	2.418×10^{20}	1 MeV gamma photon
X-ray	1.7×10^{-11}	1.8×10^{19}	75 kV tube for medical diagnosis
Ultraviolet	1.216×10^{-7}	2.465×10^{15}	First line of Lyman series (hydrogen)
Visible limit	3.8×10^{-7}	7.9×10^{14}	Approximate shortest visible wavelength
Violet	4.358×10^{-7}	6.879×10^{14}	Hg violet line
Greenish-blue	4.861×10^{-7}	6.167×10^{14}	H_β, second line of Balmer series (hydrogen)
Green	5.461×10^{-7}	5.490×10^{14}	Hg green line
Yellow	5.876×10^{-7}	5.102×10^{14}	He yellow line
Orange-yellow	5.893×10^{-7}	5.087×10^{14}	Na doublet (5,890 Å and 5,896 Å)
Red	6.438×10^{-7}	4.657×10^{14}	Cd red line
Red	6.563×10^{-7}	4.568×10^{14}	H_α, first line of Balmer series (hydrogen)
Visible limit	7.8×10^{-7}	3.8×10^{14}	Approximate longest visible wavelength
Infrared	1.5×10^{-6}	2.1×10^{14}	Peak emission of blackbody at 2000°K.
Infrared	3.431×10^{-6}	8.738×10^{13}	Vibration at C—H bond in CH_4
Microwave	2.600×10^{-3}	1.153×10^{11}	Lowest frequency in rotational spectrum of CO
Microwave	1.256×10^{-2}	2.387×10^{10}	Inversion line of ammonia
Microwave	3.0×10^{-2}	1.0×10^{10}	Radar
Microwave	2.111×10^{-1}	1.420×10^{9}	Discrete spectral line in general galactic radiation (hyperfine transition in ground state of hydrogen)
Radio	3.34	8.97×10^{7}	FM broadcast
Radio	5.3	5.7×10^{7}	TV
Radio	9.3	3.2×10^{7}	One of Hertz's original experiments
Radio	4.2×10^{2}	7.1×10^{5}	Standard broadcast

which are now employed in radio, television, and radar (Table 6–1). One can hardly ask for a better example of the benefits of "pure" research: the communications industry of today is an outgrowth of Maxwell's correction, for theoretical reasons, of the mathematical statement of Ampère's law.

The extension of the electromagnetic spectrum to wavelengths shorter than ultraviolet cannot be looked upon as a direct outcome of Maxwell's theory, although the discovery of x-rays did come at a later time. The phenomena associated with x-radiation were discovered by Roentgen in 1895 while he was making a systematic investigation of penetrating radiations associated with the cathode-ray tube. It was by no means clear initially that x-rays were to be considered a type of electromagnetic radiation, since the measurement of the associated wavelengths presented an experimental problem of some difficulty. When the identity was established, however, the Maxwell theory was immediately fruitful in describing and predicting further aspects of the behavior of x-rays.

The electromagnetic spectrum as we understand it today comprises a number of diverse phenomena under one unifying principle. Apparent distinctions among these phenomena are attributed primarily to the different orders of magnitude of the associated wavelengths. Examples of such distinctions will be considered in later chapters; they include such matters as the extent to which diffraction occurs in the propagation of the wave, and the differing requirements imposed on the instruments for production and detection.

iii) The lure of a unified field theory. For a brief moment in the history of physics, Maxwell's equations seemed tantalizingly close to being able to subsume the whole of physics within their scope. According to this appealing idea, the electromagnetic field would function as a single entity capable of accounting for the properties which we associate with "ponderable matter." The unifying link is the Einstein mass-energy relation which assigns an inertial mass to all types of energy distribution. Thus the inertial properties of an electron might be simply the inertial properties of the field surrounding the point which we consider to be the location of the electron. In this view, one could dispense with the necessity for the presence of any *material* particle. For example, "an electron at rest at $r = 0$" might consist of an electrostatic field which varies as $1/r^2$ for values of r greater than some radius a, and vanishes for r less than a. Using Eq. (6–8), the total electrostatic energy in such a field can be obtained by integration. The value of a (the so-called classical radius of the electron) would then be obtained by setting this expression equal to c^2 times the observed inertial mass of the electron. When this model is pursued further (for example, into a calculation of the momentum transported by such a field), it turns out to be *almost* correct in its prediction of the properties of an electron, but there are a few unfortunate numerical discrepancies. It is therefore still considered necessary to preserve a role for the purely mechanical concept of "matter" in addition to electromagnetic fields. This near-miss illustrates the kind of goal which is sought in current research toward a "unified field theory." The hope is that an appropriately generalized single field will be able to account for all physical phenomena; material particles would appear in this scheme merely as particular concentrations within the field.

iv) Limitations of Maxwell's theory. Developments in modern physics, such as the photoelectric effect, the Compton effect, and blackbody radiation, have shown that Maxwell's equations are not of unlimited applicability in describing the behavior of light and other electromagnetic radiations. The quantum explanation of these phenomena has introduced the photon concept, which ascribes certain corpuscular attributes to light while retaining many of the wave concepts, such as frequency and wavelength. A great deal can be accomplished by a reformulation which retains Maxwell's equations but imposes upon them a different conceptual interpretation related to the probability concepts of quantum mechanics. More advanced theories in quantum electrodynamics, however, abandon direct use of Maxwell's equations. Nevertheless, it would be incorrect to consider that Maxwell's theory is an entirely discredited and outmoded theory. Any correct advanced formulation must recognize that Maxwell's equations are an approximation which is extremely accurate over a wide range of conditions. It would be unnecessarily complicated and ridiculous, for example, to use quantum theory to deduce the radiation pattern of a radio antenna.

6–4 MAXWELL'S RELATION

We have seen that the equation $c = 1/\sqrt{\mu\epsilon}$ yields the velocity of light in free space with extreme accuracy when the values of ϵ_0 and μ_0 are substituted. Let us now consider a material medium with permittivity ϵ different from ϵ_0. We shall retain $\mu = \mu_0$, since the permeabilities of most transparent dielectric media differ only slightly from that of free space. We can express the index of refraction of such a medium as

$$n = c_0/c = \sqrt{\mu\epsilon/\mu_0\epsilon_0} = \sqrt{\epsilon/\epsilon_0} = \sqrt{K}, \tag{6–15}$$

where K is the dielectric constant of the medium. Equation (6–15) is known as *Maxwell's relation.* The implication of this equation is that an electrostatic experiment, such as the measurement of the capacitance of a parallel plate capacitor, with the given material in the space between the plates, will suffice to determine the index of refraction of the material as measured in optical experiments. In Table 6–2 we list the index of refraction of a few representative substances measured with light of wavelength $\lambda = 5893$ Å, and the square-root of the dielectric constant obtained from an electrostatic

Table 6–2. Data concerning Maxwell's relation

Substance	$\sqrt{K}$ = square root of dielectric constant	n = index of refraction for sodium yellow light
Air	1.000294	1.000293
Benzene	1.489	1.482
Hydrogen	1.000132	1.000135
Water	8.94	1.333
Ethyl alcohol	5.1	1.36
CO_2 (gas)	1.000482	1.000450

experiment. The agreement between these two numbers, predicted by Maxwell's relation, is seen to be good for the first three substances listed, but is poor for the others.

In order to understand this discrepancy we must look more thoroughly into the physical constitution of a dielectric. Let us consider, for example, what mechanism is responsible for the increase in the capacitance of a parallel plate capacitor when the space between the plates is filled with a substance such as water. The dielectric constant is measured by the ratio C/C_0, where C and C_0 are the capacitances with and without the material present. Let us suppose that the same charge Q is maintained on the plates in the two cases. The change in capacitance is then associated with different values of the potential difference for these two cases, and we obtain $K = C/C_0 = V_0/V$. We can construct a model to explain why the potential difference is smaller with the water present, by supposing that the water molecule has a permanent dipole moment. That is to say, we assume that the centers of positive and negative charge in the molecule are not coincident. The molecule as a whole can be represented by a pair of equal and opposite charges separated by a small distance. There will be a tendency for these *dipoles* to align themselves with the field between the plates. Within the medium, the positive and negative charges of adjacent dipoles result in a condition of electric neutrality. However, adjacent to either of the plates there will be a layer of uncanceled charge (Fig. 6-1). (The charge referred to in this discussion is bound to the molecules and is not free to be transferred directly to the plates.) The net electric field in the region between the plates is the superposition of the field due to the charge on the plates and the field due to the layers of bound charge. We see, therefore, that the field is weaker when the water is present, and consequently the potential difference, which is defined as the work done per unit charge in moving a charge from one plate to the other, is also smaller.

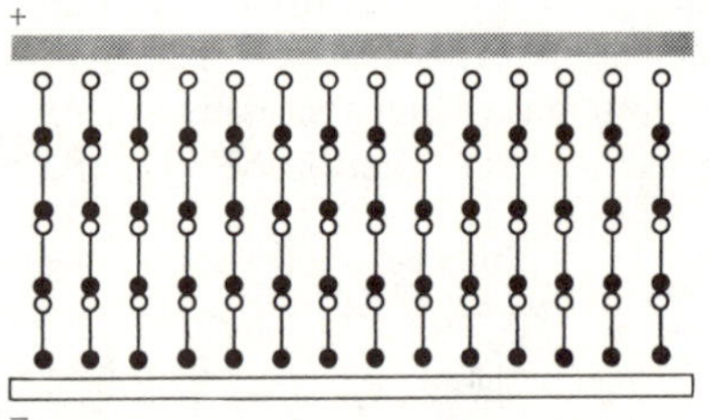

Fig. 6-1. Effective layers of bound charge produced by alignment of dipoles in the field of a parallel-plate capacitor. (The symbol ○——● represents the distribution of charge in a molecule, ○ being the center of negative charge and ●, the center of positive charge.)

Suppose now that the potential difference across the plates is reversed. The water molecules must rotate through 180° to realign themselves with the field. This process cannot be expected to occur instantaneously. The time required for the realignment is referred to as the *relaxation time*, τ. (In the qualitative discussion which follows we shall not require a more precise definition of this quantity.) If an alternating potential difference whose period is very large compared with the relaxation time is applied across the plates, ample time is provided for the dipoles to maintain alignment with the field, and we expect the molecules to contribute fully to the value of the dielectric constant. If, on the other hand, the period of the alternating potential is quite small compared with the relaxation time, the molecules have no chance to align themselves with the field in one direction before the direction of the field is reversed. This means that there is no layer of bound charge to contribute to the instantaneous electric field between the plates, and thus the instantaneous potential difference is expected to be the same as it would be in the absence of the dielectric medium. A way of putting this is to say that at sufficiently high frequencies ($\nu \gg 1/\tau$) the permanent dipoles do not contribute to phenomena associated with the dielectric constant. In terms of this model, we can interpret the data of Table 6–2 by supposing that the molecules of water, ethyl alcohol, and CO_2 are polar molecules (have permanent electric dipole moments), whereas the molecules of benzene, hydrogen, and the majority of the constituents of air are nonpolar molecules. Chemical evidence concerning the structure of these molecules lends support to this assumption. The frequency of sodium light is evidently greater than the reciprocal relaxation time of the polar materials. The model predicts that agreement will be obtained with Maxwell's relation if the index is measured for an electromagnetic wave of sufficiently low frequency. This prediction is confirmed by experiment. For example, the index of refraction of water for microwaves of wavelength $\lambda = 65$ cm is 8.88, which is in much closer agreement with $\sqrt{K}$ than is the index of refraction listed in Table 6–2.

Even if we do not consider the specific illustrations of Table 6–2, it is clear that Maxwell's relation cannot possibly be correct in the originally conceived sense, for it is well known that the index of refraction is a function of frequency, whereas ϵ and K as they appear in Eq. (6–15) were presumed to be constants measured by electrostatic experiments. If we wish to make use of Maxwell's equations without going into the details of a complete microscopic description of the material medium, we must resort to what is known as a *phenomenological approach.* This consists of defining an *effective value* for the permittivity which will make the index of refraction agree with the experimentally measured value at a particular frequency. Specifically, to work with waves of a given frequency in a given dispersive medium (one for which the index of refraction does vary with frequency), we must measure the index of refraction, n, for this frequency and then set $\epsilon = n^2\epsilon_0$ throughout Maxwell's equations. This restricts the context of the discussion of waves in a dispersive medium to the case of sinusoidal waves. Equation (6–13) has no meaning for a pulse of general form in a dispersive medium; it applies only to a sinusoidal wave for which the appropriate phenomenological value of ϵ has been inserted.

There is no single wave equation which describes the propagation of an arbitrary pulse in a dispersive medium. The situation must be handled by using Fourier analysis to determine "the frequencies contained within the pulse," and then treating each of the component frequencies separately.* Since the different frequencies propagate with different velocities, it can be foreseen that an arbitrary pulse in a dispersive medium does not propagate without distortion of form. We can, of course, continue to discuss the case of an arbitrary pulse propagating in *free space* or in a *nondispersive* medium. Also, for a case like that of a radar "blip," which is composed of frequencies confined to a band over which the index of refraction does not vary significantly, it will be appropriate to talk about the propagation of the pulse as a whole, using the effective value of ϵ for this frequency region.

6-5 DISPERSION THEORY

The model in the preceding section describes a mechanism by means of which we can understand why the index of refraction of a *polar* material is dependent on the frequency. A more complete discussion of dispersion for a general medium requires the construction of a more detailed model. We intend in this section merely to give a rough sketch of some aspects of such a model which will enable us to understand the general features of the dispersion curve over the entire range of the electromagnetic spectrum. Molecular systems may undergo internal motions of a number of different types, such as vibration at interatomic bonds, rocking of the atoms about an equilibrium bond angle, or motion of the electrons within individual atoms. Each of these changes of the configuration of the system is in general associated with a natural frequency. Consider, for example, the relative vibration of the two nuclei in a diatomic molecule. This is well-approximated by a model which assumes that the force acting to restore the atoms to the equilibrium bond distance is a linear one, and the resulting motion is simple harmonic. Among the possible internal motions of a molecule, there are some which result in a separation of the centers of the positive and negative charge distributions, i.e., in the presence of a dipole moment. It is only this type of motion which will be of concern here. When the molecule is placed in a static electric field, the equilibrium position associated with each of these types of motion will be changed; for example, the equilibrium distance between atoms in the diatomic molecule is altered. This means that associated with each of these types of motion there is an *induced dipole moment*. This induced dipole moment, along with permanent dipole moments if there are any, contributes to the effects associated with the dielectric constant. That is to say, there are layers of bound charge as shown in Fig. 6-1, associated with the *induced* dipole moment.

---

* See Chapter 16 for a fuller discussion of this topic.

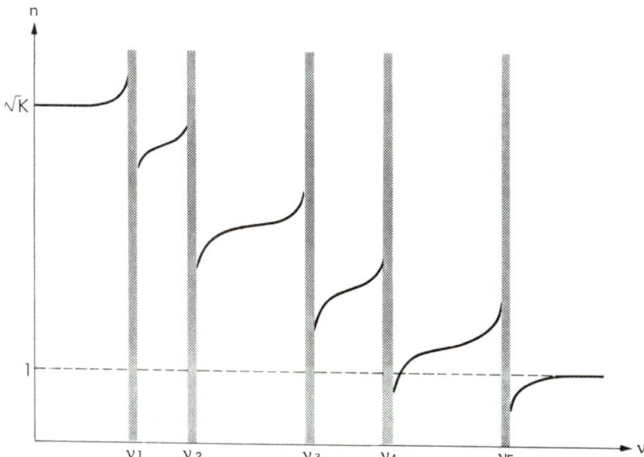

Fig. 6–2. Schematic dispersion curve for a nonpolar medium. (Resonance frequencies at $\nu_1, \nu_2, \ldots, \nu_5$; strong absorption occurs within the shaded interval about each resonance frequency.

When the applied electric field is sinusoidal with frequency ν, the question arises as to whether the induced dipole moment will continue to participate. The answer to this question depends on the relation of ν to the resonance frequencies of each type of motion. It is well known that when the frequency of a driving force is far below the natural frequency of an oscillator, the oscillator is able to follow the driving force, but when the driving frequency is well above the natural frequency, the oscillator is unable to respond and remains virtually unaffected by the presence of the driving force. This qualitative picture leads us to expect that the different types of internal motion will drop out of participation as the frequency ν is increased beyond the associated natural frequency. This means that the index of refraction n will start at the value $\sqrt{K}$ for sufficiently low frequencies and drop, by a series of more or less discontinuous steps, to the value $n = 1$ at extremely high frequencies. In the vicinity of each natural frequency we can expect that the given type of motion will respond *resonantly* and result in a strong absorption from the incident field. These predictions are in accord with the experimental findings. Although some substances are transparent over the visible region, all substances have absorption bands somewhere in the electromagnetic spectrum. The index of refraction is lower at frequencies just above an absorption band than it is at frequencies just below the band (Fig. 6–2).

A more detailed study of the model which describes the response of the system near resonance shows that the graph of $n(\nu)$ has a sharp peak below resonance and a sharp dip above resonance. Although the overall trend is for n to decrease as ν increases, the short-range behavior is just the opposite, and $dn/d\nu$ is greater than zero everywhere except in the immediate vicinity of an absorption band. A substance is said to exhibit *normal dispersion* if the index of refraction is an increasing function of frequency throughout the spectral range employed. Glass, for example, has absorption bands in the infrared and ultraviolet, and n is an increasing function of frequency throughout the visible region. This is observed experimentally by noting the order of spectral colors produced by refraction in a glass prism, blue light being deviated more than red light. If an absorption band lies in the middle of a frequency range being explored, it is still found that $dn/d\nu$ is positive on either side of the interval within which strong absorption occurs, but the index of refraction of the high-frequency side is less than that on the low-frequency side. This circumstance is referred to as *anomalous dispersion*. Consider, for example, a liquid-cell prism filled with a dye solution which absorbs the green portion of the spectrum. Blue light is deviated to a lesser extent than red by such a prism, although the order of colors from violet to green and from green to red is preserved.

6-6 A LINEARLY POLARIZED TRANSVERSE PLANE WAVE

Under the assumption that $E_x = E_z = H_x = H_y = 0$, we were led to the one-dimensional wave equation (6-13) which implies wave propagation parallel to the x-axis. The only nonvanishing fields, E_y and H_z, are *transverse* to the direction of propagation. Since each field is everywhere parallel to a fixed transverse direction, the wave is said to be *linearly polarized*. We were also led to the conclusion that E_y and H_z are functions of x and t only. This implies that at any particular instant of time the value of E_y (or H_z) is the same at all points on a plane perpendicular to the direction of propagation. We refer to this aspect by saying that the solution represents a *plane wave*. In Problem 6-3 it is shown that *any plane wave* solution to Maxwell's equations is *necessarily transverse* in character. In preparation for this problem, it should be pointed out that a trivial solution to Maxwell's equations is provided by the functions $\mathbf{H} = \mathbf{H}_0$ and $\mathbf{E} = \mathbf{E}_0$, where $\mathbf{H}_0$ and $\mathbf{E}_0$ are arbitrary vectors which do not vary with position or time. Furthermore, since Maxwell's equations are linear, a solution of the present form can always be superimposed on a given solution. Such fields which are uniform throughout all space and constant in time are obviously not associated with a wave propagation. Consequently, when we are discussing propagating solutions, we shall ignore constants of integration which are associated with fields of this trivial type.

PROBLEMS

6-1. Verify that the stated dimensions for ϵ, μ, $\mathbf{E}$, and $\mathbf{H}$ yield consistent results in the equations for energy density, energy flux, and wave propagation velocity [i.e. Eqs. (6-8), (6-9), (6-11), and (6-14)].

6–2. Eliminate E_y between Eqs. (6–6″) and (6–7″) to find a single equation which is satisfied by $H_z(x, t)$.

6–3. Starting with Maxwell's equations (Eqs. 6–4 through 6–7), show that if the instantaneous values of all field components are constant over planes perpendicular to the x-axis (i.e. none of the quantities depends on y or z), then the longitudinal components E_x and H_x must both be zero. (Integration constants are to be treated as recommended in Section 6–6.)

6–4. Show that Maxwell's equations permit a solution in which all the components of **E** and **H** vanish identically everywhere except for: (a) E_z and H_y, (b) E_x and H_z. Describe the situation represented by each of these solutions.

6–5. Integrate Eq. (6–8) over all space outside a sphere of radius a, where **E** is the inverse-square law field associated with a point charge at the center of the sphere. Use this to determine a numerical value for the "classical radius of the electron."

6–6. Consider a portion of a circuit containing a parallel plate capacitor as indicated in Fig. 6–3. Neglect edge effects. Use Gauss' law to calculate the value of **D** in the field between the plates at a time when the charge on the plates is Q. Let L be a closed rectangular loop intersecting the plane of the diagram in L_1, L_2. Let A be the surface in the plane of the loop and A' an open rectangular surface having perimeter L, with one face passing between the capacitor plates. Show that the surface integral of $\mathbf{j} + \partial\mathbf{D}/\partial t$ is the same over the areas A and A'.

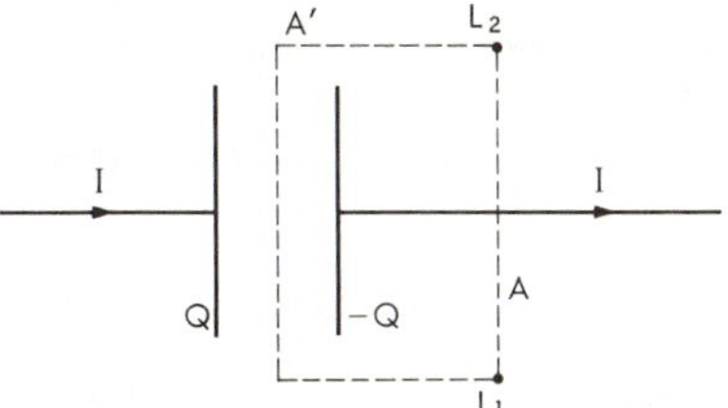

Figure 6–3

7 Analytical description of polarized electromagnetic plane waves

7-1 INTRODUCTION

On the basis of Maxwell's electromagnetic theory we can interpret light as being a propagating electromagnetic disturbance. It becomes a simple matter of reading the equations to obtain a complete physical description of the electric and magnetic fields associated with different particular solutions to the field equations. It will be of interest, for example, to examine the relation between the electric and magnetic fields. Are these two fields complementary in the sense that one is at a maximum where the other is at a minimum? Is there any necessary correlation between the directions of the E- and H-vectors at a point? How are the variations in magnitude and direction of the E-vector at one point of observation related to the spatial variations of these quantities? We shall find that Maxwell's equations immediately provide us with a variety of solutions whose character can be associated with the polarized nature of light. Although experiments for detecting polarization will not be described until the next chapter, we shall be prepared with the appropriate language for describing the observed phenomena in terms of the electromagnetic variables.

Discussion in this chapter will be confined to sinusoidal waves. The concept of "elliptical polarization" which will be developed is, in fact, only meaningful with reference to a wave of single frequency. The medium is assumed to be homogeneous, linear, and isotropic, with no free charges or conduction currents present within the region to be considered. The permittivity is assigned its effective value for the given frequency.

7-2 MORE COMPLETE DESCRIPTION OF THE LINEARLY POLARIZED SINUSOIDAL PLANE WAVE

We showed in Chapter 6 that Maxwell's equations permit a solution which represents a linearly polarized plane wave whose E-vector is everywhere parallel to the y-axis and whose H-vector is everywhere parallel to the z-axis.

We shall now consider the specific case in which $E_y(x, t)$ is a sinusoidal solution to the one-dimensional wave equation (6–13) representing a progressive wave traveling in the positive x-direction:

$$E_y = \mathcal{R}e\,\{\mathbf{A}e^{i(\omega t-kx)}\}. \tag{7–1}$$

Having made this special choice for E_y, we find that the magnetic field intensity $H_z(x, t)$ is no longer arbitrary, but becomes determined by the pair of Maxwell equations from which we previously eliminated H_z. Substitution of Eq. (7–1) in Eq. (6–6″) yields

$$\mu(\partial H_z/\partial t) = \mathcal{R}e\,\{ik\mathbf{A}e^{i(\omega t-kx)}\},$$

and upon integrating with respect to t, we have

$$H_z = \mathcal{R}e\,\{(k\mathbf{A}/\mu\omega)e^{i(\omega t-kx)}\} + \theta(x), \tag{7–2}$$

where $\theta(x)$ is an arbitrary function of x, but is a constant with respect to t.

Equation (6–7″) must also be satisfied by our solution. We find upon substitution of H_z from (7–2) into (6–7″) that

$$\mathcal{R}e\,\{(ik^2/\mu\omega)\mathbf{A}e^{i(\omega t-kx)}\} - \theta'(x) = \mathcal{R}e\,\{i\epsilon\omega\mathbf{A}e^{i(\omega t-kx)}\}.$$

Since $\epsilon\omega = k^2/\mu\omega$, the sinusoidal terms on the two sides of this equation are identical, implying $\theta'(x) = 0$. Thus θ is constant in both space and time. As was observed in Section 6–6, an arbitrary time-independent and uniform magnetic field can be superimposed upon any solution of Maxwell's equations. Hence for a discussion of the propagating wave we choose $\theta = 0$. If we replace ω by kc, Eq. (7–2) then becomes

$$H_z = \mathcal{R}e\,\{(\mathbf{A}/\mu c)e^{i(\omega t-kx)}\}. \tag{7–3}$$

We thus find that all of Maxwell's equations have been satisfied, and that if E_y represents a sinusoidal progressive plane wave of complex amplitude $\mathbf{A}$, then the associated magnetic field is of necessity also a sinusoidal plane wave traveling in the same direction, with $\mathbf{H}$ perpendicular to $\mathbf{E}$, and with complex amplitude $\mathbf{A}/(\mu c)$. Furthermore, H_z is in phase with E_y, and the instantaneous values of the two functions are proportional:

$$E_y(x, t) = \mu c H_z(x, t). \tag{7–4}$$

The graphs of E_y and H_z as functions of x at a fixed time t are given in Fig. 7–1. By plotting H_z in a plane perpendicular to that in which E_y is plotted, we obtain a schematic representation of the relation between the E- and H-vectors at each location. Because $\mathbf{E}$ and $\mathbf{H}$ have different dimensions, different scales must be chosen on the two ordinate axes.

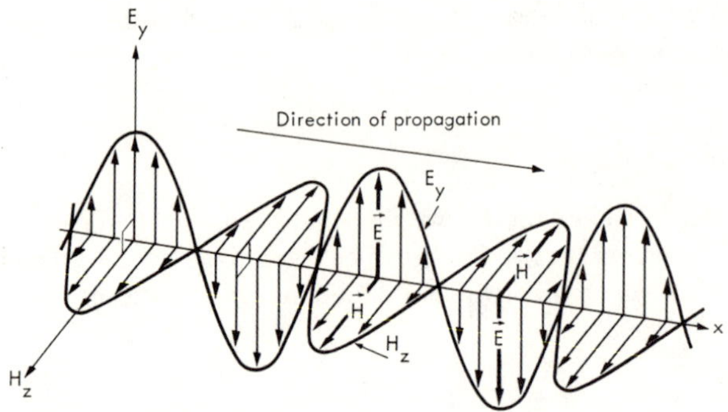

Fig. 7-1. Electric and magnetic fields at a given instant for a linearly polarized plane wave. Graphs of E_y and H_z are shown in mutually perpendicular planes. Specific E- and H-vectors are indicated at two locations.

Since Eq. (7–4) is correct only for waves traveling in the positive x-direction, we can remind ourselves of this restriction by using an appropriate superscript:

$$E_y^+ = \mu c H_z^+. \tag{7–5}$$

It is left as an exercise for the reader to show that for a plane wave (linearly polarized with its E-vector along the y-axis) traveling in the negative direction, the relation is

$$E_y^- = -\mu c H_z^-. \tag{7–6}$$

The instantaneous values of the electric and magnetic energies per unit volume are given by Eqs. (6–8) and (6–9):

$$W_{\text{el}} = (\epsilon/2)E_y^2(x, t) \qquad \text{and} \qquad W_{\text{mag}} = (\mu/2)H_z^2(x, t).$$

We observe that by virtue of Eqs. (7–5) and (7–6) these quantities are equal at any point in a progressive wave:

$$W_{\text{el}} = (\epsilon/2)E_y^2 = (\epsilon/2)[\mu c H_z]^2 = (\epsilon\mu^2 c^2/2)H_z^2 = (\mu/2)H_z^2 = W_{\text{mag}}.$$

The Poynting vector reduces to the form $\mathbf{S} = \mathbf{i}S_x$, where $S_x = E_y H_z$. In a $+$wave $S_x^+ = [E_y^+]^2/\mu c$, showing that S_x^+ is a nonnegative quantity; likewise for a $-$wave, $S_x^- \leq 0$. That is to say, in both cases the Poynting vector is parallel to the direction of propagation. Taking $\mathbf{A} = Ae^{-i\phi}$, we obtain $S_x^+ = (A^2/\mu c)\cos^2(\omega t - \phi - kx)$, which indicates that the energy flow is a pulsating quantity. It is generally only the average value, $\overline{S_x^+} = A^2/2\mu c$, which is of interest in applications. In the context of wave propagation the Poynting vector **S** will be referred to as the *intensity* of the wave.*

7-3 REFLECTION FROM A DIELECTRIC SURFACE OBTAINED BY APPEAL TO AN ISOMORPHISM

The equations of the last section suggest an isomorphism between the linearly polarized electromagnetic plane wave and the acoustic plane wave. One possible correspondence of variables is listed in Table 7–1. The reader can check that the basic equations of either system translate into true equations of the other, and hence the two systems of equations are isomorphic. Thus, for example, Eq. (6–6″) translates into $\rho_0\ddot{\xi} = -(\partial p/\partial x)$, which is Eq. (2–18), Newton's second law for the acoustic plane wave. The electromagnetic parameter μc, which is the analog of the acoustic parameter $\rho_0 c$, will be referred to as the *electromagnetic wave impedance of the medium* and will be designated by the symbol Z. It is also called the *characteristic impedance* or *intrinsic impedance* of the medium. An alternative expression is $\mu c = \sqrt{\mu/\epsilon}$. The dimensions of this quantity are the same as the dimensions of the ohm. The electromagnetic wave impedance of free space is

$$Z_0 = \mu_0 c_0 = (4\pi \times 10^{-7})(3.00 \times 10^8) = 120\pi = 377 \text{ ohms.}$$

This isomorphism is particularly useful, since it also happens that the boundary conditions for the reflection-transmission problem of an electromagnetic plane wave normally incident upon a plane interface between two dielectric media are isomorphic to the boundary conditions used in an acoustic problem whose solution we have previously obtained. Consider two homogeneous isotropic dielectric media separated by the plane $x = 0$. Let medium 1 lie in $x < 0$, and medium 2 in $x > 0$. Assume a linearly polarized plane wave (with its E-vector along the y-axis) traveling in the positive direction in

Table 7–1. Correspondence between variables in the acoustic plane wave and the linearly polarized electromagnetic plane wave

Acoustic variable	p	ξ	$\mathcal{B}_a$	ρ_0	W_{kin}	W_{pot}	$c = \sqrt{\mathcal{B}_a/\rho_0}$	$Z = \rho_0 c$	$i = p\dot{\xi}$
Electromagnetic analog	E_y	H_z	$1/\epsilon$	μ	W_{mag}	W_{el}	$c = 1/\sqrt{\epsilon\mu}$	$Z = \mu c$	$S_x = E_y H_z$

* In the subject of photometry the term "intensity" is used in a different sense. The intensity of a *source* in a given direction refers to the power radiated by the source per unit solid angle in the given direction. In this context the quantity designated by $|\mathbf{S}|$ is usually called *illumination*. The above usage, which we shall adhere to, is more in line with the terminology of wave propagation in a general context, and is commonly employed.

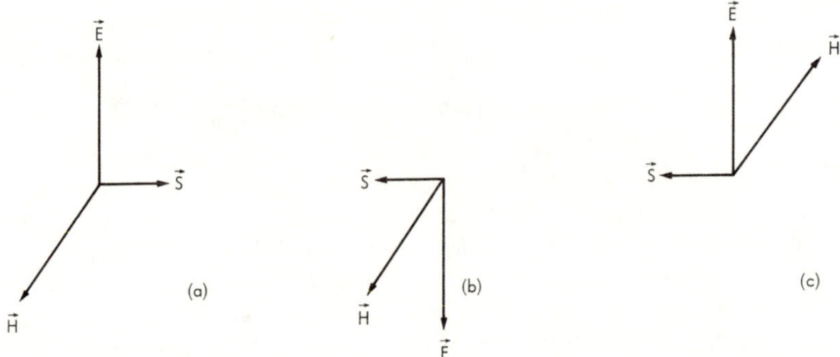

Fig. 7–2. Phase reversal upon reflection. The vectors **E**, **H**, and **S** are evaluated at a point on the interface at a particular instant (a) for the incident wave, (b) for the reflected wave when $n_1 < n_2$ (external reflection), and (c) for the reflected wave when $n_1 > n_2$ (internal reflection).

medium 1 and incident normally on the interface. For this problem the boundary conditions of Eq. (6–12) reduce to*

$$E_{y1}(0, t) = E_{y2}(0, t), \quad H_{z1}(0, t) = H_{z2}(0, t).$$

These, however, are the correct analogs of the conditions on p and ξ which were used as boundary conditions in the problem of Section 3–4. We are therefore assured that the mathematical operations of solving the boundary value problem will be exactly parallel in the two cases, and will lead to a reflection coefficient for E_y given by $R_E = (1 - Z_{12})/(1 + Z_{12})$. For applications in optics where both μ_1 and μ_2 can be replaced by μ_0,

$$Z_{12} = \mu_1 c_1/\mu_2 c_2 = c_1/c_2 = n_2/n_1 = n_{21},$$

and R_E can equivalently be written $R_E = (1 - n_{21})/(1 + n_{21})$. The other coefficients follow from the following relations (also obtained from the isomorphism): $T_E = 1 + R_E$, $R_H = -R_E$, $T_H = 1 - R_E$, $R_S = -R_E^2$, and $T_S = 1 - R_E^2$.

When the optical density of the second medium is greater than that of the first ($n_2 > n_1$), the reflection is known as an *external* reflection. In such a case $R_E < 0$, but $R_H > 0$. That is, the E-vector reverses phase upon reflection but the H-vector does not (Fig. 7–2b). Conversely, for an *internal* reflection ($n_1 > n_2$) it is the H-vector which reverses phase, whereas the E-vector does not (Fig. 7–2c). Observe that in order for Eq. (7–6) to apply to

* The continuity of the normal components of **D** and **B** is trivially satisfied since all x-components are zero.

the reflected wave, one, but not both, of E_y and H_z must reverse phase. The phase reversal of just one is also required in order to give the correct direction to the Poynting vector in the reflected wave.

7-4 REFLECTION FROM A PERFECT CONDUCTOR; DIRECT EVIDENCE OF STANDING WAVES

The reflection of a linearly polarized plane wave normally incident upon a perfect conductor is easily handled. The *perfect conductor* is defined as a medium within which **E** must be identically zero. Hence the boundary value problem is isomorphic to the reflection of an acoustic wave at a *free surface* ($E_y = 0$ at the interface corresponds to $p = 0$ at the interface). For reflection at a perfect conductor we therefore have $R_E = -1$ and $R_H = 1$. The boundary is a node for E_y and an antinode for H_z. For mnemonic purposes, one can think of the perfect conductor as a dielectric of zero impedance, infinite index of refraction, and infinite permittivity.

Standing waves played an important role in Hertz's investigation of the disturbance produced by an electric oscillator. It might be argued that refraction by a prism or focusing by a curved mirror are not properties which are specifically characteristic of waves, for this behavior can be interpreted in terms of a particle model. However, the explanation of a system of nodes and antinodes at a reflecting surface involves such concepts as *interference* and *wavelength*, which are more uniquely associated with waves. In Hertz's experiments the wavelengths were on the order of 5 to 10 m. Thus he could explore the system of standing waves produced by reflection from a metal sheet by moving the detector and noting the locations of the successive maxima in the response. (Analysis of the circuit relations shows that the receiving oscillator is responsive to the *E-vector* of the surrounding field.)

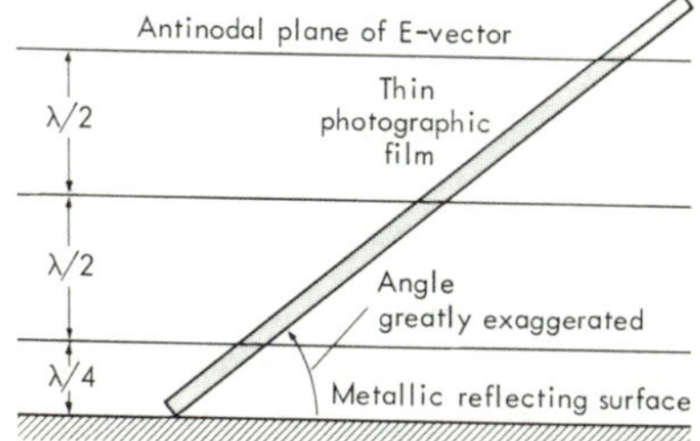

Fig. 7-3. Wiener's demonstration of standing light waves.

It is obviously much more difficult to demonstrate the existence of a standing wave pattern for light waves, since the distance between successive nodal planes is only a few thousand angstroms. This difficulty was overcome by Otto Wiener in 1890, using the arrangement indicated in Fig. 7-3. A thin film of a very fine-grained photographic emulsion is inclined at a small angle to a metallic reflecting surface. The system is exposed to plane waves of monochromatic light* incident normally on the reflecting surface. When the film is developed, it is found that a maximum deposition of silver occurs at distances corresponding to $\lambda/4$, $3\lambda/4$, $5\lambda/4$, . . . from the surface, but that the emulsion which lies at distances 0, $\lambda/2$, λ, . . . from the surface has been unaffected. That is to say, the film is blackened at the antinodes of **E**, but is unchanged at the nodes of **E** (the antinodes of **H**). This finding agrees with expectations, provided we suppose that the photochemical action which results in "exposure" of the emulsion is produced by variations in the electric field intensity, and not by variations in the magnetic field intensity. This assumption seems reasonable, since the influence of an electric field on the electronic configuration of a substance would be expected to be greater than that of a magnetic field of "comparable strength."

Consider a charge e moving with velocity v in an electric field E and a magnetic field $H = E/\mu_0 c_0$. The electric force is $F_{\text{el}} = eE$; the magnetic force is $F_{\text{mag}} = evB = ev\mu_0 H = evE/c_0$. Thus, $F_{\text{mag}}/F_{\text{el}} = v/c_0$. The magnetic force is small compared with the electric force unless the velocity of the charge is comparable to the velocity of light. This is not the case with the valence electrons responsible for chemical behavior.

A modified version of Wiener's experiment was devised by H. E. Ives. The metallic reflecting surface is coated directly with fine-grained emulsion in a film which is several hundred wavelengths thick. (In Wiener's experiment the film is only approximately one-thirtieth of a wavelength thick.) When the system is exposed to normally incident monochromatic light and the film is developed, the silver deposits in a stratified series of layers, the planes of maximum and minimum density being spaced relative to the reflecting surface as in Wiener's experiment. By analogy with the Bragg law of selective reflection it can be seen that such a series of layers reflects selectively in favor of the original color and hence takes on the corresponding color when illuminated with white light. This principle was the basis of an early form of color photography known as the *Lippmann process*.

7-5 CONSIDERATION OF A MORE GENERAL SINUSOIDAL PLANE WAVE

In this section we shall free ourselves from the restriction of talking only about a linearly polarized wave, and shall describe the relation between **E** and **H** for a general progressive sinusoidal plane wave. In the following section we discuss what types of polarization other than linear are possible. We assume as before that we are dealing with plane waves whose normal is parallel to the x-axis. Taking $H_x = E_x = 0$, and assuming that all quantities

---

* The light need not be linearly polarized. See remarks at end of Section 7-8.

are independent of y and z, we find that four Maxwell equations have to be satisfied:

$$\mu \frac{\partial H_z}{\partial t} = -\frac{\partial E_y}{\partial x}, \tag{7-7}$$

$$\epsilon \frac{\partial E_y}{\partial t} = -\frac{\partial H_z}{\partial x}, \tag{7-8}$$

$$\mu \frac{\partial H_y}{\partial t} = \frac{\partial E_z}{\partial x}, \tag{7-9}$$

$$\epsilon \frac{\partial E_z}{\partial t} = \frac{\partial H_y}{\partial x}. \tag{7-10}$$

These equations occur in pairs. The first two equations refer only to E_y and H_z, and the last two refer only to E_z and H_y. The first pair is in fact identical with Eqs. (6–6″) and (6–7″), and therefore many of the conclusions drawn from the previous discussion will apply to this more general situation. For example, elimination between these two equations shows that both E_y and H_z must satisfy the one-dimensional wave equation. Similarly, elimination between Eqs. (7–9) and (7–10) shows that E_z and H_y must also satisfy the one-dimensional wave equation with the same velocity of propagation, $c = 1/\sqrt{\mu\epsilon}$. If we try a solution in which

$$E_y = \Re e\,\{\mathbf{A}e^{i(\omega t - kx)}\}, \tag{7-11}$$

we find that Eqs. (7–7) and (7–8) will be satisfied, provided that

$$H_z = \Re e\left\{\frac{\mathbf{A}}{\mu c}\,e^{i(\omega t - kx)}\right\}, \tag{7-12}$$

regardless of the values of E_z and H_y. Similarly, if we assume

$$E_z = \Re e\,\{\mathbf{B}e^{i(\omega t - kx)}\}, \tag{7-13}$$

Eqs. (7–9) and (7–10) will be satisfied, provided that

$$H_y = -\Re e\left\{\frac{\mathbf{B}}{\mu c}\,e^{i(\omega t - kx)}\right\}. \tag{7-14}$$

The complex constant **B** can be chosen independently of **A**. We shall confine our discussion in this section to the progressive wave traveling in the positive x-direction, which is described by Eqs. (7–11) through (7–14). To simplify notation the superscript + will be omitted, though it is important to realize that some of the results will not apply to a more general field which is a superposition of waves traveling in both directions.

For arbitrary choice of the constants **A** and **B**, the relation between **E** and **H** can be characterized by the series of propositions which follow.

1) $$E_y = ZH_z. \tag{7-15}$$

This is evident from Eqs. (7–11) and (7–12).

2) $$E_z = -ZH_y. \tag{7-16}$$

This is evident from Eqs. (7–13) and (7–14). If $\mathbf{A} = 0$ and $\mathbf{B} \neq 0$, the wave is linearly polarized with its E-vector along the z-axis. The reader should convince himself that the minus sign in Eq. (7–16) yields the proper direction for $\mathbf{S} = \mathbf{E} \times \mathbf{H}$. It is also clear that this situation can be derived from the previous description of a linearly polarized plane wave by the rotation of axes specified by the equations $x' = x$, $z' = y$, and $y' = -z$.

3) **E** is at all times perpendicular to **H**.

Proof $\quad \mathbf{E}\cdot\mathbf{H} = E_yH_y + E_zH_z = Z(H_zH_y - H_yH_z) = 0.$

4) $$|\mathbf{E}| = Z|\mathbf{H}|.$$

Proof $\quad |\mathbf{E}|^2 = E_y^2 + E_z^2 = Z^2\{H_y^2 + H_z^2\} = Z^2|\mathbf{H}|^2.$

5) In summary, **H** can be expressed explicitly in terms of **E** by the relation

$$\mathbf{H} = (1/Z)\mathbf{i} \times \mathbf{E}. \tag{7-17}$$

Proof

$$\begin{aligned}(1/Z)\mathbf{i} \times \mathbf{E} &= (1/Z)\mathbf{i} \times (\mathbf{j}E_y + \mathbf{k}E_z)\\ &= (1/Z)(\mathbf{k}E_y - \mathbf{j}E_z) = \mathbf{k}H_z + \mathbf{j}H_y = \mathbf{H}.\end{aligned}$$

From these propositions we see that once a complete description has been given of **E** for a particular progressive wave, the nature of the associated H-vector is also known. In the next section we shall ignore **H** and concentrate on describing the general character of **E**.

7-6 TYPES OF POLARIZATION

We shall first describe the behavior of the E-vector as a function of time at a fixed location, which for simplicity we take to be $x = 0$. The spatial variations in the value of **E** can then be deduced from the fact that the wave is traveling in the positive x-direction with velocity c. The variety of possible cases arises from different choices of the constants **A** and **B**. Let $\mathbf{A} = Ae^{-i\phi_1}$ and $\mathbf{B} = Be^{-i\phi_2}$. Equations (7–11) and (7–13) can then be written

$$E_y = A\cos(\omega t - \phi_1), \qquad E_z = B\cos(\omega t - \phi_2). \tag{7-18}$$

The net E-vector is the resultant of these two components. The remainder of this section evolves directly from Eq. (7–18) and does not make explicit use of any relations resulting from Maxwell's equations. The resulting classification of types of polarization is identical with that given by Fresnel fifty years before Maxwell's presentation of the electromagnetic theory of light. This classification depends only on the fact that a sinusoidal plane wave can be represented by equations having the *form* of Eq. (7–18). The physical interpretation of the vector **E** is not made use of in the discussion. It is sufficient to know that a plane wave is of transverse nature and has two degrees of freedom, that is to say, it is capable of being represented by a *vector* lying in the plane perpendicular to the direction of propagation. This vector can be referred to ambiguously as the "light vector" or the "optical disturbance." In Fresnel's interpretation, this vector stands for the displacements of the particles of an elastic ether from their equilibrium positions. So far as this section is concerned, the "optical disturbance" vector can be interpreted electromagnetically by either **E** or **H**, for, by virtue of Eq. (7–17), statements made about the relative amplitudes and phases of the components of one of these vectors apply equally well to the relative amplitudes and phases of the components of the other.

Those who are familiar with Lissajous figures will recognize from Eq. (7–18) that the tip of the E-vector describes the Lissajous figure characteristic of 1:1 frequency ratio, that is to say, an ellipse, or some special case thereof. We consider certain of these special cases first. The distinguishing criteria depend upon the relative amplitudes of E_y and E_z and upon the phase difference, $\Delta\phi = \phi_2 - \phi_1$.

1) $A = B$ *and* $\Delta\phi = \pi/2$. Then

$$E_y = A\cos(\omega t - \phi_1), \qquad E_z = A\sin(\omega t - \phi_1). \tag{7–19}$$

Squaring both of these equations and adding, we find $E_y^2 + E_z^2 = A^2$, which shows that for this special case the E-vector is constant in magnitude. The angle which the E-vector makes with the E_y-axis is $\psi = \tan^{-1}(E_z/E_y)$. But from Eq. (7–19), $E_z/E_y = \tan(\omega t - \phi_1)$. Therefore $\psi = \omega t - \phi_1$, implying that the E-vector rotates in the positive direction with uniform angular velocity ω. We have thus identified the equations in (7–19) as the parametric equations of a circle. The wave is said to be *circularly polarized.* The locus of the moving tip of the E-vector is shown in Fig. 7–4(a). In this diagram the x-axis is directed perpendicularly *out* of the paper, so that the wave is approaching the reader. Looked at from this point of view the rotation is counterclockwise, and the wave is said to be *left* circularly polarized.* Figure 7–4(b) shows the E-vector of a left circularly polarized wave at various

* Some authors use the opposite convention.

locations at a fixed time t. It is evident that a left circularly polarized wave will also result if $A = B$ and $\Delta\phi = \pi/2 + 2n\pi$, where $n = \pm 1, \pm 2, \ldots$ It should also be observed that it is only the phase *difference*, $\phi_2 - \phi_1$, which is of importance in characterizing the light as being left circularly polarized. By shifting the origin of either x or t, the value of ϕ_1 could be made equal to zero.

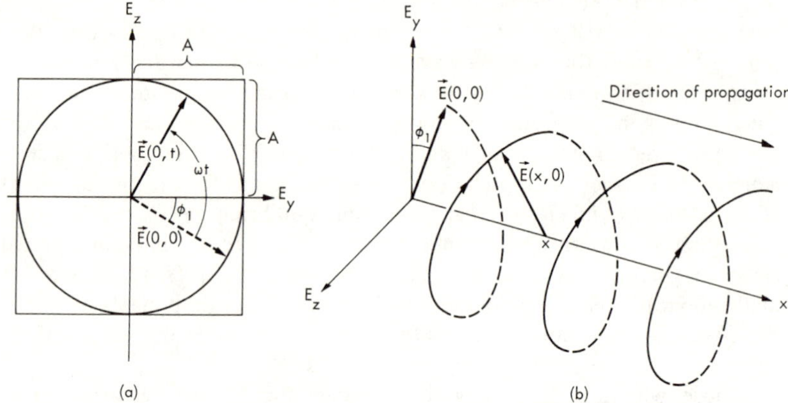

Fig. 7-4. Behavior of the E-vector in left circularly polarized light (a) as a function of time in the plane $x = 0$, and (b) as a function of x at $t = 0$.

2) *$A = B$ and $\Delta\phi = -\pi/2 + 2n\pi$*, ($n = 0, \pm 1, \pm 2, \ldots$). By analysis exactly parallel to that in case 1, the tip of the E-vector can be shown in this case to describe a circle whose direction of rotation is opposite to that above. That is to say, for case 2 the rotation is in a clockwise sense to an observer watching the beam approach. The wave is said to be *right* circularly polarized.

In many contexts the distinction between *right* and *left* circular polarization is not important. Circular polarization of one or the other type will result if $A = B$ and $\Delta\phi$ is an odd multiple of $\pi/2$. Unless one is interested in the *sense* of the rotation, it is sufficient to refer to "the phase difference between the two components" without specifying whether this is $\phi_2 - \phi_1$ or $\phi_1 - \phi_2$.

3) *$A \neq B$ and $\Delta\phi = \pi/2 + n\pi$*, ($n = 0, \pm 1, \pm 2, \ldots$). By substitution in Eq. (7-18),

$$E_y/A = \cos(\omega t - \phi_1), \qquad E_z/B = (-1)^n \sin(\omega t - \phi_1).$$

Squaring both equations and adding, we obtain

$$(E_y/A)^2 + (E_z/B)^2 = 1.$$

The tip of the E-vector therefore at all times lies on an ellipse. This ellipse has a special orientation, in that the principal axes of the ellipse coincide with the coordinate axes (see Fig. 7–5). The wave is said to be *elliptically polarized.* To distinguish this from case 6 below, it is desirable to include a statement about the special orientation of the ellipse. We may refer to the present case as a *standard* ellipse. The sense of the rotation is described by the same conventions as with circular polarization; it is left-handed if n is even, and right-handed if n is odd. The E-vector does not rotate with uniform angular velocity (see Problem 7–6).

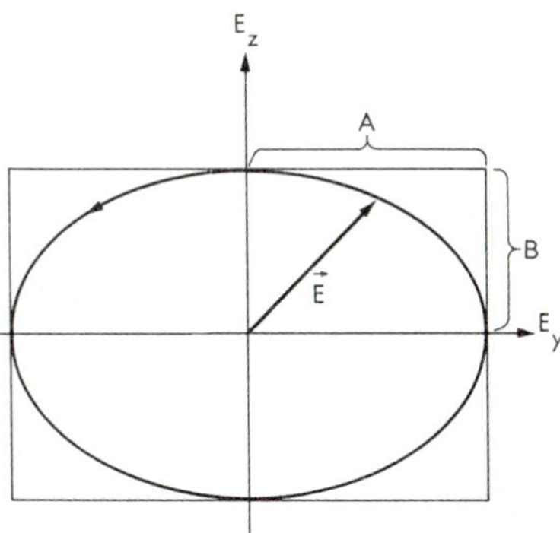

Fig. 7–5.
Elliptical polarization with standard orientation. (Principal axes and coordinate axes coincident: $\Delta\phi = \pi/2$.)

4) $\Delta\phi = 0$ (*or an integral multiple of* 2π). When there is no phase difference between E_y and E_z, these quantities vary proportionally to each other. They pass through zero value at the same time, attain their maximum values (A and B respectively) at the same time and their minimum values ($-A$ and $-B$ respectively) at the same time. From this it can be seen that the *length* of the E-vector varies in the same proportion, and, since the ratio of E_z to E_y is constant, the tip of the E-vector moves along a straight line. This result can be derived formally from Eq. (7–18) with $\phi_2 = \phi_1$. It follows immediately that $E_z/E_y = B/A$. In the plane whose coordinates are (E_y, E_z), this is the equation of a straight line passing through the origin and of slope B/A. The tip of the E-vector oscillates sinusoidally along the segment of this straight line extending from $(-A, -B)$ to (A, B) (Fig. 7–6a). This, of course, can be recognized as a linearly polarized wave whose E-vector does not happen to coincide with one of the coordinate axes. The angle between the y-axis and the line along which the E-vector lies is $\theta = \tan^{-1}(B/A)$.

5) $\Delta\phi = \pi$ (*or any odd multiple of* π). From Eq. (7–18),

$$E_y = A\cos(\omega t - \phi_1), \qquad E_z = -B\cos(\omega t - \phi_1),$$

and $E_z/E_y = -B/A$. The E-vector oscillates along the line segment joining the points $(A, -B)$ and $(-A, B)$ (Fig. 7–6b).

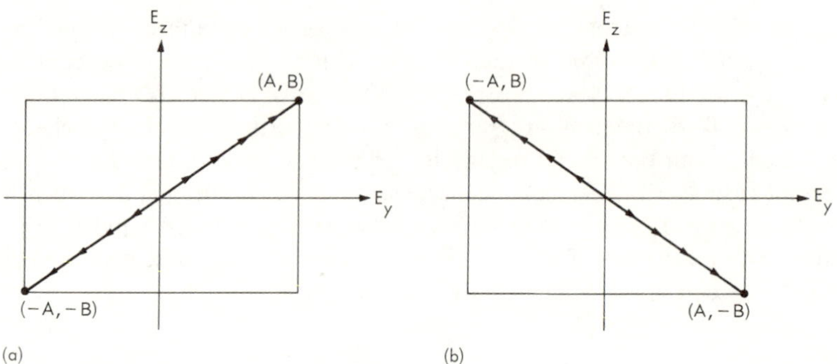

Fig. 7-6. Linear polarization: (a) phase difference of zero between E_y and E_z, and (b) phase difference of π between E_y and E_z. (The arrows show the tip of E-vector at various times.)

6) *The general case* ($\Delta\phi$ *not a multiple of* $\pi/2$). The equations in (7–18) are the parametric equations of a locus in the E_yE_z-plane which must be identified by elimination of t between the two equations. Expanding the trigonometric functions yields

$$E_y = (A \cos \phi_1) \cos \omega t + (A \sin \phi_1) \sin \omega t, \tag{7–20}$$

$$E_z = (B \cos \phi_2) \cos \omega t + (B \sin \phi_2) \sin \omega t. \tag{7–21}$$

We multiply Eq. (7–20) by $(B \cos \phi_2)$ and (7–21) by $(A \cos \phi_1)$ and subtract:

$$(B \cos \phi_2)E_y - (A \cos \phi_1)E_z = -(AB \sin \Delta\phi) \sin \omega t.$$

Likewise, we multiply Eq. (7–20) by $(B \sin \phi_2)$ and (7–21) by $(A \sin \phi_1)$ and subtract:

$$(B \sin \phi_2)E_y - (A \sin \phi_1)E_z = (AB \sin \Delta\phi) \cos \omega t.$$

Squaring and adding the last two equations, we find

$$B^2E_y^2 + A^2E_z^2 - (2AB \cos \Delta\phi)E_yE_z = A^2B^2 \sin^2 \Delta\phi. \tag{7–22}$$

This can be recognized as the equation of a conic section. The discriminant shows that for a general $\Delta\phi$ the locus is an ellipse. For special values, of course, this reduces to the cases studied above. For example, with $\Delta\phi = \pi/2$ and $A = B$, the result is a circle as in case 1. For $\Delta\phi = 0$, the equation is $(BE_y - AE_z)^2 = 0$, which is the equation of the straight line of case 4. Both the circle and the straight line may be looked upon as limiting cases of an ellipse. The principal axes of the ellipse do not coincide with the coordinate axes unless the coefficient of E_yE_z vanishes. This requires $\Delta\phi$ to be an odd multiple of $\pi/2$, which is the criterion employed in case 3 for the ellipse to have standard orientation.

In dealing with the general case it is convenient to observe that the ellipse is inscribed within the rectangle $|E_y| \leq A$, $|E_z| \leq B$. This can be seen by inspection of Eq. (7–18). For example, E_y always lies between $-A$ and $+A$ and attains each of these values once within each cycle. Figure 7–7 is a sketch of the variation of **E** for an elliptically polarized wave of nonspecial phase angle.

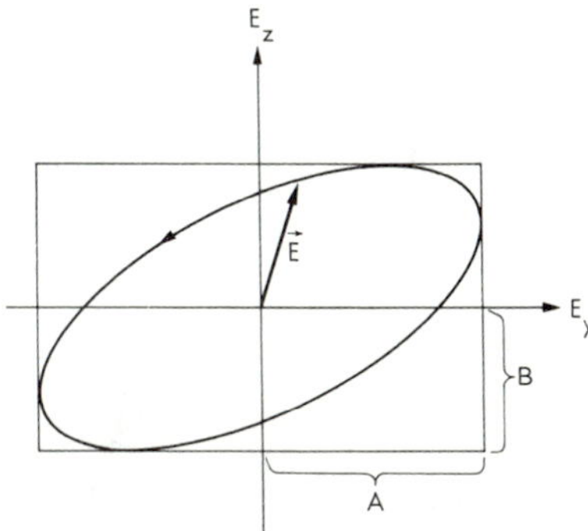

Fig. 7–7. General state of elliptical polarization.

If A and B are held fixed and the parameter $\Delta\phi$ is considered to vary, the corresponding ellipses progress through a sequence of forms. Such a sequence is shown in Fig. 9–4. The progression of patterns obtained on an

Table 7–2. Summary of criteria for type of polarization

Criterion [cf. Eq. (7–18)]	Type of polarization
$B = 0$ and $A \neq 0$	Linear, E-vector along the y-axis
$B \neq 0$ and $A = 0$	Linear, E-vector along the z-axis
$\Delta\phi = 0, \pm 2\pi, \pm 4\pi, \ldots$	Linear, E-vector making an angle $\tan^{-1}(B/A)$ with the y-axis
$\Delta\phi = \pm\pi, \pm 3\pi, \ldots$	Linear, E-vector making an angle $\pi - \tan^{-1}(B/A)$ with the y-axis
$\Delta\phi = \pm\pi/2, \pm 3\pi/2, \ldots$	Elliptical, principal axes coinciding with the coordinate axes
$\Delta\phi = \pm\pi/2, \pm 3\pi/2, \ldots$ *and* $A = B$	Circular
$\Delta\phi$ not one of the above special angles	Elliptical, principal axes inclined to the coordinate axes

oscilloscope for a nearly 1:1 frequency ratio is of this nature. [When the frequencies of the signals applied to the horizontal and vertical plates are ω and $\omega + \Delta\omega$, respectively, this may be equivalently interpreted to mean that the two signals are of the same frequency but exhibit a slow drift in relative phase, $\Delta\phi = (\Delta\omega)t$.]

Table 7–2 summarizes the criteria for the different types of polarization. It is easy to show that the listed criteria are not only sufficient conditions for the type of polarization (as we have shown above), but are also *necessary* conditions. For example, given a wave which is circularly polarized, it follows that the components of **E** along any two mutually perpendicular axes in the plane perpendicular to the direction of propagation are sinusoidal functions of equal amplitude, and that they differ in phase by an odd multiple of $\pi/2$.

7-7 NATURAL LIGHT

Table 7–2 exhausts all the possibilities for waves that are "perfectly" sinusoidal. By this we mean waves represented by Eq. (7–18), where A, B, ϕ_1, and ϕ_2 are assumed to be constant. When we come to discuss the experimental means of detecting types of polarization, we shall find that under ordinary conditions the light emitted by a source does not show the characteristics of any of these types of polarization, even though the light is what we would call "monochromatic" (e.g. a particular spectral line isolated by means of filters from the radiation of a mercury arc). By means of a symmetry argument we can convince ourselves that it would, in fact, be rather strange if a general monochromatic source did emit a wave having a clearly defined type of polarization. All the types of polarization (except circular) are characterized by certain preferred directions. If conditions within the source are isotropic, there is no mechanism which could account for the singling out of a preferred direction in the emitted radiation. If the emission were either right or left circularly polarized symmetry would again be violated, for there is no condition within the source to determine a preferred *sense* of rotation. If the emission is the superposition of right and left circularly polarized waves of equal amplitude, then the situation is equivalent to a linearly polarized wave (Problem 7–8a), which again implies a preferred direction.

We are thus faced with a dilemma: all sinusoidal waves must be polarized, yet experimentally we find that typical monochromatic light sources do not exhibit the phenomenon of polarization. One way out of this dilemma is to broaden our conception of the type of function capable of describing a wave of single frequency. We can allow any of A, B, ϕ_1, and ϕ_2 to be "slowly varying" functions of time. Consider, for example, $E_y = A\cos(\omega t - \phi_1)$.

Let us hold the amplitude A fixed, but allow the phase ϕ_1 to drift gradually. If we could examine E_y as a function of time during a few hundred of its oscillations, it would seem to be a sinusoidal function of frequency ω. If we were able to return after several thousand oscillations have taken place and examine it again for a few hundred oscillations, it would still seem to have frequency ω. However, the sine wave at this later time would not be *in phase* with the continuation of the wave observed at the earlier time. By the phrase "slowly varying" we mean that the change in ϕ_1 which occurs during one period of oscillation is insignificant: $(2\pi/\omega)(d\phi_1/dt) \ll 1$. Nevertheless, a phase change as large as $\pi/2$ can accumulate over a time interval which is large compared with the period of oscillation. Under these conditions E_y is said to be "nearly sinusoidal."*

Let us examine the consequences of supposing that the amplitudes of E_y and E_z are fixed and that the phases of both drift gradually. If ϕ_1 and ϕ_2 are slowly varying functions of time but the phase difference $\Delta\phi = \phi_2 - \phi_1$ remains constant, the resultant E-vector will at all times satisfy the criterion for a fixed type of elliptical polarization. The experimental tests for polarization would register the type of elliptical polarization associated with the amplitudes A, B and the given phase difference $\Delta\phi$. The two functions E_y and E_z are then said to be *perfectly correlated in phase* and the component waves represented by E_y and E_z are said to be *coherent* with one another. If the phases vary in such a way that $\Delta\phi$ does not remain constant, the form of the curve traced out by the resultant E-vector will drift in an irregular fashion through the sequence of ellipses inscribed in the fixed rectangle $|E_y| \le A$, $|E_z| \le B$. Since ϕ_1 and ϕ_2 are "slowly varying," the same is true for $\Delta\phi$. During any interval containing a small number of oscillations the E-vector traces out the ellipse associated with the value of $\Delta\phi$ at that time. After a large number of cycles have elapsed, the value of $\Delta\phi$ is different, and, over a short time interval starting at this later time, the E-vector will seem to be tracing out an ellipse of different orientation. If an experiment is performed to determine the type of polarization, the result will depend on whether the recording instrument (e.g. the eye or a photographic film) can respond rapidly enough to distinguish individual forms in the changing sequence. If the changes in $\Delta\phi$ are so rapid that the individual forms cannot be registered. the two functions E_y and E_z are said to be *completely uncorrelated in phase*, and the component waves are said to be *noncoherent*. Since the persistence of human vision is on the order of $\frac{1}{16}$ sec, it is easily conceivable that the changes

---

* The Fourier analysis of such a function (Chapter 15) yields a *distribution* of frequencies very highly concentrated about the value ω. This is in accord with the fact that the observed spectral line is not infinitely sharp, but is distributed over a small range of frequencies.

in $\Delta\phi$ may be "slow" in comparison with the period of oscillation of visible light ($\sim 10^{-15}$ sec), and yet too rapid to be observed. In this case we observe a time average of the responses associated with the different types of polarization.

In order to explain the fact that monochromatic light obtained directly from a source does not show the phenomena of polarization, we can therefore suppose that the component waves, E_y and E_z, are completely uncorrelated in phase. For reasons of symmetry, we also assume that the amplitudes are equal ($A = B$). The light is referred to as *natural light* and is said to be *unpolarized.* This assumption concerning natural light seems quite plausible if we consider the conditions under which a source emits light. The total emission is the sum of contributions from a large number of atoms. The wavetrain emitted by an individual atom is likely to drift in phase when the "atomic oscillator" is perturbed by the force fields of a neighboring atom. During a particularly strong interaction (collision) the phase of the oscillation may be radically changed and may then bear a random relationship to the phase before collision.* We imagine, therefore, that the type of polarized wave emitted by each individual atom is subject to continual change. Since E_y and E_z are uncorrelated in phase in each of the contributing waves, the same is true for the total.

There are alternative suppositions which are capable of serving as models to explain unpolarized behavior of light from natural sources. For example, instead of assuming that the phase shifts of the total wave are gradual, we could suppose that the phase remains constant for a number of cycles and then changes discontinuously. Another successful model is obtained if we consider natural light to consist of right and left circularly polarized waves which are equal in amplitude and uncorrelated in phase. This model is equivalent to the supposition that E_y and E_z are correlated in phase, but have slowly varying amplitudes (Problem 7–13).

A more general type of nearly sinusoidal wave is obtained if we allow all of A, B, ϕ_1, and ϕ_2 to vary, and include the possibility that these variations may be *partially correlated* with one another. A discussion of this situation requires the introduction of a number of techniques of statistics, and thus will not be gone into here. One result of this treatment which we shall have occasion to make use of is the statement that the general wave can al-

---

* The net signal produced is analogous to the sound produced by 100 violins playing middle C. Each player may produce a perfect sinusoid so long as he is bowing in one direction. When he changes the direction of bowing, the phase of the next wavetrain he emits bears a random relation to that of the previous. The note produced by the whole group will sound like middle C, and, if examined over a short period (say $\frac{1}{10}$ sec), will seem to be a perfect sinusoid. If the note is sustained for as long as 30 sec, however, the phase of the sinusoidal portion near $t = 30$ sec bears a random relation to the phase near $t = 0$.

ways be thought of as a unique combination of a certain amount of natural light and a pure polarized wave. The wave in this case is said to be *partially polarized.*

7-8 ENERGY RELATIONS FOR THE GENERAL PROGRESSIVE PLANE WAVE

The purpose of this section is to show that so far as energy relations are concerned, we can carry out the calculations on a general progressive plane wave traveling in the positive x-direction *as though* it consisted of two *separate* linearly polarized waves. Formally, the general wave

$$\mathbf{E} = \mathbf{j}E_y + \mathbf{k}E_z, \qquad \mathbf{H} = \mathbf{j}H_y + \mathbf{k}H_z$$

can be written as the superposition of the two waves

$$\begin{aligned} \mathbf{E}^{(1)} &= \mathbf{j}E_y, & \mathbf{H}^{(1)} &= \mathbf{k}H_z; \\ \mathbf{E}^{(2)} &= \mathbf{k}E_z, & \mathbf{H}^{(2)} &= \mathbf{j}H_y. \end{aligned}$$

We must show that in energy calculations no "cross terms" appear between the two component waves. Consider

$$\mathbf{S} = \mathbf{E} \times \mathbf{H} = \begin{vmatrix} \mathbf{i} & \mathbf{j} & \mathbf{k} \\ 0 & E_y & E_z \\ 0 & H_y & H_z \end{vmatrix} = \mathbf{i}S_x,$$

where $S_x = (E_yH_z - H_yE_z)$. But, for each component wave considered separately,

$$\mathbf{S}^{(1)} = (\mathbf{j}E_y) \times (\mathbf{k}H_z) = \mathbf{i}E_yH_z, \qquad \mathbf{S}^{(2)} = (\mathbf{k}E_z) \times (\mathbf{j}H_y) = -\mathbf{i}E_zH_y.$$

Consequently, the intensity of the total wave is simply the sum of the intensities associated with its two components:

$$S_x = S_x^{(1)} + S_x^{(2)}.$$

The result is more obvious in the calculation of energy densities, for

$$W_{\text{el}} = \tfrac{1}{2}\epsilon|\mathbf{E}|^2 = \tfrac{1}{2}\epsilon(E_y^2 + E_z^2) = W_{\text{el}}^{(1)} + W_{\text{el}}^{(2)}$$

and

$$W_{\text{mag}} = \tfrac{1}{2}\mu|\mathbf{H}|^2 = \tfrac{1}{2}\mu(H_y^2 + H_z^2) = W_{\text{mag}}^{(1)} + W_{\text{mag}}^{(2)}.$$

It remains for us to investigate the way in which the intensity of a wave depends on its state of polarization. In general, for any +wave the instantaneous intensity is $S_x = (1/Z)(E_y^2 + E_z^2)$. Substituting from Eq. (7–18), we find

$$S_x = (1/Z)\{A^2\cos^2(\omega t - \phi_1) + B^2\cos^2(\omega t - \phi_2)\}.$$

For a circularly polarized wave ($A = B, \Delta\phi = \pi/2$), S_x is a steady quantity. In a more general case the time variation of S_x depends on the state of polarization. However, the average value is $\overline{S}_x = (1/2Z)(A^2 + B^2)$, which depends only on the amplitudes and not on the relative phases of the two components. Given A and B, the average intensity of the wave is independent of whether the components are in phase and the wave is linearly polarized, the components have a phase difference and the wave is elliptically polarized, or the components are uncorrelated in phase and the wave is unpolarized. In Chapter 9 we shall encounter devices which alter the relative phase of the two components of a wave but do not change the amplitudes. The average intensity of the wave emerging from such a device will therefore be the same as the original intensity.

When we consider the boundary value problem concerned with the reflection of an elliptically polarized wave incident normally on a plane interface between two dielectric media, it is easy to see that the linearly polarized components (E_y, H_z) and (E_z, H_y) can be treated separately. The same type of boundary conditions apply to each of the component waves. Consequently, the reflection and transmission coefficients which were originally calculated for the linearly polarized wave (E_y, H_z) are correct without alteration for the general case. The same is true for unpolarized light, for, on a small time scale, unpolarized light has the characteristics of some kind of elliptically polarized light.

7-9 REFLECTION BY A THIN FILM

The boundary value problem for the reflection of an electromagnetic wave normally incident on the plane-parallel faces of a layer separating two semi-infinite media has a number of applications in optics. For example, thin soap films or thin films of mica or cellophane fit these conditions, the two outer media, 1 and 3, being air. The conditions are likewise met by the layer of air

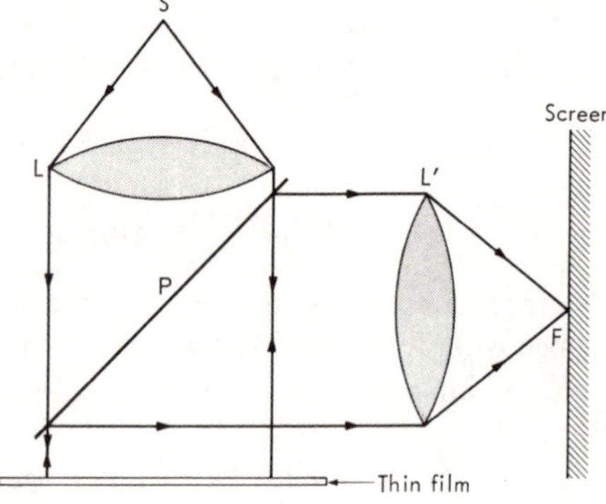

Fig. 7-8. System for viewing light reflected upon normal incidence from a thin film. S is a point source at the focal point of lens L. P is a partial reflector. The light reflected from the film and then by P is brought to a focus at F by lens L'.

between two plates of glass separated by a small space. To view the reflected light under conditions of normal incidence a reflex system such as that sketched in Fig. 7–8 is required. Since we are restricted here to the case of a *normally* incident plane wave, it is beyond the context of the present discussion to consider films of variable thickness, viewing from an oblique angle, or illumination by a broad source. The more general case is considered in Section 11–12.

An incident monochromatic plane wave of arbitrary polarization is represented as a superposition of linearly polarized component waves whose E-vectors are mutually perpendicular. The boundary conditions for each of these component waves are isomorphic to those of the acoustic problem of Section 3–7. For both components the complex amplitude of the reflected wave can be obtained through the isomorphism by translation of Eq. (3–31). Taking 1 and 3 to be identical media, we set

$$r = R_{12} = -R_{23} = (1 - n_{21})/(1 + n_{21}).$$

We then have as the reflection coefficient for the components of the E-vector

$$\mathbf{R} = r\left[\frac{1 - \cos 2k_2a + i\sin 2k_2a}{1 - r^2\cos 2k_2a + ir^2\sin 2k_2a}\right].$$

The intensity of the reflected wave relative to unit intensity of the incident wave is therefore given by

$$|\mathbf{R}|^2 = \frac{4r^2\sin^2 k_2a}{1 + r^4 - 2r^2\cos 2k_2a}. \tag{7–23}$$

The value of $|\mathbf{R}|^2$ is zero when $\sin k_2a = 0$, i.e., the film is an integral number of half-wavelengths thick. A maximum value of $|\mathbf{R}|^2 = 4r^2/(1 + r^2)^2$ is attained when the film is an odd number of quarter-wavelengths thick.* The graph of $|\mathbf{R}|^2$ versus k_2a is sketched in Fig. 7–9 for several values of r.

When the incident beam contains a number of different wavelengths, those wavelengths which are close to values for which $\lambda_2 = 2a/n$ are suppressed and the distribution of colors in the reflected beam is altered. In particular, suppose that an incandescent white light source is used. One can examine the reflected beam by allowing F in Fig. 7–8 to coincide with the entrance slit of a prism spectrometer. The spectrum will be crossed by a number of dark bands at those wavelengths for which $|\mathbf{R}|^2 = 0$. This phenomenon is known as a *channeled spectrum*. The color of the complete reflected beam is complementary to the color resulting from the superposition of the suppressed portions of the spectrum. When a is large, the suppressed colors

* The value of r depends upon wavelength in a relatively minor way through the index of refraction. This variation is neglected when we consider relative maxima with respect to wavelength changes.

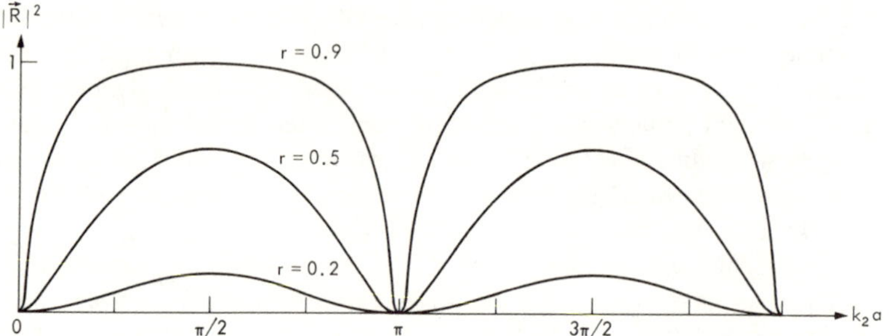

Fig. 7-9. Frequency dependence of the reflection coefficient for a thin film.

are closely spaced over the entire visible spectrum and the reflected beam appears white.

In principle, a glass plate of arbitrary thickness constitutes a layer of the type being described. In practice, however, the requirement of precise parallelism of the two faces is difficult to achieve for plates more than a few millimeters thick. If the distance between the plates varies randomly by an amount comparable to the wavelength, the light reaching the focus F must be calculated by averaging $|\mathbf{R}|^2$ over a cycle of values of k_2a. The result of this calculation is the same as that obtained by summing the *intensities* of the multiply reflected waves, rather than by first adding amplitudes and then calculating the intensity, as was done in the above solution. The result of this averaging no longer depends on the wavelength or the thickness of the plate. For this reason one can incorporate thick glass plates into optical systems without introducing highly selective reflection and transmission characteristics.

An application is made of the selective character of the reflection by a thin film in the design of *interference filters*. As is indicated in Fig. 7-9, when the single-surface reflection coefficient $r = R_{12}$ is close to ± 1, the film is an efficient reflector for all wavelengths except those very close to the critical values for which the film is transparent. Under these conditions most wavelengths are blocked from transmission by the plate; transmission occurs within a narrow range about each of the wavelengths satisfying $\lambda_2 = 2a/n$. To obtain a measure of the widths of these narrow ranges we can inquire for the values of k_2a which make $|\mathbf{R}|^2 = \frac{1}{2}$, i.e., the criterion for those wavelengths which have a transmission of one-half maximum value. The criterion yields $\sin^2 k_2a = [(1 - r^2)/2r]^2$. Thus, assuming r^2 very close to unity and k_2a close to $m\pi$, say $k_2a = m\pi + \delta\phi$, we obtain $\delta\phi = (1 - r^2)/2|r|$. The corresponding change in wavelength is given by $\delta\lambda_2/\lambda_2 = \delta\phi/m\pi \doteq (1 - |r|)/m\pi$. Thus the transmission by the layer can be made extremely selective by choosing a value of $|r|$ close to unity. Depending on the thickness of the layer, there may be several transmission bands across the visible spectrum. However, all except one of these is easily suppressed by the use of colored glass filters. In practice, values of $|r|$ close to unity are achieved by coating both sides of a dielectric layer with thin semitransparent metal films. The boundary conditions for this case are not identical with those considered above. Induced currents in the metal films introduce some dissipation of electromagnetic energy so that the maximum transmission is no longer 100%. However, the results are the same so far as the selective character of the transmission is concerned.

PROBLEMS

7–1. Estimate the rms values of electric intensity in several common electromagnetic situations and compare them with a typical electrostatic value. (Consult handbooks for experimental data on solar radiation, typical radiated power in radio waves, etc.)

7–2. Consider a linearly polarized wave normally incident on a plane interface separating two dielectric media. Derive the reflection coefficient for E_y directly by working through the boundary value problem completely in terms of electromagnetic variables.

7–3. White light is normally incident from air upon the surface of a medium of high dispersion, the index of refraction for blue light being greater than that for red. Describe the color of the reflected light.

7–4. In optical systems which involve lenses, a loss of intensity is encountered due to reflection at the lens surfaces. Assume a relative index of refraction of 1.5 and calculate the percent loss in intensity which occurs at each passage from air to glass or glass to air. (*Note:* The theory of image formation by a lens assumes that all rays are nearly parallel to the axis of the lens. Consequently it is justified to assume normal incidence in this problem.)

7–5. *Nonreflecting coating.* The index of refraction of borosilicate crown glass is 1.50 for the yellow D-lines of sodium (wavelength in air = 5893 Å). Specify the index of refraction and minimum thickness required for a coating which will make the surface nonreflecting for a normally incident wave of the given wavelength. (Assume $n = 1$ for air and $\mu = \mu_0$ for both air and glass.) When white light is incident, what hue will the reflected light have?

7–6. Show that the E-vector in elliptically polarized light rotates with nonuniform angular velocity in such a way as to sweep out equal areas in equal times.

***7–7.** When an ellipse of general orientation is inscribed within a rectangle, the eye tends to deceive us into believing that the major axis of the ellipse coincides with one of the diagonals of the rectangle (see Fig. 7–7). For the ellipse of Eq. (7–22), show that with $B/A = \tan \chi$, the angle θ between the major axis and the E_y-axis is given by the relation

$$\tan 2\theta = (\cos \Delta\phi) \tan 2\chi.$$

7–8. *The superposition of two elliptically polarized waves of the same frequency traveling in the same direction is equivalent to a single elliptically polarized wave.* The method of proof of this general proposition will be sufficiently illustrated by the following special cases:

a) Consider left- and right-handed circularly polarized waves of equal amplitude.

$$\begin{aligned} \text{Left-handed:} \quad E_y &= A\cos\omega t, \\ E_z &= A\sin\omega t; \\ \text{Right-handed:} \quad E_y &= A\cos(\omega t - \phi_1), \\ E_z &= -A\sin(\omega t - \phi_1). \end{aligned}$$

Show that the superposition is equivalent to a linearly polarized wave whose E-vector makes an angle $\phi_1/2$ with the E_y-axis. It should be clear from your proof that the converse is also true: Any linearly polarized wave may be thought of as composed of the superposition of left and right circularly polarized waves of equal amplitude.

b) Consider two linearly polarized waves which are in phase with each other but are of different amplitudes and have their E-vectors along different directions.

$$\begin{aligned} E_y^{(1)} &= A\cos\omega t, \qquad & E_z^{(1)} &= B\cos\omega t, \\ E_y^{(2)} &= A'\cos\omega t, \qquad & E_z^{(2)} &= B'\cos\omega t. \end{aligned}$$

Show that the superposition is also linearly polarized, and find the angle which the E-vector makes with the E_y-axis.

c) Show that an elliptically polarized wave can be represented as the superposition of a circularly polarized wave and a linearly polarized wave. (Without loss of generality, the axes of the ellipse may be chosen coincident with the coordinate axes.)

7–9. For each of the examples in Problem 7–8, compare the average intensity of the total wave with the sum of the average intensities of the components.

7–10. By direct appeal to Maxwell's equations show that a solution representing a standing plane wave with $E_x = E_z = 0$ and $E_y = A(\sin\omega t)(\sin kx)$ is possible. Deduce the corresponding magnetic field. Describe the energy flux. Which of the statements 1 through 5 of Section 7–5 are correct for this situation?

***7–11.** Suppose that in a given region the net electromagnetic field results from the superposition of two linearly polarized waves of equal amplitude traveling in opposite directions. Suppose further that the E-vectors in these two waves are at right angles to one another. That is, let the wave traveling in the positive x-direction have its E-vector along the y-axis, and let the wave traveling in the negative x-direction have its E-vector along the z-axis.

a) Show that in this field the net E- and H-vectors are not at right angles.

b) Show that the curve described by the tip of the net E-vector is an ellipse whose orientation depends on the value of x, and reduces to a straight line in some locations and a circle in others.

c) Show that at any given location the H-vector describes an ellipse geometrically similar to that described by the E-vector, but with a rotation in the opposite sense.

7–12. For a general case of plane waves (which may be a superposition of waves traveling in both directions), show that the relation

$$\partial W_{\text{tot}}/\partial t = -\partial S_x/\partial x$$

is an identity which follows from Maxwell's equations, Eqs. (7–7) through (7–10). What is the physical significance of this relation?

7–13. Consider a wave which is a mixture of right and left circularly polarized waves equal in amplitude and completely uncorrelated in phase. Show that E_y and E_z are perfectly correlated in phase but have amplitudes which are slowly varying functions of time. What type of polarization does the wave appear to have when observed over a short time interval? How does this change over longer intervals of time?

7–14. The channeled spectrum obtained by reflection from a thin film surrounded by air contains only two dark bands, centered at 4500 Å and 6000 Å. What is the optical thickness of the film? ("Optical thickness" is a term used for the product of the index of refraction and the geometrical thickness. Neglect the dependence of index of refraction on wavelength.)

7–15. Considered as a function of wavelength, the intensity transmitted by a thin film (relative to unit intensity for the incident beam) varies from a maximim of 1 to a certain minimum value. The reciprocal of the latter quantity may be taken as a measure of the *contrast* available in the channeled spectrum obtained upon transmission of light from an incandescent source. Express this as a function of the index of refraction of a dielectric film. Evaluate for a typical case. Comment on the possibility of constructing an interference filter without the use of metallic coatings.

7–16. A dielectric film which is one half-wavelength thick for light of wavelength 5500 Å is coated on both sides with a thin silver coating, the surface reflectivity being raised thereby to a value $|r| = 0.95$. If the film is used as an interference filter under conditions of normal incidence, what wavelengths will be passed with intensities greater than or equal to 50% of their original value? (Neglect dissipation.)

8 The production and detection of linearly polarized light

8-1 INTRODUCTION

In the next two chapters we have the task of bridging the gap between the theoretical conception of electromagnetic plane waves formulated in the last chapter and observable phenomena. It goes without saying that the rapidly varying electric field of a light wave cannot be detected by the ordinary techniques of electrical measurement. Thus, for example, we cannot ascertain that a given wave is linearly polarized by showing directly that the E-vector is parallel to a fixed direction. Instead, an indirect method must be used. We need to find observable effects produced in experiments with light which can be successfully *interpreted* by postulating that they are associated with some particular aspect of the electromagnetic plane wave.

Since indirect inference is involved in the interpretation of most optical phenomena, it is desirable to frame the initial description of an experiment in terms of a strictly operational language. We should make objective statements which can be verified without recourse to a theoretical interpretation. For example, the statement, "two disks composed of a specified material can be rotated relative to each other so that no light will pass through the combination" can be checked by direct experiment. After the association between a given effect and its theoretical interpretation has been established, we can then make statements which are based on the inferences. For example, the statement, "the light which passes through a disk of the specified material is linearly polarized" asserts that we have *inferred* from certain observations (not specified in this statement) that the light which emerges from the disk is described by equations similar to Eqs. (7–1) and (7–3). The function of this chapter and the next is to state what the "certain observations" are which we associate with linearly or elliptically polarized light and to examine the reasoning which leads to this identification.

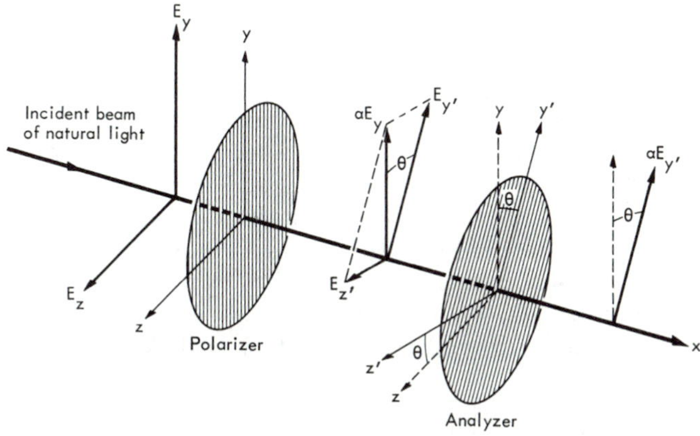

Fig. 8–1. Light passing through a polarizer and an analyzer.

8–2 THE RECOGNITION OF LINEARLY POLARIZED LIGHT

We have seen in the last chapter that a general plane wave progressing in the positive x-direction can be thought of as the superposition of two component waves E_y (with an associated magnetic field $H_z = E_y/Z$) and E_z (with an associated magnetic field $H_y = -E_z/Z$). If we can find a device which will transmit one of these two components but not the other, the emergent light will be linearly polarized. Before discussing the construction of such a device (called a *polarizer*), let us consider how we would identify one. For purposes of discussion, imagine the polarizer to be in the shape of a circular disk. We assume that the device is characterized by two mutually perpendicular directions lying in the face of the disk:

The axis of transmission. The polarizer transmits linearly polarized light whose E-vector is parallel to this direction. Since there may not be 100% transmission of this component, assume that the amplitude of E is reduced by a factor α.

The axis of extinction. The polarizer does not transmit linearly polarized light whose E-vector is parallel to this direction.

If we set the polarizer perpendicular to the x-axis, which is taken to be the direction of propagation of an incident plane wave and choose the y-axis of our coordinate system parallel to the axis of transmission of the polarizer, no matter what kind of light is incident (e.g. natural light), the emergent light will be linearly polarized with its E-vector along the y-axis (see Fig. 8–1).

Let us consider a second device (which will be called the *analyzer*) identical in construction with the first. Let the axes of transmission and extinction be y' and z' respectively. We can predict immediately that if the analyzer is placed perpendicular to the x-axis in the beam which emerges from the polarizer and set so that its axis of *extinction* is parallel to the E-vector of the incident light (i.e. z' parallel to y), *no* light will pass through the analyzer. In this position the analyzer and polarizer are said to be *crossed.* We can also predict how the intensity of the light transmitted by the analyzer will vary as the analyzer is rotated about the x-axis. To do this we must represent the light incident on the analyzer in terms of its components along the y' and z' axes. Let the angle between the y and y' axes be θ, and let the sinusoidal function describing the linearly polarized light emerging from the polarizer be given by $\alpha E_y = E_1 \cos(\omega t - \phi)$. As we can see from Fig. 8–2, this same wave is expressed relative to the y' and z' axes by the equations

$$E_{y'} = (E_1 \cos\theta)\cos(\omega t - \phi),$$
$$E_{z'} = (E_1 \sin\theta)\cos(\omega t - \phi).$$

But the analyzer transmits only $E_{y'}$, reduced in amplitude by the factor α. Therefore the wave emerging from the analyzer has an amplitude $E_2 = \alpha E_1 \cos\theta$ and an intensity

$$I_2 = \alpha^2 I_1 \cos^2\theta, \qquad (8\text{–}1)$$

where I_1 is the intensity of the beam which emerges from the polarizer. This equation is known as Malus' law.*

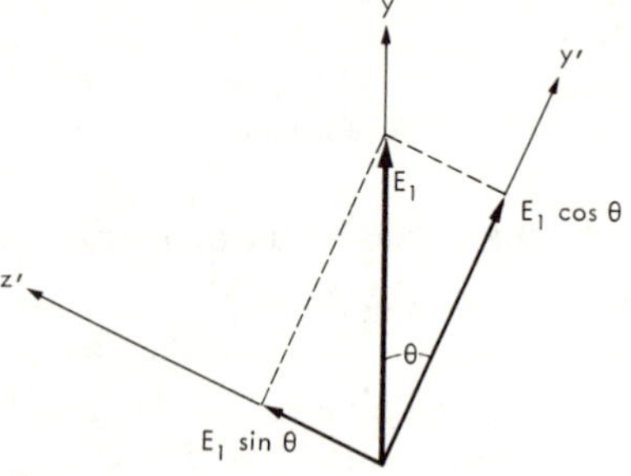

Fig. 8–2.
Resolution of the E-vector of linearly polarized light onto the axes of an analyzer.

Given a pair of devices which are asserted to have the properties described above, we can use the procedures of photometry to determine whether the intensity of light transmitted through both of them varies as the square of the cosine of the angle through which the analyzer is rotated. Note that it is not necessary to be able to single out the specific direction of the axis of transmission of either device. Since visual stimulation roughly obeys the Weber-Fechner law, it is always most accurate to make a critical adjustment at a position of minimum intensity. Therefore we assume that the analyzer is rotated until extinction occurs. The position opposite an arbitrary indicating mark can then be labeled $\theta = \pi/2$, and the rest of the scale for measurement of θ can be filled in. In later sections we describe a number of devices which are found to give good agreement with Malus' law. We may therefore con-

* First enunciated by E. Malus in 1809. Malus derived his statement of the law from a model in which light is assumed to consist of "molecules" which change their orientation under the action of mechanical forces. Following Malus' prediction, Arago made photometric measurements to show that polarizing devices give quantitative results in agreement with this equation.

clude that these devices act in such a way that one component of the incident light is removed and the other is allowed to pass, thereby producing linearly polarized light.

8-3 THE PRODUCTION OF LINEARLY POLARIZED LIGHT BY ABSORPTION

A thin slab cut from a crystal of *tourmaline* is found to have the properties of a polarizer. For slabs that are several millimeters thick, essentially perfect agreement is obtained with Malus' law. When much thinner slabs are used, it is found that complete darkness is not obtained when the analyzer is rotated to a position of minimum intensity. We infer from this last observation that one component of the incident light is progressively *absorbed* as it passes through the crystal. When the crystal is thin, there is still some of this component which gets through and the emergent beam is only *partially* polarized, i.e. the beam consists of linearly polarized light plus a small portion of natural light.

Although a slab of tourmaline which is sufficiently thick does produce perfectly linearly polarized light, there are certain disadvantages to the use of tourmaline as a polarizer. In the first place, it is relatively expensive. Second, the crystals are usually small, so that a slab of convenient cross-sectional area cannot be made. Third, the crystal is far from being transparent to the linearly polarized light whose E-vector is parallel to the axis of transmission. This component is partly absorbed. Furthermore, the absorption is much greater for certain wavelengths than for others, so that undesired coloration effects occur in applications in which white light is employed.

In 1852 the English physician Herapath discovered that the organic substance *quinine sulfate periodide* (now known as *herapathite*) has properties similar to tourmaline. When small crystals of this are viewed under a microscope, it is observed that no light is transmitted wherever two crystals overlap with their axes at right angles to each other. Herapath tried in vain to grow a crystal large enough to use as a polarizer. No application was made of herapathite until E. H. Land* conceived of the idea of leaving the crystals in

* It is interesting to learn how Land first heard about the properties of herapathite. The following quotation is from a very readable account which Land gives of the development of the different types of polarizer (*J. Opt. Soc. Am.*, **41**, 957, 1957). "Herapath's work caught the attention of Sir David Brewster, who was working in those happy days on the kaleidoscope. Brewster thought that it would be more interesting to have interference colors in his kaleidoscope than it would to have just different-colored pieces of glass. The kaleidoscope was the television of the 1850's and no respectable home would be without a kaleidoscope in the middle of the library. Brewster, who invented the kaleidoscope, wrote a book [D. Brewster, *The Kaleidoscope, Its History, Theory, and Construction*, 2nd ed (John Murray, London, 1858)] about it and in that book he mentioned that he would like to use the herapathite crystals for the eyepiece. When I was reading this book, back in 1926 and 1927, I came across his reference to these remarkable crystals and that started my interest in herapathite."

microscopic form and depositing them on a plastic sheet. To be successful as a polarizer, of course, the crystals must have their axes aligned predominantly parallel to a particular direction. In preliminary experiments Land achieved this alignment through the use of strong magnetic or electric fields. In 1932 he discovered that the alignment could be accomplished by the extrusion between long narrow slits of a viscous colloidal suspension of herapathite. The material thus obtained is known as a polarizing sheet, and is available at a relatively low cost. The process described above produces what is known as a *type J* polarizing sheet. Other types have since been developed. *Type H*, which is now most commonly employed, does not contain herapathite. An anisotropy is imposed on a film of polyvinyl alcohol by stretching it in one direction. When this film is impregnated with iodine, the iodine atoms form long polymeric chains aligned with the axis of stretch. These are responsible for the preferential absorption of one type of linearly polarized light.

The polarizing properties of this material are excellent. A crossed pair of type H polarizers transmits less than 0.01% of the energy of an incident beam of wavelength 5500 Å. The efficiency is slightly less at the blue end of the spectrum than at the red end, so that the "extinction color" is a deep blue. Type K (notable for its ability to withstand heat and intense radiation) and type J both show red as the extinction color. There is some absorption of the transmitted component, but this is fairly uniform for all wavelengths in the visible spectrum, so that incident white light still appears white after it has been passed through a single sheet of polarizing material. The factor α^2 in Eq. (8–1) is about 0.80 for a type H sheet. That is, when a linearly polarized beam whose E-vector is parallel to the axis of transmission passes through a single sheet of polarizing material, it suffers a 20% loss of intensity.

Substances such as tourmaline and herapathite, which show a preferential absorption for one kind of linearly polarized light, are known as *dichroics*. This property is evidently associated with some anisotropic feature in the structure of the substance. In dichroic crystals, for example, the axis of extinction (i.e. the direction of the E-vector in the type of linearly polarized light which is strongly absorbed) bears a definite relation to the crystallographic axes. From the electron theory of matter we can obtain an idea of the mechanism by which the absorption takes place. The variable electric fields of the light wave exert forces on the electrical charge contained within the molecules of the substance. The complete quantum-mechanical description of the distribution of charge within a molecule is obviously beyond the scope of this text. A remarkable degree of success can be achieved, however, by a simple classical model which visualizes the electrical structure of a dielectric in terms of "elementary oscillators." Consider a charge q which is associated with an equilibrium position and is subject to a linear restoring force when it is displaced from this position. So far as the dynamics of the motion of the charge is concerned, this is equivalent to supposing that the charge is bound to its equilibrium position by a system of springs like the one shown in Fig. 8–3.

In an anisotropic crystal the directions of the *y*- and *z*-axes are related to the crystal axes, and the force constants k_y and k_z are in general not equal. In a dichroic we must add a dissipative mechanism: suppose, for example, that the springs in the *z*-direction are "lossy" springs. When light whose *E*-vector is parallel to the *z*-direction passes through the crystal (the direction of propagation being perpendicular to the plane of the diagram in Fig. 8–3), the charge is set in motion parallel to the *z*-axis. This mode of oscillation continually dissipates energy in the form of heat and there is a subsequent loss of energy from the progressing wave. This type of linearly polarized light is therefore extinguished if it travels through a sufficient thickness of the crystal.

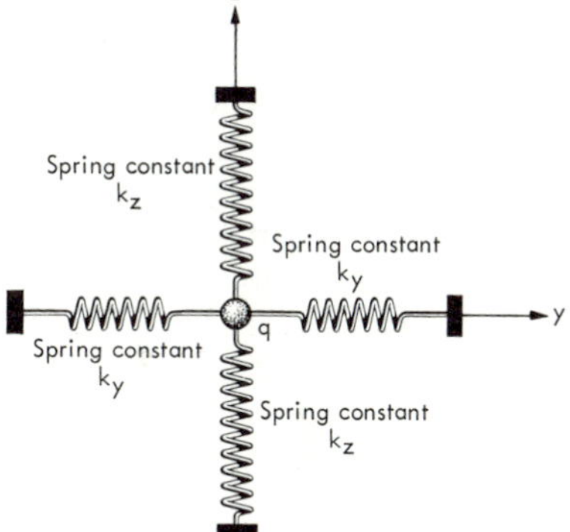

Fig. 8–3.
Anisotropic two-dimensional oscillator.

On the other hand, light whose *E*-vector is parallel to the *y*-direction induces motion parallel to the *y*-axis. If the dissipation associated with this motion is negligible, the passing wave does no net amount of work on the oscillators (once steady-state conditions have been achieved), and this type of linearly polarized light is passed by the crystal with slight attenuation. There is some energy loss due to reflection. This, in conjunction with some dissipation due to the motion of the charge, accounts for the fact that the factor α employed in Eq. (8–1) is in general less than one.

8–4 THE EXPERIMENTAL EVIDENCE THAT LIGHT IS A TRANSVERSE WAVE

A simple qualitative demonstration using an analyzer and a polarizer is sufficient to show that light exhibits behavior which is impossible for a wave of purely longitudinal character. Of course, if we believe that light waves are a special case of Maxwell's electromagnetic wave, then the theory tells us that plane waves are wholly transverse in character. We wish to show that the simple facts of observation lead to this same conclusion without the necessity of invoking any of the theoretical concepts. The demonstration has already been described in Section 8–2, but since we occasionally used some of the terminology of electromagnetic theory in that discussion, we repeat the description here in completely observational terms.

Place a slab of tourmaline crystal at a large distance from a point source of light. Let the line joining the point source to the center of the crystal be called the x-axis and set the faces of the crystal perpendicular to this axis. The light which emerges from the crystal appears quite ordinary to the naked eye. However, if we place a second slab of tourmaline in this beam with its faces perpendicular to the x-axis, we find that the light possesses an unusual characteristic: the amount of light which is transmitted by the second crystal is found to vary as this crystal is rotated about the x-axis. The beam of light being examined (that transmitted by the first crystal) therefore possesses a characteristic which distinguishes some directions perpendicular to the x-axis from others. If the beam did not possess an inherent property which varies with direction perpendicular to the x-axis, the interaction of this beam with the second crystal would be unaffected by the rotation of the latter.

A wave which is purely longitudinal cannot exhibit this type of asymmetry. In a sound wave, for example, all the associated vector quantities (e.g. particle velocity, pressure gradient) are directed along the x-axis. There is no parameter available which could single out "preferred directions" perpendicular to the x-axis. On the other hand, this type of behavior is easily associated with transverse waves. A demonstration bearing a crude resemblance to the above experiment can be performed with transverse waves on a string. A board with a slot in it, placed perpendicular to the string, will simulate the action of the tourmaline. No matter what kind of transverse vibration is present in a wave incident on this device, the vibrations in the transmitted wave are parallel to the slot. The amplitude of the disturbance which passes through two such devices is a function of the relative orientation of the slots, being a maximum when the slots are parallel and zero when the slots are perpendicular.

The above arguments show that a plane wave of light cannot be *wholly* longitudinal, but they do not rule out the possibility of a mixed longitudinal and transverse character. (Recall that the general motion of a string consists of a superposition of longitudinal and transverse vibrations.) If a longitudinal component were present in the wave transmitted by the first tourmaline crystal, it would also have to be transmitted by the second, regardless of its orientation. This contradicts the experimental fact that complete extinction is obtained when the second crystal is rotated to an appropriate position. We could still suppose that the eye does not perceive the longitudinal type of light wave or that the wave emitted by the *source* has a longitudinal component which is absorbed by the tourmaline. Experiment alone cannot rule out the possibility of a longitudinal aspect to light waves.* All we can say is that we are successfully able to interpret the results of certain experiments without making use of a longitudinal component.

We may also point out that to derive Malus' law it is necessary to assume only that the wave is associated with *some* vector perpendicular to the x-axis. In Section 8–2 no use is made of the physical interpretation of **E** as a transverse electric field. An equation having the form of Eq. (8–1) will be obtained whether **E** is interpreted as an electric field, a magnetic field, or the transverse displacement of an ether particle. Since polarizing devices give results which are in agreement with Malus' law, we see that these experiments are evidence only of the "transversality" of light and not of any of the other assumptions we have made.

8–5 POLARIZATION BY REFLECTION FROM A DIELECTRIC

Malus discovered in 1809 that light from a natural source becomes partially polarized when it is reflected at an oblique angle from the plane surface of a dielectric. By experimenting with different angles of incidence he found that for each dielectric substance there is an appropriate angle of incidence (called the *polarizing angle*) for which the reflected light is *linearly* polarized. If the surface is very clean and precisely plane, the linear polarization is nearly perfect. We can check these observations by using a disk of polarizing material as an analyzer.† For an arbitrarily chosen angle of incidence, an examination of the reflected light through an analyzer shows a variation in intensity as the analyzer is rotated. In a complete rotation through 360° two minima, 180° apart, and two maxima, at 90° from the minima, are found. This could indicate that the reflected light is elliptically polarized, or that it is a mixture of natural light and linearly or elliptically polarized light. Tests for distinguishing these cases will be described in the next chapter. They show that the reflected light is a mixture of natural light and linearly polarized

---

* Indeed, according to Maxwell's equations, waves other than the fictional case of plane waves of infinite extent *do* have a longitudinal component.

† The analyzer used by Malus was a crystal of calcite (Section 8–9). The discovery was made as he was looking through a calcite crystal at the reflection of the setting sun in a window of the Luxembourg Palace. Before retiring that night he found that the phenomenon could be duplicated using a candle and a plate of glass or a water surface. This discovery started a series of investigations which ultimately led to a recognition of the polarized nature of light. Previous to this time the curious phenomenon of double refraction in calcite had been observed, but its full implications had not been realized. For an historical account of the inductive reasoning leading from Malus' discovery to Fresnel's assertion in 1821 of the transverse character of light see Ernst Mach, *The Principles of Physical Optics, An Historical and Philosophical Treatment*, translated by J. S. Anderson and A. F. A. Young. New York: Dover Publications, 1953, Chapters 10, 11, 12.

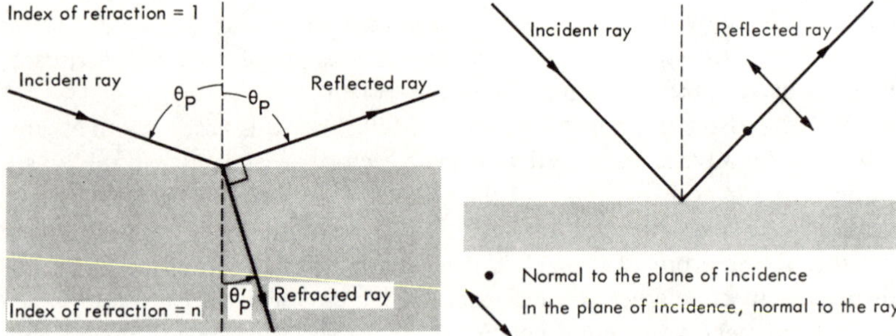

Fig. 8-4.
Relation between reflected and refracted rays at the polarizing angle.

Fig. 8-5.
Possible preferred directions associated with a wave reflected from a dielectric.

light. By trial and error an angle of incidence can be found for which no light passes through the analyzer when it is set at minimum position. This result uniquely identifies the reflected light as 100% linearly polarized, for any other type of polarized light or mixture of polarized and natural light would have a component parallel to the axis of transmission.

The relation between the polarizing angle and the index of refraction of the dielectric was investigated by Brewster. He discovered empirically that the tangent of the polarizing angle is numerically equal to the index of refraction. This relation is known as *Brewster's law.* Another way of stating this is to say that the reflected beam is 100% linearly polarized when the reflected and refracted rays are at right angles to each other. In Fig. 8-4 θ_P is the polarizing angle and θ'_P is the corresponding angle of refraction; from Brewster's law we have $\tan \theta_P = n$, and from Snell's law $\sin \theta_P = n \sin \theta'_P$. Taking the ratio of these two equations, we obtain $\sin \theta_P/\tan \theta_P = \cos \theta_P = \sin \theta'_P$. Therefore, since the angles are acute angles, $\theta_P = \pi/2 - \theta'_P$. The polarizing angle is also referred to as *Brewster's angle.*

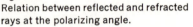

An instructive result is obtained by applying simple symmetry considerations to the experiment of polarization by reflection. Recall that the experiments of Section 8-2 provide no means of determining the specific direction of the axis of transmission of a given polarizer. This ambiguity can now be partly resolved. First we define a geometrical term used in referring to an obliquely reflected wave: An incident ray and the corresponding normal to the surface define a plane known as the *plane of incidence.* A fundamental law which we have taken for granted in drawing Fig. 8-4 is that the reflected and refracted rays both lie in the plane of incidence, i.e., the plane of the paper. Suppose now that a beam of natural light is incident at the polarizing angle. We have found that the reflected ray is linearly polarized and therefore is characterized by some preferred direction. It is clear that this preferred direction must be symmetrically related to the plane of incidence. Since the dielectric is an isotropic medium, the only features which can determine a preferred direction are the normal to the surface and the direction of the ray. If the E-vector were inclined at an angle of 20° to the plane of incidence, for example, then the mirror image of this result in the plane of incidence ought to be equally possible. There is no mechanism present to allow it to happen one way rather than the other. We therefore conclude that any vector quantity associated with the reflected wave must be either *in* the plane of incidence or *perpendicular* to it. In the former case a unique direction in the plane is determined by the requirement that the vector be transverse to the direction of propagation (see Fig. 8–5).

We cannot yet decide whether the linearly polarized reflected light has its E-vector perpendicular to the plane of incidence with its H-vector in the plane of incidence, or vice versa, but the symmetry argument has narrowed it down to these two possibilities. We can now locate "the axes" of an analyzer by placing it perpendicular to the reflected ray and setting it for extinction. In this position the directions parallel and perpendicular to the plane of incidence coincide with the axes of the analyzer, though we are not sure which is the axis of transmission and which is the axis of extinction.

Prior to the electromagnetic theory of light, when thinking about the polarization of light was done in terms of the more ambiguous "light vector," the symmetry argument was used to show that the *light vector* of the reflected light was either parallel to or perpendicular to the plane of incidence. This plane was therefore called the *plane of polarization* of the reflected ray. This is still a convenient way of specifying the transverse directionality of a linearly polarized wave. We turn now to the evidence which will decide whether it is the E-vector or the H-vector which lies in the plane of polarization.

8–6 FRESNEL'S FORMULAS

The problem of calculating both reflected and transmitted waves for a given plane wave incident at an oblique angle on the plane surface of a semi-infinite dielectric medium is a boundary value problem analogous to the ones we have solved in cases of normal incidence. The solution automatically contains a theoretical derivation of many of the laws quoted above which were originally discovered experimentally. These include Snell's law, Brewster's law, and the explanation of the partial polarization of the reflected light when natural light is incident at an angle other than the polarizing angle. The boundary value problem yields coefficients which characterize the reflected and transmitted light. We omit a derivation of the coefficients, but we shall explain their meaning and show how they can be used to study the polarization phenomena associated with reflection.

The incident beam is first represented in terms of its components onto two mutually perpendicular axes. These axes are selected so that they suit

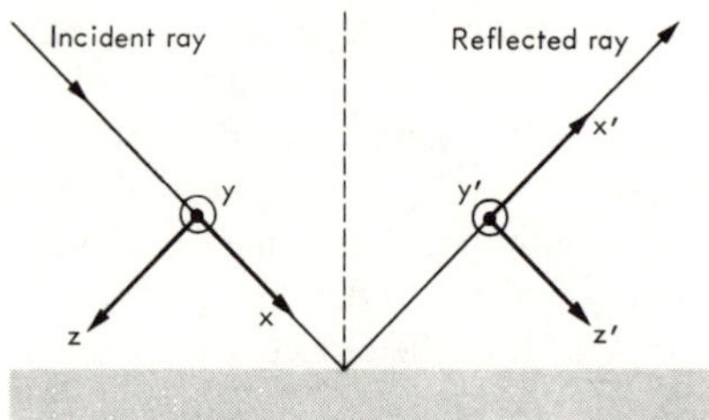

Fig. 8-6.
Coordinate systems to establish sign conventions in the Fresnel reflection coefficients.

the natural symmetry of the reflection problem, i.e., one component of the incident wave has its E-vector perpendicular to the plane of incidence and the other has its E-vector parallel to the plane of incidence. The problem reduces to a separate calculation of the reflection and transmission coefficients for each of these linearly polarized waves. We take the medium above the dielectric to be free space ($\mu = \mu_0$, $\epsilon = \epsilon_0$, $n = 1$) and assume that the permeability of the dielectric is the same as that of free space. Let the index of refraction of the dielectric be n. The angle of incidence, θ, is the angle between the direction of propagation of the incident plane wave and the normal to the surface. The derivation then shows that for either of the linearly polarized components there is a reflected plane wave with the angle of reflection equal to the angle of incidence, and a refracted plane wave for which the direction of propagation makes an angle θ' with the normal to the surface. The angle of refraction is determined by the Snell's law equation, $n \sin \theta' = \sin \theta$. The ratio of the amplitude of the E-vector in the reflected wave to the amplitude of the E-vector in the incident wave is given by the reflection coefficients: for the component whose E-vector is perpendicular to the plane of incidence,

$$R_{\perp} = -\frac{\sin(\theta - \theta')}{\sin(\theta + \theta')}; \tag{8-2}$$

and for the component whose E-vector is parallel to the plane of incidence,

$$R_{\parallel} = \frac{\tan(\theta - \theta')}{\tan(\theta + \theta')}. \tag{8-3}$$

The sign conventions associated with these coefficients are assigned as follows. Let the direction of propagation of the incident wave be the x-axis. Choose the y-axis perpendicularly out of the paper, and take the z-axis to form a right-handed coordinate system (see Fig. 8-6). Similarly, let the direction of propagation of the reflected wave be the x'-axis, choose the y'-axis perpendicularly out of the paper, and take the z'-axis to form a right-handed set. Assume that the origins of these two coordinate systems are coincident. (To avoid congestion, this has not been indicated in the diagram.)

Then the functions describing the incident wave are written

$$\begin{aligned} E_y &= A \cos(\omega t - kx - \phi_1), \\ E_z &= B \cos(\omega t - kx - \phi_2), \end{aligned}$$

and the functions describing the reflected wave are

$$\begin{aligned} E_{y'} &= A' \cos(\omega t - kx' - \phi_1), \\ E_{z'} &= B' \cos(\omega t - kx' - \phi_2). \end{aligned}$$

The reflection coefficients are then defined as

$$R_\perp \equiv A'/A, \quad R_\parallel \equiv B'/B.$$

Note that as the angle of incidence approaches zero, the z- and z'-axes are *oppositely* directed. One must take account of this to reconcile the limiting case of $\theta = 0$ with the reflection coefficients previously derived for normal incidence.

Brewster's law follows from Eq. (8–3). For that particular angle of incidence such that $\theta + \theta' = \pi/2$, the denominator is infinite and $R_\parallel = 0$. At this same angle, $R_\perp \neq 0$. Hence the reflected wave consists purely of the linearly polarized component whose E-vector is perpendicular to the plane of incidence. The previous ambiguities are resolved by this derivation. The "plane of polarization" of a linearly polarized wave is the plane containing the *H-vector*. Since we know that the E-vector in light reflected at the polarizing angle is perpendicular to the plane of incidence, the axes of an analyzer can now be labeled. When the analyzer is set for extinction, the axis of transmission (of **E**) is the intersection between the plane of the disk and the plane of incidence.

We are now in a position to discuss the reflection of natural light incident at some angle other than the polarizing angle. Since the amplitudes of the two components of incident natural light are equal, the ratio of the amplitudes in the reflected light is the same as the ratio of the reflection coefficients. From Eqs. (8–2) and (8–3), we obtain

$$\left|\frac{R_\parallel}{R_\perp}\right| = \left|\frac{\cos(\theta + \theta')}{\cos(\theta - \theta')}\right| = \frac{|(\cos\theta)(\cos\theta') - (\sin\theta)(\sin\theta')|}{(\cos\theta)(\cos\theta') + (\sin\theta)(\sin\theta')}.$$

But since $\sin\theta$, $\sin\theta'$, $\cos\theta$, and $\cos\theta'$ are all positive numbers, the numerator is smaller than the denominator and

$$\left|\frac{R_\parallel}{R_\perp}\right| \leq 1.$$

(The equals sign applies to the cases $\theta = 0$ and $\theta = \pi/2$.) Thus the amplitude of the component of the reflected wave whose E-vector is perpendicular to the plane of incidence is greater than the amplitude of the other component. If the reflected light is examined with an analyzer, a minimum is obtained when the axis of extinction is perpendicular to the plane of incidence. This provides an easy test for determining which of the two axes of an analyzer is the axis of extinction without the necessity of making an exact setting for the polarizing angle. This fact is also of practical consequence in the design of sunglasses. A polarizing sheet with its axis of extinction horizontal is effective in reducing the glare due to reflection off horizontal surfaces.

The boundary value problem also yields the following transmission coefficients for the E-vector: for the component whose E-vector is perpendicular to the plane of incidence,

$$T_{\perp} = 1 + R_{\perp} = \frac{2 \sin \theta' \cos \theta}{\sin (\theta + \theta')}; \tag{8–4}$$

and for the component whose E-vector is parallel to the plane of incidence,

$$T_{\parallel} = \frac{1 + R_{\parallel}}{n} = \frac{2 \sin \theta' \cos \theta}{\sin (\theta + \theta') \cos (\theta - \theta')}. \tag{8–5}$$

Equations (8–2), (8–3), (8–4), and (8–5) are known as *Fresnel's formulas.* They were originally derived by Fresnel on the basis of an elastic solid model of the ether. The boundary conditions which he used are isomorphic to the boundary conditions of electromagnetic theory, so that the resulting equations are similar in form, even though he had a different physical interpretation in mind for the E- and H-vectors.

Since light reflected at the polarizing angle is perfectly linearly polarized, a single glass plate can be used as a polarizer or an analyzer. This method is no longer used as a practical method of polarizing light because the intensity of the reflected beam is low. At the polarizing angle the amplitude of the reflected component is determined from Eq. (8–2) by setting $\theta' = \pi/2 - \theta$ and using $\tan \theta = n$:

$$R_{\perp} = \cos 2\theta = \frac{1 - n^2}{1 + n^2}.$$

For glass with index of refraction $n = \frac{3}{2}$ this yields $R_{\perp} = -\frac{5}{13}$. The intensity of the reflected wave can be obtained from the usual relation for a plane wave. Since $R_{\perp}^2 = 0.148$, only 14.8% of the energy of the perpendicular component is reflected. None of the parallel component is reflected; therefore only 7.4% of the total energy of the incident beam is reflected. The transmitted light is also *partially* polarized, as can be seen from the fact that $T_{\perp}$ and $T_{\parallel}$ are not equal. The light transmitted by a *single* surface contains a large proportion of natural light. If the beam passes through a number of surfaces, however, the proportion of natural light to linearly polarized light decreases (Problem 8–8).

The Fresnel formulas can be used to discuss the reflection of an incident wave that is polarized. In this case there is perfect phase correlation of the components of the incident light and consequently also of the reflected light. The type of polarization of the reflected light can be deduced from the reflection coefficients. Examples of this calculation are given in Problems 8–10 and 8–11.

*8–7 STOKES' RELATIONS

In the preceding section we considered the reflection and transmission of a wave which was incident from a medium of index 1 on a medium of index n. In general, we can take the index of the medium from which the wave is incident to be n_1 and the index of the second medium to be n_2. All the formulas discussed above hold for this general case provided that we replace n everywhere it appears by $n_{21} = n_2/n_1$. An exception must be made, however, if $n_1 > n_2$ and the angle of incidence is greater than the *critical angle* $\theta_c = \sin^{-1}(n_2/n_1)$. Under these conditions the equation of Snell's law does not lead to a meaningful angle of refraction.

The solution to the boundary value problem can nevertheless be adequately interpreted. The disturbance in the second medium does not correspond to an ordinary plane wave. The reflected wave is a plane wave having the same intensity as the incident wave. This phenomenon is referred to as *total reflection*. The two linearly polarized components of the reflected wave undergo *different* changes in phase from each other. These phase changes are functions of the angle of incidence and can be applied to produce types of polarization other than linear. We shall not go into a further description of total reflection here: In Chapter 17 the complete boundary value problem is solved for the *acoustic* case and a discussion is given of the anomalous character of the field which exists in the second medium.

There is an intimate connection between the reflection coefficients for an external reflection ($n_1 = 1, n_2 = n$) and a corresponding internal reflection ($n_1 = n, n_2 = 1$). We can derive this connection easily by inspection of the Fresnel formulas (Problem 8–7). An alternative derivation can be deduced from a simple consideration due to Sir George Stokes. The virtue of this derivation is its complete independence of the type of wave: the waves may be longitudinal or transverse. If they are transverse, the same result is obtained for any type of polarization. The argument is based on an assumption of *reversibility* of a system of rays. Consider an incident wave of unit amplitude. In consistency with our general viewpoint we do not need to specify what wave quantity this amplitude refers to. It could be any component of $\mathbf{E}$ or $\mathbf{H}$ of an electromagnetic wave, or p or ξ of an acoustic wave. Let R and T be the reflection and transmission coefficients of this quantity for a reflection taking place in medium 1 (see Fig. 8–7a). We now assume that the situation can be re-

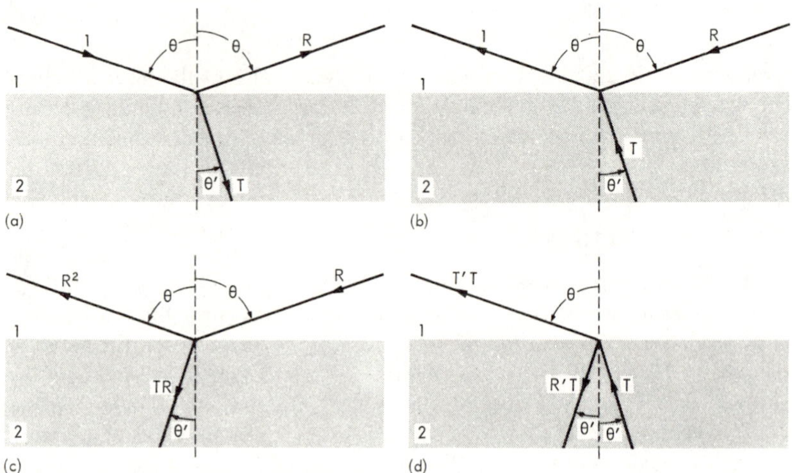

Fig. 8–7. Derivation of Stokes' relations.

versed as indicated in Fig. 8–7(b). The waves R and T must be identical with those of Fig. 8–7(a), except for a reversal of the direction of propagation. In particular, they must be perfectly correlated in phase. Independent sources could not be used to produce R and T. The superposition described is simply a thought experiment for the purpose of the derivation. As is shown in Fig. 8–7(c), a wave of amplitude R incident from medium 1 at the angle θ would give rise to a reflected wave of amplitude R^2 and a transmitted wave of amplitude TR. Similarly, Fig. 8–7(d) shows a wave of amplitude T incident from medium 2 at the angle θ' giving rise to a reflected wave of amplitude $R'T$ and a transmitted wave of amplitude $T'T$. (Primes indicate the reflection and transmission coefficients for a reflection taking place in medium 2.) For the superposition of (c) and (d) to result in the situation described in (b) it is evident that the following equations, known as *Stokes' relations*, must be satisfied:

$$R^2 + T'T = 1, \tag{8–6}$$

$$TR + R'T = 0 \qquad \text{or} \qquad R' = -R. \tag{8–7}$$

Equation (8–7) shows that in a general wave situation, a reflection at angle θ in medium 1 reflects the same percentage of the total energy as a reflection at angle θ' in medium 2. The angles θ and θ' must, of course, be *corresponding* angles in the sense of Snell's law. We also learn from the minus sign that any wave quantity undergoes opposite phase changes in internal and external reflections. Equation (8–6) is useful in calculating the transmission of a wave which passes through a layer with plane parallel sides (see Problem 8–8).

8–8 INTERPRETATION OF BREWSTER'S LAW IN TERMS OF THE ELECTRON THEORY OF MATTER

A derivation of the reflection which takes place at the surface of a dielectric follows directly from application of the appropriate boundary conditions to Maxwell's equations. No auxiliary hypotheses or suppositions are necessary in this derivation. The reflected wave simply appears as part of the mathematical solution to the problem. One does not need to think explicitly of *how* the reflected wave originates. On the other hand, the electron theory of matter takes a more microscopic view of the proceedings and enables us to visualize the production of the reflected wave in terms of a model. Briefly, the reflected wave is the result of radiation from oscillating charge within the dielectric which has been set in motion by the electric field of the incident wave. As we shall see presently, this model provides an explanation of Brewster's law. First it is necessary to make a few assertions about the nature of the radiation which is emitted by an oscillating charge.

Consider a charge q which is executing simple harmonic motion about the origin along the z-axis of a coordinate system. The nature of the electric and magnetic fields which surround a charge undergoing a periodic acceleration in this manner can be deduced from the complete set of Maxwell equations when the time-dependent charge density and current density terms are included. This problem was first solved by Hertz and is known as the *Hertzian dipole solution.* Although the details of the electromagnetic problem will not be worked out here, the analogous acoustical problem is solved in Chapter 17.

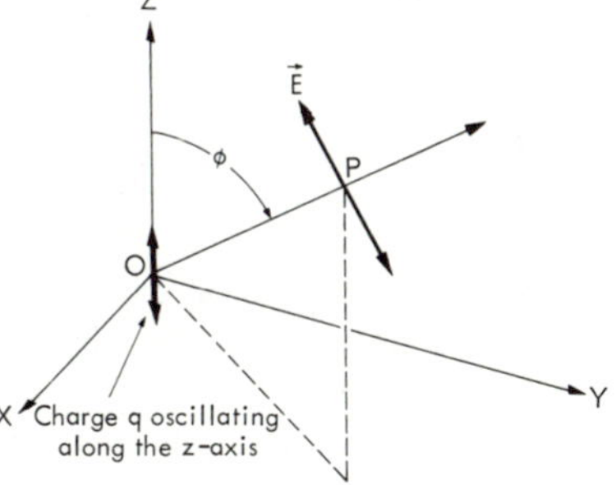

Fig. 8–8. Character of wave radiated from a charge executing linear simple harmonic motion.

The oscillating charge acts as the *source* of electromagnetic waves which are radiated in the form of spherical wavefronts. The waves are sinusoidal, having the same frequency as the frequency of oscillation of the charge. Consider a point of observation P at a large distance from the origin in a direction making an angle ϕ with the z-axis (see Fig. 8–8). A portion of the wavefront at this position will resemble a plane wave propagating in the direction OP. This wave is linearly polarized and the E-vector lies in the plane POZ. The amplitude of E is not constant over the spherical wavefront, but varies with

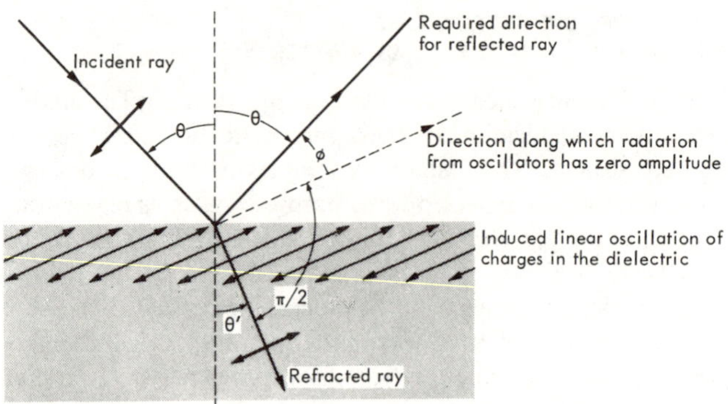

Fig. 8–9. Motion of charge near the surface of a dielectric due to a linearly polarized wave with its E-vector in the plane of incidence.

the angle ϕ. If the amplitude in the plane XOY is E_0, then the amplitude at P is $E_0 \sin \phi$. In particular, when the point of observation is on the z-axis, $\sin \phi = 0$ and the wave has zero amplitude. This means that the oscillating charge radiates no energy in the direction parallel to its direction of motion.

For a general case, consider the motion of the charge to be the superposition of two orthogonal simple harmonic oscillations of the same frequency. The charge then describes an ellipse in a plane inclined at an arbitrary angle to the coordinate axes. In this case, the character of the radiated wave observed at P depends on the projection of the motion of the charge onto a plane perpendicular to the line of sight OP. For example, if the charge *looks* from P as if it were describing a circle in the counterclockwise direction, then the wave at P is left circularly polarized. In this example the charge may actually be describing an ellipse in a plane through O which is not perpendicular to OP. The component of the motion which is parallel to OP has no effect on the radiation observed at P.

Now consider a linearly polarized plane wave incident on the surface of a dielectric with its E-vector in the plane of incidence. Let the angle of incidence be θ. The charge within the dielectric is subject to sinusoidally varying forces due to the E-vector of the *refracted* wave which propagates in a direction making an angle θ' with the normal (Fig. 8–9). The charge will therefore execute linear harmonic oscillation along a direction perpendicular to the refracted ray. Each charge is the source of a spherical wavelet. This might lead us to suppose that the "reflected" wave contains energy traveling in all possible directions. However, since the motion of all the oscillating charges is induced by the field of the refracted wave, these oscillations are perfectly phase correlated with one another. We must therefore take account of the possibility that there are certain directions for which the contributions from the individual oscillators will produce destructive interference. This situation will be discussed thoroughly in Chapter 12. It will be shown that there is only one possible direction in which the contributing wavelets interfere constructively, namely, the direction which we usually associate with the reflected ray. In all other directions the individual wavelets add up so as to produce a zero resultant. The point of observation must therefore be thought of as lying in the direction of the reflected ray. According to the Hertzian dipole solution, the amplitude of the radiation in this direction is controlled by a factor of $\sin \phi$, where ϕ, as indicated in Fig. 8–9, is the angle between the line of sight (the reflected ray) and the line along which the charge oscillates. If the angle of incidence is chosen so that $\phi = 0$, the amplitude of the radiation from the oscillators is zero in the direction of the reflected ray, and there is no reflected wave. From the figure, we see that $\phi = 0$ requires that the reflected and refracted rays be at right angles to each other, which as we have seen before is the condition for the polarizing angle. In summary, the reason for zero reflection of linearly polarized light whose E-vector is in the plane of incidence when the angle of incidence is the polarizing angle is that the charges in the dielectric are oscillating parallel to the reflected ray and produce no radiation in this direction. On the other hand, for linearly polarized light whose E-vector is perpendicular to the plane of incidence, the reflected ray is at right angles to the motion of the charge (i.e. $\phi = \pi/2$, regardless of the angle of incidence) and there is no condition under which the charges fail to radiate in the direction of the reflected ray. At the polarizing angle this component of an incident beam is reflected, whereas the other is not, and the reflected light is 100% linearly polarized with its E-vector perpendicular to the plane of incidence.

8–9 DOUBLE REFRACTION IN CALCITE

When objects are viewed through a clear piece of calcite two distinct images are seen. This phenomenon is known as *double refraction* or *birefringence*. The effect was first noticed by Erasmus Bartholinus in 1670. Huygens subjected the phenomenon to a thorough experimental study and devised an explanation (discussed in Section 8–10) which accounts accurately for the observations. The present section consists of a summary of experimental facts.

There are three different directions in a calcite crystal such that the crystal splits easily along planes perpendicular to these directions. These planes are known as *cleavage planes*. A sample of calcite all of whose faces are cleavage planes is known as a *cleavage form*. Although there are a wide variety of external forms for natural calcite crystals, they all produce the same type of cleavage form. This basic cleavage form is a rhombohedron whose faces are parallelograms in which the obtuse angle is 102° (more

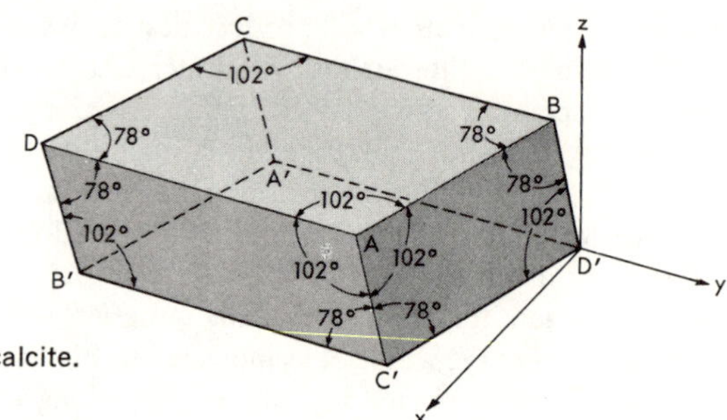

Fig. 8–10.
Cleavage form (rhomb) of calcite.

precisely 101°55′) and the acute angle is 78° (more precisely 78°5′). In the perspective drawing of Fig. 8–10 the face $A'B'C'D'$ lies in the xy-plane with $A'D'$ along the negative y-axis. The top leans to the left and *back* so that BC lies behind the yz-plane. The two vertices A and A', known as the *blunt corners*, are distinctive in that the angles in the three faces meeting at these points are all obtuse. At the other six vertices there are two acute angles and one obtuse angle.

Let us study the images formed of a point source S (e.g. an illuminated pinhole) adjacent to the lower face $A'B'C'D'$. Our point of observation P will be directly above S, that is to say, along a line passing through S which is perpendicular to the faces $A'B'C'D'$ and $ABCD$.* If S emits natural light, the two images are of equal intensity. The line joining the two images projects onto the top face as a line parallel to the angle bisectors of the obtuse angles A or C (see Fig. 8–11). We shall refer to this direction as the *principal axis of the face.* The distance of separation of the two images is proportional to the thickness (i.e. the perpendicular distance between $A'B'C'D'$ and $ABCD$). As the calcite rhomb is rotated about an axis perpendicular to $ABCD$, one of the images remains fixed while the other moves around it. The one which remains stationary is called the *ordinary image*, designated by

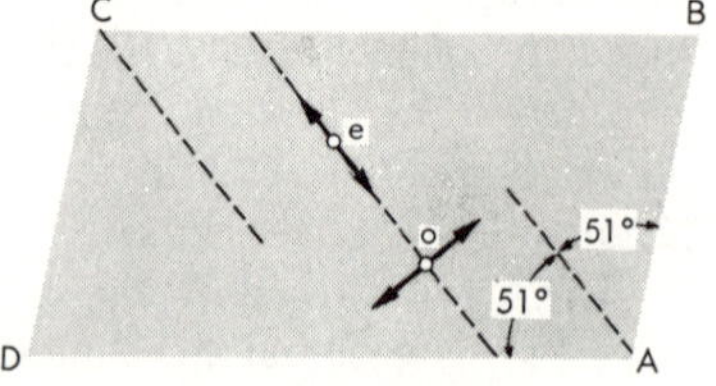

Fig. 8–11.
Top face of calcite cleavage form showing location of images of point source on bottom face.

* A number of complications arise if the point source is viewed obliquely. Even without double refraction there are such matters as astigmatism of the image and apparent change in depth. The events in double refraction are especially complicated if the plane which contains S and P and is normal to $ABCD$ is not a principal section. (See below for the definition of a principal section.)

o; the other is called the *extraordinary image*, designated by *e*. The ordinary image lies closer to the blunt corner *A* than the extraordinary image does. The apparent depth of the extraordinary image is greater than that of the ordinary image.

Since two images are observed, we infer that there are two rays emanating from *S* which arrive at the point of observation *P* (Fig. 8–12). The two rays lie in a plane which is perpendicular to the face *ABCD* and parallel to the principal axis of this face. In general, any plane which is perpendicular to a face of the cleavage form and parallel to the angle bisectors of the obtuse angles in that face is referred to as a *principal section*. Figure 8–13 shows the cross section of the cleavage form which is intercepted by a principal section passing through the vertex at the blunt corner *A*. This cross section is a parallelogram whose angles are 71° and 109°. We have assumed in drawing Fig. 8–12 that the point source *S* lies in the same principal section as that indicated in Fig. 8–13.

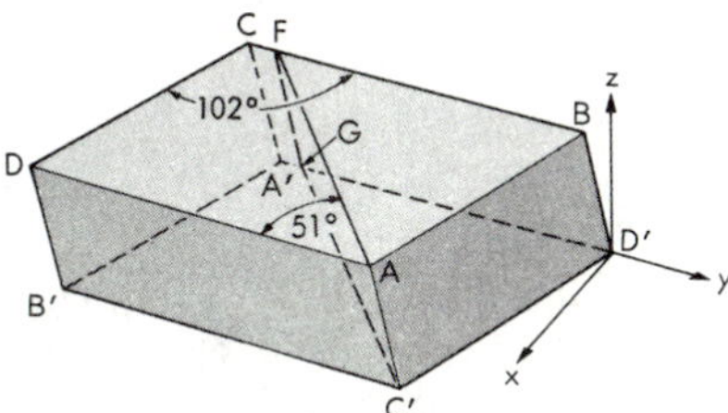

Fig. 8–13.
Principal section ($AC'GF$) of a calcite cleavage form.

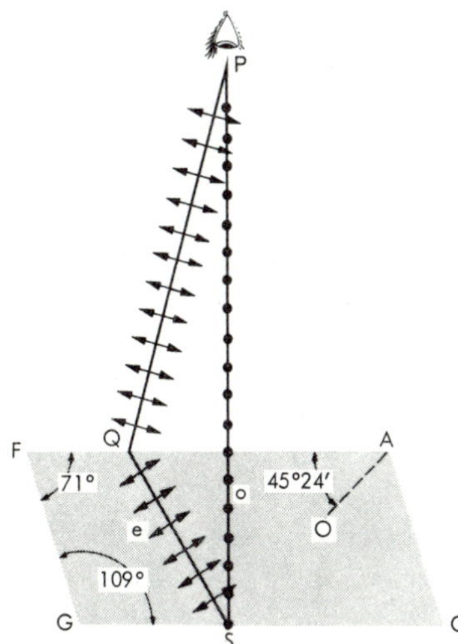

Fig. 8–12.
Ray diagram of double refraction. (The plane of the diagram is a principal section.)

The direction in which the ordinary image is observed indicates that the *o*-ray passes perpendicularly through the top surface of the cleavage form. The direction in which the extraordinary image is observed indicates that the *e*-ray has undergone a refraction at the surface which is clearly in defiance of Snell's law.

When the *o*- and *e*-images are viewed through an analyzer, it is found that both are linearly polarized. In *o* the *E*-vector is perpendicular to a principal section; in *e* the *E*-vector is parallel to a principal section. These directions are indicated by the small arrows and dots in Figs. 8–11 and 8–12.

A small dot denotes an E-vector that is perpendicular to the plane of the diagram.

The observation that the two images are polarized provides an important clue to the construction of a reasonable scheme for interpreting the phenomenon of double refraction. It seems clear that the light from the point source must be thought of in terms of linearly polarized components whose E-vectors are either parallel or perpendicular to a principal section. These components propagate independently of one another and obey different laws in the crystal medium. We can check this supposition by using a source which is linearly polarized. Let S be a pinhole covered with a sheet of polarizing material. When the axis of transmission is parallel to a principal section, only the e-image is observed; when it is perpendicular to a principal section, only the o-image is observed. A similar experiment can be performed using natural light at S and placing a second calcite rhomb on top of the first. Let $\mathbf{E}$ and $\mathbf{E}'$ represent the displacements of the extraordinary image which are effected by the respective rhombs. After passing through the first rhomb, the two images o and e are obtained with polarizations as indicated in Fig. 8–14(a). On entering the second rhomb, the light from each of these images is resolved into components parallel and perpendicular to the principal section of the second rhomb. The light from o forms an undisplaced image oo' and an image oe' having displacement $\mathbf{E}'$ (Fig. 8–14b). Similarly, e forms the two images eo' and ee'. If the angle between the principal sections of the rhombs is α, the intensities of oo', oe', eo', and ee' are in the ratio $\cos^2 \alpha : \sin^2 \alpha : \sin^2 \alpha : \cos^2 \alpha$. Thus ($oo'$, ee') and (oe', eo') are pairs of images having equal intensities. The intensities of the two pairs are complementary in the sense that the sum remains constant. When viewed through an analyzer, the pair (oe', ee') is seen to have a polarization perpendicular to that of (oo', eo').

Inspection of the cleavage form shows that there is one direction which is geometrically distinctive, namely the direction of a line which makes equal

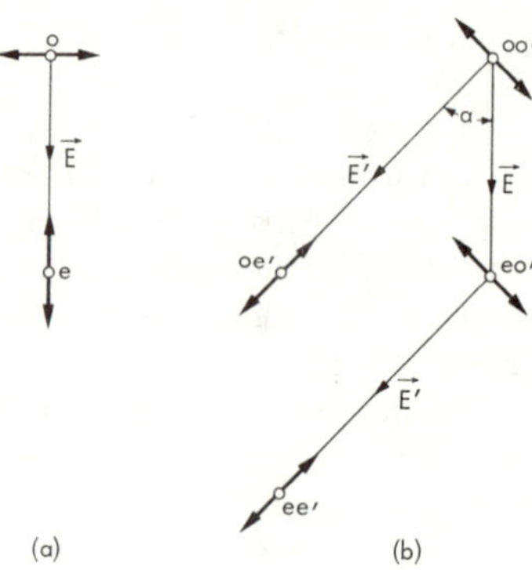

Fig. 8–14.
Images of a point source seen through a pair of calcite rhombs. The vectors **E** and **E**′ are parallel to the principal sections of the lower and upper rhombs respectively: (a) images after passage through lower rhomb, and (b) images after passage through both rhombs.

angles with the three edges meeting at a blunt corner.* This direction, which is known as the *optic axis*, is an important reference direction for the discussion of wave propagation in calcite. Every principal section of a cleavage form is parallel to this direction. In Fig. 8–12 the direction of the optic axis is indicated by the dotted line OA. It can be shown that the optic axis makes an angle of 45°24′ with all the cleavage planes (Problem 8–12). Consequently, OA makes this angle with AF, which is the intercept of a cleavage plane normal to the plane of the diagram.

Plates having a pair of plane parallel faces can be cut from calcite in such a way that the optic axis is inclined at an arbitrary angle to the faces. In general, double images are observed through these plates. A principal section is in this case defined as a plane normal to the faces and parallel to the optic axis. The displacement of the e-image and the polarization of the o- and e-images are related to the principal section in the same way as in the previous case of the cleavage form. However, if the plate is cut with the optic axis either parallel or perpendicular to the faces, only a single image is obtained of a point source when it is viewed from directly above. This distinctive property of the optic axis will be interpreted in the next section.

8–10 HUYGENS' WAVE SURFACES FOR CALCITE

The experimental observations summarized in the last section concerning double refraction in calcite were accounted for by Huygens on the basis of a few simple assumptions about the nature of the wavefronts emanating from a point source. We have seen that the light from S must be thought of in terms of its two separate linearly polarized components†—the o-wave and the e-wave. Since there are no anomalous properties associated with the o-wave, Huygens assumed that it consists of spherical wavefronts emanating from S and propagating with a velocity c_1 (see Fig. 8–15). The velocity of the e-wave is assumed to depend on the direction of propagation. Parallel to the optic axis the propagation velocity is taken to be the same as that of the o-wave. Perpendicular to the optic axis the propagation velocity is taken to be c_2, with $c_2 > c_1$. The complete wavefront is assumed to be an oblate spheroid which is generated by rotation about the optic axis of the ellipse

$$\left(\frac{x}{c_1}\right)^2 + \left(\frac{y}{c_2}\right)^2 = t^2. \tag{8–8}$$

* An example is the line joining A and A' in Fig. 8–10 if the cleavage form is picked so that the faces are *rhombuses*.

† Newton was aware of the doubly refracting properties of calcite and realized that they implied a "one-sidedness" to the character of light. He considered, however, that this ruled out the possibility that light is a wave. His thinking about waves was restricted to waves of longitudinal type, such as sound waves, with which he was most familiar.

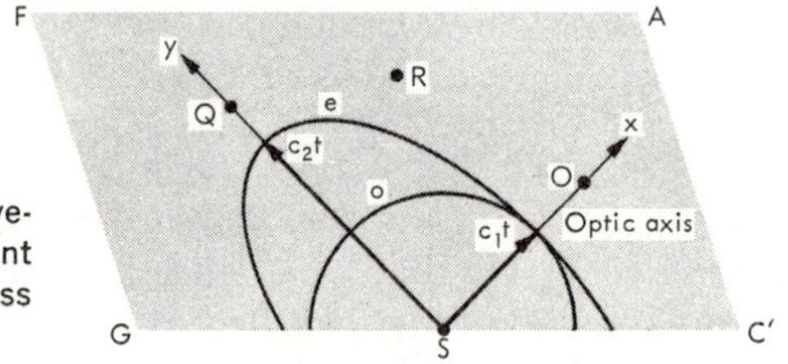

Fig. 8-15. Ordinary and extraordinary wavefronts originating from the point source S at $t = 0$. (Oblateness of spheroid exaggerated.)

In Section 9-6 we shall follow in detail some of the analytical consequences of Huygens' assumption concerning the spheroidal form of the e-wavefronts. At this point we wish to indicate in a qualitative fashion how this explains the relation between the two images of S. Let us imagine that we could take up a point of observation within the calcite rhomb. When a portion of a spherical wave is received by the eye, we perceive an image located at the center of the sphere. Thus the ordinary image of S is located at S. When the eye intercepts a small portion of the extraordinary wave, it treats this as if it were a part of a spherical wave whose center is the center of curvature of the given segment.* Hence we expect the following:

a) If the point of observation is at O (Fig. 8-15), the image produced by the e-wave is in the same direction as that of the o-wave. This corresponds to the fact that only a single image is seen when the line of sight is parallel to the optic axis.

b) If the point of observation is at Q, the directions of the o- and e-images again coincide. This corresponds to the fact that only a single image is seen when a point source is viewed from directly above through a plate whose optic axis is parallel to the surface.

c) If the point of observation is at R, the center of curvature of the e-wave passing this point is to the left of S. Thus the e-image will be displaced *away* from the blunt corner A. If c_2 were less than c_1, the displacement would be in the opposite direction. This observation can therefore be used to justify the previous assumption that $c_2 > c_1$. This choice is confirmed by the measurements described in the next chapter of the specific values of c_1 and c_2 (see also Problem 8-13).

d) The radius of curvature of the e-wave passing through R is greater than SR. The e-image will therefore seem to be at a greater distance than the o-image. This result is also dependent on the choice of $c_2 > c_1$. These predictions, based upon Huygens' assumption of the differing wave surfaces for o- and e-waves, are therefore in qualitative agreement with the observed

* Astigmatism results from the fact that the curvature is not the same in all cross-sectional planes. This will be ignored here.

relations between the *e*- and *o*-images. To obtain a quantitative prediction of the images observed at a point of observation *outside* the calcite, it would be necessary to discuss the refraction of the *e*-wave at the surface *AF*. We shall not pursue this further.

The existence of a spheroidal wavefront for an appropriately polarized wave can be derived from Maxwell's equations if account is taken of the fact that the vectors **D** and **E** are not necessarily parallel to each other in an anisotropic medium. For a discussion of Maxwell's equations in anisotropic media the reader is referred to the references cited below.*

The electron theory model, which treats the crystal as an assembly of anisotropic harmonic oscillators, enables us to understand why the direction of the *E*-vector is of importance in the determination of the propagation characteristics of linearly polarized light. Consider the oscillator in Fig. 8–3† interacting with a plane wave propagating perpendicularly out of the paper. If $k_y \neq k_z$, the response of the oscillator to an *E*-vector parallel to the *y*-axis will be different from the response to an equal stimulation parallel to the *z*-axis. In dispersion theory, the explanation of the fact that the velocity of light in a material medium is less than the velocity of light in free space is based on the interaction of the incident wave with the dipole radiation of the elementary oscillators, and therefore leads to different predicted values for the velocity in these two cases.

8-11 THE NICOL PRISM

Since light which passes through calcite is separated into two linearly polarized components which are in general displaced relative to each other, a calcite cleavage form can be used as an analyzer. If the observer concentrates his attention on one of the two images and examines the variations in intensity as the calcite is rotated, the information obtained is equivalent to the information he would obtain, for example, with a disk of polarizing material. For use as a polarizer, however, or for examination of a field of view in which there might be considerable overlap of *o*- and *e*-images, it would be desirable to eliminate one of the two images. This is accomplished by a device known as the Nicol prism, which makes use of the different refractive properties of the *o*- and *e*-waves.

A cleavage form is made with a pair of end faces that are rhombuses. Suppose, for example, that $AB = AC'$ in Fig. 8–10. The plane $ADA'D'$ will then be a principal section perpendicular to an end face. (See also Fig. 8–16 for a cross section of the cleavage form taken in this principal section.) Two

---

* R. W. Ditchburn, *Light*. Glasgow: Blackie & Son, 1952, Chapter 16. Max Born and Emil Wolf, *Principles of Optics*. New York: Pergamon Press, 1959.

† Since the absorption in calcite is small, the springs can be thought of as ideal.

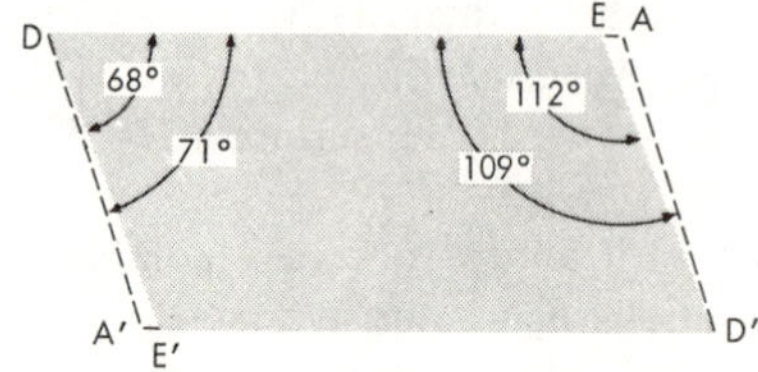

Fig. 8-16. Principal section of a calcite rhomb. $ADA'D'$ illustrates the section before grinding the end faces and $EDE'D'$, after grinding the end faces.

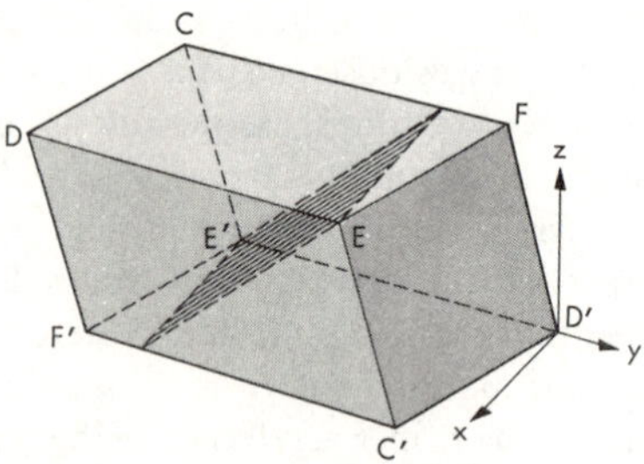

Fig. 8-17. Plane along which a calcite rhomb is cut to make a Nicol prism.

similar wedges are removed by grinding the end faces, thereby altering the angles in the parallelogram from their natural values (71°, 109°) to 68° and 112°. The rhomb is cut in two along a plane which passes through E and E' and is perpendicular to the principal section (see Fig. 8-17). The original length AD is chosen so that this plane will also be perpendicular to an end face. (In this circumstance $AD \doteq 3AC'$.) The two halves of the rhomb are then cemented together in their original position, using *Canada balsam* as the cement.

A ray which is initially parallel to a longitudinal edge of the prism is divided into an *o*-ray and an *e*-ray as it enters the prism. The refraction encountered at the surface DE' (Fig. 8-18) is different for the two rays. The *o*-ray is deviated more than the *e*-ray and is incident on the surface EE' at an angle sufficiently great so that total reflection takes place. The lateral faces of the prism are painted black, and the *o*-ray is absorbed. The *e*-ray, only partially reflected at the layer of Canada balsam, passes into the second half of the prism. The refraction at ED' renders the ray parallel to its original direction.

The Nicol prism thus results in complete elimination of the *o*-wave. The emergent beam is essentially perfectly linearly polarized. If two Nicol prisms are used as polarizer and analyzer and set in crossed position, the transmitted intensity is below the limit of detection. Even as bright an object as the sun

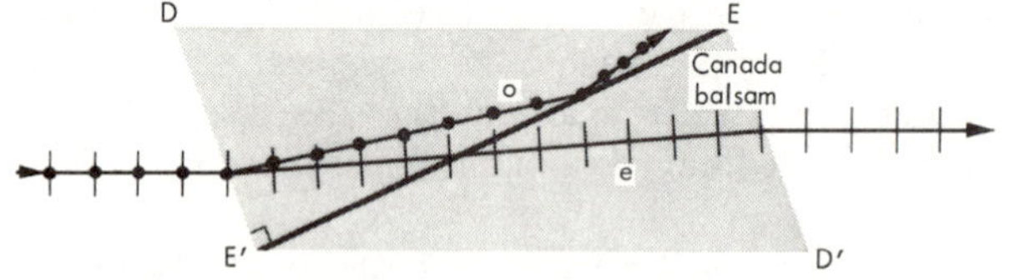

Fig. 8-18. Action of a Nicol prism. ($DED'E'$ is a principal section.)

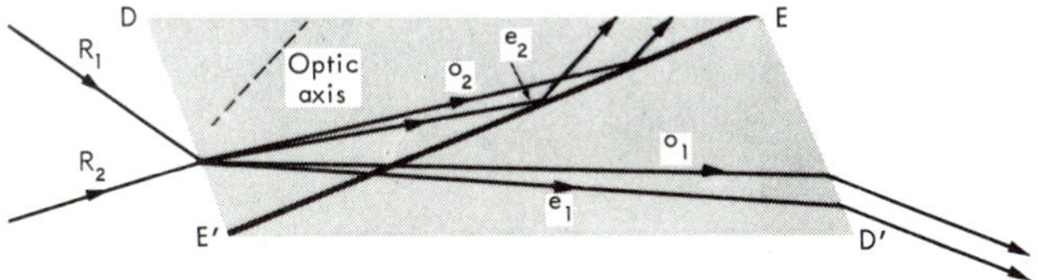

Fig. 8–19. Limited angular aperture of a Nicol prism.

can be completely blocked out if the relative orientation of the prisms is adjusted accurately. In this respect the Nicol prism is superior to other types of polarizers. One disadvantage of the Nicol prism is that it has a limited angular aperture. The *o*-ray derived from an incident ray such as R_1 in Fig. 8–19 is incident on EE' at an angle less than the critical angle and is partially transmitted. The *e*-ray derived from a ray such as R_2 is totally reflected, and no transmission occurs. These effects can be observed if one looks at a light source through a Nicol prism which is rotated about an axis perpendicular to a principal section. On one side the field is cut off sharply. If a white light source is used, the edge of the cut-off is fringed with a deep blue color. This is due to the wavelength dependence of the critical angle. On the other side, the *o*-image (bordered by a red fringe) appears at a particular angle. The minimum angle between rays of type R_1 and R_2 is about 28°. The reason for grinding down the end faces is to improve this angular aperture over that of a natural calcite rhomb of the given dimensions.

Since it is the *e*-ray that is transmitted, the direction of the E-vector in the emergent light is parallel to a principal section. This can be determined by inspection of an end face. The shorter diagonal of the rhombus (i.e. a line joining the two obtuse angles E and D' in Fig. 8–17) is parallel to a principal section.

*8–12 DOUBLE REFRACTION IN CRYSTALS OTHER THAN CALCITE

With respect to their optical properties, transparent crystals can be put into three classes.

1) Optically *isotropic*. A point source of natural light within the crystal emits only spherical waves. There is no division of the wave into two states of polarization and no double refraction takes place. Crystals of this class possess a high degree of symmetry in their crystal structure and belong to the crystallographic group known as the *cubic system*.

2) Optically *uniaxial*. There is a single preferred direction known as the optic axis. A point source of natural light within the crystal gives rise to two wave surfaces. One of these surfaces, known as the *o*-wave, is a sphere.

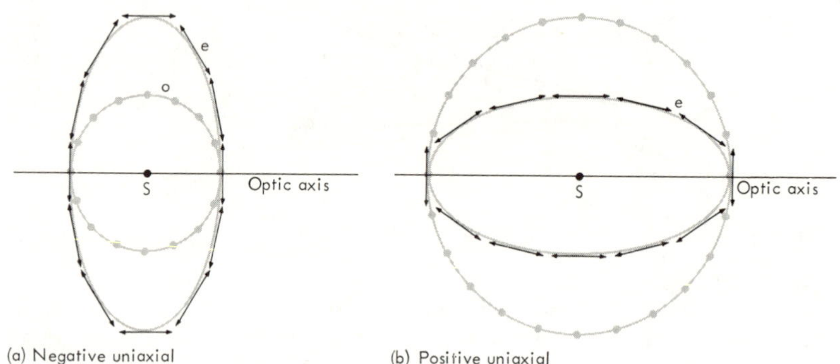

Fig. 8-20. Wave surfaces for positive and negative uniaxial crystals.

The *o*-wave obeys the normal laws of wave propagation in an isotropic medium. The other wave surface (the *e*-wave) is a spheroid, i.e. an ellipsoid of revolution, whose axis is the optic axis. Both waves are linearly polarized. The *E*-vector of the *o*-wave is everywhere perpendicular to the optic axis. The *E*-vector of the *e*-wave lies in a plane which is parallel to the optic axis.

The propagation velocities of the *o*- and *e*-waves are equal along the optic axes. Consequently, the sphere and spheroid are tangent to each other where they intersect the optic axis.* If the spheroid is *prolate*, the sphere is inside the spheroid and the crystal is said to be *negative* uniaxial (Fig. 8–20). Calcite and tourmaline are negative uniaxial. In tourmaline the *o*-wave is strongly absorbed. If the spheroid is *oblate*, the sphere is outside the spheroid and the crystal is said to be *positive* uniaxial. Ice is an example of a positive uniaxial crystal.

The properties of plane waves propagating through the crystal can be deduced from a knowledge of the wave surfaces associated with a point source. In the next chapter it will be shown that plane waves having the same linear polarization as the *e*-wave and plane waves having the linear polarization of the *o*-wave propagate with different velocities. This results in general in a change in the state of polarization of an elliptically polarized wave propagating through the crystal. However, since the propagation velocities are the same along the optic axis, a wave propagating in this direction is unchanged in its state of polarization. Thus the optic axis can be characterized from the observational point of view as the unique direction in a crystal along which a wave can propagate without a change occurring in the state of polarization.

---

* An exception occurs in the case of uniaxial crystals such as quartz which are also *optically active* (Section 10–1). Unless the direction of propagation is very nearly parallel to the optic axis, this distinction is unimportant.

Double refraction along directions other than the optic axis is a property of all uniaxial crystals. The magnitude of the effect depends on the difference between c_1, the propagation velocity of both waves along the optic axis, and c_2, the propagation velocity of the *e*-wave perpendicular to the optic axis. The effect is more pronounced in calcite than in most substances.* Calcite also has the advantage of high transparency and of being readily available in crystals of large size.

The class of uniaxial crystals is composed of members of the trigonal, tetragonal, and hexagonal systems of crystallographic symmetry.

3) Optically *biaxial.* The biaxial crystals are members of the three remaining crystallographic classifications: the monoclinic, triclinic, and orthorhombic systems. In crystals of this class there are two preferred directions, known as the *optic axes*, along which a plane wave may propagate without change in its state of polarization. The wave surface obtained from a point source is a surface of two sheets. The rays emanating from the point source and traveling along the optic axes define four points at which the two sheets coincide. That is, both sheets propagate with the same velocity along the optic axes. The two sheets are associated with opposite states of linear polarization. The uniaxial crystal may be looked on as a special case of the biaxial in which the optic axes coincide. The two sheets of the wave surface in this case become the sphere and the spheroid.

Biaxial crystals exhibit the phenomenon of double refraction. Also, an elliptically polarized plane wave passing through the crystal in a general direction has its state of polarization altered in the same way as in a uniaxial crystal. Thus in the discussion of retardation plates in the next chapter, a biaxial substance such as mica can be used in place of calcite.

It was originally supposed that Huygens' phenomenological theory would be applicable to double refraction in any anisotropic medium. Brewster's discovery that certain crystals have *two* directions along which no double refraction takes place showed that Huygens' theory was inadequate to cover all cases. The first satisfactory treatment of double refraction in biaxial crystals was provided by Fresnel. Fresnel's theory is based on more fundamental assumptions than Huygens', since Fresnel derived the form of the wave surfaces from assumed dynamical properties of the medium, whereas Huygens had introduced the spheroidal wavefronts as a purely *ad hoc* assumption. The form of the wavefronts derived from Fresnel's theory agrees with Huygens' assumption in the case of uniaxial crystals. The present-day description of the wave surfaces, which we have asserted above, is based on a derivation from electromagnetic theory. The results are essentially the same as those obtained from Fresnel's pre-Maxwellian theory.

---

* For a table of the principal indices of refraction, $n_1' = c_0/c_1$ and $n_2' = c_0/c_2$, see *American Institute of Physics Handbook*, 2nd ed. New York: McGraw-Hill, 1963, p. 6–97.

PROBLEMS

8–1. Consider a polarizer P and an analyzer A which are placed perpendicular to an incident beam and set for extinction. Let a third polarizer P' be inserted between P and A with its faces parallel to those of P and A. Show that this will allow light to pass through A. Deduce the relative intensity as a function of the angle between the axes of P' and P.

***8–2.** Suppose that $n + 1$ disks of polarizing material are stacked so that the transmission axis of the ith disk makes an angle θ/n with the transmission axis of the $(i - 1)$st disk. (Thus the first and last disks have their axes inclined at an angle θ with respect to one another.) Assume ideal polarizing material, i.e. no absorption of linearly polarized light whose E-vector is parallel to the axis of transmission. Consider a beam of linearly polarized light of amplitude E_0 incident on the first disk with its E-vector parallel to the axis of transmission. Show that when n is large the beam which emerges from the stack has the same amplitude as the incident beam but has its E-vector inclined at an angle θ to that of the incident beam.

8–3. Show that for normal incidence the Fresnel formulas reduce to values which are in accord with the reflection coefficient deduced in Section 7–3.

8–4. Consider a wave incident on a dielectric surface at an angle of incidence θ. Sketch the graphs of $R_\perp$ and $R_\parallel$, the reflection coefficients for the E-vector. Comment particularly on the phase and amplitude changes which occur on reflection at nearly grazing incidence ($\theta \doteq \pi/2$).

8–5. Consider an internal reflection in a medium of index n. Sketch the graphs of $R_\perp$ and $R_\parallel$ as functions of the angle of incidence θ for $0 \leq \theta \leq \theta_c$. Is the approach to total reflection abrupt or gradual?

8–6. Consider one of the two linearly polarized components of a general plane wave incident obliquely on a dielectric surface. Let R and T be the reflection and transmission coefficients for the E-vector and let r and t be the reflection and transmission coefficients for the magnitude of the intensity vector.

a) Show that $r = R^2$ and $t = nT^2$.

b) Show that the sum of the intensities of the reflected and transmitted waves is *not* the same as the intensity of the incident wave. That is, show that it is not correct for either the parallel or the perpendicular component that $r + t = 1$.

c) Show that (b) does not constitute a violation of the law of conservation of energy. [*Hint:* The energy flux through a given surface depends on its orientation relative to the direction of propagation. Show that for each component the total power incident on a given area on the surface of the dielectric is equal to the sum of the powers of the reflected and transmitted waves which emanate from this same area.]

8–7. Let $R_\perp$, $R_\parallel$, $T_\perp$, and $T_\parallel$ be the Fresnel coefficients for an external reflection at an angle θ from a dielectric of index n. Let $R'_\perp$, $R_\parallel$, $T'_\perp$, and $T'_\parallel$ be the Fresnel coefficients for an internal reflection at an angle θ' where $n \sin\theta' = \sin\theta$. Show that Stokes' relations can be deduced from the Fresnel coefficients for both parallel and perpendicular components.

***8–8.** Consider a plane wave incident at the polarizing angle on a glass plate with parallel sides. Let $T''_\perp$, $T''_\parallel$ be the *net* transmission coefficients of the plate for the perpendicular and parallel components respectively. That is, for each component T'' is the ratio of the amplitude of the E-vector in the wave which passes through the plate to the amplitude of the E-vector in the incident wave.

a) Show that $T''_{\parallel} = 1$. [*Note:* This follows readily if you can show that the wave transmitted by the first surface is incident on the second surface at the angle which is appropriate for Brewster's law to be applicable.]

b) Show that if multiple internal reflections are neglected, $T''_{\perp} = \sin^2 2\theta_P$. [*Note:* A simple argument will show that the reflections at the first and second surfaces are appropriately related to one another for Stokes' relations to be applicable.]

c) If a polarizer is to be constructed from a stack of glass plates, how many plates are required so that a crossed polarizer and analyzer of identical construction will transmit less than 1% of the energy of an incident beam? Assume that $n = \frac{3}{2}$.

***8–9.** In Problem 8–8 multiple internal reflections were neglected in the calculation of the transmission coefficient of the plate for linearly polarized light whose E-vector is perpendicular to the plane of incidence. In principle, the amplitude of the transmitted wave should be calculated by the methods discussed in Chapter 3, that is, either by solving the appropriate boundary value problem for oblique incidence on a dielectric layer, or by summing the contributions from successively transmitted rays, taking account of the phase differences due to additional distance traveled. Since the glass plates used in constructing a polarizer are usually not precisely plane and since the plates are a large number of wavelengths thick, it is not likely that the successively transmitted rays will have the exact phases ascribed to them. It is more reasonable to suppose that these rays are uncorrelated in phase with each other. The net intensity of the transmitted beam should then be obtained by a summation of the intensities associated with the successive rays.

a) Using this assumption, show that the net transmission coefficient for intensity is $t_{\perp} = (1 - R_{\perp}^2)/(1 + R_{\perp}^2)$. Calculate the percentage error involved in the neglect of the contributions from multiple internal reflections.

b) On a similar basis calculate $r_{\perp}$, the net reflection coefficient for intensity, and show that $r_{\perp} + t_{\perp} = 1$.

8–10 Consider a linearly polarized wave incident on the surface of a dielectric with its E-vector inclined at an angle of 45° to the plane of incidence. Apply the Fresnel formulas to show that the reflected light is linearly polarized and deduce the angle which the E-vector makes with the plane of incidence as a function of the angle of incidence.

8–11. Consider an elliptically polarized plane wave incident on the surface of a dielectric with the major axis of the ellipse parallel to the plane of incidence. Describe the polarization of the reflected wave. Show that there is one angle of incidence for which the reflected wave will be right circularly polarized, and one for which it will be left circularly polarized.

***8–12.** Given the following data concerning a calcite cleavage form:

i) The angle between any two edges meeting at a blunt corner is 101°55′.

ii) The optic axis makes equal angles with all three edges meeting at a blunt corner.

a) Find the angle between the optic axis and a face of the cleavage form.

b) Find the angles of the parallelogram intercepted by a principal section.

Suggested procedure: Let the directions of the three edges be specified by unit vectors $\boldsymbol{\sigma}_1, \boldsymbol{\sigma}_2, \boldsymbol{\sigma}_3$. Choose a coordinate system in which the z-axis is parallel to the optic axis (Fig. 8–21). If θ is the angle between the optic axis and the edges, the unit vectors can be written:

$$\begin{aligned}
\boldsymbol{\sigma}_1 &= (\sin\theta)\mathbf{i} + (\cos\theta)\mathbf{k},\\
\boldsymbol{\sigma}_2 &= -\tfrac{1}{2}(\sin\theta)\mathbf{i} + (\sqrt{3}/2)(\sin\theta)\mathbf{j} + (\cos\theta)\mathbf{k},\\
\boldsymbol{\sigma}_3 &= -\tfrac{1}{2}(\sin\theta)\mathbf{i} - (\sqrt{3}/2)(\sin\theta)\mathbf{j} + (\cos\theta)\mathbf{k}.
\end{aligned}$$

The given information is $\boldsymbol{\sigma}_1 \cdot \boldsymbol{\sigma}_2 = \boldsymbol{\sigma}_2 \cdot \boldsymbol{\sigma}_3 = \boldsymbol{\sigma}_3 \cdot \boldsymbol{\sigma}_1 = \cos\psi$, where $\psi = 101°55'$. The normal to the face whose edges are $\boldsymbol{\sigma}_2, \boldsymbol{\sigma}_3$ is specified by the unit vector $\mathbf{n}_{23} = (\boldsymbol{\sigma}_2 \times \boldsymbol{\sigma}_3)/\sin\psi$.

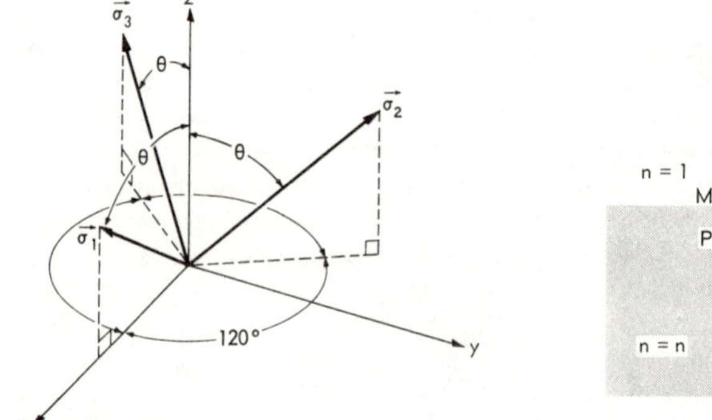

Figure 8–21

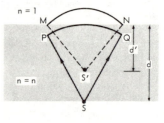

Figure 8–22

***8–13.** *Microscope method of determining the values of c_1 and c_2 for calcite.* A standard method of determining the index of refraction of a plate with plane parallel sides is to look through the top face with a microscope and observe the image location of a point source (e.g. a small particle of lycopodium powder) on the bottom face (Fig. 8–22). It is a well-known result of geometrical optics that, to first approximation, the spherical wavefront PQ is refracted as a spherical wavefront, MN, whose center lies a distance $d' = d/n$ beneath the surface. If this experiment is performed with a plate of calcite whose optic axis is perpendicular to the faces, it is found that there are two settings of the microscope at which a sharp image is obtained. Show that the apparent depths of these two images are given by $d_o' = d/n_1$ and $d_e' = n_1 d/n_2^2$. (Since the e-wave propagates with velocity c_1 along the optic axis, assume that the radius of curvature of the e-wavefront is reduced by a factor $1/n_1$ on refraction at the surface.)

9 The production and detection of elliptically polarized light

9-1 INTRODUCTION

Plane waves having an arbitrary type of elliptical polarization can be produced and analyzed by means of plates made of a doubly refracting material. It is possible to deduce the properties of such plates from Huygens' assumption concerning the form of the wave surfaces emanating from a point source within the medium. This calculation is carried out in the final section of this chapter. The earlier sections approach the problem from a simpler point of view, deducing the properties of the plates from a few elementary experimental observations. This order of presentation makes it possible to omit the material of the final section and yet have a complete understanding of the operational procedures for manipulating polarized light.

9-2 RETARDATION PLATES

For simplicity we shall confine the initial discussion to nonoptically active, uniaxial crystals, using calcite as the prototype. Consider a plate with plane parallel faces cut from a sample of calcite in such a way that the optic axis is parallel to the faces. Let a linearly polarized plane wave be normally incident on the plate. In particular, suppose that the plate is inserted between crossed polarizers. For a general orientation of the optic axis of the plate we find that light is now passed by the analyzer. However, extinction is obtained when the optic axis is either parallel or perpendicular to the E-vector of the light passed by the polarizer.

Given a plate whose optic axis is not marked, let it be placed between crossed polarizers and rotated until extinction occurs. In this position the optic axis is either parallel to or perpendicular to the transmission axis of the polarizer. Two mutually perpendicular axes, one of which is the optic axis, can be marked on the plate. For many purposes it is not necessary to know which of these two directions is the optic axis (methods for determining this will be given later). Collectively these two directions are referred to as the *axes* of the plate.

By rotating the analyzer, we can check that the light which is passed by the plate in either of these special settings has the same linear polarization as the beam which enters the plate. The observations described so far can be summarized in the following statement:

1) Linearly polarized light whose E-vector is parallel to either of the axes of a calcite plate is passed without change in its state of polarization.

Next, consider examination by means of an analyzer, of the light which emerges from the calcite plate* when one of its axes is not parallel to the E-vector of the incident linearly polarized beam. As the analyzer is rotated through 360°, the transmitted light passes through two relative maxima 180° apart and two relative minima (not zero) at 90° from the maxima. This is the behavior to be expected for either elliptically polarized light or a mixture of natural and polarized light. The possibility that the plate has converted some of the beam into natural light is ruled out by the following two experiments:

a) Except for causing a small loss of intensity due to reflection, a calcite plate has no apparent effect on a beam of natural light. Thus, in particular, a beam of natural light which has been passed through a calcite plate shows no intensity variations when examined with an analyzer.

b) If in the above experiment a second calcite plate, identical in construction with the first, is inserted next to the first plate with the optic axes of the two plates at right-angles to each other, the light which emerges from the combination is found to have the same linear polarization as the incident light. Since the second calcite plate has no ability to convert natural light into polarized light, it is evident that the beam transmitted by the first plate did not contain natural light.

In summary, we thus conclude:

2) Linearly polarized light whose E-vector is not parallel to either of the axes of a calcite plate emerges as elliptically polarized light.

We now seek a simple interpretation of the action of a calcite plate which will explain the listed observations. The incident linearly polarized light is characterized by the fact that the components of the E-vector onto any pair of mutually perpendicular axes are sinusoidal functions which are in phase. Since the emergent light is in general elliptically polarized, we infer that the components of $\mathbf{E}$ have undergone a change in relative phase. The phase of a wave which has passed through a thickness d of a material medium depends on the *optical thickness, nd,* where n is the index of refraction. This results from the phase angle in the form

$$kd = \frac{\omega d}{c} = \left(\frac{\omega}{c_0}\right)\left(\frac{c_0 d}{c}\right) = k_0(nd),$$

* The plate is assumed to have arbitrary thickness. It is shown below that special results are obtained for certain thicknesses.

where $k_0 = \omega/c_0$ is the wave number in free space. It was remarked in the last chapter that the propagation characteristics of linearly polarized light seem to depend on the orientation of the E-vector relative to the optic axis. It is natural, therefore, to hypothesize that the light which is incident on the calcite plate must be resolved into components whose E-vectors are respectively parallel to and perpendicular to the optic axis, and to assign different propagation velocities to these two waves. The first observation above is readily explained by this hypothesis, for whenever $\mathbf{E}$ is parallel to one of the axes of the plate there is only one component propagating through the plate and no question of relative phase change arises. The emergent light is linearly polarized in the same way as the incident.

For the general case, let the y-axis be chosen parallel to the optic axis and the z-axis perpendicular to it. Let the indices of refraction be n_y and n_z for linearly polarized light whose E-vector is respectively parallel to and perpendicular to the optic axis. If the E-vector of the incident light makes an angle θ with the y-axis, the incident linearly polarized light is expressed in terms of its components:

$$E_y = E_0 \cos\theta \cos(\omega t - k_0 x - \phi_1),$$
$$E_z = E_0 \sin\theta \cos(\omega t - k_0 x - \phi_1).$$

For convenience, choose $\phi_1 = 0$ and let the origin of the x-axis coincide with the position of the front surface of the calcite plate. The functions describing the wave at $x = d$ therefore have the form

$$\begin{aligned} E_y &= E_0 \cos\theta \cos(\omega t - k_0 n_y d), \\ E_z &= E_0 \sin\theta \cos(\omega t - k_0 n_z d). \end{aligned} \tag{9–1}$$

There is thus a phase difference between the two components of the emergent wave, given by

$$\Delta\phi = k_0 d(n_z - n_y). \tag{9–2}$$

A plate of doubly refracting material which introduces a relative phase shift of this sort without changing the amplitudes of the two components is referred to as a *retardation plate*.

One consequence of Eq. (9–2) which can be tested is the implied proportionality of the phase difference to the thickness of a plate. With proper selection of d, for example, it should be possible to make $\Delta\phi$ an integral multiple of 2π. This condition can be recognized easily, for with this phase difference the emergent light will again be linearly polarized, regardless of the value of θ. If such a plate is inserted between a crossed polarizer and analyzer, it is expected that no light will be transmitted by the analyzer, no matter how the plate is turned. This prediction is readily confirmed. We can find a minimum thickness d_1 for which the plate has this property. A plate of such a thickness is called a *full-wave plate*, since it retards one of the components a complete cycle in phase relative to the other. The name does not imply that the plate is only one wavelength thick, but merely that it is one wavelength *thicker* for one component than for the other.

We expect that a plate of thickness $d_1/4$ will act as a *quarter-wave plate*, that is, that it will produce a relative phase shift $\Delta\phi = \pi/2$. This is confirmed by experiments such as the following:

1) Set the axes of the plate so that the E-vector of incident linearly polarized light is equally inclined to both axes. Then, in Eq. (9–2), $\theta = \pi/4$, the amplitudes of E_y and E_z are equal, and the phase difference is $\pi/2$. The emergent wave therefore satisfies all the criteria for circularly polarized light. When it is viewed through an analyzer, no variations in intensity are observed as the analyzer is rotated.

2) Consider the elliptically polarized wave which emerges from a calcite plate of arbitrary thickness. We can find the major and minor axes of the ellipse by using an analyzer, i.e. the major axis is parallel to the extinction axis of the analyzer when it is set for minimum transmission. If a quarter-wave plate is set with its axes parallel to the axes of the ellipse, the y- and z-components of the incident light are characterized by a phase difference of $\pi/2$ (cf. Table 7–2). The quarter-wave plate introduces a phase change of $\pi/2$, so that the relative phase in the emergent wave is either 0 or π. This implies that the emergent wave is linearly polarized, which can readily be checked by using an analyzer.

The hypothesis that E_y and E_z are characterized by different wave propagation velocities in calcite is amply confirmed by its ability to predict the outcome of experiments of the type described. It is also possible to make direct measurements of the optical thickness for the two types of linearly polarized light. For example, the Michelson interferometer (Section 11–13) can be employed to determine the quantity nd for a calcite plate. Different values are obtained when linearly polarized light is used, depending on which of the two axes of the plate the E-vector is parallel to. (If the E-vector is not parallel to one of the axes, or if linearly polarized light is not used, a confused pattern is obtained in the interferometer.) If d is known, the two indices of refraction can therefore be obtained. The interferometer measurement enables us to make a distinction between the two axes of a retardation plate, since it shows which of the two is associated with the faster propagation velocity. This axis is called the *fast axis*. If one is concerned with the *sense* of rotation of polarized light (i.e. left-handed versus right-handed), it is necessary to have a retardation plate whose fast axis is labeled; otherwise, the axes can be treated indiscriminately.

A plate which is cut from calcite so that the optic axis is not parallel to the faces also acts as a retardation plate. Two mutually perpendicular axes in the face of the plate can be found such that linearly polarized light whose *E*-vector is parallel to either of these directions is transmitted without change in polarization. It turns out that one of these axes, which we shall call the *y*-axis, is determined by the *projection* of the optic axis onto the surface of the plate. This axis is always the fast axis. The value of n_y depends on the angle between the optic axis and the faces of the plate, whereas the value of n_z does not. The minimum value of n_y occurs in the case described above, where the optic axis is parallel to the faces. This minimum value, designated by n_y', is referred to as a *principal index of refraction*. For sodium light (5893 Å), the values are $n_y' = 1.4864$ and $n_z = 1.6584$. The maximum value of n_y is the same as n_z and occurs when the optic axis is perpendicular to a face. In this condition no relative phase change is introduced and the plate does not act like a retardation plate. A plate whose optic axis is perpendicular to the faces therefore transmits any incident light without change in polarization. This property is a distinguishing characteristic of the optic axis.

Other birefringent materials in addition to calcite can be used for the construction of retardation plates. The operational procedures described above will in every case determine a pair of mutually perpendicular axes in the face of the plate such that a linearly polarized plane wave normally incident on the plate with its *E*-vector parallel to either of these directions is transmitted as linearly polarized light whose *E*-vector is parallel to that of the incident beam. The indices of refraction for these two states of polarization can be determined, and the fast and slow axes labeled. This information is sufficient for dealing with the plate as a retardation plate, and we need not inquire further into the relations between these quantities and the general crystallographic structure. In uniaxial crystals the relations are similar to those for calcite: one of the axes is the projection of the optic axis on the plane of a face. For positive uniaxial crystals, this is the slow axis; for negative uniaxial crystals, this is the fast axis. Despite the fact that there are two optic axes in biaxial crystals, the propagation of normally incident plane waves is still simply characterized by decomposition into a pair of linearly polarized plane waves whose *E*-vectors are at right angles to each other. A more detailed analysis is required to relate the axes of a biaxial plate to the orientation of its crystallographic axes.

As will be seen from Problem 9–1, a quarter-wave plate of calcite would be unreasonably thin. Quartz and mica are the most commonly used substances for retardation plates. Mica is especially convenient, since we can obtain plates of desired thicknesses by splitting off layers. Calcite was taken as an example above to simplify the discussion, since it is neither optically active (like quartz) nor biaxial (like mica). The indices of refraction for mica depend on the type of mica and vary somewhat from specimen to specimen. Typical values for a sheet of *muscovite* whose faces are cleavage planes are

$n_y = 1.5941$, $n_z = 1.5997$ for sodium light. Unless the optic axis of a quartz plate is nearly perpendicular to the faces, the behavior is identical with that of a positive uniaxial crystal. For sodium light the index of refraction associated with the fast axis is $n_y = 1.5443$ and does not depend on the orientation of the optic axis. The index of refraction associated with the slow axis is dependent on the angle between the optic axis and the faces of the plate. The maximum value $n_z' = 1.5534$ is achieved when the optic axis is parallel to the faces. If the optic axis is exactly normal to the faces, linearly polarized incident light is transmitted as linearly polarized light, but the direction of the E-vector is rotated (Section 10–1).

From Eq. (9–2) the phase retardation of a plate of given thickness is seen to depend on frequency, both through the factor $k_0 = \omega/c_0$ and through the frequency dependence of the difference in indices of refraction $\Delta n = n_z - n_y$. The variation in Δn over the range of frequencies of visible light is only slight, so that $\Delta\phi$ is approximately proportional to ω. Consequently, a plate which is a full-wave plate for one frequency is not a full-wave plate for other frequencies. Suppose, for example, that a full-wave plate for sodium light is placed between a crossed polarizer and analyzer with white light used as the source. All frequencies present are converted into linearly polarized light by the polarizer. Light in the yellow region of the spectrum is transmitted by the retardation plate as *nearly* linearly polarized light (i.e. elliptically polarized with a large ratio of the major to minor axes). This light is practically extinguished by the analyzer. The light of other frequencies is transmitted by the retardation plate as a general type of elliptically polarized light, and a substantial portion of this passes through the analyzer. The light seen through the analyzer is consequently of a hue complementary to yellow: it appears as a delicate blue. If the plate is tipped slightly so that its faces are not perpendicular to the x-axis, to first approximation this can be considered as an increase in the effective thickness of the plate. This makes the plate a full-wave plate for longer wavelengths, and the color changes accordingly. A variety of spectacular effects are obtained if the light incident on a crystal plate is not a plane wave but is convergent or divergent, so that different directions of propagation and different optical path lengths are involved. The patterns obtained are useful in the study of crystal structure.*

Let us return again to the case of normally incident plane waves of white light. If a retardation plate is mounted between a polarizer and an analyzer and the analyzer is rotated, the resultant colors go through an interesting sequence of changes. If a particular color is observed for an arbitrary setting of the analyzer and the analyzer is then rotated through 90°, the color obtained is the complement of the original color. Whatever proportion of each

---

* See references cited on p. 157.

color is absorbed by the analyzer in the one position is exactly the proportion which is transmitted when the analyzer is in the other position. If the beams obtained at the two settings could be added, the distribution of frequencies would be the same as that in the original beam of white light. This implies that the colors of the two beams are exact complements of each other. By using chips of mica of different thickness, a "picture" can be constructed which is colorless in ordinary light but is brilliantly colored when viewed between polarizers.

The complementarity of colors obtained when the analyzer is rotated through 90° can be demonstrated by using a calcite rhomb as an analyzer. Place a circular aperture on the bottom face of a calcite rhomb. Illuminate this with elliptically polarized white light, i.e. place a piece of mica and a sheet of polarizing material next to the aperture and illuminate with white light. When viewed from above, the *o*- and *e*-images of the aperture have complementary colors. If the aperture is large enough so that the images partly overlap, the area of overlap is white (see Fig. 9–1). The most vivid colors are obtained when the mica is an integral wave plate for two colors in the visible region.

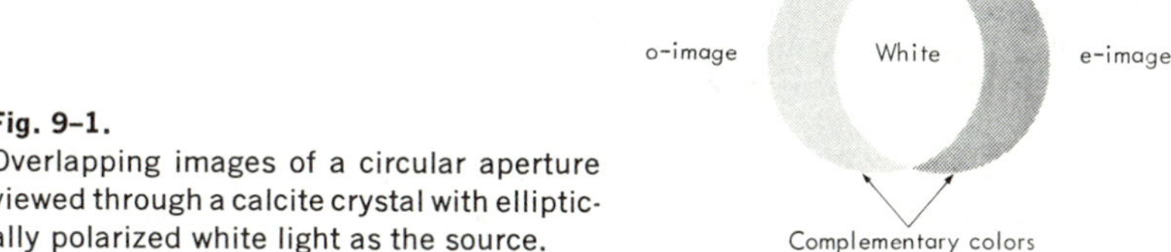

Fig. 9–1. Overlapping images of a circular aperture viewed through a calcite crystal with elliptically polarized white light as the source.

9–3 THE MANIPULATION OF POLARIZED LIGHT

It should be clear from the preceding section that any desired type of polarized light can be produced with a polarizer and a quarter-wave plate. We shall illustrate this with further examples. It will be convenient in this discussion to recall that all the ellipses described by the functions $E_y = A\cos(\omega t - \phi_1)$, $E_z = B\cos(\omega t - \phi_2)$ are inscribed within the rectangle $|E_y| \leq A$, $|E_z| \leq B$. A retardation plate changes the relative phase of E_y and E_z, but does not affect the amplitudes. Therefore what emerges from any retardation plate is inscribed in the same rectangle (with sides parallel to the axes of the plate) as that which enters. The first step in analyzing the passage of a wave through a retardation plate is to determine the rectangle by considering the relation of the incident light to the axes of the plate. Thus, if the incident light is linearly polarized with amplitude E_0 and with its E-vector inclined at an angle θ to the y-axis of the plate, this E-vector is the diagonal of the rectangle (see Fig. 9–2). A quarter-wave plate introduces a phase shift of $\pi/2$. Table 7–2 shows that the emergent wave is described by an ellipse whose axes are parallel to the y- and z-axes. The ratio of the axes of the ellipse is $B/A = \tan\theta$. We can therefore obtain an ellipse of desired eccentricity by controlling the angle θ. This angle is simply the angle between the transmission axis of the polarizer and one of the axes of the plate. If circularly polarized light is desired, we must choose $\theta = \pi/4$.

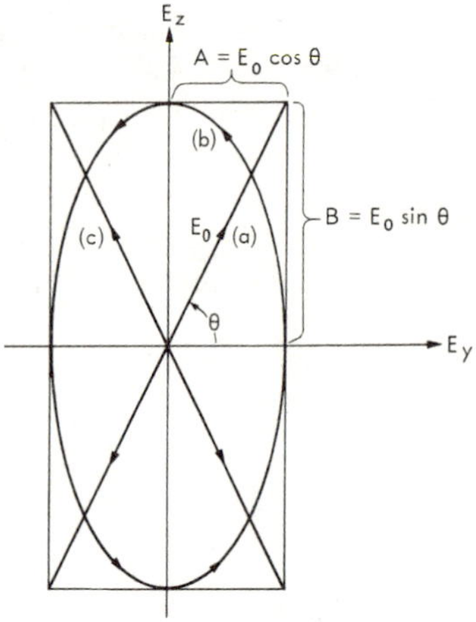

Fig. 9–2.
Rectangle associated with linearly polarized light whose E-vector makes an angle θ with the y-axis of a retardation plate. (a) Polarization of the incident wave. (b) Polarization of the wave emerging from a quarter-wave plate. (c) Polarization of the wave emerging from a half-wave plate.

A half-wave plate introduces a phase change of π, so that the emergent light is linearly polarized. Since the sign of the ratio of E_y to E_z has been changed, this linearly polarized light is represented by the other diagonal of the rectangle. Therefore, if a half-wave plate is inserted between a polarizer and an analyzer, extinction occurs for one setting of the analyzer.* Simple trigonometry shows that in this condition the angle between the extinction axis of the analyzer and the transmission axis of the polarizer is 2θ. For example, if $\theta = \pi/4$, the plane of polarization of the emergent light is at right-angles to that of the incident. That is, the rectangle is now a square and the diagonals are perpendicular to each other.

Two quarter-wave plates whose fast axes are parallel to each other are equivalent to a half-wave plate. If the fast axis of one is *perpendicular* to the fast axis of the other, whatever phase change is introduced by the first plate is canceled by the second. If both plates are made of the same material, this cancellation is exact for all wavelengths. The net effect is as if neither plate were present.

---

* Unless otherwise specified, the light referred to in this context is monochromatic light. The effects produced with white light are similar to those discussed for the full-wave plate in the preceding section.

9-4 DETERMINING THE POLARIZATION OF A GIVEN BEAM

The characteristics of a given monochromatic beam of unknown polarization can be determined with a quarter-wave plate and an analyzer. First, let us suppose that the beam is known to contain no natural light. Examination of the beam with an analyzer will yield one of three possibilities:

1) A position of the analyzer can be found at which extinction occurs.

2) No variations in intensity occur as the analyzer is rotated.

3) Variations in intensity occur as the analyzer is rotated, but no extinctions occur. (Maxima and minima are separated by $\pi/2$.) For (1) to hold, the light must be linearly polarized; for (2), it must be circularly polarized; for (3), it must be in some state of general elliptical polarization. In case (3) the axes of the ellipse are parallel to the axes of the analyzer when it is set for minimum intensity. If the analyzer is removed and a quarter-wave plate is placed in the beam with its axes parallel to the axes of the ellipse, linearly polarized light will be produced. This corresponds to a transition in Fig. 9–2 from (b) to either (a) or (c). We can determine the angle θ by examining the resultant linearly polarized light with the analyzer. From this we learn the ratio of the axes of the ellipse, $B/A = \tan\theta$.

If there is a possibility that natural light is present in the unknown beam, several ambiguities must be resolved. The beam is first examined with an analyzer as before:

If case (1) arises, the only possibility is that the beam is 100% linearly polarized.

If case (2) arises, the most general possibility is a mixture of circularly polarized and natural light. If the analyzer is removed and a quarter-wave plate is inserted, the natural light will be transmitted as natural light, but the circularly polarized light will be converted to linearly polarized light. The resultant beam is then examined with the analyzer. If case (1) *now* arises, the original beam was 100% circularly polarized; if case (2), the original beam was 100% natural light; if case (3) occurs, the original beam was a mixture for which the proportion of circularly polarized to natural light can be estimated.

If case (3) arises, the most general possibility is a mixture of natural light and elliptically polarized light, in which we include linearly polarized light as a special case. The elliptically polarized component can be converted to linearly polarized light by the same technique used before. If this process yields 100% linearly polarized light, we know that no natural light was present. If not, we can estimate the proportion of natural light. The ratio of the axes of the ellipse is given as above by the tangent of the angle between an axis of the quarter-wave plate and an axis of the analyzer when the latter is rotated to a minimum position. In particular, if this angle is 0 or $\pi/2$, the original beam consisted of natural plus *linearly* polarized light.

*9-5 THE BABINET COMPENSATOR

If retardation plates were to be used to make polarization studies of a number of different spectral lines, it would be necessary to have an accurate quarter-wave plate for each wavelength. This can be avoided by using a device known as the *Babinet compensator*. Two thin wedges of quartz are placed together to form a plate with plane parallel faces (Fig. 9–3). Each wedge is cut so that the optic axis is parallel to the faces and the optic axes of the two wedges are mutually perpendicular. The device is used with plane waves of light normally incident on one face. The angle of the wedges is sufficiently small so that we can neglect the refraction of a wave as it crosses the interface between the two wedges.

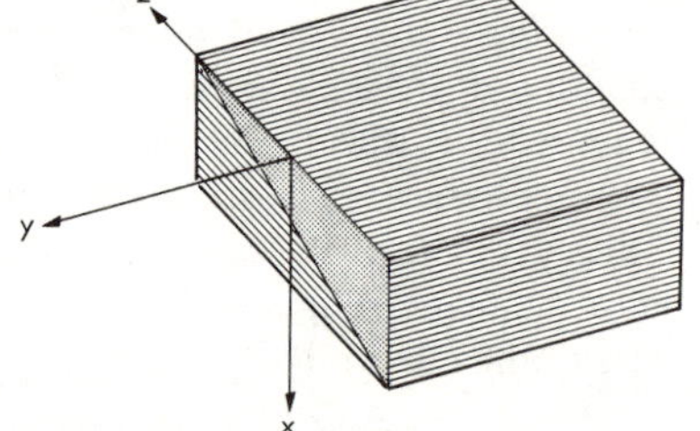

Fig. 9–3.
Babinet compensator.

Let the direction of propagation be the x-axis; choose the y-axis parallel to the optic axis of the upper wedge and the z-axis parallel to the optic axis of the lower wedge. The y- and z-axes are the preferred axes for both wedges. In the upper wedge one of the components gains in phase relative to the other, but in the lower wedge the situation is reversed. Along a given ray the net change in relative phase, $\Delta\phi$, is proportional to the difference in the thicknesses of the two wedges at the location of the ray. This quantity varies as a linear function of z. At the center of the plate where the two wedges have equal thickness, $\Delta\phi$ is zero. As far as polarization effects are concerned, it is just as though the plate had a hole at this location. For any given wavelength there will be appropriate positions on either side of the center ($z = 0$) where the plate acts like a quarter-wave plate. We shall show presently how to locate these positions experimentally.

Consider a linearly polarized plane wave incident on the compensator with its E-vector at an angle θ to the y-axis. At the center of the plate the emergent wave is linearly polarized, with its E-vector in the same direction as that of the incident wave. At other positions the polarization of the emergent wave is described by ellipses all of which are inscribed in the same rectangle (Fig. 9–4). The intensity of the emergent wave is uniform. (In Section 7–8 it was shown that the intensity of a polarized wave is independent of the phase relation between its components.) However, if the emergent wave is examined with an analyzer, we find that there are variations in in-

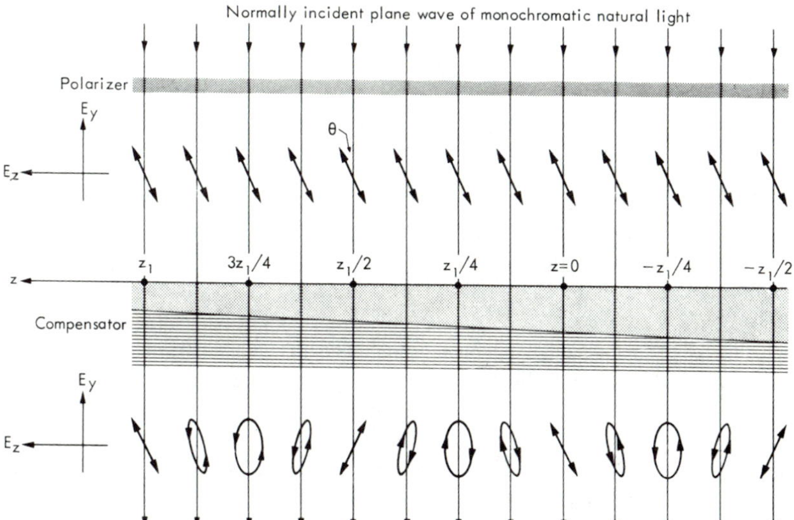

Fig. 9–4. Variation in the states of polarization of a wave emerging from a Babinet compensator. The apparatus is shown as viewed from the positive *y*-direction. The states of polarization are shown as viewed with the beam approaching the reader. The transmission axis of the polarizer makes an angle θ with the optic axis in the upper wedge of the compensator.

tensity across the field of view. Suppose, for example, that the analyzer is crossed with respect to the polarizer. Extinction occurs at the positions $0, \pm z_1, \pm 2z_1, \ldots$ Some light is passed by the analyzer at positions intermediate to these positions, so that the field of view appears as a series of equally-spaced fringes. The dark lines mark the positions where the plate acts like an integral-wave plate and the beam is unchanged in its state of polarization.

For some applications it is not essential to know which of these dark lines corresponds to *zero retardation* (i.e. $\Delta\phi = 0$). However, if this information is desired, we can obtain it by substituting white light for the monochromatic source. The pattern obtained when any combination of wavelengths is present results from the independent superposition of the patterns associated with the individual wavelengths. At the "center" of the plate, where the two wedges are of equal thickness, the condition $\Delta\phi = 0$ is met exactly for all wavelengths, and a black line appears. The position of $\Delta\phi = 2\pi$ occurs where the difference in thickness of the two wedges is d_1, which is the appropriate thickness for a full-wave plate. From Eq. (9–2), we have

$$d_1 = \frac{2\pi}{k_0\,\Delta n} = \frac{\lambda_0}{\Delta n},$$

where λ_0 is the wavelength in air. Simple trigonometry shows that the difference in thickness of the two wedges is $d = 2z \tan \alpha$, where α is the angle of the wedge. Hence we may write

$$z_1 = \frac{\lambda_0}{2\,\Delta n(\tan \alpha)}.$$

Noting that Δn varies only slightly with wavelength in the visible region, we see that the fringe patterns associated with longer wavelengths have wider spacing. It is therefore only at the center that all colors are extinguished simultaneously. Thus the position of zero retardation can be readily identified.

Let us return now to the problem of locating the position on the compensator plate at which the plate acts like an exact quarter-wave plate for a given wavelength. Since $\Delta\phi$ is a linear function of z, we see that this position lies one-quarter of the way between the position of zero retardation and the next dark line. We can find this location accurately by calibrating a screw which moves the plate past a fixed hairline. The first step is to count the number of turns of the screw required to traverse several dark lines. An appropriate fraction of this number of turns will place the hairline at a quarter-wave position. This portion of the plate can then be used as a retardation plate for analysis of polarized light of the given wavelength. In this application the position of zero retardation need not be determined. Any portion of the plate one-quarter of the way between two dark lines acts equivalently to a quarter-wave plate.

The compensator is also useful in determining the retardation of an "unknown" wave plate. Suppose that the plate is an "x-wave plate," where the value of x is to be determined for a particular wavelength. We first locate the axes of the plate by seeking the orientation which maintains extinction when the plate is placed between crossed polarizers. It is not necessary to distinguish between the fast and slow axes. The compensator is then placed between crossed polarizers and the hairline set on the dark line at the position of zero retardation. (White light can be used to identify the position of zero retardation. An accurate setting should be made using the monochromatic light alone.) When the unknown plate is inserted next to the compensator with its axes parallel to those of the compensator, the position of zero retardation is shifted. (This would be to the left in Fig. 9–4 if the optic axis of the plate is parallel to the optic axis of the upper wedge.) The position under the hairline is now an x-wave plate, and the hairline occupies a corresponding location in the fringe pattern. White light must be used to determine which dark line is the new position of zero retardation. This will show immediately what the integral part of x is. The fractional part of x can then be determined by counting the number of turns required to move the hairline to the nearest dark line in the monochromatic fringe pattern.

*9-6 HUYGENS' CONSTRUCTION FOR PLANE WAVES IN CALCITE

It is assumed that the reader is familiar from elémentary texts with *Huygens' principle* as a method of deriving Snell's law for the refraction of a plane wave at an interface between two isotropic media. The same method can be employed to derive the refraction of both *e*- and *o*-waves for a plane wave incident from air on the surface of a calcite plate. Discussion will be confined to the case of normal incidence only. The results confirm the hypotheses made in Section 9–2 concerning the action of retardation plates, and also enable us to make a quantitative study of the separation of images in the phenomenon of double refraction.

Huygens' principle provides a method of constructing the wavefront at a time $t + \Delta t$ from the given form of the wavefront at time t. Each point on the given wavefront is considered as the source of secondary wavelets. The surfaces which show how these wavelets advance during the time interval Δt are drawn. The envelope of this family of wavelets gives the location of the actual wavefront at $t + \Delta t$. We assume for calcite that the wavelets are of the type described in Section 8–10. Huygens' construction then enables us to deduce the nature of propagation of plane waves through calcite from the more basic assumptions concerning the nature of the wavefronts emanating from a point source.

First, consider a plane wave incident on a calcite plate whose optic axis is perpendicular to the face (see Fig. 9–5a). For the wavefront at time t we take a plane wave at the entering surface. A cross section of the *o*- and *e*-wavelets is drawn for three representative points. The envelope surface is the same in this case for both *o*- and *e*-waves and is a plane parallel to the original plane. This shows that both waves propagate in the calcite as plane waves and have a common propagation velocity equal to c_1. The incident wave has been drawn as a plane wave of *limited width* to show that there is no sideways deflection of the wavefront as it enters the calcite.

In Fig. 9–5(b) a similar construction is shown for a plane wave incident on a calcite plate whose optic axis is parallel to the surface. Figure 9–5(c) is a cross-sectional view of the same situation taken in a plane at right-angles to that used in Fig. 9–5(b). The envelope of the *o*-wavelets represents a plane wave propagating with velocity c_1; the envelope of the *e*-wavelets represents a plane wave propagating with velocity c_2. There is no sideways deflection of either wave. This deduction confirms the assumptions made in Section 9–2 concerning the properties of a retardation plate whose optic axis is parallel to the faces. An arbitrary incident plane wave is resolved into two linearly polarized components. The component whose E-vector is parallel to the optic axis is the *e*-wave. The velocity of this component is c_2, the maximum propagation velocity of the spheroidal wavefront. The interferometer measurement which yields the principal index of refraction n_y' can therefore be interpreted to give the numerical value of c_2. The component whose E-vector

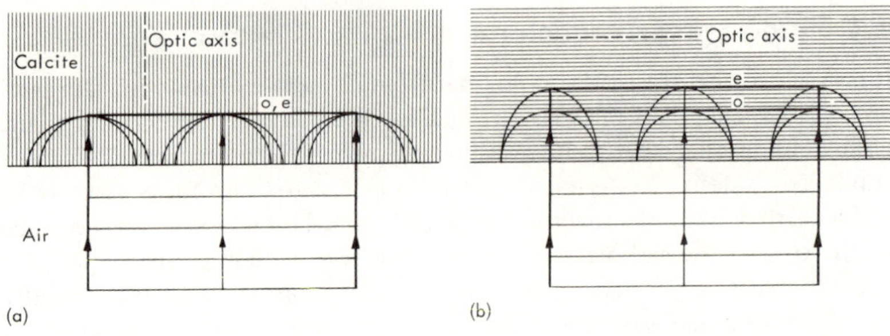

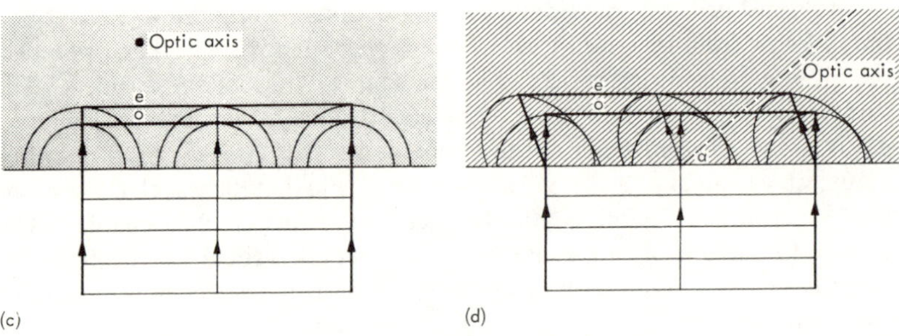

Fig. 9–5. Propagation of plane waves through a calcite plate.

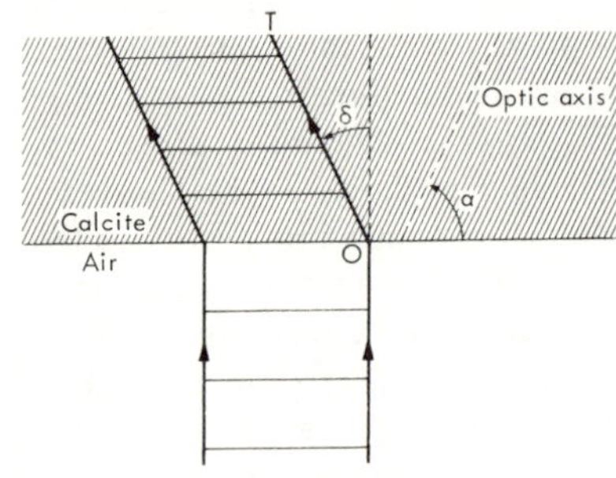

Fig. 9–6. Wavefronts in refracted e-wave produced by a limited beam of plane waves incident normally on the surface of calcite.

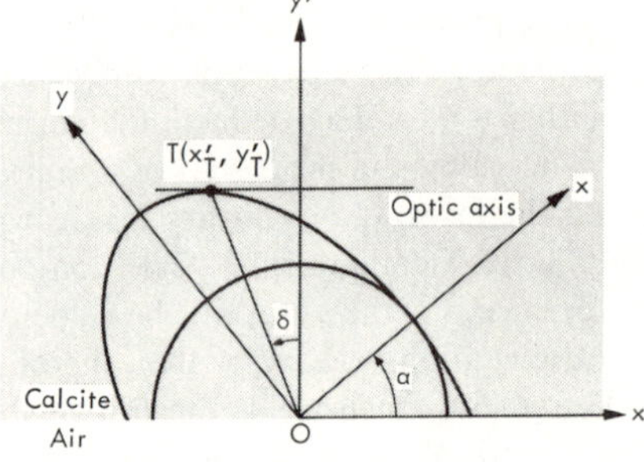

Fig. 9–7. Cross section of a spheroidal wavelet in a principal section.

is perpendicular to the optic axis is the o-wave, and we associate the index of refraction n_z with the propagation velocity c_1 of the spherical wavefronts of the o-wave.

Figure 9–5(d) shows Huygens' construction for a plane wave incident on a calcite plate whose optic axis makes an angle α with the faces. The cross section is taken in a principal section. The o-wave is an undeflected plane wave with propagation velocity c_1. The e-wave is a plane wave which remains parallel to the surface. The boundaries of this plane wave undergo a continual deflection toward the left as the wave propagates through the calcite (see Fig. 9–6). The direction of energy propagation within the calcite is OT.* By definition, this is referred to as the direction of the e-ray. In this situation the terms *ray* and *wavefront normal* are not synonymous. It is clear that the refraction of the e-ray does not obey Snell's law, i.e. a law of the form $\sin\theta/\sin\theta' =$ constant, for in this case $\theta = 0$ and $\theta' = \delta \neq 0$.

The angle of refraction of the e-ray can be calculated from Fig. 9–7, which shows an enlarged view of one of the spheroidal wavelets. By analytic geometry we find the coordinates x'_T and y'_T of the point T where the ellipse has a horizontal tangent. From Eq. (8–8), the equation of the ellipse is

$$\left(\frac{x}{c_1}\right)^2 + \left(\frac{y}{c_2}\right)^2 = (\Delta t)^2,$$

or, by rotation of axes:

$$\frac{(x'\cos\alpha + y'\sin\alpha)^2}{c_1^2} + \frac{(-x'\sin\alpha + y'\cos\alpha)^2}{c_2^2} = (\Delta t)^2. \qquad (9\text{–}3)$$

Differentiating implicitly with respect to x' and setting $dy'/dx' = 0$, we obtain one relation connecting x'_T and y'_T:

$$x'_T\left[\left(\frac{\cos\alpha}{c_1}\right)^2 + \left(\frac{\sin\alpha}{c_2}\right)^2\right] + y'_T(\sin\alpha)(\cos\alpha)\left[\frac{1}{c_1^2} - \frac{1}{c_2^2}\right] = 0. \qquad (9\text{–}4)$$

From Fig. 9–7 we see that $\tan\delta = -x'_T/y'_T$. Therefore

$$\tan\delta = \frac{\dfrac{1}{c_1^2} - \dfrac{1}{c_2^2}}{\dfrac{\operatorname{ctn}\alpha}{c_1^2} + \dfrac{\tan\alpha}{c_2^2}} = \frac{n_z^2 - (n'_y)^2}{n_z^2\operatorname{ctn}\alpha + (n'_y)^2\tan\alpha}, \qquad (9\text{–}5)$$

where $n'_y = c_0/c_2$, and $n_z = c_0/c_1$. Applying this to the calcite cleavage form for sodium light, we have $\alpha = 45°24'$ (see Problem 8–12), $n'_y =$

---

* This follows in our present derivation by considering the *boundaries* of the envelope of the spheroidal wavelets. It can also be shown from Maxwell's equations that the Poynting flux vector is parallel to OT.

1.4864, and $n_z = 1.6584$, from which $\delta = 6°14'$. (Because we have exaggerated the oblateness of the spheroidal wave surface, this angle is not shown accurately in the diagram.)

The refraction of the *e*-ray can be observed experimentally by using a narrow beam of plane waves incident normally on the face of a calcite cleavage form* (see Fig. 9–8). The *e*-wave passes through the top surface as a plane wave which then proceeds to propagate in air without further deflection. This experiment can be performed quantitatively by measuring the separation of the *o*- and *e*-beams projected on a screen.

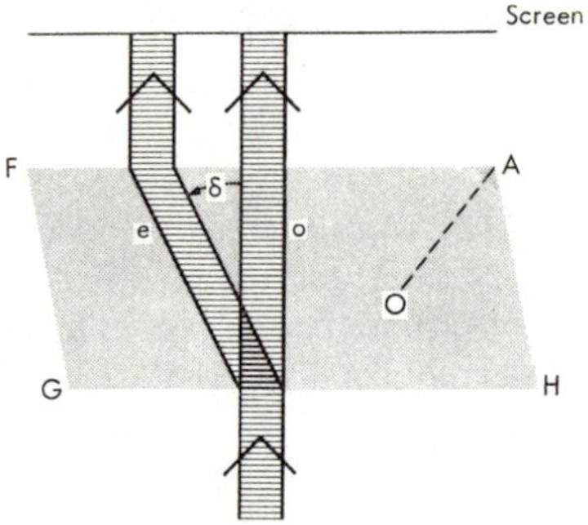

Fig. 9–8. Principal section of a calcite rhomb showing the splitting into o- and e-waves of a narrow beam of plane waves incident normally on a face.

Figure 9–7 also enables us to make a calculation of the way in which the index of refraction n_y associated with the *e*-wave depends on the orientation of the optic axis. The velocity with which the plane wavefront of the *e*-wave propagates through the calcite is $c_y = y'_T/\Delta t$. By elimination of x'_T between Eqs. (9–3) and (9–4), we can solve for the ratio of y'_T to Δt:

$$y'_T/\Delta t = \sqrt{c_1^2 \sin^2 \alpha + c_2^2 \cos^2 \alpha} = c_y.$$

The associated index of refraction of the *e*-wavefront is therefore

$$n_y = c_0/c_y = \frac{n'_y n_z}{\sqrt{(n'_y \sin \alpha)^2 + (n_z \cos \alpha)^2}}. \tag{9-6}$$

From this result it follow that $n'_y \leq n_y \leq n_z$. The minimum value $n_y = n'_y$ corresponds to $\alpha = 0$ when the optic axis is parallel to a face. The maximum value $n_y = n_z$ occurs when $\alpha = \pi/2$ and the optic axis is perpendicular to a face.

---

* If the beam is not incident normally and if the plane of incidence is not a principal section (i.e. is not parallel to the optic axis), the refracted *e*-beam does not lie in the plane of incidence. Thus another of the usual laws of refraction (plane of refraction = plane of incidence) is not obeyed by the *e*-wave.

PROBLEMS

9–1. Calculate the thicknesses required to make quarter-wave plates for sodium light from calcite, quartz, and mica. For calcite and quartz take the optic axis parallel to the faces of the plate. For mica assume that the faces are natural cleavage planes.

9–2. A half-wave plate is placed between a polarizer and an analyzer. The analyzer is set for extinction.

a) If the polarizer is rotated clockwise through an angle ϕ, through what angle and in which direction must the analyzer be rotated to maintain extinction?

b) If the half-wave plate is rotated through an angle ϕ, through what angle and in which direction must the analyzer be rotated to maintain extinction?

9–3. A polarizing sheet is attached to one face of a quarter-wave plate with the axis of transmission at 45° to the axes of the plate. A beam of linearly polarized light is looked at through this device held with the polarizing sheet toward the observer. Describe the intensity variations as the device is rotated.

9–4. Show how a quarter-wave plate whose fast axis is labeled can be used to produce a beam of right circularly polarized light. Conversely, show how such a plate can be used to tell whether a given circularly polarized beam is right- or left-handed.

9–5. *Polarizing filter for reducing surface reflections.* A film, available commercially, consists of a doubly refracting layer of quarter-wave thickness superimposed on a sheet of polarizing material whose axes are at 45° to the axes of the former. If the film is placed on a shiny surface with the polarizing material on the top face, reflections from the surface are eliminated, although light originating behind the surface can still be seen. Consider a plane wave normally incident on such a system. Analyze the changes in polarization and show that, except for reflections off the top surface of the film, the beam is absorbed by the polarizing material.

9–6. Consider a plane wave of natural light normally incident on a calcite surface containing the optic axis. Assume that the reflection coefficients for o- and e-type polarizations depend in the usual way on the respective indices of refraction.

a) Calculate the ratio of the intensities of the two types of polarization in the reflected light.

b) Suppose that it were possible to make a calcite plate thin enough for a quarter-wave plate. Taking *reflection losses* into account (but neglecting internal reflections), calculate the angle between the transmission axis of a polarizer and the axes of the plate which would be required to produce circularly polarized light.

9–7. Two quarter-wave plates are cemented together with the fast axis of one plate inclined at an angle of 45° relative to that of the other. None of the axes is labeled. An observer comes upon the device and supposes that it may be a simple retardation plate of unknown retardation. Describe an experiment which he can perform to deny this hypothesis.

***9–8.** A Babinet compensator is set between crossed polarizers with its axes making an angle of 45° with the axes of the polarizers. An observer is planning to measure the retardation of a plate which happens to be an exact quarter-wave plate. In doing so, he inserts the quarter-wave plate between the analyzer and the Babinet, but forgets to align the axes of the plate parallel to those of the Babinet. Given that the angle between these axes is ϕ, calculate the value he will obtain for the retardation of the plate.

9–9. Linearly polarized white light is normally incident on a mica plate with the E-vector at an angle of 45° to the axes of the plate. Given that the thickness is such that the plate is a 3-λ plate for light of wavelength 5000 Å, calculate the wavelengths in the visible spectrum which will emerge circularly polarized.

9–10. A channeled spectrum is obtained when light from an incandescent source is passed through a retardation plate between crossed polarizers and then examined with a spectroscope. Assume that the plate is uniaxial with the optic axis parallel to the faces and that it is of measured thickness d. Show how a measurement of the wavelengths of two adjacent dark lines can be used to determine the difference between the principal indices of refraction of the crystal.

***9–11.** *The principle of a polarizing monochromator.* Consider a set of polarizing sheets which are lined up with their axes of transmission parallel to each other. Retardation plates are inserted between each pair of polarizing sheets, with their axes at 45° to those of the sheets. If each retardation plate is an integral-wave plate (M-λ) for some particular wavelength λ_0, then this wavelength is transmitted by the entire device. For any given set of plates that are integral-wave plates for λ_0, there will in general be other wavelengths λ_0' which also fulfill this condition. However, by suitable choice of the thicknesses of the plates, the separation between λ_0 and the various λ_0' can be made sufficiently large so that colored glass filters can be used to absorb the wavelengths λ_0'. This leaves isolated a narrow portion of the spectrum in the immediate vicinity of λ_0. To illustrate the preceding, consider the following calculations. Let the plates be made of quartz with the optic axis parallel to the faces; let the thicknesses be such that the plates are integral-wave plates for $\lambda_0 = 5893$ Å.

a) Approximately what thickness is required if the first plate is to be an integral-wave plate with $M = 1024$?

b) What is the wavelength separation between λ_0 and the closest wavelength which is completely suppressed by the first plate?

c) If the second plate is exactly half as thick as the first, show that every other maximum in the channeled spectrum of the first is suppressed.

d) If there are six plates altogether whose thicknesses are in the ratio $1:\frac{1}{2}:\frac{1}{4}\cdots\frac{1}{32}$, describe the wavelengths λ_0' in the visible spectrum which are fully transmitted by the device. That is, what are the nearest wavelengths to λ_0, and how many values of λ_0' are there across the entire spectrum?

e) Is this device capable of isolating one of the two components in the sodium doublet? [*Note:* The pass wavelength λ_0 can be altered by increasing the optical path length through slight tilting, or by changing the temperature and exploiting the temperature dependence of the indices of refraction. This procedure allows a small frequency region to be *scanned.*]

9–12. Consider a Babinet compensator placed between crossed polarizers. Let θ be the angle between an axis of the Babinet and an axis of one of the polarizers. If a plane wave of wavelength λ is incident, show that the average intensity of the wave at position z on the Babinet (Fig. 9-3) is given by the expression $I = I_0 \sin^2 2\theta \sin^2 \beta z$, where I_0 is independent of θ and z, and $\beta = (\pi/\lambda_0)(\tan \alpha)\,\Delta n$. Describe the appearance of the field. What advantage is there in choosing $\theta = \pi/4$?

9–13. A Babinet compensator is placed between crossed polarizers with the y-axis in Fig. 9–3 vertical and the axes of the polarizers at an angle of 45° to it. A plane wave of white light is incident on the polarizer. The light emerging from the analyzer is blocked off with a narrow vertical slit so that only the portion of the wave between z and $z + \Delta z$ is passed. This light is then refracted by a prism and examined with a telescope. Suppose that the position of the slit is gradually moved through increasing values of z, starting at $z = 0$, the center of the Babinet. Describe the successive changes in the appearance of the spectrum. (Assume that the spectrum of the white light source contains a uniform frequency distribution.)

9–14. Consider a plane wave obliquely incident on a plane surface of calcite, with the optic axis perpendicular to the plane of incidence. Show that both the o- and e-rays obey Snell's law with respective indices of refraction n_z and n'_y. [*Note:* Since the use of the angle of minimum deviation to determine indices of refraction is based solely on Snell's law, it is clear therefore that n_z and n'_y can be measured in the ordinary way with a calcite prism whose optic axis is parallel to the refracting edge.]

9–15. At what angle to the optic axis should the faces of a calcite plate be cut to render the anomalous angle of refraction δ of a normally incident e-ray a maximum?

9–16. Fermat's principle asserts that the ray connecting a source point S with a point of observation P will be such as to minimize the time of travel of the wave between these two points. Thus we observe in Fig. 8–12 that since the velocity of the e-wave is greater in the direction SQ than in direction SP, the minimum time occurs along path SQP, despite the fact that the geometrical length of this path is greater than the straight-line distance SP. For simplicity consider P to be at infinity so that the problem reduces to finding the position of the point Q which renders the time of travel along SQ a minimum. Choose $\theta = \measuredangle PSQ$ as a parameter and calculate $t(\theta)$, the time required for a spheroidal wave emitted at $t = 0$ to arrive at the point Q. Show that the value of θ which renders $t(\theta)$ a minimum is the same as the angle δ given by Eq. (9–5).

10 Additional optical properties of matter

10-1 OPTICAL ACTIVITY

As an example of the experimental observations which are associated with *optical activity*, consider a flat plate which is cut from a quartz crystal in such a way that the optic axis is perpendicular to the faces of the plate. Let a beam of linearly polarized light be incident on the plate. The emergent beam is then analyzed by the techniques of the preceding chapter to determine its polarized character. It is found that in all cases the emergent beam is linearly polarized, but that the direction of the E-vector in the emergent beam is, in general, different from that of the incident beam. We may speak of this as a rotation of the plane of polarization, the angle between the incident and emergent E-vectors being designated as the rotation angle χ. Substances which exhibit this effect are said to be *optically active*. This phenomenon is clearly distinct from that of birefringence; in the latter case the emergent beam is, in general, elliptically polarized and is linearly polarized only if special conditions are met.

Further study of the effect of the quartz plate shows that the rotation angle χ is directly proportional to the thickness of the plate, the rate being 21.7° per mm thickness for sodium light. This leads us to depict the propagation of the wave through the quartz as being essentially that of a linearly polarized wave whose plane of polarization rotates steadily as the wave advances. In one type of quartz the rotation of the plane of polarization is in a clockwise sense to an observer who is facing the oncoming beam. Substances which rotate the plane of polarization in this sense are said to be *dextrorotatory*. For a given quartz plate the sense of rotation is independent of which face is turned toward the incident beam. Other samples of quartz crystal can be found which rotate the plane of polarization at the same rate but in the opposite sense. Substances of this nature are referred to as *levorotatory*.

There is no chemical distinction between dextrorotatory and levorotatory quartz. Since *fused* quartz is not optically active, the property appears to be associated with different patterns of crystal structure. Pairs of quartz crystals can be found which resemble each other identically in external form except that one is the mirror image of the other. These crystal forms are said to be *enantiomorphs* of each other. Plates cut from such a pair rotate the plane of polarization in opposite senses. All transparent substances which exhibit enantiomorphic crystal forms are found to be optically active.

Optical activity is also exhibited by various liquids, such as turpentine, and certain substances dissolved in nonoptically active solvents. The organic compounds which are classified as sugars are an important example of optical activity in solution. To observe the effect, we replace the quartz plate by a liquid cell with plane parallel faces set perpendicular to the incident beam. To a high degree of accuracy the rotation angle χ is found to be proportional to the concentration of the solute. Since χ is also proportional to the thickness of the cell, we can combine the results by stating that for solutions, χ depends only on the number of solute molecules swept out by a unit cross-sectional area of the beam. This experimental result suggests very strongly that the optical activity is associated with the structure of an individual solute molecule. Since the substance is capable of distinguishing "right" from "left," it is evident that the molecule possesses a mirror asymmetry. That is, a mirror image of the molecule cannot be rotated to coincide in all respects with the original. The human hand is an example of this type of asymmetry. The mirror image of a right hand is a left hand; right and left hands cannot be made to coincide by any combination of translations and rotations. Another example is a corkscrew. Note that a right-handed corkscrew has the property of right-handedness no matter from which end it is looked at. This analogy is helpful in appreciating the fact that even though the solute molecules are oriented in random directions in a solution, there is no tendency for the molecules "facing in one direction" to cancel out the effect of those "facing in the opposite direction." The liquid cell has no "front" and "back" faces, but rotates the plane of polarization in the same sense no matter what the direction of propagation of the light.

The subject of *stereochemistry* originates from Pasteur's proposal in 1860 that optical activity is associated with asymmetry in molecular structure. The subject deals with the construction of three-dimensional models which locate the atoms within molecules of an optically active substance in a structure which has mirror asymmetry. As a simple example, though one which would be impractical to prepare, consider a methane molecule, CH_4, in which three of the hydrogen atoms have been replaced by different halogens, Cl, Br, and I. Symbols lying in the plane of the page, as in the conventional formula

```
     Cl
     |
  H—C—Br
     |
     I
```

are not capable of showing the nature of the asymmetry. However, Fig. 10–1(a) indicates a possible spatial configuration for the resulting molecule with the carbon atom placed at the center of a tetrahedron formed by the other four atoms. The mirror image of this configuration is shown in Fig. 10–1(b). These two structures can be superimposed so that any two of the atoms H, Cl, Br, or I fall in the same positions, but the positions of the other two will always be interchanged. The molecule thus has a mirror asymmetry, and the carbon atom is said to be *asymmetric*. The asymmetry does not hold if any two of the groups attached to the carbon atom are equal, as would be the case, for example, with the compound CH_2BrI. We would expect a substance whose molecules are all like that in Fig. 10–1(a) to be optically active. We would likewise expect a substance whose molecules are like that in Fig. 10–1(b) to be optically active and to produce a rotation of the plane of polarization equal in magnitude to that of Fig. 10–1(a), but of opposite sense. An equal concentration of molecules of both types (called a *racemic mixture*) would constitute a solution exhibiting no optical activity.

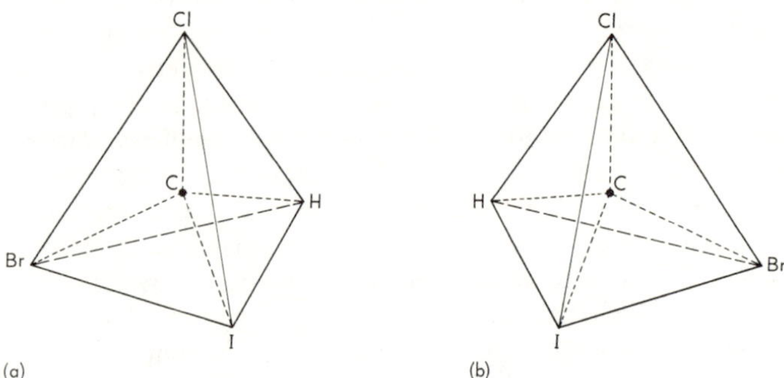

Fig. 10–1. Stereoisomers of CHClBrI.

Compounds which have the same empirical formulas but which differ with respect to their spatial configuration in the manner indicated above are referred to as *stereoisomers*. In the example of Fig. 10–1 there are just two isomers, but in more general organic compounds having several asymmetric carbon atoms the number of possibilities is increased. Any two isomers which are exact mirror images of each other are referred to as *enantiomorphs;* their external crystal structures show *enantiomorphism*.

The laboratory synthesis of a potentially optically active organic compound usually produces a racemic mixture containing all possible stereoisomers. In view of this fact it is interesting to note that biological organisms show a distinct preference for the manufacture and utilization of specific isomers. For example, when a mold feeds on a nutrient solution which initially contains equal concentrations of two isomers, one of the isomers is more rapidly consumed, leaving an effectively pure sample of the other. A striking example is the exclusive occurrence of one particular isomer of each of the amino acids which are an important part of proteins. Thus, for example, the central carbon in the amino acid *alanine*,

$$\begin{array}{c} \mathrm{COOH} \\ | \\ \mathrm{H{-}C{-}CH_3} \\ | \\ \mathrm{NH_2} \end{array}$$

is asymmetric, and the compound has two possible stereoisomers. Nevertheless, the alanine which is found in plants and animals consists in all cases of the isomer which is dextrorotatory. It is likewise found that all naturally occurring samples of the higher amino acids are of the type which would result from the replacement of the CH_3 group in alanine by more complex groups, without changing the configuration surrounding the asymmetric carbon atom. Not all are dextrorotatory, since the net optical rotation is determined by the molecule as a whole. However, since they have identical asymmetric configurations at the end of the molecule, these acids form a series designated as the L-*series*. (The L refers to the levorotatory property of a related compound, glyceraldehyde.) Except in very rare instances, the biological universe seems to make no use whatever of the D-series isomers, which have the opposite configuration about the terminal asymmetric carbon atom. In fact, organisms are unable to utilize the D-isomer of an amino acid which they are normally capable of metabolizing in L-form. One can visualize the possibility of a set of D-organisms whose amino acids are all of the D-series coexisting with the present L-organisms. No organism of one type would be capable of feeding on organisms of the other type. We can understand the noninteraction of the two series if we assume that a kind of lock-and-key principle applies to the production and utilization of amino acids. However, it is surprising that all existing organisms are exclusively of one type. It is not known whether there is some principle which favors the development of the L-system, or whether it is a matter of pure chance that the earliest organisms were of the L-system.

Irrespective of the models of crystal or molecular structure suggested by optical activity, a phenomenological interpretation devised by Fresnel yields a simple description of the effect. This is based on the formal equivalence of a linearly polarized beam with the superposition of two oppositely circularly polarized beams of equal amplitude. The azimuth of the linearly polarized beam is determined solely by the *phase relation* between the two circularly polarized components. (This proposition was considered in Problem 7–8, but the results will be fully derived in the following analysis.) Fresnel observed that if the two circularly polarized beams are endowed with different propagation velocities in an optically active substance, their phase relation will be altered as a result of passage through the medium, leading to a consequent change in the plane of polarization. Thus consider a wave of frequency ω associated with propagation velocities c_+ and c_- for right and left circularly polarized light. Corresponding to c_+, the wave number is

$$k_+ = \frac{\omega}{c_+} = \frac{\omega}{c_0}\frac{c_0}{c_+} = k_0 n_+, \tag{10–1}$$

where k_0 is the wave number in free space and n_+ is the index of refraction for right circularly polarized light. A similar relation applies to the wave number for the left circularly polarized beam. In this notation the functions describing circularly polarized beams of equal amplitude A propagating along the x-axis through the optically active medium are:

$$\text{Right circularly polarized} \quad \begin{cases} E_y^+ = A\cos(\omega t - k_+ x), \\ E_z^+ = -A\sin(\omega t - k_+ x); \end{cases} \tag{10–2}$$

$$\text{Left circularly polarized} \quad \begin{cases} E_y^- = A\cos(\omega t - k_- x), \\ E_z^- = A\sin(\omega t - k_- x). \end{cases} \tag{10–3}$$

The superposition is represented by the fields $E_y = E_y^+ + E_y^-$ and $E_z = E_z^+ + E_z^-$. Through the use of trigonometric identities, we can reduce these to

$$\begin{aligned} E_y &= 2A\cos\left[(k_+ - k_-)\frac{x}{2}\right]\cos\left[\omega t - (k_+ + k_-)\frac{x}{2}\right], \\ E_z &= 2A\sin\left[(k_+ - k_-)\frac{x}{2}\right]\cos\left[\omega t - (k_+ + k_-)\frac{x}{2}\right]. \end{aligned} \tag{10–4}$$

As the beam enters the medium at $x = 0$, the net field is given by $E_y = 2A\cos\omega t$ and $E_z = 0$, which implies linearly polarized light with the E-vector along the y-axis. At any fixed position in the medium the sinusoidal functions of the time in Eq. (10–4) are in phase, which shows that the beam has the character of linearly polarized light. The ratio of E_y to E_z is time-independent and specifies the tangent of the angle through which the E-vector has been rotated relative to the incident beam. Since the rotation angle χ is usually considered positive for rotations which are clockwise to an observer facing the oncoming beam, it is given by $\chi = -\tan^{-1}(E_z/E_y)$. In particular, if the emergent face is located at $x = d$, we obtain

$$\tan\chi = \frac{-E_z}{E_y} = \tan\left[(k_- - k_+)\frac{d}{2}\right],$$

or, in radians,

$$\chi = (k_- - k_+)\frac{d}{2} = \frac{\pi d}{\lambda_0}(n_- - n_+). \tag{10–5}$$

The model correctly predicts the fact that the emergent beam is linearly polarized and that the rotation angle is proportional to the thickness traversed. Since we would expect $(n_- - n_+)$ to be proportional to the concentration of the solute for solutions, the model also correctly predicts the dependence of χ on the concentration. From Eq. (10–5) we expect a wavelength dependence of χ both through the factor $1/\lambda_0$ and through $(n_- - n_+)$. Early experimental observations by Biot showed that optical activity is a highly dispersive phenomenon which is reasonably well represented by $\chi \propto \lambda_0^{-2}$ (Biot's law). We infer therefore that the difference in index of refraction for the two types of circularly polarized light is itself inversely proportional to the wavelength.

It is not difficult to understand qualitatively why the optically active medium interacts differently with right and left circularly polarized light. The easiest way to get an idea of the type of mechanism involved is to consider an analogy which can be constructed as a demonstration experiment with microwaves. Let the "optically active medium" consist of an assemblage of right-handed helices of wire whose dimensions are comparable with the wavelength of the microwave to be employed. If the helices are parallel to each other, the analogy is to a quartz crystal whose optic axis is parallel to the axes of the helices. Each helix represents a number of quartz molecules assembled in a corkscrew structure. If the helices have random orientation, the analogy is between a single helix and the molecule of an optically active solute. Now consider a circularly polarized wave traveling parallel to the axis of a helix. The currents induced in the helix will depend on whether the sense of rotation of the E-vector is the same as or opposite to the sense of the helix. The superposition of the incident radiation with the secondary radiation from the induced currents determines the net field within the medium and accounts for the altered propagation velocity. Since right and left circularly polarized waves interact differently with the medium, we therefore expect a difference between the two propagation velocities. Once it has been established that the two propagation velocities are different, it follows for microwaves, as it did for light, that the plane of polarization of an incident linearly polarized wave is rotated through an angle given by Eq. (10–5). Since polarized detectors are used with microwave equipment, the effect can be demonstrated by rotating the receiving waveguide until a maximum response is obtained.

10–2 THE FARADAY EFFECT

Physicists in the early part of the nineteenth century did not have the benefit of an understanding of the electrical structure of matter and the electromagnetic character of light. Nevertheless, Michael Faraday was guided by a strong intuitive feeling that the optical properties of matter would be influenced by the presence of electric or magnetic fields. He pursued a systematic series of investigations to discover such an interaction. A number of effects which are known today escaped Faraday's notice because of their small magnitude. However, Faraday's researches were rewarded in 1846 by the discovery of the strongest of the electro- and magneto-optical effects, which is now known as the *Faraday effect*. Faraday found that when a transparent substance which is not normally optically active is placed in a magnetic field it acquires the ability to rotate the plane of polarization of a beam of linearly polarized light which is traveling parallel to the direction of the magnetic field. The angle of rotation is proportional to the path length through the substance and to the strength of the magnetic field.

It is frequently stated that the magnetic field induces optical activity in the medium. This is somewhat misleading, however, because the sense of rotation cannot be characterized by the statement that the medium is either dextrorotatory or levorotatory. For example, in a diamagnetic medium the rotation of the plane of polarization is counterclockwise to an observer who is facing the oncoming beam if the magnetic field is parallel to the direction of propagation, but is clockwise if these two directions are antiparallel. If the direction of propagation of the beam is along the positive x-axis, the experimental findings can be summarized by the formula

$$\chi = -VH_x d, \tag{10-6}$$

where d is the path length in the medium, H_x is the x-component of the applied field, V is a constant (Verdet's constant) characteristic of the medium, and χ is the rotation angle in the clockwise sense to an observer facing the oncoming beam. For certain *paramagnetic* media the rotation occurs in the reverse direction, leading to assignment of negative values to V.

The phenomenological interpretation of the Faraday effect is that the magnetic field produces a change in the indices of refraction of right and left circularly polarized beams. Just as in the last section, the incident beam of linearly polarized light is formally equivalent to the superposition of equal beams of right and left circularly polarized light. The rotation of the plane of polarization is then given by Eq. (10–5), where $(n_- - n_+)$ is the difference in indices of refraction induced by the magnetic field.

It is beyond the scope of this text to derive the Faraday effect from a model based on the electron theory of matter. A brief idea of the mechanism involved in the case of *diamagnetic* media can be obtained through Larmor's theorem. As far as the electronic motion in atoms is concerned, the effect of an external magnetic field can be accounted for by transformation to a rotating coordinate system which rotates with angular velocity $\omega_L = \mu eH/2m$ in a positive sense about the direction of the magnetic field (μ is the permeability of the medium, e the magnitude of the electronic charge, and m the mass of the electron). Viewed from this coordinate system the charge behaves dynamically the same as it would in the absence of the magnetic field. But in this coordinate system a right circularly polarized beam of frequency ω traveling in the direction of the magnetic field appears to have frequency $\omega + \omega_L$, and is therefore associated with an index of refraction

$$n_+ = n(\omega + \omega_L) \doteq n(\omega) + \omega_L \frac{dn}{d\omega},$$

where $n(\omega)$ is the function defining the regular index of refraction as a function of frequency. The apparent frequency of a left circularly polarized beam is $\omega - \omega_L$, and hence

$$n_- \doteq n(\omega) - \omega_L \frac{dn}{d\omega}.$$

Thus from Eq. (10–5) the rotation angle in radians is

$$\chi = -\left(\frac{\pi d}{\lambda_0}\right)\left(\frac{\mu e H}{m}\right)\frac{dn}{d\omega}.$$

When we compare this with Eq. (10–6) we find that we have derived the proportionality of the rotation angle to H and d and have obtained as a representation of Verdet's constant the expression

$$V = \left(\frac{\pi \mu e}{m\lambda_0}\right)\frac{dn}{d\omega}.$$

Since $(dn/d\omega) > 0$ in nonabsorbing regions of the spectrum, the sign of V is positive. This formula, known as *Becquerel's formula*, relates the Verdet constant to $(dn/d\omega)$ as obtained from measurement of the index of refraction in an ordinary experiment in the absence of a magnetic field.

The ceramic materials called *ferrites* exhibit a strong Faraday effect at microwave frequencies. This has permitted the design of a device for stabilizing the output of a generator by isolating it from the influence of waves which are reflected at waveguide junctions or by obstacles in the radiated beam. The principle by which the reflected wave is eliminated is considered in Problem 10–2(b). The analog of the dichroic which serves to attenuate the reflected beam is a sheet of poorly conducting material placed parallel to the direction of propagation and perpendicular to the E-vector of the radiated wave. Since no currents are induced in the sheet under these circumstances, this wave is propagated without attenuation. Through action of the Faraday effect any reflected wave returns with its E-vector parallel to the sheet. This causes induced currents in the sheet, which dissipate the energy.

10-3 INDUCED BIREFRINGENCE

The normal lack of birefringence in gases, liquids, glasses, plastics, etc. is associated with isotropy of the medium. Thus the propagation velocity of a linearly polarized plane wave is indifferent to the orientation of the E-vector. Several external influences which introduce an anisotropy in these media and produce a birefringence similar to that of the uniaxial crystals are listed below. In all cases the anisotropy is associated with the establishment of a preferred direction which plays a role similar to that of the optic axis in calcite. The most effective method of studying these effects is to arrange to have the axis of anisotropy perpendicular to the direction of propagation of an incident linearly polarized beam. The specimen then behaves like a retardation plate, and the emergent beam is some type of elliptically polarized light. A Babinet compensator can be used to study the emergent beam and to measure the relative phase retardation introduced by the specimen. This, in conjunction with Eq. (9–2), enables us to make a calculation of Δn, the difference in the indices of refraction for linearly polarized beams whose E-vectors are respectively perpendicular to and parallel to the axis of anisotropy.

a) The Kerr Effect

A strong applied electric field $\mathbf{E}$ causes the molecules of certain substances to have an induced electric dipole moment. These dipoles tend to align parallel to $\mathbf{E}$. The optic axis of the induced birefringence is thus parallel to the applied electric field. The effect is second order in $\mathbf{E}$, that is, $\Delta n \propto E^2$. The orientation of the dipoles is a relaxation type phenomenon, requiring a characteristic length of time for its establishment. This time varies from a few seconds for certain solids to the order of 10^{-11} sec for certain liquids. The extremely short relaxation time of liquids enables construction of a rapid shutter for producing an intermittent light beam. The electric field is produced by placing a parallel plate capacitor in the liquid cell, this part of the device being known as a *Kerr cell.* The cell is placed between crossed polarizers. For maximum contrast the axes of the latter are set at 45° to the electric field. When no voltage is applied across the capacitor plates the field is dark. When a voltage is applied the intensity of the beam passed is controlled by variations in the potential difference. This device, known as a Kerr electro-optical shutter, has received important application as an accurate method of determining the velocity of light.

b) The Cotton-Mouton Effect

This is a magneto-optical effect which, though relatively weaker, is directly analogous in principle to the Kerr electro-optical effect. An applied magnetic field $\mathbf{H}$ produces an anisotropy of the medium. The associated birefringence has an optic axis parallel to $\mathbf{H}$ and $\Delta n \propto H^2$.

c) Stress Birefringence (The Photoelastic Effect)

In some substances the presence of mechanical stress is accompanied by an alignment of anisotropic molecules which are otherwise randomly oriented. Cellophane and certain transparent tapes show such an alignment due to a drawing process in their manufacture; they can be used conveniently as retardation plates in qualitative experiments. The axes are parallel and perpendicular to the axis along which the material has been stretched.

A common application of stress birefringence is in the study of structural weaknesses by examination between crossed polarizers of plastic models under various loading conditions. Glasswork can be inspected for evidence of poor annealment by a similar procedure. If all the stress in a sample is parallel to a particular direction, no pattern is obtained when the transmission axis of either the polarizer or the analyzer is parallel to this direction. In general however, different portions of the sample have different directions for the axis of stress, and there is no orientation of the polarizers for which the pattern disappears. If such a change in direction of the axis of stress occurs across the thickness of the sample, the effect is similar to that which would be obtained from a stack of retardation plates whose axes are in different directions. Under these conditions the pattern changes which occur as the analyzer is rotated can be quite complex. Due to its practical importance, the analysis of photoelastic stress patterns has been the subject of considerable study.*

d) Lamellar Flow in Liquids

Long chain molecules in liquids tend to align themselves perpendicular to velocity gradients in steady nonturbulent flow. The associated birefringence can be used to study the velocity distribution.

10–4 SCATTERING (THE TYNDALL EFFECT)

Consider a light beam which passes along the axis of a cylindrical tube T with transparent walls (Fig. 10–2). If the tube is evacuated, an observer who looks in at a point O from the directions QO, RO, or SO receives no energy from the beam. However, if the tube is filled with any material medium, the observer at Q, R, or S receives some light. This is referred to as light which has been *scattered* from the incident beam. An obvious illustration of the effect is the appearance of shafts of sunlight through breaks in a cloud-covered sky.

After studying the scattering by colloids (suspensions of finely divided particles), Tyndall discovered that scattering takes place even in a pure

---

* M. M. Frocht, *Photoelasticity*. New York: John Wiley & Sons, Vol. I, 1941, Vol. II, 1950.

substance such as a sample of air which has been completely cleared of suspended water droplets and dust particles. We shall see below that this scattering is associated with a lack of perfect homogeneity of the medium. That is, even though the sample contains no impurities, the random thermal motion of the molecules results in temporary local fluctuations in density.

The scattering by large particles (e.g. specks of dust) is easily understood as being caused by the reflection of the incident beam from the surfaces of the particles. Since the angle of reflection is equal to the angle of incidence in specular reflection, this type of scattering is not isotropic. In this circumstance different intensities are observed in the scattered light at Q, R, and S.

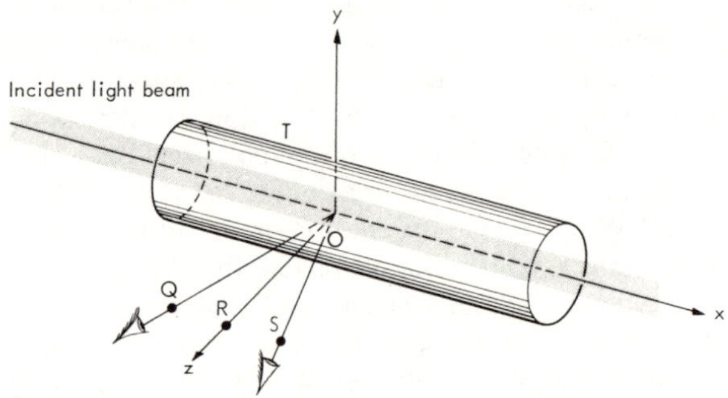

Fig. 10-2. The Tyndall effect.

Another type of scattering, about which we can learn a great deal from a simple model, is the scattering by a random distribution of particles whose dimensions are small compared with a wavelength. The assumption that the distribution is *random* is made so that the scattered radiation from individual particles can be considered independently. The total intensity is then simply the sum of the intensities from all the scatterers. If this assumption is not satisfied, it is necessary to consider the possibility of interference effects among the individual contributions to the net scattered radiation. If the dimensions of the particles are small compared with a wavelength, the variable electric field of the incident light causes an oscillation of charge within the molecules of the particle. The scattered radiation is the secondary radiation produced by this oscillation of charge, governed by the laws of Hertzian dipole radiation which were asserted in Section 8-8. The direction of propagation of the incident light is designated as the x-axis, the line OR as the z-axis. If the particle itself is isotropic, the induced motion of the charge takes place in the yz-plane. One inference which can be drawn from the previous discussion of dipole radiation is that the intensity of the scattered radiation will be symmetric about this plane. That is, the apparent motion (perpendicular to the line of sight) is the same when looked at from Q or S, and the scattered intensity will be the same in these two directions. We can also infer that the scattered radiation in the direction OR will be linearly polarized, since the projection of the motion of the charge perpendicular to the line of sight OR is straight line motion parallel to the y-axis. Suppose additionally that the incident beam is linearly polarized with its E-vector inclined in the yz-plane at an angle θ with the y-axis. The projection of the motion of the oscillating charge as seen from the fixed point of observation R is affected by a factor $\cos\theta$, and the intensity by a factor of $\cos^2\theta$. In particular, with θ chosen to be $\pi/2$, no scattered radiation is expected at the point R, the induced motion of charge then being parallel to the line of sight.

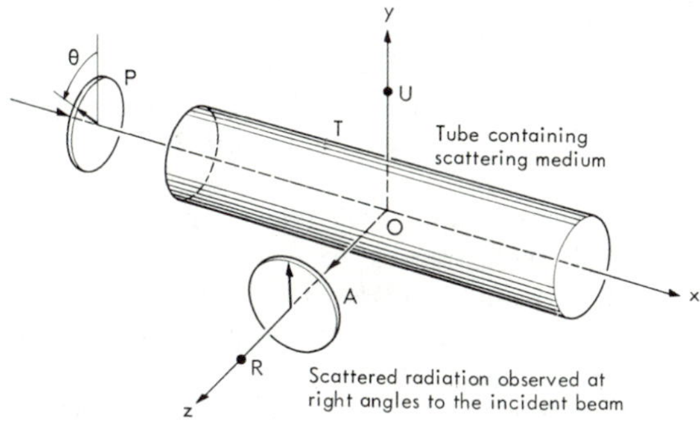

Fig. 10–3. Demonstration of polarization effects with scattered radiation. (The arrows indicate the transmission axes of the polarizer P and analyzer A.)

EXAMPLE 1. Let tube T (Fig. 10–3) be filled with water which has been made turbid by the addition of a drop of milk. With natural light incident and the polarizer P removed, the polarization of the scattered radiation can be demonstrated by rotating the analyzer A. A distinct minimum occurs when the transmission axis of the latter is parallel to the axis of the tube. Or, remove A and insert P. The observer at R sees the tube filled with light which diminishes in intensity as the transmission axis of P is rotated toward the position $\theta = \pi/2$, at which point a distinct minimum is obtained. An observer at U who looks in from above witnesses maximum intensity while the observer at R witnesses a minimum.

EXAMPLE 2. The brightness of the sky on a clear day is produced by the scattering of sunlight by air molecules and shows a partial polarization of the type described above. Choose a portion of the sky such that the line of sight is perpendicular to a ray from the sun. Minimum intensity is transmitted by an analyzer when its axis of transmission is parallel to the latter direction. Further analysis of the light from the sky in such a direction shows that it is a mixture of linearly polarized and unpolarized light. The lack of complete linear polarization is primarily due to the fact that the molecules of air are not spherically symmetric. When a light beam is incident on an asymmetric molecule, the motion of induced charge is not confined to the plane perpendicular to the direction of propagation. As an extreme example, the currents induced in a thin needle are along the axis of the needle and are determined by the component of **E** onto this direction.

The type of scattering which we have been describing qualitatively (that due to a particle whose dimensions are small compared with a wavelength) was described theoretically by Lord Rayleigh and is known as *Rayleigh scattering*. The calculation predicts that the effect will be strongly dependent on the wavelength, the scattered intensity being inversely proportional to the fourth power of the wavelength. Thus blue light is much more strongly scattered than red. This accounts for the blue color of the sky and of the scattered radiation which is observed in Example 1. Furthermore, the incident beam of white light becomes successively more depleted of blue light as it progresses through the medium, so that its spectral distribution favors the red end of the spectrum after some distance of travel. Thus it can be observed that the light which emerges from the end of the tube in Example 1 has a red tinge. The red color of the sun as seen through a thick layer of air at sunset is an obvious analog.

In an ideal homogeneous medium each molecule may be thought of as a source of scattered radiation. If these sources were all independent, the intensity of the scattered radiation would increase proportionally to N, the number of scattering molecules per unit volume. However, the superposition of the radiation from a number of closely spaced sources which maintain definite phase relations with each other (their motions being induced by the same incident wave) results in a high degree of destructive interference at a point such as R. A comparable situation is discussed in the introduction to Chapter 12. Further pursuit of this line of reasoning shows that density fluctuations are essential for the existence of a net scattered wave. Thus the transition from gaseous to liquid phase does not show an increase in scattering proportional to the increase in N, for the density fluctuations are significantly less in the liquid than they are in the gas. Marked statistical fluctuations in density occur near the critical temperature of a substance and result in strong scattering, the effect being known as *critical opalescence*.

10-5 ABSORPTION

It has been mentioned previously that all media are strongly absorbing in some region of the electromagnetic spectrum. In addition we should recognize that a certain amount of absorption occurs even in regions where the medium is considered to be relatively transparent. Two mechanisms are chiefly responsible for this:

a) Scattering. Corresponding to the energy which is redirected in the form of scattered radiation there is a withdrawal of energy from the incident beam.

b) Dissipation in the form of heat. As the molecules of the substance undergo collisions with one another, some of the energy which has been given up by the field to set the charge within the molecules in oscillation is transferred to the form of random kinetic energy of the molecules. This process is sometimes referred to as *true absorption*, as distinguished from absorption due to scattering.

Regardless of the mechanism involved, a description of the attenuation of a plane wave can be deduced from the simple assumption that the net power loss per unit volume, P, is proportional to the local intensity I: $P = \alpha I$. The frequency-dependent parameter α is characteristic of the medium and is referred to as the *absorption coefficient*. The quantities I and P denote time average rather than instantaneous values. Consider then a volume bounded by the planes x and $x + \Delta x$ with area A perpendicular to the x-axis. The net power flux into this volume is

$$AI(x) - AI(x + \Delta x) = -A\left(\frac{dI}{dx}\right)\Delta x.$$

This is equal to the power absorbed, which is $AP\,\Delta x$. Thus $-(dI/dx) = \alpha I$. This equation integrates to yield *Lambert's law*:

$$I = I_0 e^{-\alpha x}. \tag{10-7}$$

Hence the attentuation occurs exponentially, requiring a distance $1/\alpha$ for the intensity to be reduced to $1/e$ of its original value. The assumptions made in this derivation are equally applicable to acoustic or electromagnetic plane waves.

If the light absorption by a liquid solution is attributable primarily to the solute molecules, it can be anticipated that for low concentrations the absorption coefficient α should be proportional to N, the number of solute molecules per unit volume. The *molecular absorption coefficient* β is defined by $\beta = \alpha/N$. A solution is said to obey *Beer's law* over a certain range of concentrations if β has a constant value over this range. The fact that departures are obtained from Beer's law at large concentrations indicates that the absorption by an individual molecule is influenced by its environment.

10-6 SPECTROSCOPIC EFFECTS

The emission and absorption spectra of different substances are topics which are normally considered in a course on atomic physics. We include here only a brief description of certain effects which are associated with discrete line spectra.

Fluorescence: A substance absorbs light of a particular frequency from an incident beam and re-emits light of the same or lower frequency. The case in which the emitted and absorbed frequencies are the same is more particularly known as *resonance radiation*. The observation that the re-emitted light is not of higher frequency than the incident is known as *Stokes' law*. The spectral composition of the incident beam must contain energy at a frequency which coincides with a resonance frequency of the fluorescing substance, as this condition is necessary for the absorption process.

EXAMPLE 3. Although the spectrum of iodine vapor is extremely complex, it happens that the green line produced in a low pressure mercury arc is sufficiently narrow to overlap just one of the absorption lines of iodine. When illuminated under these conditions, the iodine vapor emits a number of frequencies all of which are lower than or equal to the frequency of the green line. These frequencies all belong to the normal emission spectrum of iodine vapor but constitute a select subset thereof.

The quantum explanation of fluorescence assumes that a molecule is first raised to an excited state by absorption of an incident photon. The return to the ground state can then occur through emission of a photon of the same frequency (resonance radiation) or through transition to a series of intervening levels, each of the photons in this case having less energy than the photon absorbed, and hence a lower frequency. If the absorbing atom happens to be initially in an excited state (a rare event at room temperature) and then emits all its energy after absorption by a single transition to the ground state, the resulting photon has more energy than that absorbed and thus leads to an exception to Stokes' law.

Raman effect: A substance is irradiated with monochromatic light which does not coincide with one of its resonance frequencies. The scattered radiation consists predominately of light which is identical in frequency with the incident light. In addition, however, the scattered radiation contains a series of lines of both higher and lower frequency than the original. The *spacing* of these lines is characteristic of the *scattering* substance and is intimately related to its regular infrared spectrum.

EXAMPLE 4. Filters are used to isolate the mercury violet line (4358 Å), and the scattering of this light by liquid benzene is examined. In addition to 4358 Å, the scattered radiation contains a number of nearby lines. A prominent one of these has the wavelength 4555 Å. The frequency difference between these two lines is $2.972 \times 10^{13}\ \text{sec}^{-1}$, for which the associated wavelength is 1.0088×10^{-5} m. This wavelength is in the infrared range and corresponds to an observed line in the benzene spectrum.

The quantum interpretation of scattering is different from that of fluorescence in that separated processes of absorption and later re-emission are not involved. Scattering is attributed to a *collision* between an incident photon and a molecule. The majority of these collisions are *elastic* and the scattered photon has the same energy (and hence frequency) as the incident. *Inelastic collisions* occur if the internal energy of the molecule is changed during the collision process. This usually involves a change in the discrete vibrational or rotational kinetic energies of the molecule. If the molecule is raised to an excited state by the collision, the colliding photon loses an amount of energy equal to the energy level difference of the molecule. In this case, the scattered photon has a lower frequency than the incident. Since this is similar to Stokes' law for fluorescence, the corresponding lines are said to be Stokes lines. On the other hand, a number of the molecules may be in excited vibrational or rotational states before the collision; these transfer a discrete amount of energy to the colliding photon. This gives rise to anti-Stokes lines, i.e., scattered photons of frequency higher than the frequency of the incident photons. Since the infrared spectrum is associated with *direct* rotational and vibrational transitions, the frequency differences observed in the Raman effect correspond to lines in the infrared spectrum.

Zeeman effect: A source emitting a discrete spectrum is subjected to a strong homogeneous magnetic field. Each spectral line is observed to split into a number of closely spaced, polarized components. The spacing of the components is proportional to the field strength. A number of lines conform to a law which can be deduced classically from Larmor's theorem (Section 10–2). The remainder are said to exhibit the *anomalous* Zeeman effect, which is successfully accounted for only by quantum mechanics.

Stark effect: A splitting of spectral lines into polarized components, similar to the Zeeman effect, occurs when the source is subjected to a strong *electric* field. A part of the broadening of spectral lines in a discharge tube is explained by the Stark effect produced in a given atom due to the electric fields of neighboring ions. The atoms are subject to a variety of field strengths, with subsequent different amounts of splitting. The superposition therefore contains a broad distribution of frequencies.

PROBLEMS

10–1. An optically active substance obeying Biot's law is mounted between crossed polarizers. The thickness is such that light of wavelength 5000 Å undergoes a rotation of 4π radians. Describe the appearance of the spectrum of the emergent light when white light is incident.

10–2. A monochromatic plane wave is incident on a polarizer and passes through a thickness of material which rotates the plane of polarization through 45°. The beam is then reflected at a mirror and passes back through the material. (a) Given that the material is optically active, determine whether or not the returning beam will be passed by the polarizer. (b) Do the same for a material exhibiting the Faraday effect.

10–3. Given the value $\chi = 21.7°$ per mm, calculate the value of $(n_{-} - n_{+})$ for sodium light in quartz.

10–4. For many substances the index of refraction as a function of wavelength can be fitted empirically to an equation of the form $n = A + B/\lambda^2$ (Cauchy's formula). Use this information and Becquerel's formula to determine the wavelength dependence of the angle of rotation in the Faraday effect.

10–5. Assume that Becquerel's formula holds for water (a diamagnetic material). Given that $n = 1.3372$ for $\lambda = 4860$ Å and $n = 1.3312$ for $\lambda = 6560$ Å: (a) Estimate the value of Verdet's constant for sodium light. (Actual measured value: $V = 4.8 \times 10^{-6}$ radians per ampere.) (b) Approximately what field strength would be required to make the rotation angle per centimeter for the Faraday effect in water the same as the rotation angle per centimeter for optical activity in quartz?

***10–6.** A variable potential difference $V(t)$ applied across the capacitor plates in a Kerr electro-optical shutter produces an amplitude modulation of the transmitted beam. Assume that the changes in the applied potential difference are slow compared with the period of the incident sinusoidal light wave. The polarizer and analyzer are crossed and are at 45° to the field between the capacitor plates. (a) Express the E-vector of the emergent beam as a function of the time for given $V(t)$. (b) Obtain a simplified expression for the amplitude of the emergent light beam when the maximum value of $V(t)$ corresponds to a retardation which is small compared with one radian. (c) Is a sinusoidal variation in $V(t)$ faithfully reproduced in the amplitude modulation of the light beam?

10–7. In the experimental demonstration of Example 1 sodium light is used and the polarizer P is inserted. The tube is one meter long and contains a solution of one gram of sucrose (cane sugar) for each cubic centimeter of water. A solution of this concentration rotates the plane of polarization of linearly polarized sodium light through an angle of 66.5° in 10 cm. Describe the appearance of the tube as viewed from the side.

11 Interference pattern from a pair of point sources

11-1 INTRODUCTION

Consider two monochromatic point sources S_1 and S_2 as indicated in Fig. 11-1. The sources are assumed to be of the same frequency, though not necessarily in phase with each other. The net disturbance at a point of observation P is given by the sum of the sinusoidal functions representing the disturbances arriving at P from S_1 and S_2. The amplitude of the resultant is critically dependent on the phase difference $\Delta\phi$ between these two contributions. At those locations where $\Delta\phi = 2n\pi$, $(n = 0, \pm 1, \pm 2, \ldots)$ the contributing waves tend to re-enforce each other; where $\Delta\phi = (2n + 1)\pi$, $(n = 0, \pm 1, \pm 2, \ldots)$ they tend to cancel each other. Whether or not this cancellation is exact depends on the relative amplitudes, a circumstance we shall examine in more detail presently.

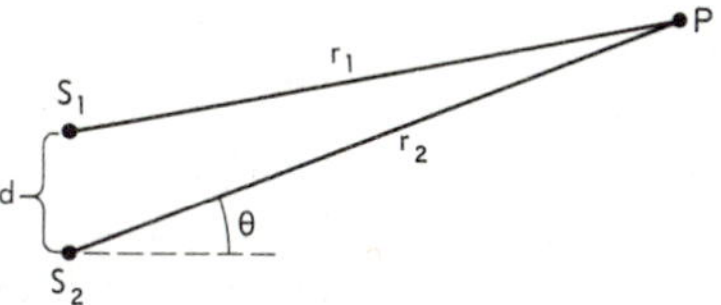

Fig. 11-1. Signals arriving at point of observation P from two point sources S_1 and S_2.

If we represent the time dependence of the signal emitted at S_1 by the function $\cos\,[\omega t - (\phi_1)_0]$, then, after propagating a distance $S_1P = r_1$, the time dependence of the signal arriving at P from S_1 is

$$\cos\left[\omega\left(t - \frac{r_1}{c}\right) - (\phi_1)_0\right] = \cos\,(\omega t - \phi_1),$$

where $\phi_1 = (\phi_1)_0 + kr_1$. Thus the phase difference between the signals arriving at P from S_1 and S_2 is

$$\Delta\phi = \phi_2 - \phi_1 = (\Delta\phi)_0 + k(r_2 - r_1), \tag{11-1}$$

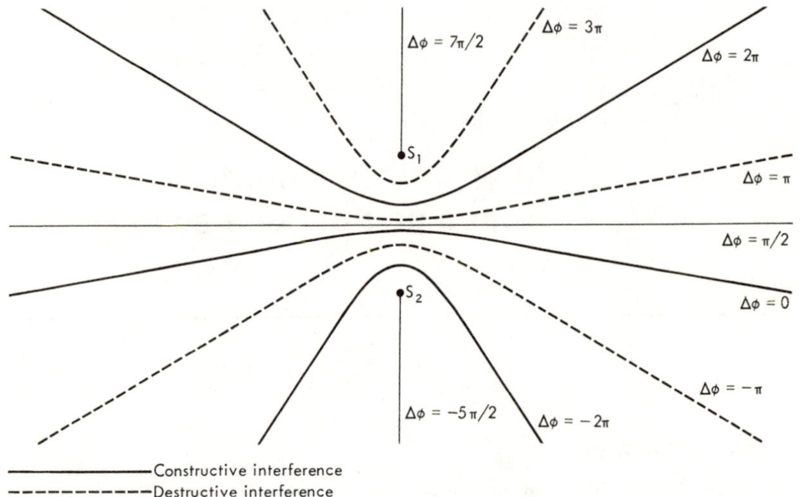

Fig. 11-2. Hyperboloidal loci of constant $\Delta\phi$ for $(\Delta\phi)_0 = \pi/2$ and $kd = 3\pi$.

where $(\Delta\phi)_0 = (\phi_2)_0 - (\phi_1)_0$ is the phase difference between the two sources.

The set of points for which $r_2 - r_1$ has a constant value defines a locus along which $\Delta\phi$ is constant. It is clear from the general definition of an hyperbola that the locus $r_2 - r_1 =$ constant is a figure of revolution about the axis S_1S_2, whose intercept in the plane of the diagram is one branch of an hyperbola. Thus the loci of constructive and destructive interference constitute a family of hyperboloids of revolution of which S_1 and S_2 are the common foci. If the distance between S_1 and S_2 is d, the value of $r_2 - r_1$ lies between $-d$ and $+d$. Consequently, a limited number of loci of constructive or destructive interference are present. Figure 11-2 shows a typical case where $(\Delta\phi)_0 = \pi/2$ and $kd = 3\pi$.

The precise nature of the "disturbance" which emanates from each point source depends on the physical variable being described. (See Chapter 17 for a specific discussion of the acoustic case.) For general purposes we shall assume that we are dealing with a scalar field variable ψ whose square determines the intensity of the radiated wave. The dependence of the amplitude of ψ on r, the distance from the point source, follows from the assumption that a point source radiates uniformly in all directions. Thus writing $\psi = A(r)\cos(\omega t - \phi)$, we see that the average flux through a sphere of radius r concentric with the source is proportional to $A^2(r)$ and to the area of the sphere, $4\pi r^2$. In a steady state (as is the case with our sinusoidal disturbance) the energy content of the region between two spheres of different radii remains constant. This implies that the average power flux through the two spheres must be the same. We therefore conclude that $r^2A^2(r)$ is constant and hence $A(r) \propto 1/r$ for the steady-state radiation from a spherically symmetric point source.

If the sources are of equal strength, the same function $A(r)$ represents the amplitudes of the contributions from S_1 and S_2, and the net disturbance at P is given by the superposition

$$\psi = \psi_1 + \psi_2 = A(r_1)\cos(\omega t - \phi_1) + A(r_2)\cos(\omega t - \phi_2). \quad (11\text{–}2)$$

If we assume that P is in a location such that the amplitudes can be considered approximately equal, $A(r_1) \doteq A(r_2) = A$, this expression becomes

$$\psi = A\{\cos(\omega t - \phi_1) + \cos(\omega t - \phi_2)\}.$$

With $\bar{\phi} = (\phi_1 + \phi_2)/2$ this reduces by trigonometric identity to

$$\psi = 2A\cos(\Delta\phi/2)\cos(\omega t - \bar{\phi}). \quad (11\text{–}3)$$

According to this formula, ψ vanishes wherever $\Delta\phi = (2n + 1)\pi$. Note, however, that according to the more accurate expression in Eq. (11–2), the condition $\phi_2 = \phi_1 + (2n + 1)\pi$ yields

$$\psi = [A(r_1) - A(r_2)]\cos(\omega t - \phi_1).$$

Unless $A(r_1) = A(r_2)$, the two contributions do not exactly cancel each other. In more specific terms, the approximation $A(r_1) \doteq A(r_2)$ requires

$$1 \gg \left|\frac{A(r_1) - A(r_2)}{A(r_1)}\right| = \left|1 - \frac{r_1}{r_2}\right| = \left|\frac{r_2 - r_1}{r_2}\right|.$$

The largest possible value of $|r_2 - r_1|$ is d, hence the approximation is assured if $1 \gg d/r_2$, i.e., if the distance from the point of observation to the sources is large compared with the distance between sources. If we cannot assume that $A(r_1) \doteq A(r_2)$, we must obtain the net disturbance at P by direct addition of the sinusoidal functions in Eq. (11–2).

Using the approximate form given by Eq. (11–3) we observe that the amplitude of the net signal at P differs from that due to a single source by the factor $2\cos(\Delta\phi/2)$. Therefore the average intensity I differs by a factor $4\cos^2(\Delta\phi/2)$ from the average intensity I_0 we would have from a single source:

$$I = 4I_0\cos^2(\Delta\phi/2). \quad (11\text{–}4)$$

By virtue of the dependence of $\Delta\phi$ on the location of P this formula expresses the intensity variation as a function of position for distances which are large compared with the distance between the sources. On a locus of destructive interference, $\Delta\phi = (2n + 1)\pi$ and $I = 0$. On a locus of constructive interference, $\Delta\phi = 2n\pi$ and $I = 4I_0$. Thus the intensity is four times what would be obtained from one of the sources without the presence of the other source.

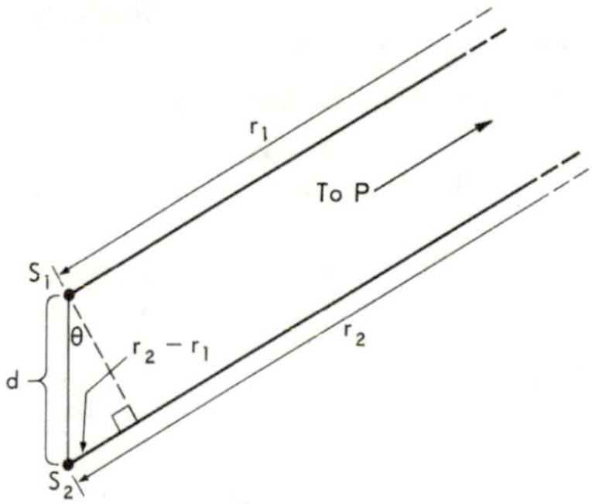

Fig. 11-3.
Path difference for rays going from the point sources S_1 and S_2 to a distant point of observation P.

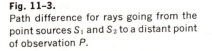

To study the interference pattern given by Eq. (11-4) it is convenient to introduce an approximation for the quantity $r_2 - r_1$ as it appears in the phase difference given by Eq. (11-1). Since we have assumed that r_1 and r_2 are much larger than d, we can treat the lines S_1P and S_2P as nearly parallel to each other. Then Fig. 11-3 shows that $r_2 - r_1 \doteq d \sin\theta$. More precisely, on applying the law of cosines to Fig. 11-1 it follows that

$$r_1^2 = r_2^2 - 2r_2 d \sin\theta + d^2,$$

or

$$\frac{r_1}{r_2} = (1 - 2x \sin\theta + x^2)^{1/2} = f(x),$$

where $x = d/r_2$. Expanding in a Maclaurin series, we have

$$\frac{r_1}{r_2} = f(0) + xf'(0) + \frac{x^2}{2} f''(0) + \cdots$$

$$= 1 - x \sin\theta + \frac{x^2}{2} \cos^2\theta + \cdots$$

or

$$r_1 = r_2 - d \sin\theta + \frac{d^2}{2r_2} \cos^2\theta + \cdots$$

It will be legitimate to drop the term in d^2, provided this has a negligible effect in Eq. (11-4). The contribution that this term makes to $\Delta\phi$ is, from Eq. (11-1), $-(kd^2/2r_2)\cos^2\theta$. This produces a negligible change in $\cos(\Delta\phi/2)$, provided $(kd^2/4r_2)\cos^2\theta \ll \pi/2$. Setting $k = 2\pi/\lambda$ and $\cos\theta \sim 1$, we find that the approximation $r_2 - r_1 \doteq d \sin\theta$ requires $r_2 \gg d^2/\lambda$. It can be shown that this approximation is equivalent to the assumption that the point of observation P is sufficiently far away from the sources so that the hyperbolas $\Delta\phi$ = constant are well approximated by their asymptotes. It is to be noted that the crude replacement $r_1 = r_2$ which was used in approximating $A(r_1) \doteq A(r_2)$ is not adequate for the phase factor $\Delta\phi$ since significant phase changes may be associated with the additional term $d \sin \theta$. Substituting into Eq. (11–4), we find that the final expression for the intensity distribution under the conditions $d \ll r_2$ and $d \ll \sqrt{r_2\lambda}$ is

$$I = 4I_0 \cos^2\left[\frac{(\Delta\phi)_0}{2} + \frac{kd}{2}\sin\theta\right]. \tag{11–5}$$

The remainder of this chapter is concerned with applications of Eq. (11–5). These are treated under the categories $d \ll \lambda$, $d \sim \lambda$, and $d \gg \lambda$.

*11–2 SOURCES CLOSE TOGETHER COMPARED WITH A WAVELENGTH; THE DIPOLE SOURCE

If the two sources S_1 and S_2 are in phase $[(\Delta\phi)_0 = 0]$ and closely spaced compared with a wavelength ($kd \ll 1$), Eq. (11–5) becomes

$$I = 4I_0 \cos^2\left[\frac{kd}{2}\sin\theta\right] \doteq 4I_0. \tag{11–6}$$

The angular dependence of the intensity pattern is slight. For all intents and purposes the two sources act together like a single source of double strength.

On the other hand, for a *dipole source* the two sources are taken to be exactly out of phase with each other and closely spaced compared with a wavelength. Thus setting $(\Delta\phi)_0 = \pi$ and $kd \ll 1$, Eq. (11–5) becomes

$$I = 4I_0 \cos^2\left[\frac{\pi}{2} + \frac{kd}{2}\sin\theta\right] = 4I_0 \sin^2\left(\frac{kd}{2}\sin\theta\right) \doteq (kd \sin\theta)^2 I_0. \tag{11–7}$$

From this formula $I \ll I_0$, which means that the radiation from the dipole source is inefficient compared with the radiation from a single source. The inefficiency is the more pronounced the smaller kd is. For a given distance of separation, d, lower frequencies radiate less efficiently (in comparison with a single source of the same frequency) than higher frequencies.

Complete destructive interference takes place only in the direction $\theta = 0$ (along the line perpendicular to the axis of the dipole). There is no position at which constructive interference occurs. The maximum intensity occurs at $\theta = \pi/2$. These results are summarized in the polar plot of intensities in Fig. 11–4.

In acoustics an isotropic source of spherical waves can be visualized as a sphere whose radius varies sinusoidally about a mean value. (For the excess acoustic pressure p to have the properties assumed for the general variable ψ, the point of observation must be at a large distance compared with a wavelength; see Chapter 17.) An acoustic dipole could be constructed by separating two such spheres, whose vibrations are π out of phase, by a distance which is small compared with a wavelength. Alternatively, we obtain the equivalent of a dipole source from a single sphere of fixed radius whose center executes simple harmonic motion of amplitude which is small compared with the radius and small compared with a wavelength. For qualitative purposes any oscillating body, for example, a single tine of a tuning fork, can be considered as an assemblage of dipole sources (see Fig. 11–5).

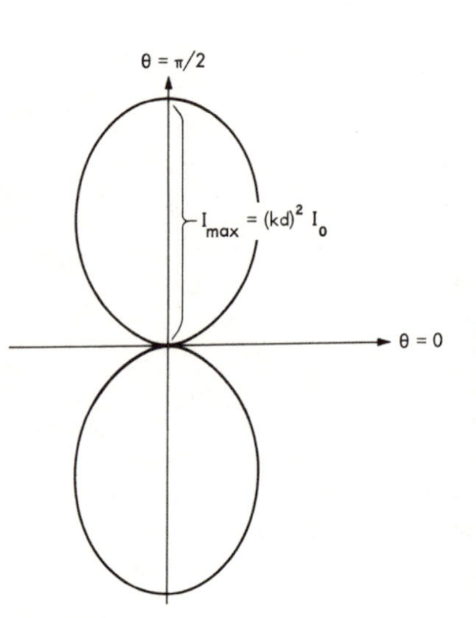

Fig. 11–4.
Intensity vs. angular position at a fixed distance from a dipole source. (Axis of dipole is $\theta = \pi/2$.)

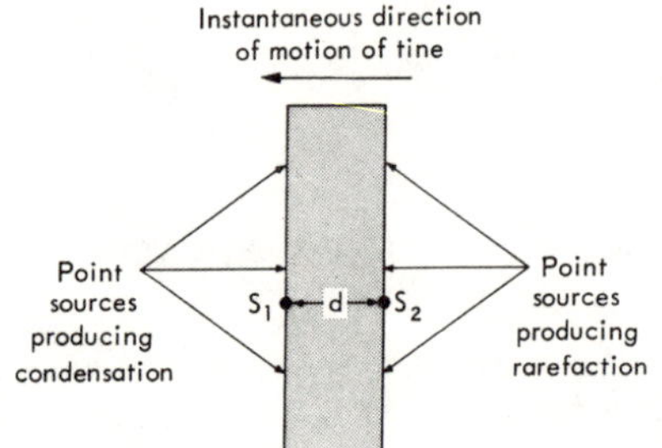

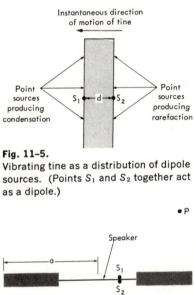

Fig. 11–5.
Vibrating tine as a distribution of dipole sources. (Points S_1 and S_2 together act as a dipole.)

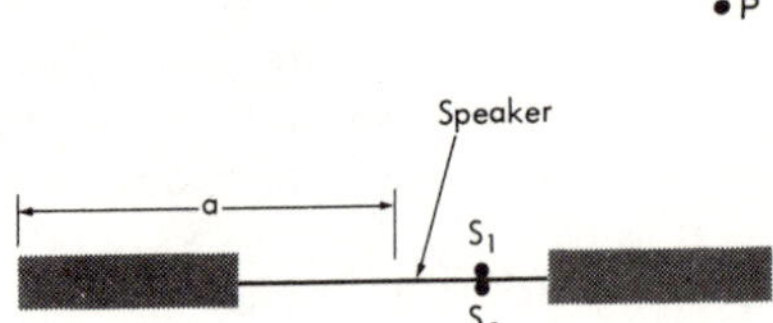

Fig. 11–6.
Speaker mounted in a baffle.

If you hold a vertical tuning fork (not mounted on a resonator) near one ear, and rotate it about a vertical axis, the sound you hear has distinct minima for certain orientations. However, there are four minima for one revolution of the tuning fork, and not two as in Fig. 11–4. The reason is that both tines are present. This cannot be avoided by placing a card between the tines, since there would then be an "image source" due to reflection. The tines move in opposite directions and there is a tendency for the signals they produce to cancel each other. Two closely-spaced dipole sources of opposite phase constitute what is known as a *quadrupole* source. The only thing we can be sure of from the present analysis is that there will be a null on the $\theta = 0$ axis: if neither dipole radiates in that direction, the combination cannot do so.

It is easy to visualize why an acoustic dipole is an inefficient radiator. When the pressure near S_1 is a maximum, the pressure near S_2 is a minimum. There is thus a large pressure gradient which tends to accelerate particles from S_1 to S_2. The predominant motion is a "local flow," or an oscillation of the particles back and forth in the vicinity of the sources. When S_1 alone is present, the pressure gradients and particle velocities are entirely radial and give rise to a greater outward flux of energy.

A loudspeaker is an inefficient radiator of low frequency sounds if both faces of the vibrating cone are open to the room, because it then acts like a collection of dipole sources. This inefficiency can be avoided by mounting the speaker so that waves generated at the back surface cannot reach the point of observation. This wave may be suppressed by enclosing the back of the speaker in a cabinet lined with absorbent material, or by having the speaker mounted in the wall of a room. If the speaker is mounted in the center of a board of radius a (see Fig. 11–6), the dipole effect is in part removed. Without the baffle the geometrical path difference $S_2P - S_1P$ is small, and since S_1 and S_2 are π out of phase, the two signals cancel to a large degree. When the baffle is present, any disturbance emanating from S_2 cannot go directly to P, but must travel an additional distance the order of magnitude of a. The effect may be considered a virtual separation of S_1 and S_2 by the distance a. For those wavelengths for which $\lambda \gg a$ we are still dealing with the dipole situation, and the radiation efficiency is not improved. But the more efficient radiation patterns to be described in the next section apply to those wavelengths for which $\lambda \lesssim a$. That is, local flow has been thwarted for the vibrations of wavelength shorter than a. A convincing demonstration of the effectiveness of a baffle is provided by listening to the change in the low frequency response of a small speaker when a baffle is alternately inserted and removed.

11-3 VARIOUS INTERFERENCE PATTERNS FOR $d \sim \lambda$

In Fig. 11–7 the polar plot of intensities is given for three examples of Eq. (11–5) when d is comparable to λ. These graphs give the angular distribution of intensity in the interference patterns which would be obtained from two equal sources of sound which radiate isotropically (e.g. two pulsing spheres). We should recall that the circle along which observations are made must have a radius r such that $r \gg d$ and $r \gg d^2/\lambda$. For example, in case (a) with $d = 2\lambda$, both conditions are satisfied with $r \gg 4\lambda$.

We can apply these formulas to a study of the radiation pattern of arrays of radio antennas. Suppose that S_1 and S_2 are *vertical* antennas and that the observations are made at various angular positions in a horizontal plane. It is true that S_1 and S_2 are not point sources and hence the formulation of Section 11–1 is not strictly applicable. The individual antennas do not

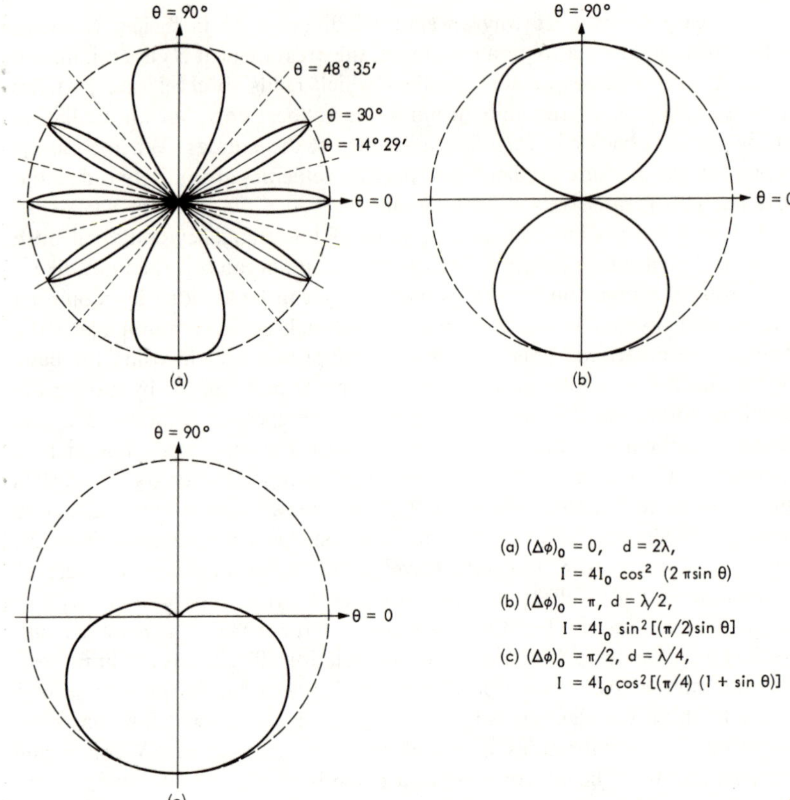

Fig. 11-7. Polar plots of Eq. (11-5). The radius of the dotted circle is $4I_0$. The line joining the sources is $\theta = 90°$.

radiate isotropically in all directions. However, if the antennas are vertical, the radiation is uniform so far as directions in the horizontal plane are concerned. It is doubtful now that the amplitude factor in Eq. (11-2) should be taken proportional to $1/r$. However, we ended up by ignoring the distinction between r_1 and r_2 in the amplitude factors. The factor I_0 can be taken to stand for the intensity which would be obtained from a single source at the given distance regardless of how this may depend on r. The arguments concerning the variation of r in the important phase factors are all correct for this application. We may therefore interpret the graphs of Fig. (11-7) as directional radiation patterns in a horizontal plane from a pair of vertical antennas having the given spacing and phase relation. Case (c) is a "directional array" which would be efficient if the receiving station is in the direction $\theta = 3\pi/2$.

Case (b) can also be applied to a single *horizontal* antenna placed a quarter-wavelength above the surface of the earth. Since the earth is a good conductor at radio frequencies, the net pattern obtained will be due to the direct wave from the antenna S_1 and the reflected wave which can be associated with an image source S_2 a distance $\lambda/4$ below the surface. Due to the phase change upon reflection, S_1 and S_2 are nearly π out of phase, and the distance separating the sources is $d = \lambda/2$. We must suppose in this case that observations are made along a vertical semicircle at a large distance from S_1. We cannot predict the variations of intensity in a horizontal direction unless we know the angular distribution of radiation from the antenna. However, if the plane of observation is the perpendicular bisector of the horizontal antenna, the radiation is isotropic in this plane. The conditions of case (b) then apply and can be used to judge the efficiency of reception as a function of the height of the receiving antenna, which determines the angle θ. (Many factors have been neglected: curvature of the earth, the existence of a "ground wave;" the interference effect considered is an important one at radar frequencies.)

*11-4 TOTAL POWER RADIATED FROM A PAIR OF POINT SOURCES

In the elementary discussion of interference (Section 4–8) we found that the intensity of the superposition of two waves which interfere constructively is four times the simple sum of the intensities of the component waves. When the waves interfere destructively, the intensity is zero. At first sight it seems that we are able to create or destroy energy merely by adding two waves with the appropriate phase relation. The examples of the last two sections are concrete illustrations of how waves are added in a physical situation. This suggests the possibility that the energy excesses in certain regions may be compensated for by energy deficiencies elsewhere. In Fig. 11–7(a), for example, we might expect that the energy is channeled along the loci of constructive interference but that the total energy integrated over a sphere would be the sum of the energies associated with the two sources individually. We can easily perform this calculation. The interference pattern is a figure of revolution about the axis $\theta = \pi/2$. We can therefore choose an element of area on the surface of a sphere of radius r:

$$dA = 2\pi r^2 \cos\theta \, d\theta.$$

The total energy crossing this sphere per unit of time is

$$P = \int_{-\pi/2}^{\pi/2} I(\theta) 2\pi r^2 \cos\theta \, d\theta,$$

where $I(\theta)$ is given by Eq. (11–5). An elementary integral results upon making the substitution $\mu = \sin\theta$. The actual power radiated is to be compared with $2P_0 = 2(4\pi r^2 I_0)$, which is the sum of the powers associated with the individual sources. Defining ϵ as the ratio of these two quantities, we find

$$\epsilon \equiv \frac{P}{2P_0} = 1 + \left(\frac{\sin kd}{kd}\right)\cos(\Delta\phi)_0. \tag{11-8}$$

In general, therefore, we do not obtain $\epsilon = 1$, which is what we would expect if the total power radiated by the sources acting together were the sum of the powers of the two acting independently. We see that linearity of superposition does not apply to the total power radiated any more than it does to the intensities at a given point. We have assumed that the signals from S_1 and S_2 add linearly. Thus, for example, S_2 produces a contribution to the disturbance at large distances which is the same as that which would be produced if S_1 were not present. This does not imply, however, that S_2 is doing the same amount of work on its surroundings as it would if S_1 were not present. Suppose, for example, that S_2 is a sphere whose surface is vibrating in simple harmonic motion. The power delivered by a surface moving with given velocity depends on the pressure at the surface. We can see in the case of the dipole source that as S_2 expands, the pressure surrounding S_2 is less than normal because S_1 is contracting at this time. (There is no time lag since S_1 and S_2 are close together compared with a wavelength.) Therefore less work is done by S_2 as it expands than would be the case if S_1 were not present.

It should also be pointed out that there may be a more direct interaction of the two sources. For example, it may not be possible for the driving mechanism of S_2 to maintain exactly the same motion of the surface as it would in the absence of S_1. Similarly, if S_1 and S_2 are radio antennas, the field due to S_1 will tend to alter the current distribution which would otherwise exist in S_2. This form of coupling of the two sources is a practical matter that can be minimized by appropriate design of the driving mechanisms and is not directly connected with matters of principle concerning the law of conservation of energy.

It is interesting to note that $\epsilon = 1$ if the two sources are in quadrature, that is, if they are $\pi/2$ out of phase. Also, for $kd \gg 1$ the magnitude of the second term in Eq. (11-8) is insignificant, and $\epsilon \doteq 1$. For $kd \ll 1$,

$$\frac{\sin kd}{kd} \doteq 1 - \frac{(kd)^2}{6} \quad \text{and} \quad \epsilon \doteq 1 + \left[1 - \frac{(kd)^2}{6}\right]\cos(\Delta\phi)_0.$$

If the two sources are in phase, $\epsilon \doteq 2$, and, as we have noted before, the two sources act together like a single source having twice the amplitude. For the dipole source $\cos(\Delta\phi)_0 = -1$ and $\epsilon \doteq (kd)^2/6 \ll 1$. We can see this result from Eq. (11-7) if we note that the average value of $\sin^2\theta$ over the surface of a sphere is $\frac{1}{3}$.

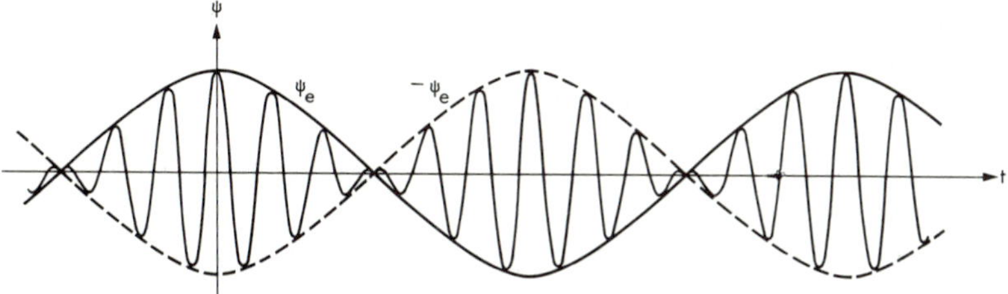

Fig. 11–8. The sum of two sinusoidal functions of different frequency. (Graph of Eq. 11–9.)

*11–5 THE PHENOMENON OF BEATS

Consider two waves of different frequencies, ω_1 and ω_2, which are superposed at a given point of observation where their amplitudes are equal. We can represent the resultant by the following identity:

$$\begin{aligned} \psi &= \mathcal{R}e\,\{Ae^{i\omega_1 t} + Ae^{i\omega_2 t}\} \\ &= \mathcal{R}e\,\left\{Ae^{i(\omega_1+\omega_2)t/2}\left[e^{i(\omega_1-\omega_2)t/2} + e^{-i(\omega_1-\omega_2)t/2}\right]\right\} \\ &= 2A\cos\Omega t\cos\bar{\omega}t, \end{aligned} \tag{11–9}$$

where $\Omega = (\omega_2 - \omega_1)/2$ and $\bar{\omega} = (\omega_1 + \omega_2)/2$. If $\Omega \ll \bar{\omega}$, the function $\cos\Omega t$ varies slowly in comparison with $\cos\bar{\omega}t$, and it is convenient to conceive of the graph of Eq. (11–9) in terms of the envelope concept. Thus, if $\psi_e = 2A\cos\Omega t$, the graph of ψ lies between the graphs of ψ_e and $-\psi_e$. When $\cos\bar{\omega}t = +1$, ψ touches the graph of ψ_e, and when $\cos\bar{\omega}t = -1$, ψ touches the graph of $-\psi_e$. This is illustrated in Fig. 11–8. Let the instantaneous intensities associated with ω_1 and ω_2 be $I_1 = KA^2\cos^2\omega_1 t$ and $I_2 = K'A^2\cos^2\omega_2 t$. The proportionality factors K and K' may be functions of the frequency, but since we are applying this to the case $\omega_2 - \omega_1 \ll \bar{\omega}$, we shall assume $K' = K$. The time average intensities associated with the individual signals are therefore equal:

$$\bar{I}_1 = \bar{I}_2 = \tfrac{1}{2}KA^2 \equiv \bar{I}_0.$$

The instantaneous intensity associated with the superposition is

$$I = 4KA^2\cos^2\Omega t\cos^2\bar{\omega}t.$$

On averaging over one cycle of $\cos\bar{\omega}t$, and treating $\cos\Omega t$ as approximately constant over this relatively short time interval, we obtain the average intensity

$$\bar{I} = 4\bar{I}_0\cos^2\Omega t. \tag{11–10}$$

If the response time τ of a receiving instrument is small compared with $2\pi/\Omega$, the instrument will be sensitive to the fluctuation in intensity; this is referred to as the phenomenon of beats. Maximum intensity occurs when $\Omega t = n\pi$, that is, once every π/Ω seconds. The frequency of the beats is therefore $\Omega/\pi = (1/2\pi)(\omega_2 - \omega_1) = \nu_2 - \nu_1$ and is numerically equal to the difference in frequency of the two signals. If $\tau \gg 2\pi/\Omega$, the receiving instrument perceives only the time average of Eq. (11–10) over a period of $\cos \Omega t$: $\overline{\overline{I}} = 2\overline{I}_0$. In this case the intensity of the superposition is just the sum of the intensities of the components. There is no "long term interference" between two sources of different frequency.

The phenomenon of beats can be heard when two musical instruments sounding the same note are slightly out of tune. If the frequency difference is relatively large, a rasping "difference tone" is heard. If the two frequencies are close, the beats are experienced as a succession of swells whose frequency decreases as the frequency difference decreases. Recorder players can tell how much the pitch of their instrument depends on how hard it is blown by observing the beats produced against an instrument blown steadily.

As an alternate method, the phenomenon of beats can be derived by application of Eq. (11–4). The two sources of slightly differing frequency can be viewed as equivalent to two sources of the same frequency but with a slowly varying relative phase. Since $\Omega = (\omega_2 - \omega_1)/2$, we have $\omega_2 = \omega_1 + 2\Omega$ and

$$\cos \omega_2 t = \cos [(\omega_1 + 2\Omega)t] = \cos (\omega_1 t + (\Delta\phi)_0),$$

where $(\Delta\phi)_0 = 2\Omega t$ is the phase difference between the two sources. At a distant point of observation where $\Delta\phi = (\Delta\phi)_0 + kd \sin \theta$, the intensity then varies according to Eq. (11–4):

$$I = 4I_0 \cos^2 \frac{\Delta\phi}{2} = 4I_0 \cos^2 \left\{\Omega t + \frac{kd}{2} \sin \theta\right\} \cdot \qquad (11\text{–}11)$$

This is equivalent to Eq. (11–10) except for the arbitrary choice of the origin of time. (Observers at different values of θ will hear the beat maxima at different times.) Effectively this means that the beat phenomenon can be interpreted as a slow migration of the pattern of Fig. 11–2, the minima occurring as the loci of destructive interference sweep past the ear.

11-6 INTERFERENCE PATTERNS WHEN $kd \gg 1$

When the distance between sources is large compared with a wavelength, the pattern is similar to that of Fig. 11–7(a), except that the number of lobes is now large. The receiving instrument may be incapable of registering the fine structure in the interference pattern, in which case it responds only to a spatial average of the intensity. We can obtain this from Eq. (11–5) by setting the average value of the cosine-squared equal to $\frac{1}{2}$. (This assertion is justified by the more meticulous calculation of Problem 11–9.) The average intensity is thus $2I_0$, the simple sum of the intensities which would be obtained from the sources acting independently. The result is equivalent to ignoring interference between the two sources and treating them as if they were noncoherent. The only interesting cases will be ones in which kd is large, but not so large that the interference pattern cannot be discerned.

a) Consider first a case in which the pattern is examined along a section of a sphere (or the tangent plane to it) located near the axis $\theta = 0$. The intersections of the hyperboloids of revolution with this surface are approximately straight lines perpendicular to the plane of the diagram in Fig. 11–9.

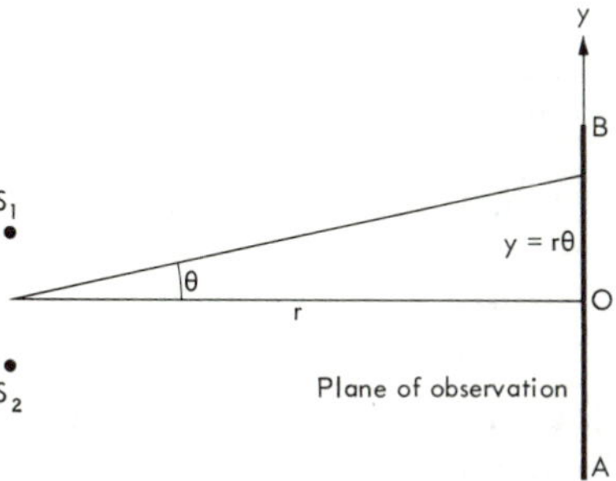

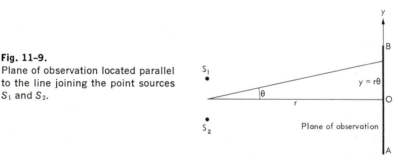

Fig. 11–9. Plane of observation located parallel to the line joining the point sources S_1 and S_2.

When $kd \gg 1$, a large number of interference loci will be encompassed in the region for which θ is small. We therefore assume that observations are restricted to such a region and make the replacement $\sin\theta \doteq \theta$. From Eq. (11–5) the intensity variation along the line AB is then given by

$$I = 4I_0 \cos^2\left[\frac{(\Delta\phi)_0}{2} + \frac{kd}{2}\,\theta\right] = 4I_0 \cos^2\left[\frac{(\Delta\phi)_0}{2} + \frac{kd}{2r}\,y\right]$$

$$= 4I_0 \cos^2\left[\frac{kd}{2r}\,\bar{y}\right], \tag{11–12}$$

where $\bar{y} = y + r(\Delta\phi)_0/kd$. The "center" of the pattern occurs at $\bar{y} = 0$ or $y = -r(\Delta\phi)_0/kd$ and is a position of constructive interference at which $\Delta\phi = 0$. The graph of Eq. (11–12) is shown in Fig. 11–10(a). Because of the approximate logarithmic response of the eye, it is perhaps more appropriate in an optical context to plot the logarithm of the intensity (Fig. 11–10b). If there is a threshold intensity I_{thr}, the regions $I < I_{\text{thr}}$ will appear completely dark. These dark bands are equally spaced along the screen and are separated by bright fringes which are relatively uniform except near the edges. Because of this uniformity it is difficult for the eye to judge the position of maximum intensity; more accurate settings can be obtained at the center of a dark fringe.

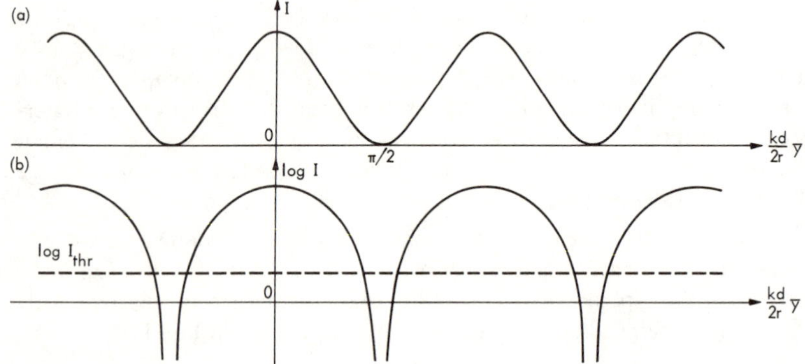

Fig. 11-10. Intensity distribution in a straight fringe pattern. (a) Intensity vs. position, (b) Logarithm of intensity vs. position.

The distance between adjacent dark bands is given by

$$\Delta\left(\frac{kd}{2r}\,\bar{y}\right) = \pi, \qquad \text{or} \qquad \Delta\bar{y} = \Delta y = \frac{\lambda}{d}\,r. \tag{11-13}$$

The *angle* subtended at S_2 by a complete fringe (from the center of one dark band to the center of the next) is independent of r:

$$\Delta\theta = \frac{\Delta y}{r} = \frac{\lambda}{d}\cdot \tag{11-14}$$

b) The second case of special interest is observation along a small section of a plane set perpendicular to the axis $\theta = \pi/2$. The intersections of the hyperboloids of revolution with this plane are circles. In the context of optics the pattern consists of a series of dark circular bands separated by bright circular fringes.

11-7 YOUNG'S EXPERIMENT

In seeking to demonstrate the interference patterns obtained between two light sources we are immediately confronted with the problem of coherence. As long as the phase difference between the sources remains constant a definite interference pattern will be obtained; the position of the pattern will depend on the value of $(\Delta\phi)_0$ (cf. Eq. 11-12). However, we have already had evidence in Section 7-7 that the phase of a light source does not remain constant over a period of time comparable to the resolving time of the eye. Therefore, if two independent sources are used, the associated interference pattern shifts position more rapidly than the eye can follow. The eye then responds at a given location to the intensity given by Eq. (11-12), averaged

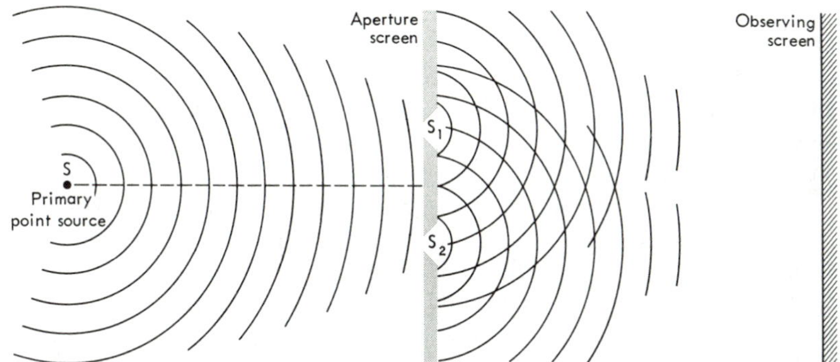

Fig. 11-11. Young's double-pinhole experiment.

with respect to a random set of values of $(\Delta\phi)_0$. This average is given by

$$\bar{I} = \frac{4I_0 \int_0^{2\pi} \cos^2\left(\frac{u}{2} + \frac{kd}{2r}\,y\right) du}{2\pi} = 2I_0\,. \tag{11-15}$$

This average value is independent of position and is the same as the value that would be obtained if we ignored any interaction between the signals from the two sources.

A common method of obtaining a pair of coherent sources in optics is based on the use of *secondary* sources S_1 and S_2 derived from a single primary source S. This method has the result that the phase difference between S_1 and S_2 remains constant even though the phase of S varies. Thomas Young devised a simple arrangement for obtaining this condition experimentally. Two pinholes S_1 and S_2 are made in a screen (see Fig. 11-11). A monochromatic point source S lies on the perpendicular bisector of S_1S_2. Since the path lengths SS_1 and SS_2 are equal, the light signals at the two pinholes remain in phase with each other at all times. We anticipate from general observations concerning the waves passing through a small opening that diffraction will take place at the pinholes. A more complete discussion of this effect will be deferred to the next chapter. For the moment we shall assume by analogy with ripple tank experiments that the extent to which diffraction occurs depends on the ratio of the radius of the pinhole to the wavelength. In particular, for a small enough pinhole, diffraction is complete, and the diffracted wave consists of spherical wavefronts diverging from the pinhole as if it were a point source. We will assume this condition of complete diffraction in the present discussion. In this way the pinholes S_1 and S_2 can be considered as a pair of point sources which remain in phase with each other.

When this experiment is performed it is found that straight fringes are obtained on the observing screen. It can be checked that the spacing of the fringes depends on the distance between the pinholes and on the distance between the two screens in the manner predicted by Eq. (11–12). These measurements can then be used to infer the value of the wavelength of the monochromatic source.

In 1807, when Young originally performed this experiment, Huygens' wave theory of light had long since been abandoned in favor of a corpuscular theory. To interpret the results of his experiment, Young found it fruitful to use the wave theory with a few additional features. Huygens' discussion of waves had been primarily concerned with the propagation of "wavefronts" with certain characteristic velocities and made no essential use of sinusoidal waves. The important notion added by Young was the concept of periodicity and phase which introduces the possibility of interference.

Young received his training in medicine and wrote his doctoral dissertation on the human voice. An important investigation on the lens of the eye won him election to the Royal Society at the age of 21. Having a thorough knowledge of both optics and acoustics, he was intrigued by the analogies between them. It was this comparison which led him to devise the interference experiment. (Young also did important work in the study of elasticity (Young's modulus), the theory of color vision, and is noted for his contribution in deciphering the Rosetta Stone.)

Young's interpretation of the results of his experiment as a wave-interference phenomenon was not greeted with the kind of impartiality we like to think characterizes the scientific community. As the result of a particularly violent attack by one individual, Young's reputation was considerably damaged and he withdrew from the field of optics. Acceptance of his ideas did not occur until after they had been amplified and put to brilliant use by the French physicist Augustin Fresnel. In 1818 the French Academy offered a prize for the best paper concerning certain experimental results which had been termed "diffraction effects." This was done with the full expectation that these effects could be explained on the basis of the corpuscular theory. One of the members of the Commission of the Academy was the mathematician Poisson who was a staunch supporter of the corpuscular theory. While reading the paper submitted by Fresnel, Poisson worked out, as a consequence of Fresnel's theory, the curious prediction that the shadow of a circular disc should have a bright spot at the center. Poisson suggested that his friend Arago perform the experiment to demonstrate the absurdity of the theory. What Arago found, however, was exactly what the theory had predicted. Fresnel was awarded the prize and from that point on the wave theory of light was taken more seriously. In later work Fresnel made an important addition to the theory by recognizing the transverse character of the waves. On this basis he was able to answer some of the legitimate objections which had been raised to the original form of Young's theory.

The next two chapters are devoted to applications of the concepts introduced by Young and Fresnel. For reasons of space we must omit discussion of some of the interesting special effects such as the Arago bright spot.*

11-8 SOME PRACTICAL CONSIDERATIONS IN YOUNG'S EXPERIMENT

The discussion of the interference experiment in the last section was based on the assumption that the primary source S is an ideal point source and that the pinholes S_1 and S_2 are of negligible size. Various ways in which these conditions can be relaxed will be discussed in the present section. In the ideal experiment the interference pattern can be observed on any screen placed at an arbitrary distance satisfying the restrictions $r \gg d$ and $r \gg d^2/\lambda$. The smallness of the pinholes of course implies low intensities, which may require that the fringes be detected by a time-exposure photograph. On the other hand, if the size of the pinholes is increased, complete diffraction does not take place, and the secondary sources do not radiate isotropically. We shall see that this imposes the necessity of using a lens in order to "focus" the interference pattern on an observing screen in a fixed location, and that this pattern is then limited to the region in which the diffraction cones of S_1 and S_2 overlap. Since the pattern consists of straight fringes, we can further increase the intensity by replacing the primary source with a line source parallel to the direction of the fringes. Finally, this line source need not be of infinitesimal width; a criterion will be developed for the breadth of source which still permits the observation of a clear set of interference fringes.

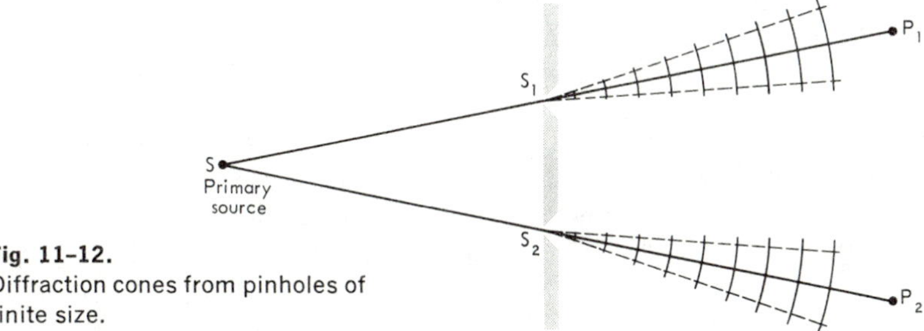

Fig. 11-12. Diffraction cones from pinholes of finite size.

a) Limited Diffraction Due to Finite Aperture Size; The Use of a Lens

As indicated in Fig. 11-12, the diffracted wave produced by an aperture whose size is not small compared with a wavelength consists of spherical wavefronts along which the intensity is not uniform. If S_1P_1 is the extension of the central

---

* For the story of the public reception of Young's work and a detailed account of Fresnel's contributions see pages 144–163 and Chapter XIV of E. Mach, *The Principles of Physical Optics, An Historical and Philosophical Treatment*, translated by J. S. Anderson and A. F. A. Young, New York: Dover Publications, 1953.

ray SS_1, the energy in the diffracted wave from S_1 is concentrated in a cone whose axis is the central ray. (It is assumed that the aperture size remains sufficiently small so that the *geometrical* pencil of rays joining S with all of the points in S_1 can still be treated like a single ray.) At any point on an observing screen placed to the right of the aperture screen the amplitudes of the signals arriving from S_1 and S_2 will be radically different and the interference pattern derived for the case of two equal sources is not to be expected.

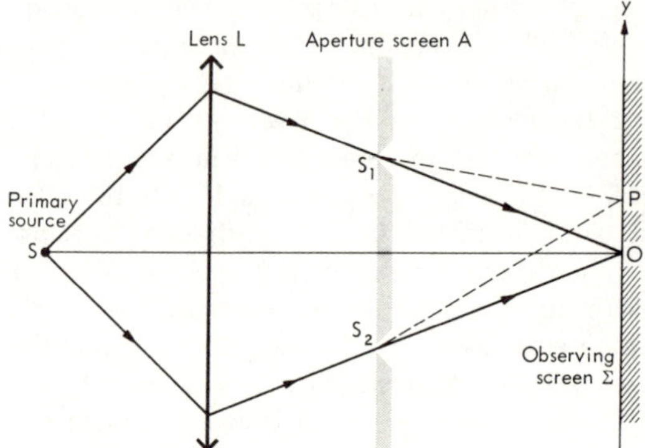

Fig. 11–13. Central rays from a point source on the axis of a lens.

Suppose, however, that a lens L is introduced as in Fig. 11–13, and an observing screen is placed in a location such that the central rays are brought together at point O. This implies that O and S are conjugate image points with respect to the lens. In other words, the lens is placed so that a focused image of the primary source would be obtained on the observing screen if the aperture screen were absent. This arrangement surely produces a considerable overlap of the diffraction cones from S_1 and S_2. Furthermore it is clear from the symmetry of the figure that the secondary sources S_1 and S_2 are in phase, since the rays joining them to S by refraction in the lens pass through equal thicknesses of glass. Thus two rays S_1P and S_2P within the respective diffraction cones meet at P with a relative phase determined only by the geometrical path difference $PS_2 - PS_1$. The straight fringe pattern of Eq. (11–12) is obtained in the region of overlap.

b) Primary Source Not on the Axis of the Lens

To treat the effect of an increase in size of the primary source we need to consider the phase relation between the secondary sources S_1 and S_2 due to a point source S' not on the axis of the lens (Fig. 11–14). The argument by symmetry of the preceding section is no longer applicable. However, we note that the effect of the lens is to produce a spherical wavefront AB whose center is the image point O' of S' in L. This means that all points on AB are in phase.

Therefore the phase difference between S_2 and S_1 is

$$(\Delta\phi)_0 = k(BS_2 - AS_1) = -k(S_2O' - S_1O').$$

This result can be summarized by considering that the source S' is replaced by its image O'. Since O' is to the right of the aperture screen it is "virtual." Thus the phase difference can be written in the form $(\Delta\phi)_0 = k(r_2 - r_1)$, where $r_i = O'S_i$ is considered negative when O' is to the right of the aperture screen.

If the image of S' falls to the left of the aperture screen it is quite obvious that O' acts like a primary source in determining the phase relation between S_1 and S_2, since rays then diverge from O' exactly as if it were a real point source. For the present application we are interested specifically in a case where O' acts as a virtual source for the diffraction, since maximum overlap of the diffraction cones requires O' to lie on the observing screen.

The important results are the obvious ones: the phase relation between S_1 and S_2 is such as to make O' the center of the interference pattern; other than that, the width of the fringes is given by Eq. (11–13), $\Delta y = r\lambda/d$, where r is the distance between the aperture screen and the observing screen.

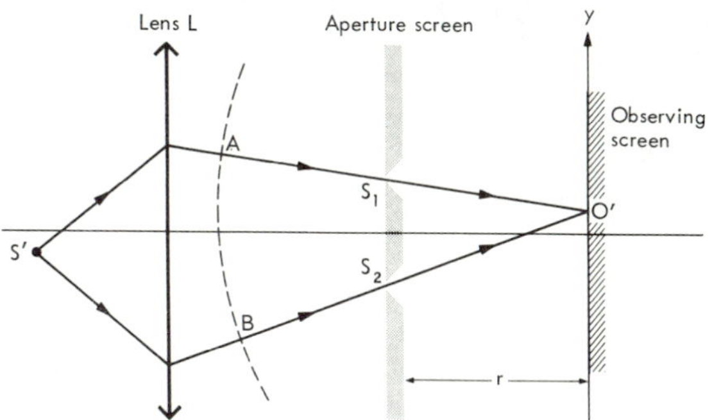

Fig. 11–14. Central rays from a point source S' not on the principal axis of a lens.

c) Use of a Line Source as the Primary Source

Let S and S' be two independent point sources, with S lying in the plane of the diagram in Fig. 11–13 and S' below S in a direction perpendicular to the plane of the diagram. By an argument similar to that of the last section, both S and S' produce straight fringes perpendicular to the plane of the diagram. The fringe width is the same in both cases and the centers are located at $y = 0$. The fringe system associated with S' is shifted relative to that of S in a direction parallel to the fringes. This is shown in Fig. 11–15, where C' and C represent the intersections of the respective diffraction cones with the observing screen. Since S and S' are noncoherent, the intensity on the screen is obtained by addition of the separate intensities; no additional interference takes place.

It is therefore clear that the primary source can be replaced by a line source S oriented in a direction perpendicular to the line joining the pinholes. The result is a series of straight fringes parallel to the line source. In the y-direction this fringe system is still limited by the size of the diffraction cone, but in the z-direction the fringes extend a distance essentially determined by the angle subtended at the lens by the two ends of the line source.

In practice, we can obtain a line source by placing a narrow slit (the source slit) adjacent to a discharge tube. The different molecules of the source which lie behind each point of the slit radiate incoherently of each other. More conveniently, a focused image of the discharge tube can be projected onto the source slit.

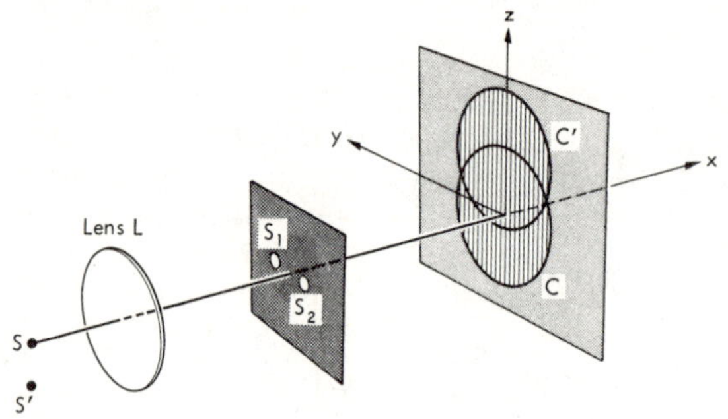

Fig. 11–15. Independent fringes produced by noncoherent sources S and S'.

The breadth of the source slit is restricted by a condition which we shall discuss in the next section. It turns out that the source slit can be made relatively broad in the sense that diffraction need not be taken into account on passage through it. This is especially true since the sources are effectively located in the plane of the slit itself.

It is tempting to suppose that the effect of lengthening the apertures S_1 and S_2 into "diffracting slits" in the z-direction of Fig. 11–15 could be considered a simple extension of the present discussion. Although the resulting pattern indeed resembles the patterns considered so far, the interpretation requires a more complete understanding of the diffraction by an aperture of finite size, and will be deferred until Chapter 12. To understand that there is an added complication, consider a second pair of diffracting pinholes, S_1' and S_2', displaced from S_1 and S_2 in the z-direction. This pair would independently produce a fringe system which partly overlaps the fringes produced by S_1,S_2. However, the light transmitted by S_1',S_2' *is* coherent with that from S_1,S_2, since they are both derived from the same primary source. Therefore it is necessary to consider additional interference effects to have a proper interpretation of the resulting pattern.

d) Breadth of Source Condition

Consider now the effect of using a primary source S which is not of infinitesimal width in the direction perpendicular to the straight fringes. Such a situation is to be considered a collection of independent points S' radiating incoherently of one another. Each source point S' produces a set of straight fringes whose center of symmetry is determined by the geometry of Fig. 11–14. When the total width of the source is small, these patterns lie in approximately the same location on the observing screen and the superposition consists of a clear set of fringes. For a broad source, however, the bright bands produced by some of the points S' will fall at the location of the dark bands produced by other points on the source, and the net pattern tends to wash out. To study this process let the source consist of a line segment bisected by the axis of the lens and parallel to the line S_1S_2 joining the two pinholes. When the aperture screen is not present, the image of the source on the observing screen is designated in Fig. 11–16 by the line segment AB of length b. On the introduction of the aperture screen, the light which fell on the original image between y_0 and $y_0 + dy_0$ is now distributed in a straight fringe pattern centered about y_0. According to Eq. (11–12), the contribution to the intensity along the screen due to this portion of the source is

$$dI = (\alpha\, dy_0)\cos^2\left[\frac{kd}{2r}(y - y_0)\right],$$

where α is a constant, provided that we assume no variation in brightness across the source. From Eq. (11–13) we write

$$\frac{kd}{2r} = \frac{\pi d}{\lambda r} = \frac{\pi}{w},$$

where $w = r\lambda/d$ is the width on the observing screen of one complete fringe in the simple two-point interference pattern.

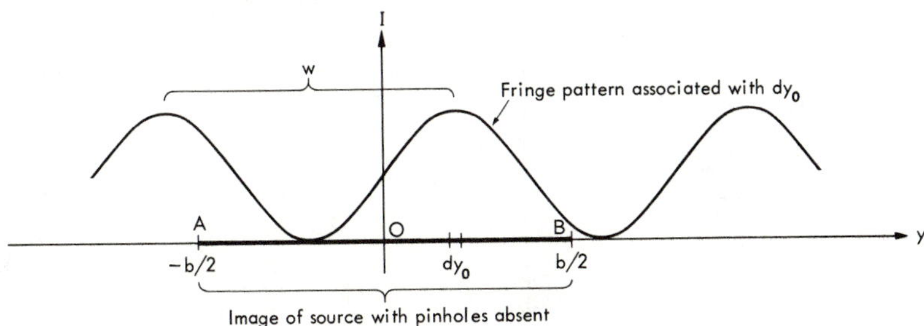

Fig. 11–16. Contribution to total pattern associated with portion dy_0 of a broad source.

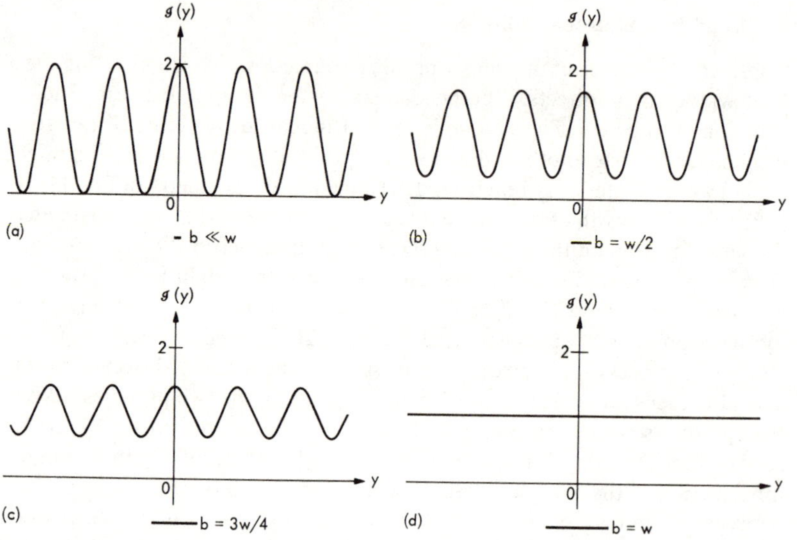

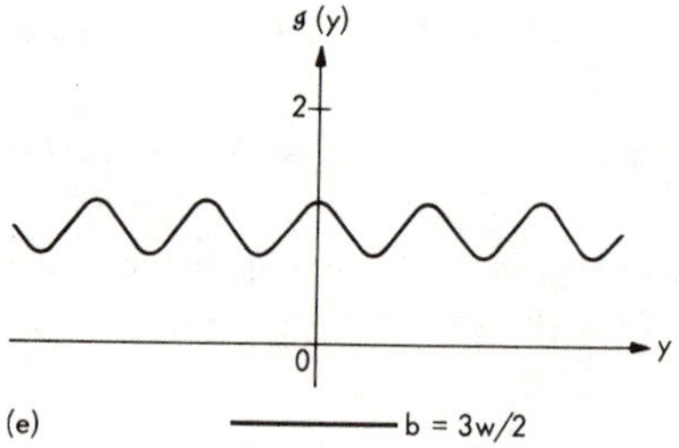

Fig. 11-17. Relative intensity distributions for different breadths of source slit. (The line beneath each graph represents the image of the source slit when the aperture screen is removed.)

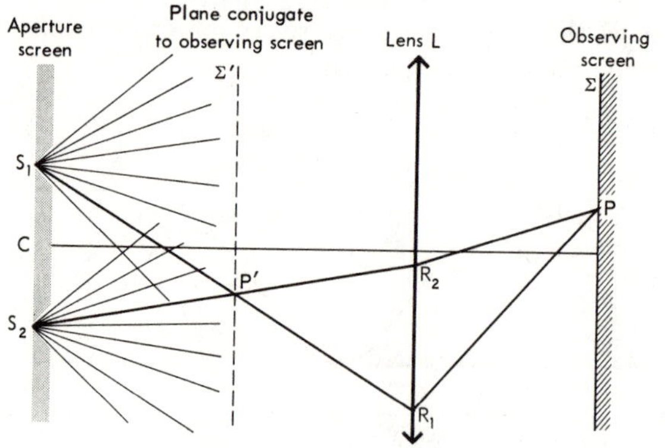

Fig. 11-18. Refraction of the rays arriving at the same point of observation P from pinholes S_1 and S_2.

The total intensity at any point y is therefore obtained by addition of the contributions from all the segments of the primary source:

$$\begin{aligned} I(y) &= \alpha \int_{-b/2}^{b/2} \cos^2 \left[\frac{\pi(y - y_0)}{w} \right] dy_0 \\ &= \frac{b\alpha}{2} + \frac{\alpha w}{2\pi} \sin \frac{\pi b}{w} \cos \frac{2\pi y}{w} . \end{aligned} \tag{11–16}$$

This result shows that the intensity is periodic in y with a period equal to the elementary fringe width w. The average intensity over a complete fringe is $b(\alpha/2)$, which increases in direct proportion to the breadth of the source. For convenience in presenting the result graphically, we divide $I(y)$ by its average value to obtain the relative intensity distribution:

$$\mathcal{I}(y) = \frac{2I(y)}{b\alpha} = 1 + \frac{\sin (\pi b/w)}{(\pi b/w)} \cos \frac{2\pi y}{w} . \tag{11–17}$$

This function is graphed in Fig. 11–17 for progressively increasing values of the breadth of source in relation to the fringe width of the interference pattern. We observe that the pattern loses contrast as b is increased and that the fringes completely disappear when $b = w$. Beyond this point an intensity variation reappears, but the contrast remains poor. We may therefore adopt as an order-of-magnitude criterion for a "broad" source the condition $b \gtrsim w$. Thus it is advantageous to let more light through a source slit by widening it until the width of the image of the slit becomes comparable to the width of a complete fringe in the two-source interference pattern.

e) Additional Use of Lenses

It is frequently desirable to use collimators, telescopes, microscopes, or the lens of the eye as part of the optical system to observe a two-point interference pattern. To treat these cases we must supplement the preceding discussion by considering the effect of a lens interposed between the aperture and observing screens. We shall find that this lens has the effect of "projecting" onto the observing screen the interference pattern which in principle exists in a different plane of observation.

For simplicity let us assume that the pinholes are sufficiently small to allow complete diffraction to take place at each. (If this is not the case, the optimal viewing conditions are obtained as before when the entire combination of lenses in the system is arranged so that a focused image of the primary source is obtained on the observation screen when the aperture screen is removed.) In Fig. 11–18 rays are shown emanating in all directions from the pinholes S_1 and S_2. Assume that the lens and observing screen are placed in arbitrary locations, with the sole provision that a focused image of the aperture screen does not fall on the observing screen. Under these conditions one and only one ray S_1R_1P emanating from S_1 is refracted by the lens to intercept the observing screen at a given point P. Likewise, there is one ray S_2R_2P arriving at P from S_2. These rays will interfere at P according to their phase relation, which it is now our task to determine.

Consider the intersection P' of the two rays in question. (A virtual point of intersection can be considered if the two rays do not meet before passing through the lens.) It is clear from the imaging properties of the lens that P' lies in the plane Σ' conjugate through L to the plane of the observing screen Σ. This implies that a point source located at P' would be imaged at P and hence, by an argument similar to one employed in (b) above, the phase retardations are the same along all rays joining P' and P by refraction through the lens. This means in particular that the phase relation of the two rays uniting at P is the same as the phase relation between these two rays when they intersect at P'. Hence the interference pattern obtained on the observing screen corresponds point by point with the pattern which would be obtained on a screen located in the plane Σ'.

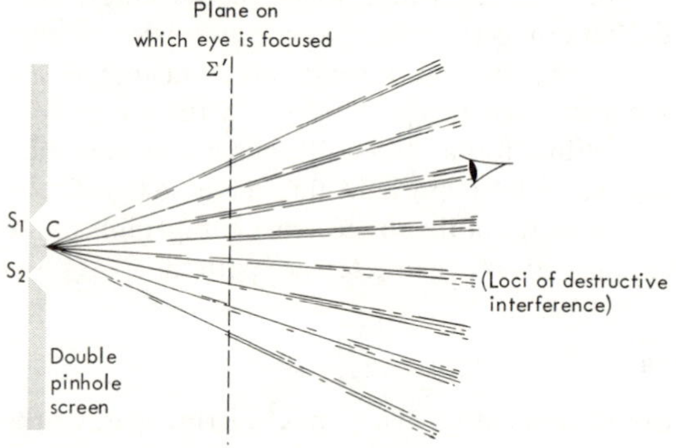

Fig. 11-19. Does the eye see the bright fringes in Σ'?

We will now point out by means of a paradox one assumption that has been implicit in the preceding discussion. Suppose that lens L is the lens of an eye which is focused on Σ'. We have implied that the image on the retina will consist of a projection of the interference pattern across Σ'. Suppose, however, that the pupil of the eye is located on a locus of destructive interference within the total interference pattern from the double source S_1S_2 (see Fig. 11-19). If there is no energy within the field at the location of the pupil, how can the eye register the bright bands across Σ'? The answer is, of course, that it cannot. In the preceding discussion we have tacitly assumed that lens L is of indefinite extent. It can be seen, for example, that if the lens is slightly smaller than indicated in Fig. 11–18, the ray S_1P' will not be refracted to point P and there will be no interference. It should be understood, therefore, that in general the interference pattern projected on the observing screen is only a limited portion of the pattern in Σ'. Considering the closeness of S_1 to S_2, this portion is effectively the circular area intercepted by the cone subtended at C by the circumference of the lens. We can resolve the above paradox by noting that the portion of the interference pattern in Σ' which the eye can perceive is limited to a dark band if the pupil itself lies within a dark band—the relevant cone from C coincides with the asymptote to a locus of destructive interference. (It is to be recalled that our discussion of interference patterns is restricted to the asymptotic region.)

APPLICATIONS

i) *Direct visual observation.* A pair of pinholes can easily be prepared by piercing a small card with a fine-pointed needle. The diameter of the holes must be kept small so that the diffraction disk from each hole will be perceptibly broad. The distance between centers must be relatively small or the fringes in the interference pattern will be excessively close together. The pinholes are placed directly in front of the pupil of the eye. In this case the lens L is the lens of the eye and the "observing screen" is the retina. A distant street lamp will serve as the primary source S. To obtain maximum overlap of the diffraction disks produced by the two pinholes, the eye adjusts itself for focus on an object at a distance equal to that of the source. This means that the plane Σ' conjugate to the retina coincides with the location of S. (This is a case in which the intersection point P' is virtual.) Thus the virtual object for the image produced on the retina can be considered a set of interference fringes located in the plane of the source S. (Note that the conditions $r \gg d$ and $r \gg d^2/\lambda$ are adequately met even though the pinholes are close to the retina. The effect of the lens of the eye is to project an interference pattern for which r is the distance between the observer and the primary source.)

The angular width of the fringes is the quantity which is significant in visual observation. This is given by λ/d and is independent of the distance from the source to the observer. As the card is rotated, the fringes remain perpendicular to the line joining the pinholes.

A variety of line sources are available in any night scene: a tubular lamp, the crack in a lighted doorway, an illuminated column, the reflection streak of a light bulb on wet pavement or the straight filament of an unfrosted incandescent bulb. If the card is turned properly, interference bands can be seen across the diffraction images of such sources. We can observe the breadth of source condition by noticing that the interference bands are not obtained from sources which subtend a wide angle.

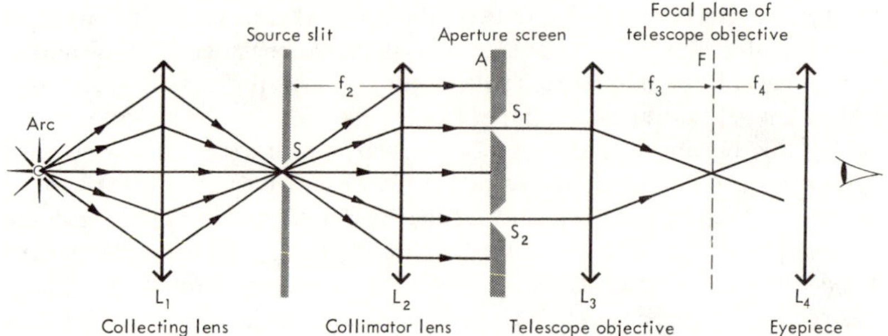

Fig. 11-20. Observation of interference fringes with collimator and telescope.

ii) *Collimator and telescope.* Figure 11–20 indicates an arrangement suitable for quantitative measurement of the two-source interference pattern. The source slit S is located in the focal plane of the collimator lens L_2. From the point of view of the aperture screen this is equivalent to having a real source at infinity. The central rays (passing through S_1 and S_2 from S) are brought together in F, the focal plane of the telescope objective, and this is the plane in which the best focus will be obtained on the interference fringes. Since the plane conjugate through L_3 to F is at infinity, the interference pattern in F is equivalent to the pattern at an infinite distance from the aperture screen, thus satisfying ideally the conditions for the approximations made in the calculations. The fringes in F are observed by means of the eyepiece L_4 which produces a virtual image at infinity so that the eye can operate in its most relaxed condition. The angle subtended at L_3 by the fringes in F is λ/d, corresponding to an actual width $f_3\lambda/d$. These fringes therefore subtend an angle $(f_3/f_4)(\lambda/d)$ from L_4. Hence the telescope produces an angular magnification of f_3/f_4 over what would be obtained by omitting L_3 and L_4 and observing directly with the eye.

f) Interference Fringes with White Light

The context of the preceding discussion has been that of monochromatic light of specified wavelength. If the light contains a mixture of wavelengths, such as with white light, each component gives rise to an interference pattern independent of the others. The perpendicular bisector of S_1S_2 is a locus of zero path difference for all wavelengths. This position is therefore the center of a bright fringe for all the components and the resultant color is identical with that of the original source. Progressing outward from the center of the pattern, we reach a point of extinction for shorter wavelengths sooner than for longer wavelengths, and therefore progress through a series of colors complementary to the extinguished colors. At large distances from the center of the pattern a number of colors across the visible spectrum are missing. The addition of the closely spaced remaining colors gives rise to the sensation of white light.

g) The Laser as a Source for Interference Experiments

The preceding parts of this section describe a number of artful dodges by means of which the intensity of a system of fringes can be boosted while still maintaining the characteristics of the double-source interference pattern. The necessity for seeking such compromises results from the nature of conventional light sources. Prior to 1960, the only sources available involved the emission of light by a collection of atoms which radiate incoherently of one another. The atoms are raised to excited states either by thermal processes for incandescent solids or by collision with electrons of the tube current for arcs and gaseous discharges. The atoms then radiate energy by the process of *spontaneous emission*. In this process an atom can remain for a while in an excited state, but eventually emits a photon whose frequency corresponds to the energy lost in a transition to one of the lower energy levels. Associated with the fact that there is statistical variation in the time at which spontaneous emission occurs, there is no phase correlation among the waves emitted by different atoms. Given this noncoherency, the only way to obtain clear fringe patterns is to employ a collection of atoms whose independent fringe patterns will coincide in position.

Aside from spontaneous emission, there is another process by which an excited atom can radiate when it is in the electromagnetic field of an incident light wave. This process, known as *stimulated emission*, can be partially understood from the following classical explanation. Consider an elementary oscillator consisting of an electron bound by an elastic restoring force associated with angular frequency ω_0. Assume that the oscillator receives energy from collisions occurring at random time intervals. In the intervals between collisions the electron is subject to a driving force due to the external field of strength E_0 and frequency ω'. The differential equation which determines the displacement $x(t)$ of the electron is thus an inhomogeneous equation; the general solution is the sum of a particular solution and a solution to the homogeneous equation. From this it is possible to calculate the instantaneous rate of dipole radiation, which is proportional to $|\ddot{x}|^2$, the square of the acceleration. Now the homogeneous, or transient, solution is sinusoidal with the natural frequency ω_0 and has no phase correlation with the incident field. The associated radiation is the incoherent spontaneous radiation. On the other hand, the particular, or steady-state, solution is sinusoidal at the driving frequency ω', has a definite phase relation to the incident field, and has an amplitude proportional to E_0. The contribution which this steady-state solution makes to the emitted radiation corresponds to the stimulated, or induced, emission. Its most important characteristics are:

i) Coherence with the incident wave.

ii) Frequency equal to that of the incident wave. (More generally, if the incident wave is not perfectly sinusoidal, the stimulated emission has the same line shape as the incident wave and is not affected by line broadening due to collisions, as is the spontaneous radiation.)

iii) Since $|\ddot{x}|^2 \propto E_0^2$, the radiated power is proportional to the power of the incident wave.

iv) The ratio of the radiated power to the incident power is a function of the frequency difference $\omega' - \omega_0$. An especially strong (resonant) response is obtained only if this frequency difference is smaller than the line breadth of the spontaneous radiation. (When ω' is not close to ω_0, the stimulated emission is identical with the scattered radiation of Section 10–4. The Rayleigh scattering formula can be obtained by carrying out the calculations outlined above.)

Several important inferences can be drawn from the fact that the stimulated emission is coherent with the incident wave. First, this means that a collection of atoms radiating under the inducement of the same incident wave are no longer to be treated independently. This invites the possibility of producing high power levels of coherent radiation through the "cooperation" of a large number of atoms. Second, consider the atoms which lie in a plane perpendicular to the direction of propagation of an incident plane wave. Each of these emits a spherical wave in phase with all the rest. If the atoms are closely spaced, the superposition of these waves can be obtained by application of Huygens' construction to the contributing spherical waves. The common tangent is a plane wave parallel to the wavefronts of the incident wave. For application to simple interference experiments this is the most important feature to be exploited. It means that a high-intensity plane wave can be obtained with almost no angular dispersion. Compare this with the conventional means of producing a "plane wave": a "small" source is placed at the focal point of a lens—to the extent that the source *is* small, low power is obtained; to the extent that the source is *not* small, a set of plane waves is obtained whose normals define a range of angles equal to the angle subtended by the source from the center of the lens. Consider also the focusing by a lens of a plane wave produced by the cooperative action of atoms under stimulated emission. If we ignore diffraction effects associated with finite aperture of the lens or of the device in which the stimulated emission takes place, the image of the plane wave is a perfect point. All the energy emitted by a large number of atoms converges to this single point. Hence, we can realize the conditions of an ideal point source by using either the plane wave directly (a point source at infinity) or the image produced by the lens. No hedging is required to boost intensities.

Although the concept of stimulated emission was available in classical physics, it required the advent of quantum mechanics before a practical means could be found of meeting certain requirements which are obvious from the above discussion:

a) The plane wave produced by stimulated emission from a collection of atoms has the same characteristics as the incident wave and can be considered an "amplification" of it. But the incident wave is subject to absorption as it passes through the collection of atoms. Unless the energy added by stimulated emission more than makes up for the absorption, the net emergent wave will be more feeble than the incident wave, and no useful amplification will be obtained.

b) Spontaneous emission occurs simultaneously with stimulated emission. Under conditions in which we can expect stimulated emission to be significant ($\omega' \sim \omega_0$), these two emissions have nearly the same frequency. Thus the emergent wave is adulterated with noncoherent radiation unless the power level in the incident wave can be raised high enough to allow the stimulated emission to predominate.

A device which successfully meets these requirements for electromagnetic waves in the visible or near-visible range of frequencies is known as a *laser*, this word being an acronym for light amplification by the stimulated emission of radiation. The principles on which such a device can be constructed were outlined in a paper by Schawlow and Townes in 1958 and were first put into successful operation by Maiman in 1960. The idea was suggested by analogy with the maser (a device for microwave amplification by the stimulated emission of radiation), which had been developed in 1954 by Gordon, Zeiger, and Townes. For this reason lasers are sometimes referred to as *optical masers*.

The details of laser construction are varied, and it is not our purpose to digress too far into the field of quantum electronics.* To give an idea of how requirements (a) and (b) can be met we will discuss the principles of one successful arrangement. Consider an atomic system of ground state energy E_0 and a selected pair of energy levels E_1 and E_2 with $E_2 > E_1$. The frequency $\nu_{21} = (E_2 - E_1)/h$ is to be the frequency at which laser action takes place. That is, an incident wave of this frequency is to be amplified by stimulating the emission of additional photons of the same frequency from atoms which happen to be in an initial state of energy E_2. Of a given number of atoms having this energy, some undergo spontaneous transition to E_1 with the emission of *incoherent* photons of frequency ν_{21} and some undergo

---

* Reference texts on the subject of lasers are G. Birnbaum, *Optical Masers*. New York: Academic Press, 1964; B. A. Lengyel, *Lasers, Generation of Light by Stimulated Emission*. New York: John Wiley & Sons, 1962; O. S. Heavens, *Optical Masers*. New York: John Wiley & Sons, 1964.

induced transition to E_1 with the emission of *coherent* photons of frequency ν_{21}.* The coherent process can be made to predominate over the incoherent process if the incident wave is of sufficiently high intensity, since the rate of spontaneous emission is independent of the incident intensity, whereas the rate of stimulated emission is proportional to it. The intensity of the incident wave is built up to a high level by the device of trapping it in a resonant cavity. For example, suppose that plane mirrors are set up at two ends of a cylinder within which the laser action is taking place. If the mirror surfaces are optically flat (i.e. have no irregularities larger than a small fraction of a wavelength), any photons which are emitted in a direction perpendicular to the mirror surfaces will be reflected back and forth within the cavity. In their travel through the laser medium these photons stimulate the emission of additional photons coherent with themselves and with directions of propagation which are also parallel to the axis of the cylinder. If the spacing between mirrors is a precise integral multiple of a half-wavelength for ν_{21}, a set of standing waves is formed. If their intensity can be allowed to build up sufficiently, a condition in which stimulated emission predominates over spontaneous will have been achieved and requirement (b) will have been met.

Turning now to the question of loss of photons due to absorption, we encounter the most ingenious aspect of the principle of laser operation. Photons of frequency ν_{21} are highly susceptible to absorption by atoms in energy state E_1 since they have precisely the right energy to raise the atom to state E_2. The rate at which this process occurs is proportional to the intensity of the incident wave; consequently, the high power levels required to enhance stimulated emission also increase the rate of absorption. However, the rate of absorption is proportional to n_1, the number of atoms in state E_1, and the rate of stimulated emission is proportional to n_2, the number of atoms in state E_2. It can be shown that the constants of proportionality are the same in the two cases; hence the ratio of the rates is given by the ratio of the population numbers, n_2/n_1. What is required, therefore, to have a net gain of coherent photons produced over photons absorbed, is to have $n_2 > n_1$. In thermal equilibrium the opposite condition prevails, and in fact $n_2 \ll n_1$.† For this reason the required condition is referred to as a *population inversion*. Such a situation can be achieved by using an additional energy level $E_3 > E_2$.

---

* Also some undergo spontaneous transition to other energy levels with $E < E_2$. These emissions have low intensity in comparison with the laser-amplified beam and are furthermore irrelevant since they occur at frequencies separated from ν_{21}.

† From thermodynamics the equilibrium ratio is given by the ratio of the Boltzmann factors:

$$\frac{n_2}{n_1} = \frac{e^{-E_2/kT}}{e^{-E_1/kT}} = e^{-(E_2-E_1)/kT} = e^{-h\nu_{21}/kT}.$$

With ν_{21} an optical frequency and T the order of room temperature, $h\nu_{21}/kT \sim 100$.

The atom can be excited to E_3 by a variety of processes. In the method known as *optical pumping* an arc lamp is selected which has a spectral line of frequency $\nu_{30} = (E_3 - E_0)/h$. When the laser material is irradiated with this source, atoms in the ground state E_0 are raised to E_3 by absorbing photons of energy $h\nu_{30}$. Since the vast majority of atoms are originally in the ground state, a large number of such transitions are induced. Some of the atoms which have been raised to E_3 undergo spontaneous transitions to E_2 and E_1 as well as to other levels which are irrelevant. Now if E_2 has been chosen so that there is a *slow* rate of spontaneous transition out of E_2 to any lower level (E_2 is said to be a metastable, or long-lived state), and if E_1 is chosen to have a *high* rate of spontaneous transition to lower states, a relative accumulation occurs in E_2, and the population inversion is achieved.

With $n_2/n_1 > 1$ the two "stimulated" processes, emission and absorption, show a net gain. The ratio must be made sufficiently large to make up for losses due to imperfections in the mirror surfaces (some of the reflected light does not travel along the axis and passes out the sides) and, of course, the energy extracted in the output.

The useful laser beam is obtained by having one or both of the mirror surfaces slightly transmissive. A small proportion of the photons which constitute the standing-wave system are transmitted rather than reflected. The emergent beam is a plane wave with nearly perfect collimation. In principle, the only reason for angular divergence of the beam is that the plane wave is not of infinite extent in the transverse direction. We can think of the emergent beam as a plane wave which has passed through a circular aperture whose radius is that of the end face of the cylinder. The diffraction associated with such a wide aperture gives rise to a slight degree of lateral spreading of the beam. Beam divergences as low as one milliradian have been achieved in practice with a ruby laser whose end face is 1 cm in diameter. This corresponds to a beam spread of only 10 ft at a distance of one mile.

11-9 INTERFERENCE PATTERNS OBTAINED BY THE USE OF VIRTUAL SOURCES

In Young's interference experiment we rely on two pinholes in an aperture screen to serve as a pair of coherent point sources. The phenomenon of diffraction plays an essential role, since the pinholes act like point sources radiating in many directions only if they are sufficiently small for a significant amount of diffraction to take place. There are a number of methods for producing the optical double-source interference pattern which avoid this complication and make no use of diffraction. The key idea is to produce virtual sources by means of reflection or refraction in mirrors, lenses, and prisms. The rays from such sources are mutually coherent, since they are derived from the same primary source, and interference is obtained in any region from which two sources are simultaneously in view.

Experiments of this type played a critical role in the controversy aroused by Young's revival of a wave theory of light. Skeptics were inclined to attribute the pattern which Young obtained to some "modification" of the light as it passed through the aperture screen. These alternative methods are not open to the same objection.

A few examples of the use of virtual sources are given in this section. The remaining sections of this chapter consist of further examples of the same principle, although in somewhat more elaborate form.

It is to be understood in the remainder of this section that "the primary source S" can be produced in any of the ways previously mentioned. For example, if plane waves from a laser source are used, S lies at infinity in a specified direction; if lenses have been interposed, S stands for the image of the original source.

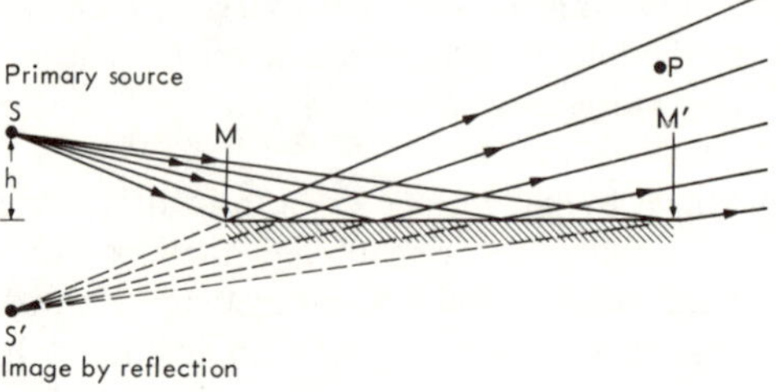

Fig. 11–21. Lloyd's mirror.

a) *Lloyd's mirror* (Fig. 11–21). The primary source is located at a distance h above the plane of a reflecting surface MM'. Since the angles of incidence are nearly grazing, the surface of any dielectric will do and need not be silvered. The rays between SM and SM' are reflected at the surface and appear to be diverging from a point S' at a distance h below the surface. Since grazing incidence is involved, the primary source S and the virtual source S' are of nearly equal strength. The source S' is coherent with S but π out of phase (cf. Problem 8–4). At any position within the angle $MS'M'$ rays are received from both S and S' and the double-source pattern associated with $d = 2h$ and $(\Delta\phi)_0 = \pi$ is obtained. The existence of the phase change on reflection can be demonstrated by focusing a microscope on a plane containing M', such as $M'P$. At M', where the edge of the mirror is in focus, there is a black band. The bright fringes between M' and P are of equal width. (If there were no phase change, M' would be at the edge of a bright fringe, half as broad as the others.) Since h must be small, the angle $MS'M'$ is also small. If one attempts to obtain a wider field of fringes by increasing h, the spacing of the fringes becomes smaller and they are more difficult to observe. Optimum conditions are obtained by putting S near M and using as long a surface MM' as is available.

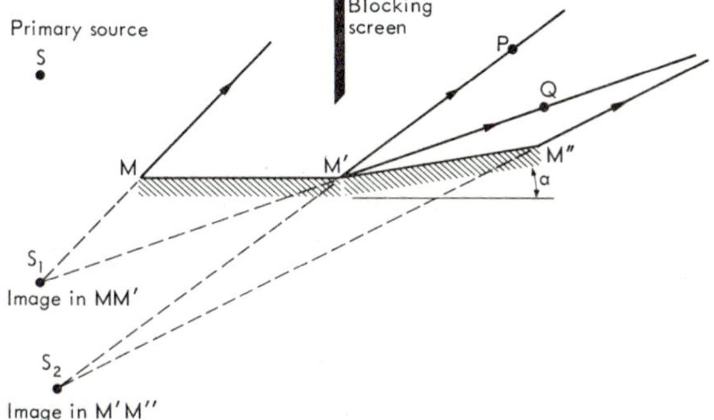

Fig. 11–22. Fresnel's double mirror.

b) *Fresnel's double mirror* (Fig. 11–22). Two reflecting surfaces MM' and $M'M''$ are inclined at a slight angle α. The respective images S_1 and S_2 are both visible within the angle $PM'Q$, which is the overlap region between angles MS_1M' and $M'S_2M''$. An obstacle is interposed so that direct rays from the primary source are blocked off. The virtual sources S_1 and S_2 are of equal strength and in phase with one another. Fringes are seen across a plane of observation such as PQ. The angular separation of the two sources from M' is $S_1M'S_2 = 2\alpha$. To achieve broad fringes, we would like α to be small. However, the "useable angle" $PM'Q$ is also 2α, so that when the fringes are broad there are few of them. Since the pattern near the edges of the field, P and Q, is complicated by diffraction effects, a reasonable compromise is one in which 15 or 20 fairly narrow fringes lie between P and Q.

c) *Fresnel's biprism* (Fig. 11–23). The triangle ABA' represents a cross section of a prism with equal angles α at A and A'. For simplicity, we consider the source S to be at infinity along the perpendicular bisector of AA'

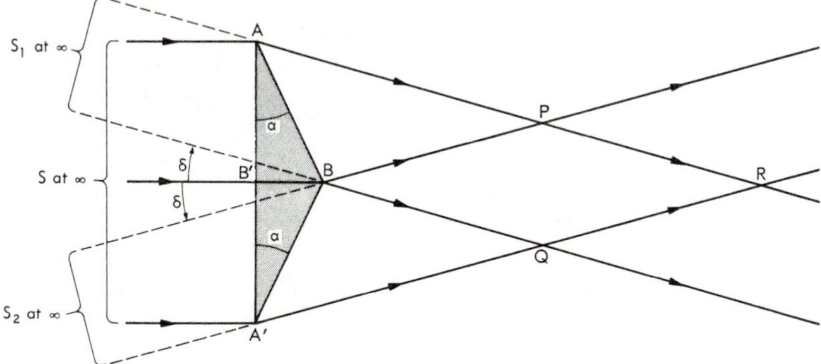

Fig. 11–23. Fresnel's biprism.

so that plane waves are incident normally on AA'. If α is a small angle, the deviation of a ray as it passes through the prism is $\delta = \alpha(n-1)$, where n is the index of refraction of the prism substance. The light which passes through the top half of the prism emerges as a parallel bundle equivalent to that obtained from a virtual source S_1 at infinity at an angle δ from the axis BB'. Similarly, the light passing through the bottom half is equivalent to a source S_2 where the angular separation of S_1 and S_2 is 2δ. Rays from the two virtual sources overlap within the region $BPRQ$ and a straight fringe interference pattern can be observed along a plane such as PQ.

The interference of electrons after diffraction by a pair of slits is often considered as a thought experiment in quantum mechanics, but it is the analog of the Fresnel biprism which has proved to be more practical to perform. Such an experiment was performed by Möllenstedt and Düker in 1956.* The clear pattern of straight fringes obtained is strikingly direct evidence for the wave nature of the electron.

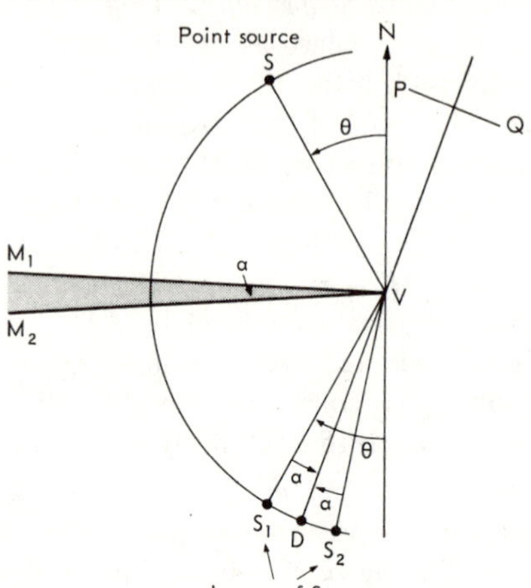

Fig. 11–24. Images by reflection in the two surfaces of a thin wedge.

11–10 REFLECTION FROM A THIN WEDGE

Let M_1V and M_2V be the plane surfaces of a wedge of dielectric material represented in cross section in Fig. 11–24. The angle $\alpha = M_1VM_2$ must be a small angle if the effects considered below are to be significant. Let VN be constructed normal to the upper surface M_1V, and consider a point source S whose angular position from the vertex V is $\theta = SVN$. We are interested in considering interference effects in the light reflected by the wedge. It will be assumed that the reflection coefficients at the two surfaces are sufficiently small so that multiple internal reflections need not be taken into account. At every point in the field of reflected light we therefore consider only the superposition of rays reflected directly from M_1V and rays which have entered the wedge

* See O. Klemperer, *Electron Physics.* New York: Academic Press, 1959, Section 9.5.

and been reflected from M_2V. For qualitative purposes it is convenient to associate these sets of rays with virtual sources S_1, the image of S in M_1V, and S_2, the image of S in M_2V. Sources S_1 and S_2 are to be thought of as π out of phase, since one is produced by an external and the other by an internal reflection.

In those regions from which both S_1 and S_2 can be seen, we therefore expect to find the interference patterns associated with a pair of point sources. This simple picture is adequate for an understanding of the general characteristics of the resulting interference pattern, but must be replaced when it comes to a matter of specific calculation. This is because there is in reality no single image point which can be associated with the reflection at M_2V of rays which pass at varying angles through the dielectric. Furthermore, the phase of such a reflected ray is not simply related to the geometrical distance from an image point. We shall ignore these difficulties in the preliminary discussion and bypass them in the calculations of the next section.

Since we assume that the lines SS_1 and SS_2 are perpendicular to the surfaces M_1V and M_2V respectively, it follows that S, S_1, and S_2 are equidistant from V, and hence lie on a circle with center at V. Simple geometry then determines the angles indicated in Fig. 11–24. The interference pattern from the double source S_1S_2 consists of hyperboloids of revolution which can be seen as straight fringes along a plane such as PQ, or circular fringes along a plane perpendicular to the line joining S_1S_2. (The straight fringes are seen only along that portion of PQ which is serviced by rays from both S_1 and S_2. There is no need to complicate the present discussion by further reference to this fact. We shall speak of the fringes as if they exist all along PQ.) The center of symmetry of the straight fringe pattern lies on the perpendicular bisector of S_1S_2. Since S_1S_2 is a chord on the circle, the bisector passes through the center, and is represented on the diagram by the line DV. Taking account of the phase difference of π, we note that this line is a locus of destructive interference.

In all preceding discussions of straight fringes, the plane of observation has been taken to be perpendicular to the bisector of the two sources. In preparation for the next section we note the effect of tilting the plane of observation about an axis parallel to the fringes. It is clear that the hyperboloids of revolution still intersect the plane in straight lines, but that the spacing between fringes is increased. Thus, for example, if PQ is originally perpendicular to DV and is then rotated an angle ϵ about an axis perpendicular to the plane of the figure, the spacing of the fringes along PQ is increased by a factor of $\sec \epsilon$. The latter will be referred to as the *inclination factor*.

When ϵ is a small quantity, $\sec \epsilon \doteq 1 + \epsilon^2/2$ differs from unity by a second-order small quantity. It is for this reason that the setting of the plane of observation is not a highly critical adjustment in quantitative experiments with a straight fringe pattern.

The circumstance under which circular fringes are most important is that of a dielectric plate with parallel faces. This can be obtained from Fig. 11–24 by holding the position of S and the thickness of the wedge below S fixed and moving V off to infinity on the right. This makes the angle α approach zero, i.e., the two surfaces M_1 and M_2 become parallel. Then SS_1S_2 becomes a circle of infinite radius, which corresponds to the obvious fact that S_1 and S_2 both lie perpendicularly below S. Circular fringes are obtained in planes parallel to the faces of the plate and the centers lie on the perpendicular from S.

When S is truly a point source, there is no critical requirement on the location of the plane of observation to obtain clear fringes of either the straight or circular type. For example, PQ in Fig. 11–24 can be close to or far from the wedge. Fringes are observed either by inserting a screen or by focusing a telescope on any such plane. Because of this the fringes are said to be *nonlocalized*.

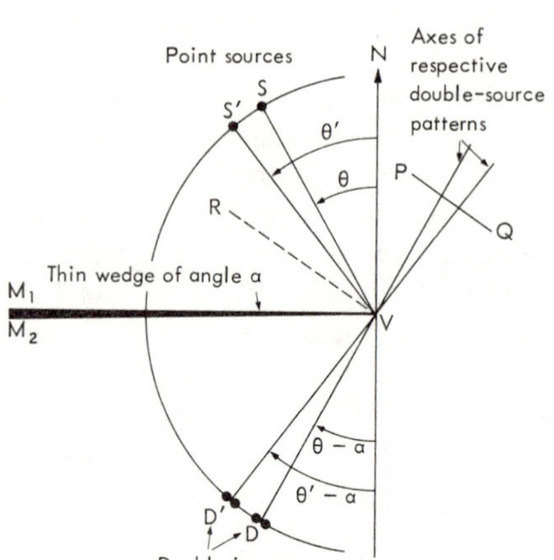

Fig. 11–25. Double images of independent point sources by reflection in a thin wedge.

11-11 BROAD SOURCES AND LOCALIZED FRINGES IN REFLECTION FROM A THIN WEDGE

In Fig. 11–25 we consider two incoherent primary sources S and S', representative of two points on a broad source. (The angle α is taken to be very small in this figure.) The pair of virtual sources obtained by reflection of S in M_1V and M_2V is indicated at D; DV is the axis of this double source. Similarly, $D'V$ is the axis of the double source obtained by reflection of S' in the two surfaces of the wedge. The intensity distributions associated with D and D' superimpose without interference. In a plane such as PQ the two fringe systems do not coincide. On the other hand, the vertex of the wedge is on a locus of destructive interference for both D and D', and the fringe systems in a plane such as VR have a common center of symmetry at V.

For a general orientation of VR the periodicities of the fringes associated with D and D' will not be the same, due to the inclination factor, sec ϵ, discussed in the last section. If there are just two sources S and S', the periodicities can be made equal by aligning VR perpendicular to the angle bisector of $D'V$ and DV. (This makes $M_1VR = [(\theta + \theta')/2] - \alpha$.) However, if a broad source extends from S to S', the fringe widths associated with points between S and S' do not all have the same inclination factors. Nevertheless, it should be evident that the indicated orientation of VR represents an optimum condition. All of the contributing fringe systems are "in phase" at V. If the source is not excessively broad, the fringes closest to V coincide approximately in position, and the superposition consists of a series of relatively clear fringes. Eventually the contributing fringes get out of step with each other and at some distance from V the pattern becomes less clear.

It is implied therefore that there is a *preferred* plane of observation for the straight fringes produced by reflection of a broad source in a thin wedge. The fringes are said to be *localized*. If they are to be observed on a screen, this must be placed along VR; if they are to be observed with a telescope, this must be focused on VR. When looked at directly by eye, the eye is drawn into focusing along VR since interesting things are seen there.

We now proceed to a calculation of the localized fringes obtained in the specific case of a broad source located at infinity in a direction close to the normal VN. This makes the preferred angle $M_1VR = [(\theta + \theta')/2] - \alpha$ close to zero, which means that the plane of localization of the fringes is approximately coincident with the upper surface of the wedge. Assume therefore that the observing instrument is focused on the plane M_1V. It is first necessary to analyze the phase relation between the rays which combine after reflection from the two surfaces when S is a point source at a specific angle θ. This result is then used to consider the visibility of the fringes obtained from a broad source.

Since the observing instrument is focused on the upper surface of the wedge, all rays which emerge from a point P on this surface are brought together at one point in the image plane. From a single point source at infinity there are just two such rays (see Fig. 11–26): APQ resulting from reflection in M_1V and $BCEPR$ which has been refracted in the wedge and reflected in M_2V. The phase difference between these two rays when they unite in the image plane is the same as their phase difference at P. A perpendicular erected from P to CE is a wavefront inside the dielectric (normal to a ray); the phases are equal at P and D. Therefore a phase difference arises from the extra distance $DE + EP$ along the refracted ray. It is convenient to express this distance in terms of $d_P = PF$, the thickness of the wedge at P. From the geometry of the figure, a simple calculation shows that

$$DE + EP = 2d_P \cos(\theta' - \alpha).$$

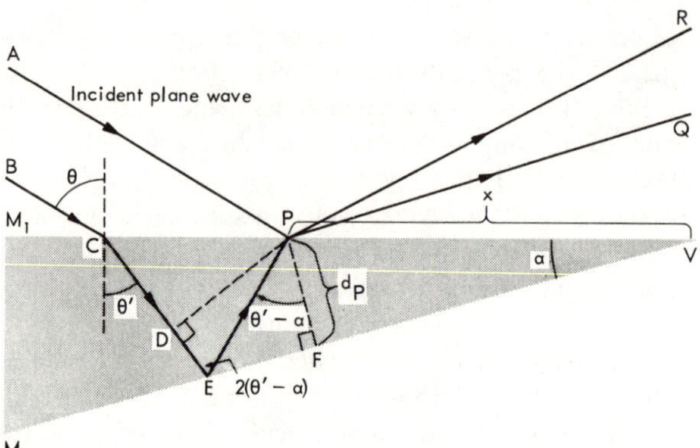

Fig. 11-26. Interfering rays for condition of focus on the top surface of wedge.

Since α is assumed to be a small angle, we obtain the approximation

$$DE + EP = 2d_P \cos\theta'. \tag{11-18}$$

The phase difference associated with this path length is

$$\Delta\phi = k'(DE + EP) = 4\pi(d_P/\lambda')\cos\theta',$$

where $\lambda' = \lambda/n$ is the wavelength in the dielectric having index of refraction n. If we take account of the phase difference of π between internally and externally reflected rays, the criterion for destructive interference is $\Delta\phi = 2m\pi$, where $m = 0, 1, 2, \ldots$ From this it follows that black interference bands parallel to the vertex of the wedge are obtained at those locations where the local thickness is

$$d_P = m(\lambda'/2)\sec\theta'. \tag{11-19}$$

The thickness of the wedge increases linearly with distance along the upper surface, hence the fringes are equally spaced. If we write $x = d_P/\sin\alpha \doteq d_P/\alpha$, the width of one fringe is $\Delta x = \lambda'(\sec\theta')/2\alpha$. In particular for $\theta = 0 = \theta'$, $\Delta x = \lambda'/2\alpha$, which is what is to be expected from the fact that the virtual sources subtend an angle 2α from the vertex of the wedge. The extra factor of $\sec\theta'$ under nonnormal incidence is the inclination factor to be expected if the ideal plane of the fringes is inclined at an angle θ' with the upper surface of the wedge.

Consider now the superposition of the independent fringe patterns associated with a broad source extending between $\theta = 0$ and $\theta = \Delta\theta$. The problem is similar to that of Section 11-8(d) except that in this case we are dealing with differently spaced fringes which have a common origin rather than equally spaced fringes which have different origins. In the present case there is no angular width of source which makes the pattern wash out completely.

Even when $\Delta\theta$ is large, the visibility of the fringes is good near the vertex and gradually becomes worse with increasing fringe order. As an arbitrary measure of the number of "good" fringes, we adopt the condition which makes the fringes associated with $\Delta\theta$ exactly out-of-step with the fringes of $\theta = 0$. Thus, if a particular location is an mth order black fringe for $\theta = 0$, $d_P = m(\lambda'/2)$, we desire this to be at the center of a bright fringe for $\Delta\theta$:

$$d_P = (m - \tfrac{1}{2})(\lambda'/2) \sec(\Delta\theta').$$

Equating these two expressions for d_P and using the approximation $\cos(\Delta\theta') \doteq 1 - (\Delta\theta')^2/2$, we can solve the relation for m:

$$m = (1/\Delta\theta')^2 = (n/\Delta\theta)^2.$$

This represents an order-of-magnitude estimate of the number of clear fringes obtainable with a broad source subtending an angle $\Delta\theta$ under conditions of nearly normal incidence. For example, if $n = 1.5$ and $\Delta\theta = 10° \sim \frac{1}{6}$ rad, we find $m \sim 81$ fringes. It is thus evident that sources are permitted to be considerably broader in the present context than they were in Young's experiment.

For direct visual observation, or other cases in which the lens of the viewing instrument subtends a small angle from a point on the wedge, the relevant angular width $\Delta\theta$ is not determined by the actual width of the source itself. As indicated in Fig. 11–27. when the eye is focused on P, only the segment EF contributes rays which pass through P and are intercepted by the lens of the eye. To improve the visibility of the fringes, the effective $\Delta\theta$ can be decreased by placing a pinhole directly in front of the eye. Another condition must be observed, however. Since both rays PQ and PR in Fig. 11–26 must be intercepted by the eye, the latter must subtend an angle at least as wide as RPQ. This angle is the order of 2α, the angular separation of the virtual sources in Fig. 11–24.

If the broad source terminates at S and S' in Fig. 11–27, fringes will not be seen outside the segment QR on the wedge. It is thus essential to have a source of considerable breadth to view localized fringes directly by eye.

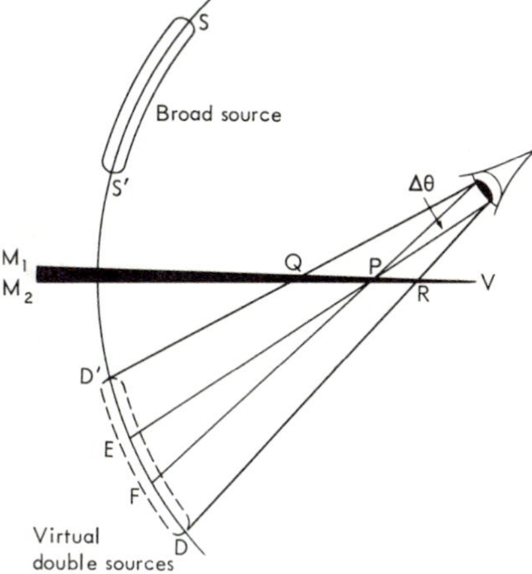

Fig. 11–27.
Effective breadth of source under direct viewing by eye.

11-12 NONUNIFORM THIN FILMS

We consider now the interference between the reflections at the two surfaces of a thin dielectric film which is not a perfect wedge. If the surfaces are relatively smooth and nearly parallel, then any local portion of the film can be treated like a wedge of small angle. The condition of viewing for *nonlocalized* fringes is not of interest in this context. If, for example, a lens collects light which has passed through different portions of the film, the situation cannot be analyzed in terms of a simple pair of virtual sources. However, under the viewing condition for *localized* fringes, with the viewing instrument focused approximately on the upper surface of the film, the phase relation of rays uniting at a given point of observation P is solely determined by the local characteristics of the surfaces. If the curvature of the surfaces is sufficiently small so that they can be approximated by planes which are nearly parallel, we can apply the analysis leading to Eq. (11–19). Thus for the case of nearly normal incidence, with $\sec\theta' \doteq 1$, destructive interference is expected wherever the local thickness of the film is an integral number of half-wavelengths. A given locus of destructive interference maps out a line along which the thickness of the film is constant; the fringes are referred to as *fringes of equal thickness*. The ideal wedge is a special case in which the fringes are equally spaced and are parallel to the vertex of the wedge. The following are additional examples:

i) *A thin film of air between nearly plane surfaces of glass*. The appearance of the fringes can be used as a test of flatness. If the departure from an ideal plane is nowhere larger than a tenth-wavelength, a surface is said to be *optically flat*. Figure 11–28 shows the fringes of equal thickness obtained between a supposedly plane surface and an optical flat.

ii) *Newton's rings*. When the convex surface of a lens rests on an optical flat, an air film is formed for which the loci of constant thickness are circles concentric with the point of contact. A set of localized circular fringes is obtained when light from a broad source is reflected at nearly normal incidence. Unless the lens surface has a large radius of curvature, the fringes are closely spaced and may not be resolved (see Problem 11–26).

iii) *Soap films*. A vertical soap film tends to settle under gravity so that the top portion is exceedingly thin (much less than a wavelength). This portion of the film is a locus of destructive interference and is dark when observed under reflection of light from a broad source. Note that it is essential to include the phase change of π distinguishing internal and external reflections to obtain the desired continuity of behavior as the thickness of the film tends to zero. Without taking this into account we could not say that a film which is "sufficiently thin" is equivalent to no film at all.

A region which has the characteristics of a thin wedge develops below the thin portion of the film. In monochromatic light the fringes are approximately equally spaced and horizontal. Since the spacing is wavelength dependent, the fringes for different colors get out of step, and a series of subtraction colors are obtained under illumination with white light. The reflected light appears white at large thicknesses, but yields a channeled spectrum upon examination with a spectroscope. The iridescence of soap bubbles is explained by the fact that the loci of constant thickness corresponding to destructive interference are different for the different colors.

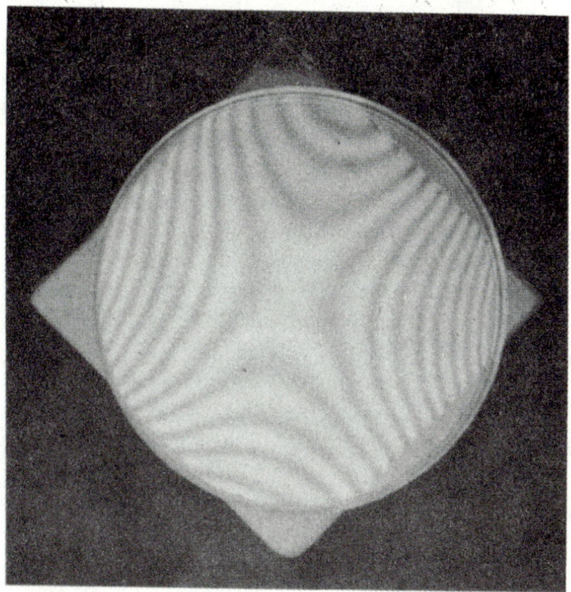

Fig. 11-28. Testing an optical flat. The disk resting on top has a bottom surface which is known to be optically flat. The localized fringes of equal thickness show that the top surface of the square plate is not flat. A study of the contours shows where additional polishing is necessary. (Photograph by courtesy of Kollmorgen Corp.)

iv) *Oil slicks.* Thin films of oil (not more than a few wavelengths thick) give rise to fringes of constant thickness with monochromatic illumination, and to a range of subtraction colors when illuminated by white light. The spectacular coloring of an oil slick on wet asphalt is a common example. A layer of water between the oil and the asphalt provides a smooth lower surface to the oil film. The light which passes through the film is absorbed by the black asphalt and hence is prevented from overwhelming the interference colors produced by reflection in the oil film.

v) *Oxide coatings on metals.* Dark metals such as cast iron often display iridescent colors due to thin oxide films of variable thickness.

11-13 THE MICHELSON INTERFEROMETER

A schematic diagram of the Michelson interferometer is given in Fig. 11-29. In the diagram S is a primary point source and R is a semireflector which we shall suppose reflects one-half the incident light. The eye at E receives light from two virtual images of the original point source. One of these is associated with rays transmitted through R, reflected at M_1, and reflected at R. The other is produced by reflection at R and M_2 and transmission through R. The relative location of the two virtual images can be controlled by changing the location and orientation of the mirrors M_1 and M_2. For example, if M_1 and M_2 are equidistant from R, and M_2 is very nearly perpendicular to M_1, the two virtual images lie in a plane perpendicular to the line of sight from E. Their apparent separation can be changed by tilting one of the mirrors slightly. Nonlocalized straight fringes can then be observed by focusing a telescope on any plane of observation sufficiently far from the plane of the virtual sources. Similarly, nonlocalized circular fringes can be obtained by setting M_1 and M_2 exactly perpendicular to each other and at slightly different distances from R. The two virtual sources are then separated in depth along the line of sight, and the hyperboloids of revolution intersect the plane of observation in circles.

The Michelson interferometer is customarily used with a broad source in place of the primary point source S, and the resulting fringes are localized. It is convenient to analyze this situation in the following manner. The geometry of ray reflections in Fig. 11-29 can be seen to be equivalent to that indicated in Fig. 11-30. As far as rays of type two are concerned, the reflection in R is equivalent to replacing S by a virtual source S', which is the image of S in R. The rays from S' are then reflected in M_2 to the point of observation. The geometry of rays of type one is equivalent to the reflection of rays from S' in a plane surface M_1', which is the image of M_1 in R. Therefore the entire arrangement is equivalent to the reflection of light from a point source S' in a thin wedge $M_1'VM_2$. We can therefore apply the results of the preceding sections. As a matter of fact, several of the simplifying assumptions previously made are ideally satisfied in the present case. There are no multiple reflections between M_1' and M_2 to be ignored, since the rays associated with M_1' do not in fact encounter the surface M_2. It is not an approximation to assume that the reflections in M_1' and M_2 correspond to sources of equal strength, since this is governed by making R a semireflector. Furthermore, the complications associated with *refraction* in a dielectric wedge, where $n \neq 1$, do not occur in the present case; the two surfaces of the virtual wedge serve to produce reflections, but the index of refraction can be set equal to unity in the preceding formulas.

Suppose then that a broad source is used and the mirrors are adjusted so that M_1' and M_2 form a small angle α and their line of intersection V lies within the field of view. Under these conditions straight fringes are observed

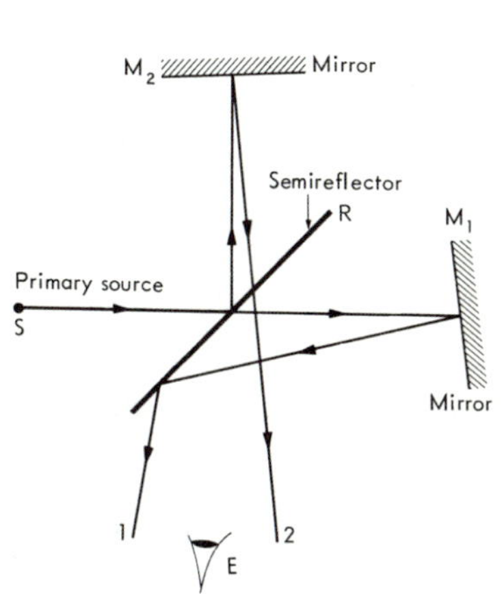

Fig. 11-29.
Michelson interferometer (schematic).

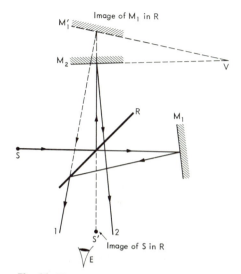

Fig. 11-30.
Michelson interferometer related to reflection in a thin wedge.

parallel to V. These fringes are localized approximately in the plane of M_2. The fringe width can be controlled by varying α by tilting-screws on the mirrors. The mirror M_2 is mounted on an accurate track so that it can be moved parallel to itself. Since the fringes are loci of equal thickness, moving M_2 through a certain distance changes the thickness everywhere along the virtual wedge by the same amount; the fringes move across the field of view.

Since the fringe width is proportional to the wavelength, the spacing of the fringes is different for light of different colors. Thus when white light is used, the contributing fringe systems gradually become out of step with increasing distance from V. At a given location near V a small number of colors across the visible spectrum are extinguished and a subtraction color is perceived. At some distance from V the light has a channeled spectrum, but appears to be almost white. Hence the so-called "white light fringes" consist of a series of colored bands located near V; the bands fade off into less saturated colors until they are indistinguishable from the white of the original source.

In the acoustical version of the Michelson interferometer, the semireflector R is a perforated sheet of metal, the area of the holes being equal to half the total area. The source is customarily either a loudspeaker or a Galton whistle and is not a "broad source" in the sense of being an assembly of noncoherently radiating points. It is consequently only the nonlocalized fringes which are of interest. The role of the eye is played by a microphone which explores the field point-by-point.

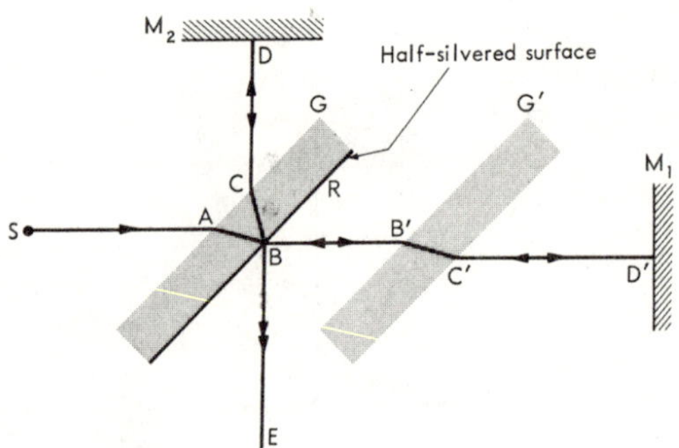

Fig. 11-31. Passage of rays through glass plates in a Michelson interferometer.

We can create a semireflector in optics by silvering one surface of a plane-parallel plate of glass to raise the reflection coefficient to 50% (this is referred to as *half-silvering*). The diagram of Fig. 11-29 is therefore only schematic, since it does not take account of the passage of the light through the glass. This is indicated in Fig. 11-31, where $SABCDCBE$ is a ray of type two and $SABB'C'D'C'B'BE$ is a ray of type one. The *compensator plate* G' is a glass plate identical to G, but without half-silvering, and parallel to it. The plate G' serves no essential purpose when only monochromatic light is used, but it is necessary in applications requiring the white light fringes. The latter can be obtained only if there is some setting which makes the difference in optical path length zero for all colors simultaneously. From Fig. 11-31, we find that the optical path difference between the two rays is

$$\begin{aligned}\Delta L &= 2[(nBC + CD) - (nB'C' + BB' + C'D')] \\ &= 2[CD - (BB' + C'D') + n(BC - B'C')]. \qquad (11\text{-}20)\end{aligned}$$

If G' were not present ($B'C' = 0$), it would be possible to move M_2 to make $\Delta L = 0$ for any *particular* wavelength λ_1, but for a different wavelength λ_2 the optical path difference would then be

$$\Delta L = 2[n(\lambda_2) - n(\lambda_1)]\,BC.$$

For a typical glass plate $BC \sim 0.5$ cm; if we take λ_1 in the middle of the visible spectrum and λ_2 at either end, we obtain $[n(\lambda_2) - n(\lambda_1)] \sim 0.01$. Thus $\Delta L \sim 0.01$ cm. The corresponding phase difference is

$$\Delta\phi = k\Delta L \sim 2\pi(0.01/\lambda_2),$$

which is many multiples of 2π. This shows that λ_2 is far from the condition of zero retardation. When the compensator plate *is* present, $B'C' = BC$, and ΔL in Eq. (11-20) is independent of wavelength. The condition of zero retardation is achieved for all wavelengths when the geometrical distances CD and $BB' + C'D'$ are equal.

The presence of the glass plate G introduces a distinction between the internal reflection ABC and the external reflection $B'BE$. When R is half-silvered, the corresponding relative phase shift is, in general, neither zero nor π. The position of zero retardation ($\Delta L = 0$) is not a position of either constructive or destructive interference, and there is a subsequent skewing of the symmetry of the white light fringes. If the surface R is unsilvered, the two virtual sources are not equal in strength. The instrument can nevertheless be used under these conditions, for this means only that the contrast in the fringe pattern will be reduced inasmuch as complete extinction does not occur at the minima. The relative phase change for this case is π and the center of the white light pattern is a dark fringe. The color sequence is then similar to that of a thin dielectric wedge, or a Babinet compensator between crossed polarizers.

In principle, Fig. 11–31 ought also to include a ray reflected at M_2 after external reflection at A and another returning from M_1 and internally reflected at A. The first of these rays passes through a glass plate only once, whereas the second passes through glass five times. This pair cannot produce white light fringes, but does add uniform illumination to the field and reduces the contrast. When monochromatic light is used, the fringe pattern from this pair may at times be noticed but is easily ignored. A nonreflecting coating on the front side of G can be used to eliminate both members of this undesired pair of reflections.

11-14 EXPERIMENTS USING THE MICHELSON INTERFEROMETER

To illustrate basic optical principles we shall consider several laboratory experiments which can be performed with a Michelson interferometer. For specific practical applications this instrument has by and large been superceded by other types.* A short text by Michelson describes some of the important experiments which were originally performed with a Michelson interferometer.† These include a study of light propagation relative to the ether, determination of the fine structure of spectral lines by examining visibility curves, and the determination of the number of times a selected monochromatic wavelength is contained in the standard meter.

To be definite, we shall assume the relative phase change of π for an unsilvered glass plate. The final results are the same regardless of the phase change, since differences are always involved in which an arbitrary constant

---

* S. Tolansky, *An Introduction to Interferometry*. New York: Longmans Green and Co., 1955.

† A. A. Michelson, *Light Waves and Their Uses*. Phoenix, Chicago (1961) PSS 508.

phase angle will be canceled out. For the localized straight fringes we use Eq. (11–19) corresponding to a thin wedge with the source at infinity at $\theta \doteq 0$. The dark bands are given by wedge thicknesses

$$d = m\lambda/2, \qquad m = 0, \pm 1, \pm 2, \ldots \tag{11–21}$$

If the mirror M_2 is moved parallel to itself, the dark fringe of order m moves to a different location within the field of view. Keeping a fixed point of observation, we find that the dark fringe of order $(m + 1)$ arrives at a point originally occupied by the fringe of order m when the distance moved by the mirror is $\Delta d = \lambda/2$.

It is sometimes advantageous to use the system of circular fringes localized at infinity which arise when M_1' is set parallel to M_2. From Problem 11–29 we learn that the radii of the dark fringes subtend angles θ_m given by the equation

$$\cos \theta_m = m\lambda/2d, \tag{11–22}$$

where d is the distance between M_1' and M_2. The integer m has a maximum value corresponding to the dark fringe of smallest radius. Keeping m fixed and increasing d causes the value of θ_m to increase; thus the circles expand as the thickness of the virtual plane parallel plate formed by M_1' and M_2 is increased. A new fringe of higher order appears at the center each time $\lambda/2d$ passes through an exact integral value. It is therefore convenient to use the center of the circular fringe pattern as a reference point. We can then assert, as we did in the case of straight fringes, that the motion of M_2 through a distance $\Delta d = \lambda/2$ corresponds to the motion of one complete fringe past the reference point.

a) Determination of a Monochromatic Wavelength

Using either the circular or straight fringes, let M_2 be moved through a total distance Δd which can be read from a vernier scale. If m fringes pass the reference point in the process, the wavelength is given by $\lambda = 2\,\Delta d/m$. The accuracy is increased by counting a large number of fringes, for if δd is the least count on the vernier, $\delta\lambda \sim 2\,\delta d/m$ is the uncertainty in the wavelength. There is, of course, a practical limit imposed by limited precision in the construction of the instrument (accurate parallel transport of M_2, temperature control of the vernier scale, etc.). A limitation of a different sort is imposed by the fact that no source is perfectly monochromatic. This is evidenced by a loss in distinctness of the fringes when d is increased to large values. This effect is discussed in Section 11–15.

b) Determination of the Wavelength Separation of a Doublet

Two closely spaced spectral lines, λ_1 and $\lambda_2 = \lambda_1 + \Delta\lambda$, give rise to independent straight fringe systems of slightly different spacing. There will be regions of good visibility within which the bright fringes of λ_1 are nearly coincident with the bright fringes of λ_2, and regions of poor visibility within which the bright fringes of one system occur at nearly the position of the dark fringes of the other. We define the "order number" at a position where the wedge thickness is d as $m \equiv 2d/\lambda$. The centers of dark and bright fringes for a given wavelength occur where m is integral and half-integral respectively. A region of poor visibility is characterized by $m_1 - m_2 \sim p + \frac{1}{2}$, where p is an integer. The centers of these regions are therefore given by

$$m_1 - m_2 = 2d\left(\frac{1}{\lambda_1} - \frac{1}{\lambda_2}\right) \doteq \frac{2d\,\Delta\lambda}{\lambda_1^2} = p + \tfrac{1}{2}. \qquad (11\text{–}23)$$

Starting at minimum visibility corresponding to some value of p, let the mirror M_2 be moved through a distance Δd until the next region of minimum visibility moves to the point of observation. This is associated with going from p to $p + 1$ in Eq. (11–23) and we have

$$2(d + \Delta d)\frac{\Delta\lambda}{\lambda_1^2} = p + \tfrac{3}{2}. \qquad (11\text{–}24)$$

Eliminating p between Eqs. (11–23) and (11–24) and solving for $\Delta\lambda$, we find that

$$\Delta\lambda = \frac{\lambda_1^2}{2\,\Delta d}. \qquad (11\text{–}25)$$

For the two lines of the sodium yellow doublet the positions of minimum visibility can be estimated with reasonable accuracy; they yield a wavelength separation of $\Delta\lambda = 6$ Å.

We can verify the measurement on the sodium doublet by determining the position of the line centers using a spectroscope of sufficient resolution. The present method is of interest because of its use of indirect inference. The simple discussion given here is suitable for the case of two component lines of approximately equal strength, where the structure of the individual lines does not need to be taken into account. More complicated cases can be handled by studying the variation in visibility of the fringes as a function of d, using methods described in the references cited in the next section.

c) Measurement of the Index of Refraction of a Thin Plate

If a thin sheet of a transparent dielectric is inserted normal to the beam in one of the arms, the optical path difference ΔL is changed by the amount $2(n - 1)l$ through the replacement of l by nl, where l is the thickness of the sheet and n is its index of refraction. This change can be compensated for by moving mirror M_2. Since $(n - 1)$ is wavelength dependent, monochromatic light should be used. White light fringes must be used, however, to "tag" a particular monochromatic straight fringe. If the sheet is sufficiently thin, the zero-order monochromatic fringe is the one which appears nearest in position to the center of the white light fringe system. Starting with this opposite a fixed reference point, the sheet is then removed. If the mirror must be moved a distance Δd to return the zero-order fringe to the reference point, the associated change in path length is $2\,\Delta d$. Thus $(n - 1)l = \Delta d$. The value of Δd can be obtained most accurately by counting fringes and using the known wavelength of the monochromatic light, taking account of the fact that each fringe results from a change in d of one-half wavelength. The value of n or l separately can be deduced if the other is known.

11-15 THE PURITY OF A SPECTRAL LINE

A perfectly sharp monochromatic line would in principle enable us to see circular fringes with the Michelson interferometer when M_1' and M_2 are parallel, regardless of their distance of separation d. In practice we find that for every spectral line there is an upper limit on d beyond which the interference patterns cannot be obtained. This can be interpreted by supposing that even the best monochromatic sources emit a continuous distribution of wavelengths in some narrow interval between λ and $\lambda + \Delta\lambda$. When d is small, the circular fringes for all the contributing wavelengths are effectively coincident. But as d increases, the rate of expansion of the circles and the rate of production of new fringes at the center is different for each wavelength between λ and $\lambda + \Delta\lambda$. There is no definite cutoff point at which the fringes suddenly lose distinctness. However, we can adopt as an order-of-magnitude criterion the separation d_i, which causes the extremes λ and $\lambda + \Delta\lambda$ to produce fringe systems which are exactly out of step at the center. If the center is a black fringe for λ and a bright fringe for $\lambda + \Delta\lambda$, by consulting Eq. (11–22) we have

$$1 = \frac{m\lambda}{2d_i} = (m - \tfrac{1}{2})\,\frac{(\lambda + \Delta\lambda)}{2d_i}.$$

Eliminating m and solving approximately for $\Delta\lambda$, we find that

$$\Delta\lambda = \lambda^2/4d_i. \qquad (11\text{–}26)$$

Thus if the fringes become noticeably indistinct when $d \gtrsim d_i$, we can interpret this to mean that a spread of wavelengths is present having a "line width" given by the above formula.*

* The present discussion is highly qualitative and does not enable precise definitions to be given to such terms as d_i and $\Delta\lambda$. The arguments have been constructed at this intuitive level to introduce the concept of a distribution of wavelengths associated with a spectral line, without requiring more advanced mathematical techniques. For a more satisfactory treatment see J. Strong, *Concepts of Classical Optics*. San Francisco: W. H. Freeman & Co., 1958, Appendix F. J. M. Stone, *Radiation and Optics*. New York: McGraw-Hill, 1963, Chapter 13. M. Born and E. Wolf, *Principles of Optics*. New York: Pergamon, 1959, Section 7.5.8.

Michelson found that the cadmium red line (6438 Å) is one of the most ideally monochromatic sources available, allowing fringes to be discerned to values of $d \sim 15$ cm. Setting $d_i = 15$ cm, we obtain from Eq. (11–26) a wavelength spread of $\Delta\lambda \sim 0.007$ Å.

An alternative interpretation can be considered for the indistinctness when d is large of the circular fringe pattern obtained from supposedly monochromatic illumination. The difference in optical path length between the rays of types one and two (reflected from M_1 and M_2 respectively) is $\Delta L = 2d$, where d is the thickness of the virtual plate $M_1'M_2$. But a difference of path length implies that the signals arriving at the point of observation were emitted at different times from the source. The question arises as to whether the source can maintain coherence over indefinitely long time intervals. In Section 7–7 it was suggested that the signal from a customary monochromatic source can be considered to be sinusoidal over relatively short time intervals, but does not maintain a persistent phase over long time intervals. If τ is the *interval of coherence*, such that phase correlation exists over time intervals $\Delta t \ll \tau$ but does not exist for $\Delta t \gg \tau$, the fading out of the circular fringe patterns can be interpreted to mean that the time difference $2d/c$ has become comparable to τ. For example, $d_i = 15$ cm for the cadmium red line implies a coherence interval $\tau \sim 2d_i/c = 10^{-9}$ sec. Although this is a short time interval in macroscopic terms, the number of oscillations of the signal is $\tau/T = 2d_i/\lambda \sim 5 \times 10^5$ cycles.

These alternative interpretations of the washing out of the interference pattern use different language, but can be shown to be equivalent through the technique of Fourier analysis. A sine wave with slowly varying phase is only "nearly sinusoidal." The analysis of such a function shows that it may be regarded as a superposition of perfect sinusoidal waves having a variety of wavelengths. The important contributions to this superposition come from wavelengths close to the apparent wavelength of the nearly sinusoidal wave. The frequency spread $\Delta\nu$ turns out to be inversely proportional to the coherence interval τ and obeys an order-of-magnitude relation $\tau\Delta\nu \sim 1$. This relation can be derived from the qualitative arguments above if we assume the equivalence of the two interpretations. From Eq. (11–26),

$$\Delta\nu = \frac{c\,\Delta\lambda}{\lambda^2} = \frac{c}{4d_i},$$

and

$$\tau\,\Delta\nu = \left(\frac{2d_i}{c}\right)\left(\frac{c}{4d_i}\right) = \tfrac{1}{2}.$$

This relation implies that an infinitely sharp spectral line ($\Delta\nu = 0$) is associated with an infinite interval of coherence ($\tau = \infty$). That is, the ideal sinusoidal wave has no frequency spread and remains coherent over indefinitely long periods of time. A decrease in the interval of coherence is associated with an increase in breadth of a spectral line.

Among the mechanisms which may be responsible for the broadening of spectral lines are the following:

i) *Collision broadening.* Under certain conditions of temperature and pressure the coherence of the radiation emitted in a discharge tube is limited by interruptions due to collisions. In this case τ is the order of the mean time between collisions. Increasing the pressure shortens τ and causes the spectral lines to become broader. This explains the fuzzy appearance of the spectral lines from a high pressure Hg arc as compared with one of low pressure.

ii) *Doppler broadening.* To a stationary observer the molecules of a gas have a distribution of velocities associated with thermal motion. Thus, even if we suppose that each molecule is emitting perfect sinusoidal waves of given frequency, the Doppler effect implies that these will be received as waves of increased or decreased frequency depending on the component of the velocity of the molecule along the line of sight. The expected distribution of frequencies can be calculated from a knowledge of the statistics of the velocity distribution. Since the mean-square velocity of the molecules is proportional to the absolute temperature and inversely proportional to the molecular mass, this effect is most important for high temperatures and light molecules.

iii) *Natural broadening.* The law of conservation of energy implies that the amplitude must decrease as an oscillating charge radiates energy spontaneously. If collisions are relatively infrequent, the signal emitted is therefore an attenuated sine wave which drops effectively to zero before the next excitation. The Fourier analysis of such a signal shows that it is composed of a spread of frequencies close to the undamped natural frequency. Hence, even if conditions are such that collision and Doppler broadening are not important, the spontaneous radiation is prevented from being a perfect sinusoid by this mechanism of *intrinsic* or *natural* line broadening.

In addition to the fact that a laser produces high power levels in a highly collimated beam, another of its outstanding characteristics is a degree of spectral purity much higher than that possible with conventional light sources. Experiments using a helium-neon gas laser with a Michelson interferometer arrangement involving a path difference $\Delta L = 9$ m obtained clear circular fringes, showing that the interval of coherence had not yet been exceeded. Experiments of a different type show that the spectral purity is of the order of one part in 10^{14} for a wavelength of $\lambda = 1.153\,\mu$. From Eq. (11-26) we find that the value of d_i is

$$d_i = \left(\frac{\lambda}{\Delta\lambda}\right)\left(\frac{\lambda}{4}\right) = \left(\frac{10^{14}}{1}\right)\left(\frac{1.153 \times 10^{-6}}{4}\right) \sim 3 \times 10^7 \text{ m}.$$

This corresponds to a coherence interval of $\tau = 2d_i/c \sim \frac{1}{5}$ sec.

The theoretical causes of line broadening of the output of a laser are complicated. A major practical limitation is the variation in the distance between the end mirrors due to mechanical vibration and thermal drift.

PROBLEMS

11–1. a) Consider a point source of sound above a perfectly reflecting plane. By means of a figure similar to Fig. 11–2, discuss the progressive changes in the distribution of loci of constructive and destructive interference as the source moves perpendicularly away from the plane. What effect will be observed at a fixed point of observation? b) The effect considered in (a) is sometimes demonstrated by moving a tuning fork in front of a reflecting wall. Assuming that the fork moves with constant velocity, derive an expression for the frequency at which new constructive loci appear along the axis $\theta = \pi/2$. (Make the simplifying assumption that the tuning fork is a point source of uniform spherical waves.) c) An observer who is on the axis $\theta = \pi/2$ further away from the plane than the moving tuning fork receives signals from the approaching source and from its receding image by reflection. Calculate the Doppler shift in the frequencies of these two waves and the beat frequency of their superposition. Compare with (b).

11–2. Set up a cartesian system in Fig. 11–2 and determine the equation in standard form for the family of hyperbolas defined by

$$(\Delta\phi)_0 + k(r_2 - r_1) = \text{const} = \Delta\phi.$$

Write the equation for one of the asymptotes and show that when this is solved for $\Delta\phi$ the result is equivalent to the approximation $\Delta\phi = (\Delta\phi)_0 + kd \sin\theta$.

11–3. Sketch a polar plot of the intensity in the radiation pattern at large distances corresponding to the conditions of Fig. 11–2 (equal point sources in quadrature with a spacing of $3\lambda/2$).

11–4. A horizontal antenna is placed a half wavelength above the surface of the earth, the latter being assumed to act like a perfect reflecting plane. Plot the radiation pattern in a vertical plane perpendicular to the antenna at its midpoint.

11–5. Sketch the polar plot of the intensity in the radiation pattern at large distances from two equal sinusoidal sources, π out of phase, having a distance of separation $d = 3\lambda/4$, $d = 5\lambda/4$, and $d = 3\lambda/2$. Interpolate into this sequence the sketches for $d \ll \lambda$ (Fig. 11–4), $d = \lambda/2$ (Fig. 11–7b), and $d = \lambda$ (Problem 11–4). Compare with the discussion of Problem 11–1(a).

***11–6.** Suppose the assumption is not made that the amplitude factors are equal in combining the signals from two sinusoidal point sources (Eq. 11–2). Show that the time average of the square of the field variable can be written in the form

$$\overline{\psi^2} = \tfrac{1}{2}(A_1 - A_2)^2 + 2A_1A_2 \cos^2(\Delta\phi/2).$$

b) Sketch a polar plot of the intensity in the radiation pattern at large distances from two point sources of sinusoidal waves for the case in which they are π out of phase, are spaced a half-wavelength apart, and in which one of the sources has twice the strength of the other. [Conditions are the same as for Fig. 11–7(b) except that the sources are not of equal strength.]

11–7. a) Verify the integration leading to Eq. (11–8) for ϵ, the relative total power radiated by a pair of point sources. b) Given that the two sources are in phase, sketch the graph of ϵ as a function of the distance of separation of the two sources. [The material of Section 12–4 will be helpful in considering the function $((\sin kd)/kd)$.]

11–8. Verify that the average value of $\sin^2 \theta$ over the surface of a sphere is equal to $\frac{1}{3}$.

11–9. Using $I(\theta)$ given by Eq. (11–5), set up the expression for the intensity averaged spatially over a zone lying between θ_1 and θ_2. Show that if the integration is carried out over a complete fringe in the interference pattern ($\Delta\phi$ changes by 2π), the average intensity is $\bar{I} = 2I_0$.

11–10. a) If θ_1 and θ_2 are the angles subtended by adjacent loci of destructive interference in the radiation pattern from a pair of point sources, show that $\sin \theta_2 - \sin \theta_1 = \lambda/d$. b) For $kd \gg 1$, consider a general point of observation (not necessarily close to $\theta = 0$). Show that the angular width of a fringe is the same as that which would be obtained if one projected the actual point sources onto a line perpendicular to the line of sight.

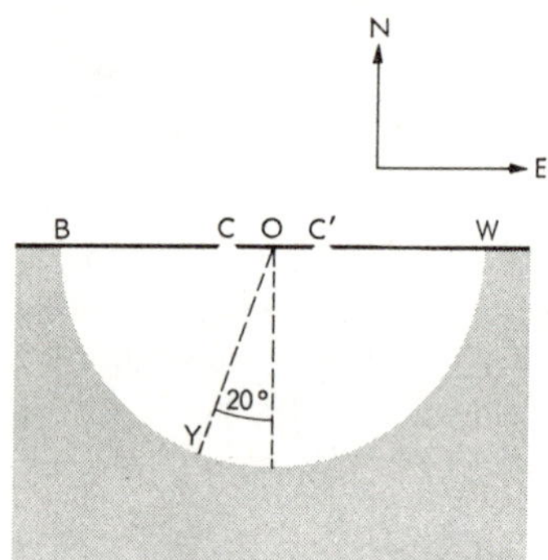

Figure 11–32

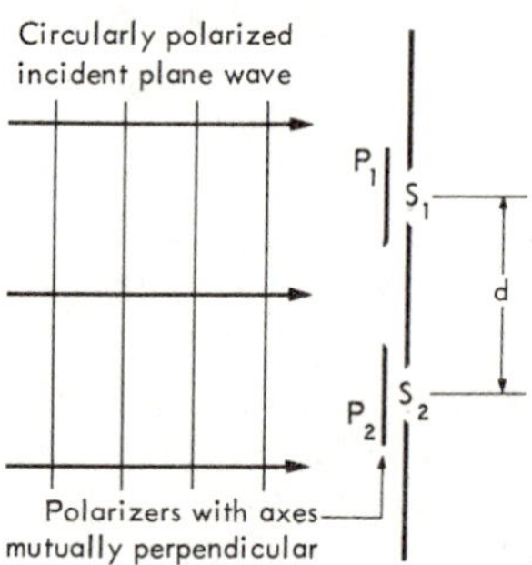

Figure 11–33

11–11. A yacht club Y is located on the semicircular shoreline of a harbor protected by a breakwater BW (Fig. 11–32). It is found that the most damaging storms are associated with swells which come from due north and have a wavelength of 20 ft. To give the yacht club maximum protection from these storms, it is decided to have two openings in the breakwater at C and C' equidistant from the center O. (Consistent with the fictitious nature of the problem, simplifications should be made as desired.) a) What should the spacing be between the openings? b) Along how much of the shoreline on either side of Y will the wave amplitude during a storm be less than half of what it would be with only a single opening? (Take the radius $OY = 1500$ ft.)

11–12. A plane wave of circularly polarized monochromatic light is normally incident on a screen containing two pinholes S_1 and S_2 separated by a distance d (Fig. 11–33). The diameters of the individual pinholes may be assumed to be very small compared with the wavelength, so that each acts as a source of uniform spherical waves. A polarizer P_1 is placed directly in front of S_1, and a second polarizer P_2, whose axis is perpendicular to that of P_1, is placed in front of S_2. Assume that the wavelength is reasonably small compared with d and let a screen be placed at a large distance from the pinholes. What is the character of the light arriving at different points of the screen? Describe the appearance of the screen. What will happen if an analyzer is placed directly in front of the screen?

11–13. If it is desired to have $r \geq 10d^2/\lambda$ in a double pinhole interference experiment, what is the smallest permissible value of r when $d = 0.2$ mm? $d = 1$ mm? (Assume that $\lambda = 5500$ Å.)

11–14. Verify the integration in Eq. (11–16) leading to the intensity distribution in the Young's pattern produced by a broad source.

11–15. a) If the eye is unable to distinguish fringes which subtend an angle less than about 1 minute of arc, what is the maximum distance allowable between pinhole centers for direct visual observation of a Young's interference pattern? (Assume a wavelength of 5500 Å.) b) How far away must a frosted incandescent bulb be located to act as a point source for the experiment in (a)? (The diameter of the bulb $\sim$ 2.5 in.)

11–16. If an unfrosted light bulb with a straight filament is viewed directly through a pair of pinholes having a separation $d = 0.7$ mm, the interference fringes become indistinct at a distance of about 2 m. Estimate the width of the filament.

***11–17.** *Principle of the stellar interferometer*
The angular breadth of a star is too small to be resolved directly with the best telescopes. However, we have seen that the breadth of a source in a double pinhole experiment affects the visibility of the interference fringes. As an approximation, assume that the source strength is uniform across the disk of the star, i.e., if the star is divided into vertical strips parallel to the fringes produced by the pinholes, assume that all these strips make equal contributions. Assume also that a filter is used to pass a single wavelength. Let a double pinhole be placed in front of the lens of a telescope focused on a particular star. a) Start with a small value of d, the distance of separation of the pinholes. What happens to the interference pattern as d is increased? b) If d_0 is the smallest value of d at which the fringes disappear (uniform intensity), show how this could be used in principle to determine the angular diameter of the star. c) If the nearest stars (at distances the order of five light years) are about the same size as the sun (diameter $= 8.6 \times 10^5$ miles), would you expect this to be a feasible method of measuring their diameters? d) Michelson devised the arrangement indicated schematically in Fig. 11–34 for exaggerating the effects due to broadness of the source without requiring the pinhole separation to be excessively large. Mirrors inclined at 45° to the axis of the instrument, as defined by the perpendicular bisector of S_1S_2, are labeled M_1, M_2, M_3, and M_4. The pinhole separation d can be fixed at any desired value according to the telescope lens to be used, the latter being placed to the right of the aperture screen. The value of D can be varied by moving the mirrors M_1 and M_4 symmetrically away from the axis.

Consider an incoming plane wave whose normal makes an angle α with the axis, corresponding to a point source at infinity. From the geometry of the ray

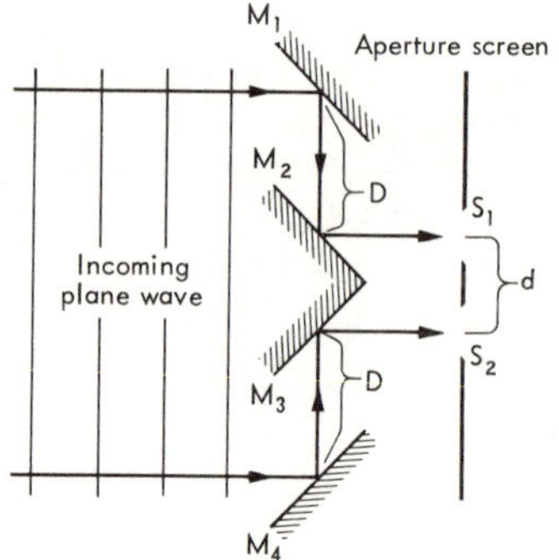

Fig. 11–34. Michelson stellar interferometer.

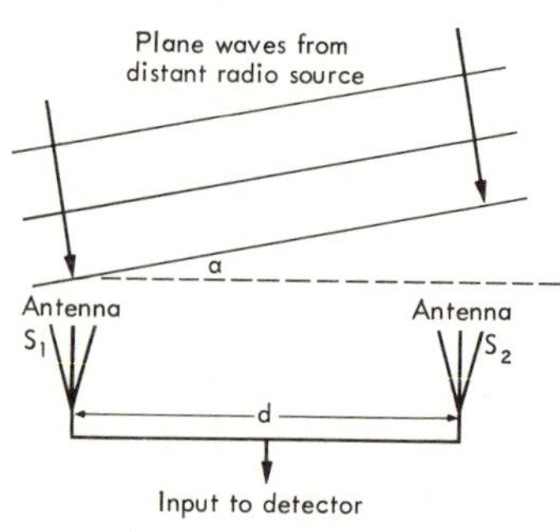

Fig. 11–35. Radio-frequency stellar interferometer.

paths show that for small α the phase difference between the signals arriving at S_1 and S_2 is $k\alpha(d + 2D)$, and therefore that the mirror arrangement can be thought of as producing a multiplication of the angle α by the factor $[1 + (2D)/d]$. e) Which parameters affect the breadth of the interference fringes in the arrangement of (d) and which affect the location of the center of the fringe system? f) If a broad source subtends a total angle α, what will be observed as D is increased? Does the measurement desired in (c) now seem feasible? g) Consider a double star consisting of two equal components each of which is small enough to be considered a point source, the angle between centers being α. If this is used as a source with the apparatus of Fig. 11–34, what will be observed as D is increased? How could you distinguish experimentally between this case and that of a single broad source?

11–18. *Radio-frequency stellar interferometer*

Using radio telescopes, the angular breadth of a cosmic source of radio waves can be determined by a method basically the same as that of the Michelson stellar interferometer (Problem 11–17). The apparatus is indicated schematically in Fig. 11–35. For simplicity the radio antennas S_1 and S_2 will be considered to be nondirectional. These are analogous to the two pinholes of the optical experiment; the assumption of nondirectionality is equivalent to the assumption that the pinholes are sufficiently small for complete diffraction to take place. The signals are detected by a receiver which is sensitive to a narrow band of frequencies, thus permitting the incident wave to be treated as having a single wavelength. The radio signals from the antennas are connected, by cables of equal length, to the same input of the detection system, thus making the instantaneous voltage at this point the sum of the instantaneous voltages of the signals from the two antennas. This is analogous to adopting a *fixed* point of observation on the axis $\theta = 0$ of a double pinhole interference pattern. The output of the detection system records the rms voltage of the combined signals, which is the analog of the average intensity I in the optical case. a) Given an incoming plane wave equivalent to a point source at infinity at a small angle α. As the earth turns, α varies linearly with time. Describe the expected changes in the output signal. b) Discuss how a variable baseline d might be used to determine the angular diameter $\Delta\alpha$ of a broad source. c) When it was first established that the discrete radio source called Taurus A was associated with the optically observed Crab Nebula, it was not known whether the radio emission takes place throughout the nebula, or from a single star embedded in it. If we use a frequency of 10^8 sec^{-1}, what baseline would be required to measure the angular width of the radio source in each of these two cases? (The visual angular width of the nebula is $\Delta\alpha \sim 5'$; Betelgeuse, the star of largest angular width in visible light has $\Delta\alpha = 0.05''$.) Experimentally it is found that loss of modulation of the output signal occurs for a baseline somewhere between 1 and 5 km. What can be inferred about the relation of the radio source to the nebula?

11–19. The image of a point source S is focused by means of a lens L onto an observing screen Σ. a) An aperture screen A containing a pair of pinholes is placed between L and Σ as in Fig. 11–13. How does the linear width of the fringes vary as A is moved back from the observing screen toward the lens? b) When A reaches L it is switched from one side of the lens to the other. What change does this produce in the fringes on Σ? How does the pattern change as A is then moved back from L toward S?

11–20. In a double pinhole interference experiment let a lens L_2 be placed between the aperture screen A and the observing screen Σ. (Assume that appropriate other arrangements are made so that the overall focusing condition is satisfied.) Let A' be the image (real or virtual) of A in L_2. Show that the width of the interference fringes on Σ is equivalent to that which would be obtained by the removal of A and L_2 and the insertion of an aperture screen having the size and location of A'.

11–21. A point source is fixed at one end of an optical bench and an observing screen at the other; the distance of separation is 3 m. Given a converging lens L of focal length $f = 50$ cm and an aperture screen A with a pair of pinholes whose centers are 0.5 mm apart. How should L and A be placed to yield interference fringes of maximum width? What is this maximum fringe width for the Hg green line?

11–22. In the collimator-telescope arrangement of Fig. 11–20 the aperture screen A is displaced sideways by a small amount. (The screen A remains parallel to itself; the displacement is perpendicular to the axes of the lenses.) It is found that the fringe pattern is *unaffected* by this change. Why? What happens if A is moved closer to the lens L_3?

11–23. Consider the effect of using a broad source in a Lloyd's-mirror experiment (Section 11–9a). If S and $\overline{S}$ are two points at slightly different distances above the plane of the mirror, how do the associated fringe systems compare? Why does broadening the source cause the *higher*-order fringes to be lost first? Does the same thing happen in the Fresnel double-mirror experiment?

11–24. A sound generator which produces a signal of $\nu = 500$ sec^{-1} is placed at a depth of 20 ft below the surface of a body of water. (Take $c = 5000$ ft/sec.) Assume that the water surface is perfectly smooth and that there is a phase change of π upon reflection. Ignore variation in the velocity of sound throughout the body of water and ignore reflections from the bottom. A hydrophone is lowered vertically at a distance of 2000 ft from the location of the source. At what depths will pressure minima be found? What experiment in optics and what situation in radio transmission are analogous to this acoustical experiment?

11–25. *Fresnel biprism experiment with a collimated laser beam*
A biprism is illuminated with a collimated laser beam normally incident on the face AA' (Fig. 11–23). (*Practical note:* To produce a significant overlap region of the two refracted beams the laser beam spot can be widened by passing the light backward through a telescope focused for infinity.) a) Show that in any two-point interference experiment the width of the fringes on a screen set parallel to the line joining the sources S_1 and S_2 can be expressed as $\Delta x = \lambda/\beta$, where β is the angle subtended by S_1S_2 from the observing screen. b) Argue therefore that in the above experimental arrangement the width of the fringes is independent of the distance of the plane of observation from the biprism. Calculate the fringe width for $n = 1.6$, $\alpha = 15''$, and $\lambda = 6328$ Å. c) To demonstrate the presence of the fringes without using a microscope, the observing screen can be rotated until the beam strikes it at nearly grazing incidence. Show that rotation of the screen through an angle ϵ about an axis parallel to the fringes multiplies the fringe width by a factor of sec ϵ.

11–26. Let the point of contact of a spherical lens surface of radius R and an optical flat be taken as the origin of a cartesian system with the y-axis perpendicular to the flat. a) Show that near the origin the intercept of the lens surface with the xy-plane can be represented approximately by the parabola $y = x^2/2R$. (This approximation is sometimes known as the *sagitta formula.*) b) Show therefore that the radii of Newton's dark rings obtained under reflection at normal incidence are proportional to the square roots of the integers. Calculate the lens radius required to make the radius of the tenth dark fringe equal to 1 cm for $\lambda = 5500$ Å.

11–27. It is desired to exhibit localized interference fringes in sodium yellow light by forming a thin air wedge between two optical flats. The flats are 5 cm long and the wedge is formed by inserting a thin piece of paper between the flats along one edge. How thick must the paper be if the fringes are to be 1 mm wide?

11–28. In an acoustic version of the Michelson interferometer S is a point source of sound and E is a microphone at a fixed position (Fig. 11–29). The angles between the semireflector R and each of the reflectors M_1 and M_2 are set at 45°. The analog of counting fringes as they appear at the center of a circular fringe pattern is obtained by moving M_2 back parallel to itself and counting the number of minima in the signal detected by the microphone. a) Sketch the location of the virtual sound sources produced by reflection and deduce the condition required for a minimum at E. b) Given that the frequency of the source is 15 kc and room temperature is 24°C, calculate the distance M_2 must be moved to go from one minimum to the next.

11–29. In Fig. 11–36 S is a point source of monochromatic light and M_1M_2 is a plane parallel plate having index of refraction n. The observing instrument is focused for infinity, i.e. the observing screen Σ is located in the focal plane of lens L. If multiple internal reflections in the plate are neglected, there are two rays from S arriving at any given point of observation P, resulting from reflections in M_1 and M_2 respectively. a) Given

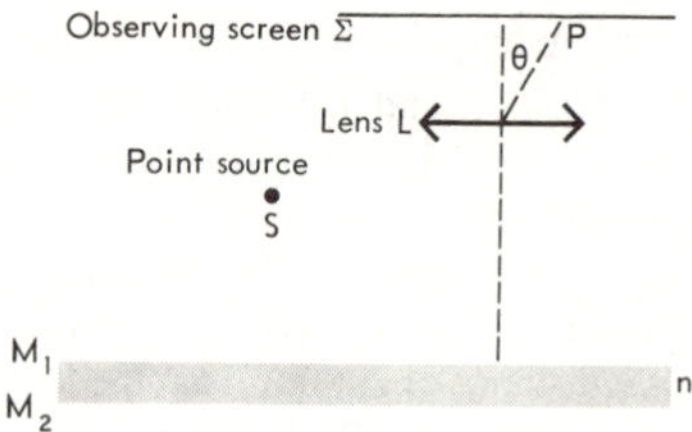

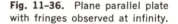

Fig. 11–36. Plane parallel plate with fringes observed at infinity.

that the thickness of the plate is d, show that the phase difference between these two rays is

$$\Delta\phi = \pi + 2kd\sqrt{n^2 - \sin^2\theta},$$

where θ is the angle corresponding to the chosen point of observation. (*Note:* Since this result is completely independent of the location of S, the point source can be replaced by a source which is indefinitely broad without causing the fringe pattern to lose distinctness. That is, the loci $\Delta\phi$ = const have precisely the same location for all points of a broad source. A necessary condition for this independence of the position of S is that the observing instrument be focused for infinity. The fringes are said to be *localized at infinity*.) b) For application to the Michelson interferometer, take $n = 1$ and show that the loci of destructive interference are circles given by $\cos\theta = m\lambda/2d$, where m is an integer. c) In direct visual observation, where L is the lens of an eye focused for infinity (i.e. relaxed), what determines where the center of the circles will lie?

11–30. Using a Michelson interferometer, the position of zero retardation is located by means of white light fringes. The source is then replaced by a high pressure Hg arc, using a filter for the green line $\lambda = 5461$ Å. As mirror M_2 is moved back, visibility of the fringes gradually becomes poorer. After approximately 1000 fringes have passed, the pattern is difficult to perceive and does not improve with increasing path difference. Calculate the order of magnitude of the wavelength spread in the collision broadened line, and the associated coherence interval, τ.

12 Continuous distributions of coherent sources; the Fraunhofer approximation

12-1 INTRODUCTION

Christian Huygens is responsible not only for a general method of constructing successive wavefronts for any kind of propagating wave, but also for the specific hypothesis that light is to be considered as a wave. As shown in Chapter 9, these ideas enable us to give a satisfactory explanation of the phenomenon of double refraction. Newton was familiar with Huygens' work, but rejected the notion that light is a wave primarily on the basis of the distinction between light and sound with regard to the production of shadows.*

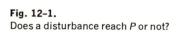

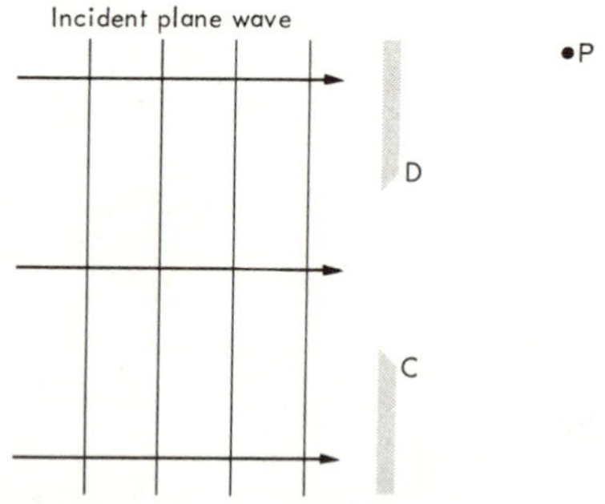

Fig. 12-1.
Does a disturbance reach P or not?

Consider, for example, a plane wave incident from the left on a screen with opening CD (Fig. 12-1). According to Huygens' principle, every point on the wavefront between C and D may be considered to be a source of

* Sir Isaac Newton, *Opticks*, Book Three, Part I, Question 28. New York: Dover Publications, p. 362.

spherical wavelets. Thus if P is any point to the right of the screen, we would expect these wavelets to contribute to a disturbance at P. This is an acceptable conclusion for sound, for common experience verifies that sound spreads out in all directions after passing through an open window. Light, on the other hand, casts a relatively sharp shadow and no disturbance reaches P if it is within this shadow region. Huygens' construction of secondary wavelets would be identical for these two cases and contains no suggestion as to why the results should be so different.

The wave hypothesis of light was rescued from this dilemma by Fresnel who combined the ideas of Huygens and Young in a statement which we shall refer to as the *Huygens-Fresnel principle*. According to Fresnel we consider the points along the wavefront between C and D as secondary sources, but in adding the contributions at P we must take account of the fact that the contributing wavelets have traveled different distances and consequently are not in phase with one another as they arrive at P. If the maximum path difference $|PC - PD|$ contains many wavelengths, the phases of the contributing wavelets will range over many multiples of π. Thus most of the wavelets can be paired with other wavelets which differ in phase by π; the contributions tend in large measure to cancel each other. However, if $|PC - PD|$ is small compared with a wavelength, the contributions are all nearly in phase and the interference is constructive. According to Fresnel, Newton's qualitative distinction between light and sound is spurious. The important distinction is the comparison of the wavelength λ with the dimensions CD of the opening. If $CD \ll \lambda$, there will be no "shadow," since $|PC - PD| \leq CD$ and the criterion $|PC - PD| \ll \lambda$ is satisfied everywhere. On the other hand, if $CD \gg \lambda$, there will be a shadow region within which $|PC - PD| \gg \lambda$ and, as we shall show in detail below, the destructive interference of contributing wavelets is almost complete. Thus the "short" wavelength electromagnetic waves do cast sharp shadows, whereas the "long" wavelength sound waves do not.

It is mainly in connection with the Huygens-Fresnel principle that we will be concerned with the radiation pattern from a continuous distribution of coherent point sources. That is, when a plane wave passes through an opening in a screen, the Huygens-Fresnel principle identifies the distribution of intensities in the diffracted wave with the radiation pattern which would be obtained from a number of point sources closely distributed over the area of the aperture. We can, however, imagine a situation in which an actual distribution of sources would be involved. An example is a number of coherently excited radio antennas in a closely spaced linear array. In the following sections the radiation patterns from certain source distributions are discussed abstractly before they are applied in the context of a diffraction pattern.

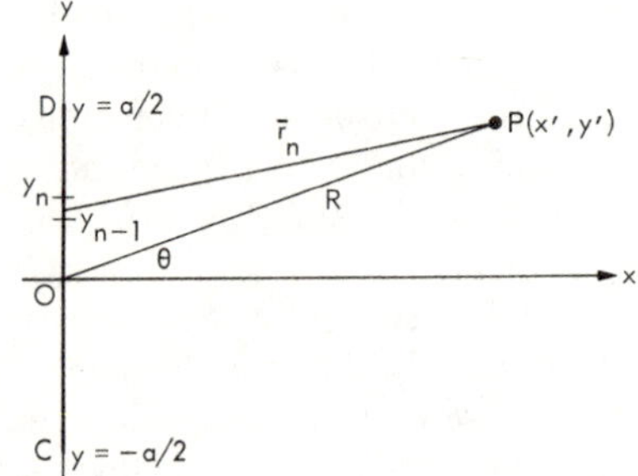

Fig. 12-2. Linear distribution of coherent point sources.

12-2 RADIATION PATTERN FROM COHERENT SOURCES CONTINUOUSLY DISTRIBUTED ALONG A LINE SEGMENT

In Fig. 12-2 let *CD* represent a linear array of uniformly distributed point sources which are in phase and of equal strength. It is clear that the radiation pattern will be symmetrical about the *y*-axis; hence, it will be sufficient to calculate the net disturbance at a point *P* in the *xy*-plane. Divide the *y*-axis between *C* and *D* into *N* intervals by picking points of subdivision $y_0 = -a/2 < y_1 < y_2 < \cdots < y_n < \cdots < y_N = a/2$. If we think of the total segment *CD* as containing a uniform distribution of a large number *M* of individual point sources, the number contained within the segment $\Delta y_n = y_n - y_{n-1}$ is $(\Delta y_n/a)M$. Let each of these point sources contribute a spherical wave of the form

$$\mathcal{R}e\left\{A\,\frac{e^{i(\omega t-kr)}}{r}\right\}$$

to the disturbance at *P*. Now if Δy_n is sufficiently small, all the point sources within Δy_n are close together compared with a wavelength and the net contribution from the *n*th segment can be represented as

$$\psi_n = \mathcal{R}e\left\{A\,\frac{e^{i(\omega t-k\bar{r}_n)}}{\bar{r}_n}\cdot\frac{M\,\Delta y_n}{a}\right\}, \tag{12-1}$$

where $\bar{r}_n$ is measured from some intermediate position in the interval. As we let *M* approach infinity we must think of the strength of each individual point source as approaching zero, for clearly an infinite number of point sources of finite strength would give rise to an infinite resultant. Thus let

$$\lim_{M\to\infty}\left(\frac{AM}{a}\right) = B.$$

Equation (12-1) is then equivalent to the assumption that each small segment of width Δy_n is the source of a spherical wavelet of strength $B\,\Delta y_n$. That is, each small segment contributes a wavelet whose strength is proportional to the width of the segment. The proportionality constant *B* is the source strength per unit length. The net disturbance at P is obtained by adding the contributions from all the segments:

$$\psi = \sum_{n=1}^{N} \psi_n = \mathcal{R}e \left\{ Be^{i\omega t} \sum_{n=1}^{N} \frac{e^{-ik\bar{r}_n}}{\bar{r}_n} \Delta y_n \right\}. \tag{12–2}$$

If we now let N approach ∞ in such a way that each Δy_n tends to zero, from the definition of the definite integral we may write

$$\psi = \mathcal{R}e \, \{Be^{i\omega t} \Psi(x', y')\},$$

where

$$\Psi(x', y') = \int_{-a/2}^{a/2} \frac{e^{-ikr}}{r} \, dy \tag{12–3}$$

and $r = r(y) = \sqrt{x'^2 + (y' - y)^2}$. The problem of calculating the radiation field at $P(x', y')$ has been reduced to the problem of calculating an integral of the form specified by $\Psi(x', y')$. The reader should convince himself that despite the apparent simplicity of this integral, it is not reducible to any of the forms listed in standard integral tables. We must content ourselves with an examination of the radiation pattern within regions where an approximate expression for Ψ is an adequate representation, or else carry out an explicit numerical evaluation of the integral.

12-3 THE FRAUNHOFER APPROXIMATION

The first approximation we make in evaluating Eq. (12–3) is the assumption that the "amplitude factor," $1/r$, does not vary significantly over the range of integration. Let the distance OP in Fig. 12–2 be designated by

$$R = \sqrt{x'^2 + y'^2}\,.$$

We then replace $1/r$ by $1/R$ and obtain

$$\Psi(x', y') = \frac{1}{R} \Phi(x', y'), \qquad \text{where} \qquad \Phi = \int_{-a/2}^{a/2} e^{-ikr} \, dy. \tag{12–4}$$

The approximation will be valid only at relatively large distances from the source, such that $R \gg a$. The integral in Eq. (12–4) is still not a standard integral, and further approximation is required in the exponential factor. (Recall from Section 11–1 that more subtle approximations are required in the phase term than can be tolerated in the amplitude factor.) If we expand $r(y)$ in a Maclaurin series we obtain

$$r(y) = R - y \sin\theta + (y^2/2R) \cos^2\theta + \cdots, \tag{12–5}$$

where we have set $y' = R\sin\theta$, θ being the angle subtended by the point of observation P from the perpendicular bisector of CD. If the third term in this expansion can legitimately be neglected, $r(y)$ is approximated by a linear function of y, and the integral is of elementary form. The third term can be neglected in the expansion if $(ky^2/2R)\cos^2\theta \ll \pi/4$, for then $e^{-i(ky^2/2R)\cos^2\theta}$ is indistinguishable from unity. Since the largest value of y^2 is $a^2/4$, the criterion becomes

$$(ka^2/8R)\cos^2\theta \ll \pi/4, \qquad \text{or} \qquad (a^2/\lambda R)\cos^2\theta \ll 1.$$

If R is sufficiently large, this condition is satisfied for all values of θ and the approximation obtained is valid throughout the whole pattern. If, however, $\lambda R/a^2 \sim 1$, the approximation will be valid only for angles θ such that $\cos^2\theta \ll \lambda R/a^2$, but will not be valid near the axis where $\theta = 0$. The replacement $r = R - y\sin\theta$ is called the *Fraunhofer approximation*. In the next chapter we shall consider the *Fresnel approximation*, which is a more sensitive approximation obtained by retaining all three terms in Eq. (12–5).

Under the conditions of the Fraunhofer approximation we may then write

$$\mathbf{\Phi} = \int_{-a/2}^{a/2} e^{-ik(R-y\sin\theta)}\,dy = e^{-ikR}\left(\frac{e^{ik(a/2)\sin\theta} - e^{-ik(a/2)\sin\theta}}{ik\sin\theta}\right)$$

$$= ae^{-ikR}(\sin\beta)/\beta, \tag{12–6}$$

where $\beta = (ka/2)\sin\theta$. Collecting terms we have finally

$$\psi = \mathcal{R}e\left\{\frac{aB}{R}\left(\frac{\sin\beta}{\beta}\right)e^{i(\omega t-kR)}\right\} = \frac{aB}{R}\left(\frac{\sin\beta}{\beta}\right)\cos(\omega t - kR)\,. \tag{12–7}$$

This enables us to study the intensity distribution along a circle of radius R as a function of angular position. If we take the time average and consider only the angle-dependent factors, we have

$$I(\theta) \propto \overline{(\psi)^2} \propto \left(\frac{\sin\beta}{\beta}\right)^2.$$

Since $\beta = 0$ when $\theta = 0$ and

$$\lim_{\beta\to 0}\left(\frac{\sin\beta}{\beta}\right) = 1,$$

we write

$$I(\theta) = I_0\left(\frac{\sin\beta}{\beta}\right)^2, \tag{12–8}$$

where $I_0 = I(0)$ is the intensity on the axis $\theta = 0$.

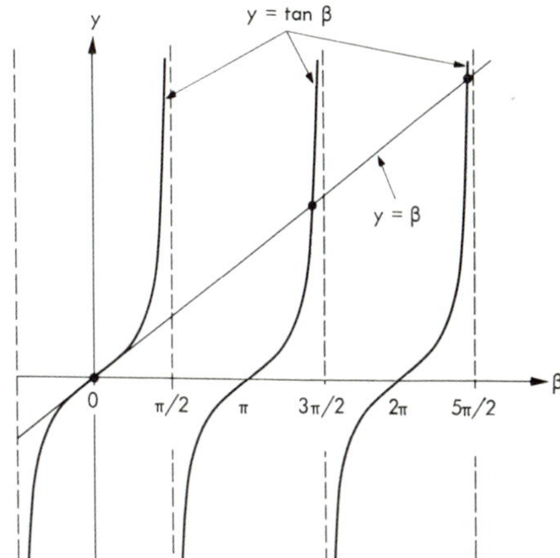

Fig. 12-3.
Locating the roots of $\beta = \tan \beta$.

12-4 STUDY OF THE FRAUNHOFER PATTERN

The function $f(\beta) = (\sin \beta)^2/\beta^2$ has the following properties:

i) $f(\beta) \geq 0$.
ii) $f(0) = 1$.
iii) $f(\beta) = 1/\beta^2$, where $\sin \beta = \pm 1$, $\beta = (2n + 1)\pi/2$, $n = 0, \pm 1, \pm 2, \ldots$
iv) $f(\beta) = 0$, where $\beta = n\pi$, $n = \pm 1, \pm 2, \ldots$
v) $f(\beta)$ is an even function of β.
vi) The possible extrema of $f(\beta)$ occur where $f'(\beta) = 0$ or

$$\frac{2 \sin \beta \cos \beta}{\beta^2} - \frac{2 \sin^2 \beta}{\beta^3} = \frac{2 \sin \beta}{\beta^3} (\beta \cos \beta - \sin \beta) = 0.$$

The roots $\sin \beta = 0$ correspond to minima at the zeros of $f(\beta)$. Setting $(\beta \cos \beta - \sin \beta) = 0$, since $\cos \beta = 0$ is clearly not a root, we obtain

$$\beta = \tan \beta. \tag{12-9}$$

We can obtain the roots of Eq. (12-9) graphically by considering the intersections between the graphs of $y = \tan \beta$ and the straight line $y = \beta$ (Fig. 12-3). One root is $\beta = 0$. If we consider $\beta > 0$, the graph shows that for every $m = 1, 2, 3, \ldots$ there is one intersection β_m such that $m\pi < \beta_m < m\pi + \pi/2$. Furthermore, when m is large $\beta_m \sim m\pi + \pi/2$. Since there is only one such root between each of the minima determined above, the function must be maximum at these values. The values of β_m can also be obtained by inspection of tables of $\tan \beta$. The lowest root other than $\beta = 0$ is $\beta_1 = 1.43\pi$.

Properties (iii) and (iv) show that $f(\beta)$ oscillates between the β-axis and the envelope curve $y = 1/\beta^2$. Comparing (iii) and (vi) we see that the places

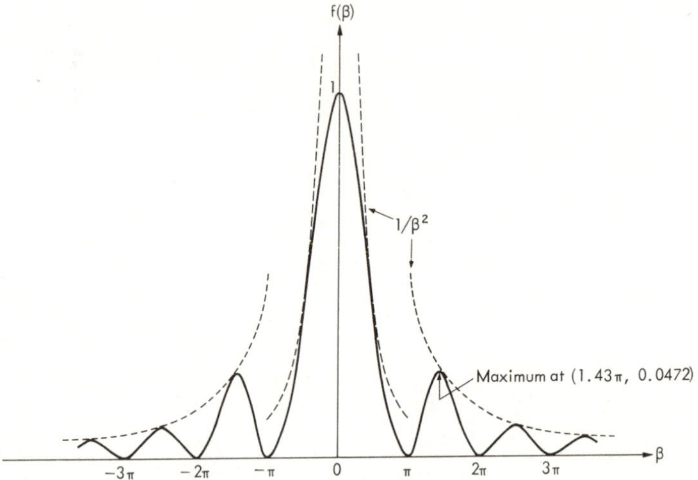

Fig. 12–4. Graph of $f(\beta) = (\sin\beta)^2/\beta^2$. [*Note:* In order to show subsidiary maxima on the same graph as the principal maximum the value of $f(\beta)$ has been multiplied by five in the range $|\beta| > \pi$.]

where $f(\beta)$ touches the envelope are not identical with the maximum points but are nearly coincident for large m. The graph of $f(\beta)$ is shown in Fig. 12–4.

The maximum at $\beta = 0$ is referred to as the *principal maximum;* the maxima at $\beta = \pm\beta_m$ are *subsidiary maxima.* The first subsidiary maximum occurs at $\beta_1 = 1.43\pi$, $f(\beta_1) = 0.0472$. For large m,

$$f(\beta_m) \sim \frac{1}{\beta_m^2} \sim \frac{1}{\pi^2(m + \frac{1}{2})^2}.$$

Since $\beta = (ka/2)\sin\theta$, the graph of $I(\theta)$ from Eq. (12–8) is related to the graph of $f(\beta)$ through a nonlinear stretching parallel to the axis of abscissas. Only the portion of $f(\beta)$ lying between $\pm\beta_{\max} = \pm ka/2$ is relevant. The zeros of $I(\theta)$ correspond to the zeros of $f(\beta)$ and occur at $\sin\theta = n\lambda/a$ where $n = \pm 1, \pm 2, \ldots \pm n_{\max}$. The extrema on the graph of $I(\theta)$ are given by

$$I'(\theta) = I_0 f'(\beta)(d\beta/d\theta) = I_0 f'(\beta)(ka/2)\cos\theta = 0$$

and correspond to extrema on $f(\beta)$ with the possible addition of a maximum or a minimum at $\theta = \pi/2$. A polar plot of $I(\theta)$ is given in Fig. 12–5 for the case $a = 1.6\lambda$.

In general, we see that the secondary lobes in the radiation pattern carry only a small portion of the total energy; the important contributions come from angles such that $-\pi < \beta < \pi$, that is, $-\lambda/a < \sin\theta < \lambda/a$. It is shown in Problem 12–2 that this principal maximum contributes not less than 91% of the total energy flux.

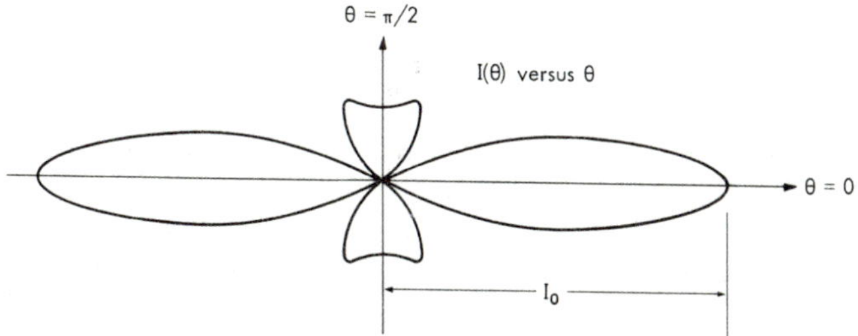

Fig. 12-5. Polar plot of $I(\theta) = I_0[\sin^2(1.6\pi \sin\theta)]/(1.6\pi \sin\theta)^2$. [*Note:* Values of $I(\theta)$ have been multiplied by a factor of five outside of the principal lobes.]

Two extreme cases are worthy of consideration:

a) $ka \ll 1$. Then $\beta \ll 1$ for all θ. Hence $\sin\beta \sim \beta$ and

$$I(\theta) = I_0 \frac{\sin^2\beta}{\beta^2} \sim I_0.$$

This is to be expected, since in this case all of the points along the linear source are close together compared with a wavelength and act together constructively to produce a spherical wave with uniform intensity in all directions.

b) $ka \gg 1$. In this case $I(\theta)$ is insignificant except near $\theta = 0$, and we can replace $\sin\theta$ by θ throughout the region of interest. Since β and θ are proportional in this approximation, the graph of $I(\theta)$ vs. θ has the same form as the curve in Fig. 12–4. We conclude from this that when a linear source is many wavelengths long the radiation pattern is highly directional, the energy flux being concentrated about the direction $\theta = 0$. This statement applies, of course, only to distances such that $R \gg a$; it does not describe the field near the source.

12-5 VIBRATION CURVE FOR THE FRAUNHOFER APPROXIMATION

In Section 12–3 we wrote the net disturbance at point P in the radiation pattern from a line source as

$$\psi = \mathcal{R}e\left\{\frac{Be^{i\omega t}}{R}\,\mathbf{\Phi}\right\}$$

where $\mathbf{\Phi}$ is the so-called *diffraction integral*,

$$\mathbf{\Phi} = \int_{-a/2}^{a/2} e^{-ikr}\,dy.$$

The only dependence on the angular position θ occurs in $\mathbf{\Phi}$, through the dependence of r on θ. The angular distribution of intensity is consequently determined by taking $I(\theta) \propto |\mathbf{\Phi}|^2$. When the Fraunhofer approximation is made, r is replaced by a linear function in y and the integration of the diffraction integral is immediate. Nevertheless it is instructive to observe that this definite integral, by virtue of its being the limit of a sum of complex numbers, can be interpreted by means of a vector diagram. We shall derive this interpretation from general arguments which apply not only to the present case of the Fraunhofer approximation, but also to the case of the Fresnel approximation to be discussed in Chapter 13. Thus, consider the integral

$$\mathbf{Z}(u) = \int_0^u e^{i\phi(u)}\, du, \tag{12-10}$$

where u is a real parameter and $\phi(u)$ is a real function of u. The complex number $\mathbf{Z}$ can be written in rectangular form

$$\mathbf{Z}(u) = X(u) + iY(u),$$

where

$$X(u) = \int_0^u \cos[\phi(u)]\, du, \qquad Y(u) = \int_0^u \sin[\phi(u)]\, du. \tag{12-11}$$

As u is varied, the vector associated with $\mathbf{Z}$ traces out a locus in the complex plane referred to as the *vibration curve* (see Fig. 12-6). The cartesian coordinates of the moving point are (X, Y) and the locus is specified with u as parameter by the parametric equations

$$X = X(u), \qquad Y = Y(u).$$

The origin $X = 0$, $Y = 0$ is a point on the locus corresponding to $u = 0$.

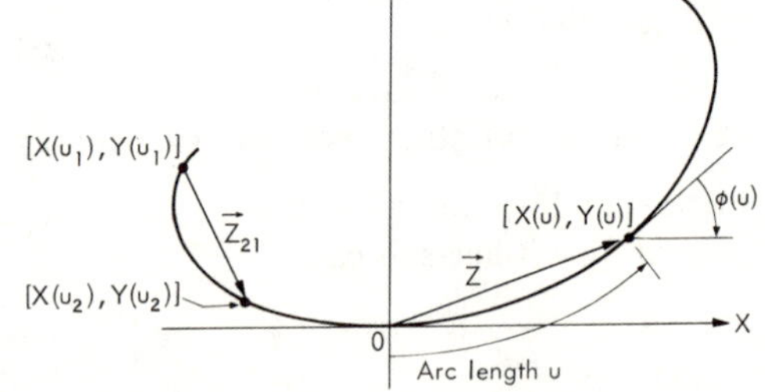

Fig. 12-6. Vibration curve determined by $\mathbf{Z} = \int_0^u e^{i\phi(u)}\, du$.

The geometrical significance of the parameter u is that it represents the arc length measured along the vibration curve from the origin $(0, 0)$ to the general point (X, Y).

Proof: The differential of arc length is

$$ds = \sqrt{dX^2 + dY^2} = \sqrt{(X')^2 + (Y')^2}\, du,$$

where $'$ designates d/du. But from Eq. (12–11), $X' = \cos\phi$, $Y' = \sin\phi$ and $(X')^2 + (Y')^2 = 1$. Therefore

$$ds = du. \tag{12–12}$$

The geometrical significance of the function $\phi(u)$ is that it represents the inclination of the vibration curve at the point corresponding to parameter u.

Proof: The slope is

$$\frac{dY}{dX} = \frac{Y'}{X'} = \frac{\sin\phi}{\cos\phi} = \tan\phi. \tag{12–13}$$

The curvature K is defined from analytic geometry as the rate of change of inclination with respect to arc length, $K \equiv d\phi/ds$. But from Eq. (12–12) this can be written

$$K = d\phi/du = \phi'(u). \tag{12–14}$$

The complex number

$$\mathbf{Z}_{21} = \int_{u_1}^{u_2} e^{i\phi(u)}\,du = \int_0^{u_2} e^{i\phi(u)}\,du - \int_0^{u_1} e^{i\phi(u)}\,du = \mathbf{Z}_2 - \mathbf{Z}_1$$

is represented by the difference between the vectors $\mathbf{Z}_2$ and $\mathbf{Z}_1$; hence $|\mathbf{Z}_{21}|^2$ is the square of the length of the chord on the vibration curve joining the two points corresponding to values u_1 and u_2 of the parameter.

Applying these results to the case of the diffraction integral in the Fraunhofer approximation, we take $r = R - y\sin\theta$ and

$$\mathbf{\Phi} = e^{-ikR}\int_{-a/2}^{a/2} e^{iky\sin\theta}\,dy.$$

Let

$$\phi(y) = ky\sin\theta \qquad \text{and} \qquad \mathbf{Z}(y) = \int_0^y e^{iky\sin\theta}\,dy. \tag{12–15}$$

Then

$$\mathbf{\Phi} = e^{-ikR}\mathbf{Z}_{21},$$

where $y_1 = -a/2$, $y_2 = a/2$. But $|\mathbf{\Phi}|^2 = |\mathbf{Z}_{21}|^2$; therefore $I(\theta)$ is proportional to the square of the length of the chord joining points $\mathbf{Z}_1$ and $\mathbf{Z}_2$ on the vibration curve of $\mathbf{Z}(y)$. The points $\mathbf{Z}_1$ and $\mathbf{Z}_2$ lie in opposite directions along the curve from the origin and are at equal distances $a/2$ measured along the curve.

The geometry of the vibration curve is determined by the fact that $K = \phi'(y) = k\sin\theta$ is independent of y. The locus is therefore a *circle* of radius $1/K = 1/k\sin\theta$. Since $\phi(0) = 0$, the circle is tangent to the real axis at the origin.

A sequence of vibration curves is shown in Fig. 12–7 corresponding to progressively increasing values of θ, the angular position of the point of

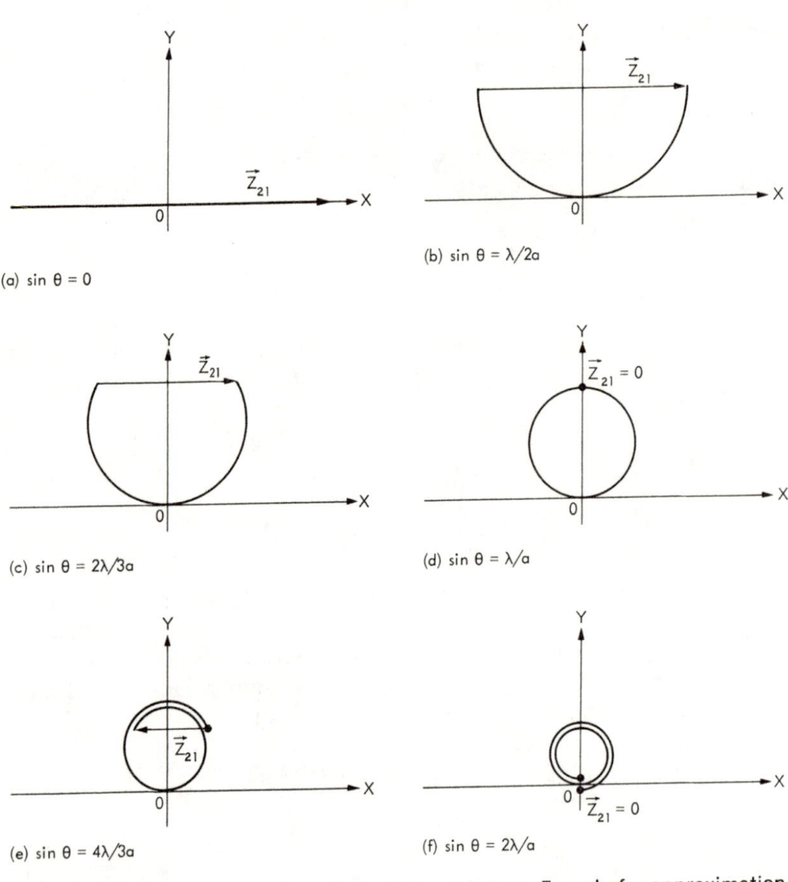

Fig. 12-7. Vibration curves for coherent linear source, Fraunhofer approximation.

observation. When $\theta = 0$ all contributions to the vector sum are in phase (according to the Fraunhofer approximation), the vibration curve is a straight line ($K = 0$), and the maximum chord length is obtained. As θ increases the vibration curve curls up, but maintains the fixed arc length $\Delta y = a$. The sequence is similar to that which we would obtain by wrapping a string of fixed length around cylinders of progressively smaller radius. The resultant $\mathbf{Z}_{21}$ is zero whenever a is an integral multiple of the circumference:

$$a = m \cdot 2\pi \frac{1}{K} = \frac{2\pi m}{k \sin \theta} = \frac{m\lambda}{\sin \theta}\,.$$

This criterion for the nulls in $I(\theta)$ is of course the same as $\beta = m\pi$ obtained from the previous analytical treatment.

12–6 DIFFRACTION BY AN EXTREMELY NARROW SLIT

We will now apply the formulas obtained under the conditions of the Fraunhofer approximation for the radiation pattern from a coherent line source in the context of the Huygens-Fresnel principle to determine the *diffraction pattern* obtained when a plane monochromatic wave is incident on a screen containing a rectangular aperture. The most direct application of the preceding formulas is the case of an *extremely* narrow slit. Let the y and z dimensions of the slit be a and b respectively (Fig. 12–8). If we assume that b is very small compared with a wavelength, then the entire length of the slit can be divided into small segments each of which can be treated effectively as a point, and the slit consists of a linear array of coherently radiating point sources. We should note that the usual laboratory slit does not fulfill the condition $b \ll \lambda$. In fact, this is not a condition which can be achieved practically in an optical context. The present discussion is being given as a matter of principle; it will serve as a basis for the discussion in the next section of the laboratory slit, which satisfies the opposite extreme condition $b \gg \lambda$.

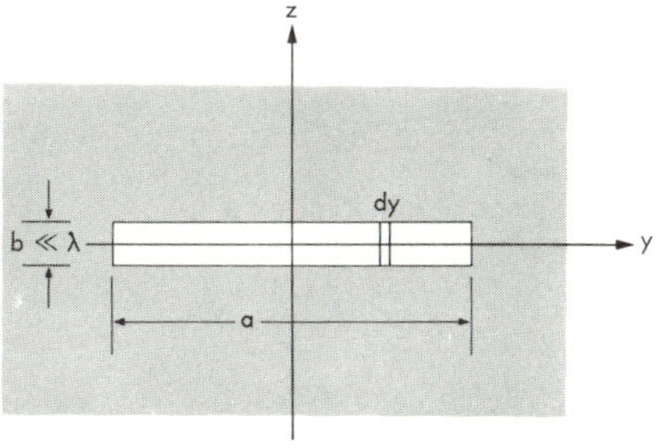

Fig. 12–8.
Dimensions of an extremely narrow slit.

To determine the disturbance at any point P beyond the screen, divide the rectangle into segments dy as indicated in Fig. 12–8. Since $kb \ll 1$ all the points within a given segment dy are close together compared with a wavelength and hence act cooperatively like a single point source. The entire slit is then equivalent to a linear distribution of coherent sources and the preceding results apply directly, provided that the point of observation is sufficiently remote to justify the Fraunhofer approximation.

Suppose that the original plane wave is produced by a point source at the focal point of a lens. If a second lens is placed beyond the screen, the conditions of the Fraunhofer approximation apply ideally to the intensity distribution in the focal plane of this lens. If the aperture screen is removed, a single point image is produced of the original source. When the aperture screen is present, this image is spread out indefinitely in a direction parallel to the narrow dimension of the slit; that is, corresponding to the axial symmetry of the radiation pattern from a line source, the intensity at large distances is uniform along any line parallel to the z-axis of Fig. 12–8. The intensity variation in the y-direction is described by Eq. (12–8). Thus a series of fringes are obtained which are parallel to the z-axis and whose spacing is governed by the relation between a and λ.

We can devise microwave or acoustic experiments which demonstrate diffraction by an extremely narrow slit. For example, for a sound wave with $\lambda = 20$ cm, $b = 1$ cm $= \lambda/20$, and $a = 32$ cm, the slit is narrow compared with a wavelength. The condition $a = 1.6\lambda$ corresponds to the data of Fig. 12–5. If the slit is mounted horizontally and a microphone is moved over the surface of a sphere of 5 or 10 m radius from the center of the slit (so that $R \gg a$ and $R \gg a^2/\lambda$), no variations in intensity are expected when the microphone is moved vertically; the variations described by Fig. 12–5 are expected when the microphone is moved in any horizontal plane.

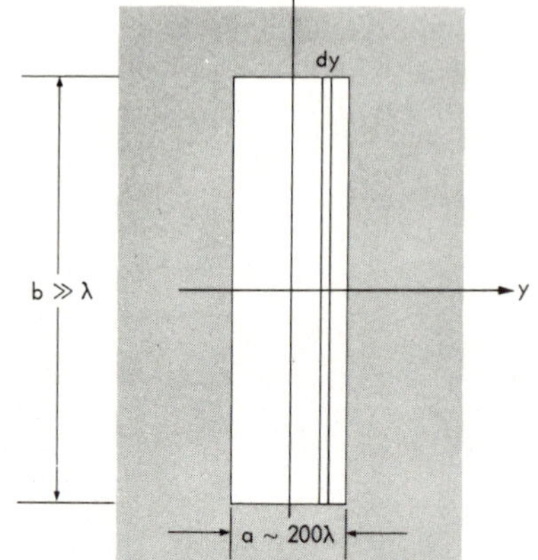

Fig. 12–9.
Dimensions of a laboratory slit.

12-7 DIFFRACTION BY AN EXTREMELY LONG SLIT

The dimensions of the usual laboratory slit are depicted in Fig. 12–9. The width a is not extremely small compared with the wavelength but is more commonly on the order of hundreds of wavelengths. On the other hand, the length b is an extremely large multiple of the wavelength. To determine the diffraction by such a slit when a plane wave is normally incident, we divide the rectangle into a large number of strips parallel to the z-axis. Each one of these strips dy is equivalent to a coherent line source which is extremely long compared with a wavelength. We observed in Section 12–4 that the superposition of the contributions from all the points along such a "long" source is vanishingly small except near $\theta = 0$, i.e., except for points of observation lying very near the plane $z = 0$. If P lies in the plane $z = 0$, the factor $(\sin \beta)/\beta$ in Eq. (12–7) is unity. This formula also shows that the *phase* of the resultant is equivalent to that of a signal arriving at the point of observation P from the midpoint of the line segment. In summary, the net signal from each elementary strip within the slit is only significant in the plane $z = 0$; within this plane we can associate the signal (as far as phase relations are concerned) with a point lying on the y-axis. All that remains is to add the contributions from all the strips ranging from $y = -a/2$ to $y = a/2$. But this summation is equivalent to the determination of the radiation pattern from a linear array of length a. We can therefore apply Eq. (12–8) directly. Consider the diffraction pattern obtained when the incident wave is a plane wave from a laser source and the plane of observation is the focal plane $O'Y'Z'$ of a lens placed on the opposite side of the slit. The pattern is sharply concentrated about the line $z' = 0$. Intensity along this line varies according to Eq. (12–8). Thus a series of "dashes" are seen (Fig. 12–10) whose spacing depends on the relation between a and λ.

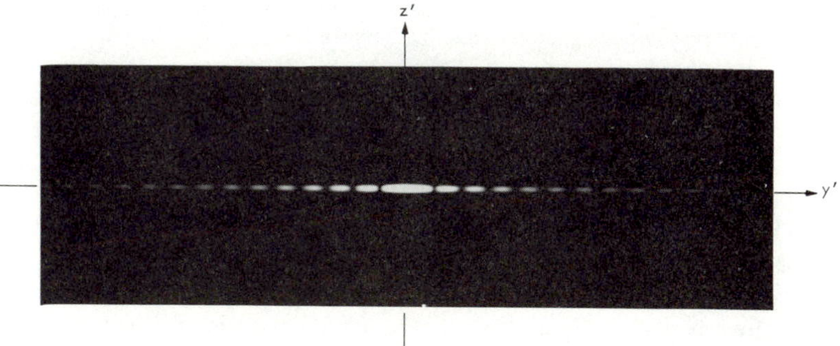

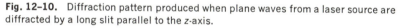

Fig. 12–10. Diffraction pattern produced when plane waves from a laser source are diffracted by a long slit parallel to the z-axis.

This same pattern is obtained if the incident plane waves are produced by a single point source S at the focal point of a lens. It is significant to observe that the center of the pattern is the position at which a single point image of S would be obtained in the absence of the aperture screen. We can then foresee the result of replacing S by an *incoherent* line source L consisting of a linear distribution of independent point sources which emit signals bearing random phase relations to each other. Each point of L produces its own independent pattern in the plane of observation. The resultant pattern consists of bands parallel to the line source L. In particular, if L is parallel to the z-axis, the result is that indicated in Fig. 12–11.

In Young's double pinhole experiment it was advantageous to use a line source since this increases the illumination of the bright bands in the plane of observation. The same advantage does not exist in the present case. The use of a line source merely produces a repetition of the desired pattern,

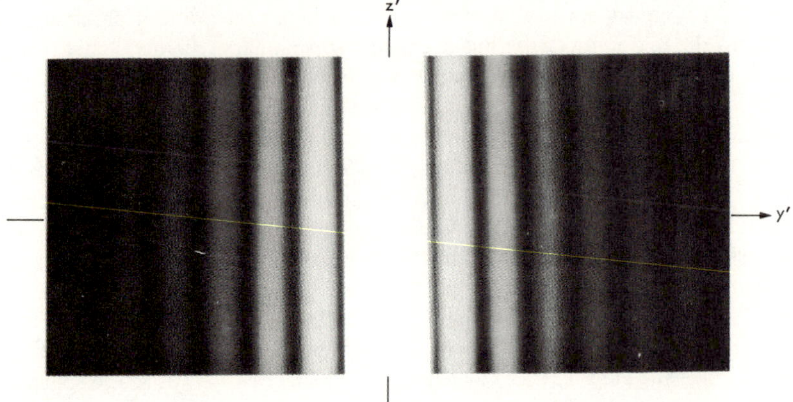

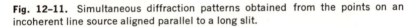

Fig. 12-11. Simultaneous diffraction patterns obtained from the points on an incoherent line source aligned parallel to a long slit.

without overlap. Caution must be used in interpreting this pattern in a quantitative experiment: The *perpendicular* distance between two dark bands does not represent the desired angular separation unless the source slit has been aligned parallel to the diffracting slit.

The effect of broadening the primary source in a direction perpendicular to the slit will be the same as that discussed in Section 11-8(d). The illumination of the diffraction pattern can be increased without serious loss of clarity by broadening the source until the point is reached at which the angle subtended by the source becomes comparable to the angular width of the fringes.

12-8 THE FRAUNHOFER APPROXIMATION APPLIED TO A RECTANGULAR DISTRIBUTION OF COHERENT POINT SOURCES

Let a uniform planar distribution of sources be characterized by a source strength B per unit area. In Fig. 12-12 the rectangular array of sources is divided into elements of area $dA = dy\,dz$. Such an element radiates a spherical wavelet producing a signal at P of the form

$$d\psi = \mathcal{R}e\left\{\frac{Be^{i(\omega t - kr)}}{r}\right\} dy\,dz,$$

where

$$r^2 = (x')^2 + (y - y')^2 + (z - z')^2.$$

The total signal at P is therefore given by

$$\psi = \mathcal{R}e\left\{e^{i\omega t}\int_{-b/2}^{b/2}\int_{-a/2}^{a/2}\frac{Be^{-ikr}}{r}\,dy\,dz\right\}. \qquad (12\text{-}16)$$

Since B is assumed to be independent of y and z, we obtain

$$\psi = \mathcal{R}e\,\{Be^{i\omega t}\Psi(x', y', z')\},$$

where

$$\Psi(x', y', z') = \int_{-b/2}^{b/2}\int_{-a/2}^{a/2} \frac{e^{-ikr}}{r}\,dy\,dz. \tag{12-17}$$

The Fraunhofer approximation consists in the replacement of r by $R = \sqrt{(x')^2 + (y')^2 + (z')^2}$ in the amplitude factor and by an expression linear in y and z for the phase factor:

$$\begin{aligned} r &\doteq [(x')^2 + (y')^2 + (z')^2 - 2yy' - 2zz']^{1/2} \\ &= R[1 - 2yy'/R^2 - 2zz'/R^2]^{1/2} \doteq R - yy'/R - zz'/R. \end{aligned} \tag{12-18}$$

The last step is the result of omitting the higher-order terms in a binomial expansion of the square root. Thus we write

$$\Psi = \frac{e^{-ikR}}{R}\int_{-b/2}^{b/2}\int_{-a/2}^{a/2} e^{ik(yy'+zz')/R}\,dy\,dz, \tag{12-19}$$

and, upon integrating, we obtain

$$\begin{aligned} \Psi &= \frac{e^{-ikR}}{R}\left(\int_{-a/2}^{a/2} e^{iky'y/R}\,dy\right)\left(\int_{-b/2}^{b/2} e^{ikz'z/R}\,dz\right) \\ &= \frac{abe^{-ikR}}{R}\left(\frac{\sin\beta_{y'}}{\beta_{y'}}\right)\left(\frac{\sin\beta_{z'}}{\beta_{z'}}\right), \end{aligned} \tag{12-20}$$

where $\beta_{y'} = kay'/2R$ and $\beta_{z'} = kbz'/2R$. If we confine our observations to the surface of a sphere or to a small region on a tangent plane, so that variations in R are negligible, the intensity distribution is given by

$$I(y', z') = I_0\left(\frac{\sin\beta_{y'}}{\beta_{y'}}\right)^2\left(\frac{\sin\beta_{z'}}{\beta_{z'}}\right)^2. \tag{12-21}$$

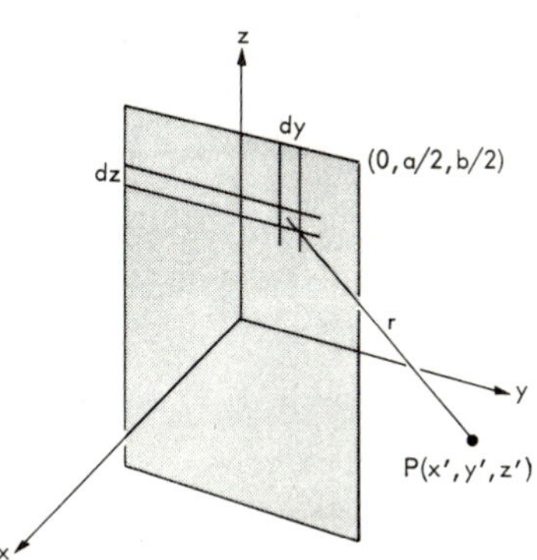

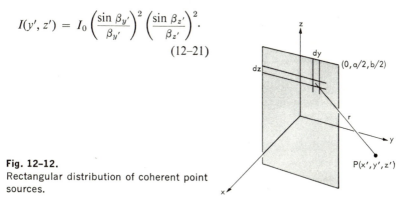

Fig. 12-12.
Rectangular distribution of coherent point sources.

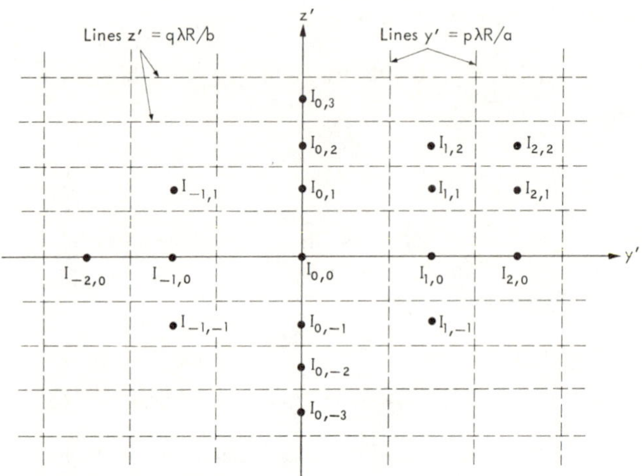

Fig. 12–13. Rectangular grid associated with the radiation pattern from a rectangular distribution of coherent point sources.

We can discern the character of this radiation pattern by observing that the tangent plane is divided into a rectangular grid defined by the loci of destructive interference $y' = p\lambda R/a$ and $z' = q\lambda R/b$ where p, $q = \pm 1$, $2, \ldots$ (zero not included). Along these straight lines $\sin \beta_{y'} = 0$ or $\sin \beta_{z'} = 0$ respectively. Note that the dimensions of the rectangular grid vary inversely as the dimensions of the rectangular source. A relative maximum of intensity is reached within each cell at approximately the center. Let these maximum values be designated by $I_{m,m'}$ as indicated in Fig. 12–13. Using $\beta_m \doteq (m + \frac{1}{2})\pi$ as the approximate value of β for which $f(\beta) = (\sin \beta)^2/\beta^2$ attains its mth subsidiary maximum and $f(\beta_m) \doteq 1/\beta_m^2$, we find that the intensities of the cells along either the y' or z' axes vary as

$$I_{0,0} : I_{1,0} : I_{2,0} : I_{3,0} : \cdots \doteq 1 : \tfrac{4}{9}\pi^2 : \tfrac{4}{25}\pi^2 : \tfrac{4}{49}\pi^2 : \cdots$$
$$\doteq 1 : \tfrac{1}{22} : \tfrac{1}{62} : \tfrac{1}{122} : \cdots$$

The off-axis cell with $m = 1$, $m' = 1$ has an intensity given by $I_{1,1}/I_{0,0} = 1/(22)^2 = \frac{1}{484}$. The reader can check that this is less than the intensity $I_{m,0}$ of the on-axis cells out to $m = 6$. Hence the cells extending along the coordinate axes are the most prominent features of the pattern.

12-9 DIFFRACTION BY A RECTANGULAR APERTURE

Consider plane waves from a laser source incident on a rectangular aperture, and an observing screen placed at a reasonably large distance from the aperture. According to the Huygens-Fresnel principle the aperture acts like a

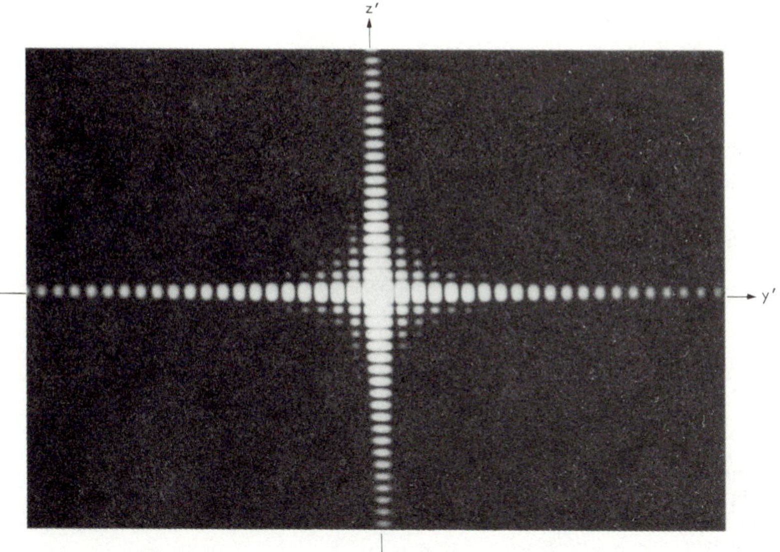

Fig. 12–14. Diffraction pattern produced when plane waves from a laser source are diffracted by a rectangular aperture.

rectangular distribution of coherent point sources; the resultant pattern is described by Eq. (12–21). Figure 12–14 is a photograph taken of a Fraunhofer pattern obtained under these conditions. The effect can be demonstrated to a large group by inserting a ground glass screen and viewing from the opposite side.

A pair of razor blades can be used to define one pair of edges of the aperture and a second pair at right angles to the first laid on top. Alternatively, an aperture whose dimensions can be changed can be obtained from a closely spaced pair of variable slits with their long dimensions at right angles to each other.

The formulas of the preceding section offer a more satisfactory way of describing the usual laboratory slit in optics than the discussion in Section 12–7. If the value of b is increased until $b \gg \lambda$, the rectangles in Fig. 12–14 converge onto the y'-axis, yielding the series of dashes of Fig. 12–10. From our present standpoint we are in a position to appreciate the structure of this pattern perpendicular to the y'-axis. The width of the narrow dimension of the line of dashes is determined by the criterion $\beta_{z'} \sim \pi$.

In the general case (whether or not $\lambda \ll b$), the effect of using an incoherent line source (e.g. a source slit) as the primary source is to smear out the pattern of Fig. 12–14 parallel to the source slit. If the source slit itself is parallel to one of the edges of the rectangular diffracting aperture, the resulting pattern is the same as that in Fig. 12–11.

12-10 OBLIQUE INCIDENCE

If the Huygens-Fresnel principle is to be applied to determine the diffraction pattern of a rectangular aperture when a plane wave is obliquely incident, the secondary sources may no longer be considered in phase with each other. The phase variation is, however, a linear function of position within the aperture, and is easily incorporated into the equations developed so far.

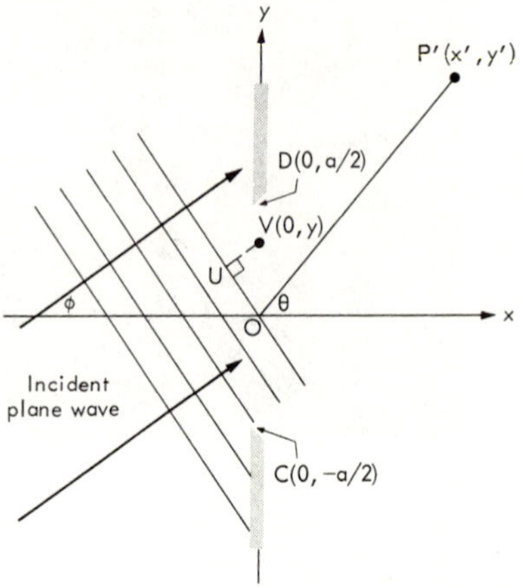

Fig. 12-15. Plane wave obliquely incident on a rectangular aperture. [The plane of the aperture is perpendicular to the plane of the diagram, the vertices of the rectangle being (0, $\pm a/2$, $\pm b/2$).]

Figure 12-15 represents a cross section, taken in the plane $z = 0$, of a rectangular slit of dimensions a, b. Consider an incident plane wave whose normal lies in the xy-plane and makes an angle ϕ with the x-axis. The phase difference between the signals at O and V is represented by the geometrical distance $UV = y \sin \phi$. Therefore if we take $\mathbf{B}_0$ to represent the complex amplitude per unit area of the sources located at $y = 0$, we have $\mathbf{B} = \mathbf{B}_0 e^{-iky \sin \phi}$ as the source strength per unit area at y. If we insert this into Eq. (12-16) and make the usual Fraunhofer approximations, the total signal at the point of observation is

$$\psi = \mathcal{R}e\,\{\mathbf{B}_0 e^{i\omega t} \mathbf{\Psi}(x', y', z')\},$$

where

$$\mathbf{\Psi}(x', y', z') = \frac{e^{-ikR}}{R} \int_{-b/2}^{b/2} \int_{-a/2}^{a/2} \exp\left\{ik\left[y\left(\frac{y'}{R} - \sin\phi\right) + \frac{zz'}{R}\right]\right\} dy\, dz.$$

Comparing this with Eq. (12-19) we find that the only change is the replacement of y'/R by $[(y'/R) - \sin\phi] = \sin\theta - \sin\phi$, where θ is the angle between OP' and the x-axis, P' being the projection onto the xy-plane of the point of observation P. Thus choosing a constant of proportionality I_ϕ we can write

$$I(y', z') = I_\phi \left(\frac{\sin \bar{\beta}_{y'}}{\bar{\beta}_{y'}}\right)^2 \left(\frac{\sin \beta_{z'}}{\beta_{z'}}\right)^2, \qquad (12\text{-}22)$$

where $\bar{\beta}_{y'} = (ka/2)[\sin\theta - \sin\phi]$ and I_ϕ is the maximum intensity, which occurs now at $\theta = \phi$, $z' = 0$ instead of at $\theta = 0$, $z' = 0$ as before.

If the angle of incidence ϕ is sufficiently small so that $\sin\phi \doteq \phi$, and if observations are confined to small angles θ so that $\sin\theta \doteq \theta$, the only effect of the oblique incidence is to shift the center of symmetry of the pattern to the line $\theta = \phi$. This fact was presupposed in Sections 12–7 and 12–9 when the use of an incoherent line source was mentioned.

If ϕ is not small, and if $ka \gg 1$, the only significant portion of the pattern occurs where $\theta \doteq \phi$. Using the identity

$$\sin\theta - \sin\phi = 2\sin\frac{(\theta - \phi)}{2}\cos\frac{(\theta + \phi)}{2} \doteq (\theta - \phi)\cos\phi,$$

we find that

$$\bar{\beta}_{y'} = \frac{ka\cos\phi}{2}(\theta - \phi).$$

The effect consists therefore not only in shifting the center of the pattern to $\theta = \phi$ but also in reducing the "effective width" of the slit from a to $a\cos\phi$. The latter is the width of the projection of the slit onto a plane perpendicular to the line of sight. This apparent narrowing of the slit can be observed in a single slit diffraction pattern by rotating the slit about its long axis.

12–11 REFLECTION OF A PLANE WAVE FROM A RECTANGULAR SURFACE

Consider a dielectric slab having a plane rectangular surface of dimensions a, b. Choose an x-axis normal to the surface and y-, z-axes parallel to the two sides of the rectangle (see Fig. 12–16). Let a sinusoidal plane wave whose normal is in the xy-plane be incident on the surface at an angle ϕ. We may think of the incident wave as exciting motion of the charges in the surface of the dielectric, thus causing them to act as secondary sources. It is easy

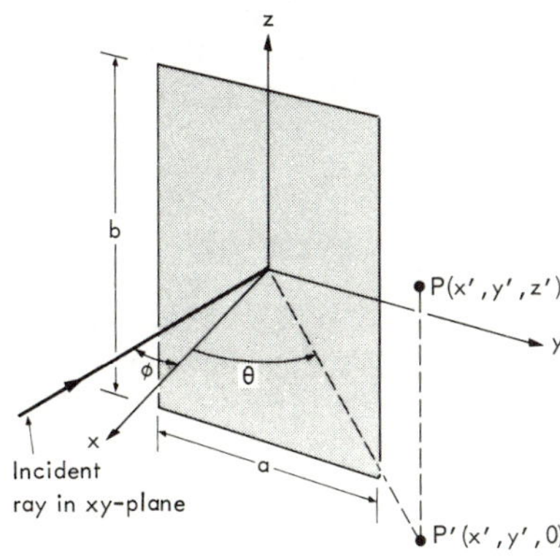

Fig. 12–16.
Plane wave obliquely incident on a rectangular surface.

to see that the relative phase of these sources varies linearly with y in the same way as was deduced in the last section. The combined effect of the radiation from all the charges in the surface can then be considered as another application of the radiation pattern of a rectangular distribution of coherent point sources having a linear phase variation.

Consider first the extreme case of a broad surface, $ka \gg 1$ and $kb \gg 1$. Equation (12–22) suitably interpreted in the present context shows that the radiation is insignificant unless the point of observation P lies in the plane $z = 0$ in a direction such that $\theta = \phi$. This is the way the electron theory of matter interprets the law of reflection of geometrical optics. When the point of observation P is at a large distance, there is only one direction, that associated with the geometrically reflected ray, which is such that all the contributions from the elementary oscillators within the surface arrive in a condition of constructive interference. For all other locations of P the contributions vary in phase by many multiples of π; the result is substantial destructive interference and a negligible net signal. The present discussion shows only that $\theta = \phi$ is a *necessary* condition for the existence of a reflected wave. This fact was asserted in the discussion of Brewster's law (Section 8–8) where the nonisotropic character of the radiation from the individual oscillators was also taken into account.

The notion of an image source is useful in clarifying the equivalence of the formulas describing the intensity distribution in the reflection from a rectangular surface and the diffraction by a rectangular slit. In the reflection problem the real physical oscillators in the dielectric surface behave exactly like the virtual sources which the Huygens-Fresnel principle would associate with diffraction, by an aperture having the same dimensions as the reflecting surface, of a wave originating from a mirror image of the actual source.

In the Lloyd's-mirror and the Fresnel double-mirror experiments, reflection at nearly grazing incidence is involved. Under these conditions the "effective width" of the mirror becomes relatively small. In our previous discussion of these experiments, we assumed, in effect, that the mirrors were of infinite length. In an actual experiment, diffraction effects, particularly those of Fresnel type (to be discussed in Chapter 13), are observed in addition to the fringes of the ideal double source pattern.

12-12 FRAUNHOFER DIFFRACTION BY A CIRCULAR APERTURE

Consider a sinusoidal plane wave normally incident on a screen containing a circular aperture of diameter a (Fig. 12–17). It is clear that the diffraction pattern will have rotational symmetry about an axis normal to the screen at the center of the circle. Let the point of observation P be at a distance R from the center, at an angle θ from the axis. If R is sufficiently large the Fraunhofer approximation can be applied to a calculation based on the Huygens-Fresnel principle to obtain the net disturbance at P. The calculations are the same as for the rectangular aperture, up to the point of inserting the limits of integration in Eq. (12–16). When the limits of integration range over the circular aperture the resulting integral is not of elementary type and requires the use of Bessel functions.*

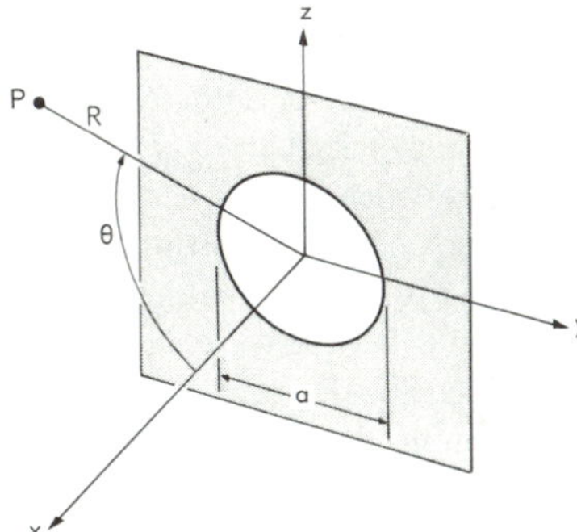

Fig. 12–17.
Geometry of a circular aperture.

Because of the practical importance of circular apertures, a few of the significant features of the resulting diffraction pattern will be asserted. In an optical context the pattern consists of a central bright disk (known as the *Airy disk*) and a series of annular bright rings separated by dark rings (see Fig. 12–18). The intensity in the radiation pattern as a function of angular position is qualitatively similar to the result for the coherent line source given in Eq. (12–8) and Fig. 12–4.

The radius of the *m*th dark ring is specified by an equation of the form $\sin\theta_m = \mu_m(\lambda/a)$, where $\mu_1 = 1.220$, $\mu_2 = 2.233$, $\mu_3 = 3.238$, and $\mu_m \approx (m + \frac{1}{4})$ for large m. (By way of contrast, the zeros in the case of the

Fig. 12–18. Diffraction rings produced when plane waves from a laser source are diffracted by a circular aperture.

* See Appendix VIII.

coherent line source are given by an equation of the same form where $\mu_m = m$ for all m.)

In the extreme case $a \ll \lambda$ there is no locus of destructive interference, since $1.220(\lambda/a) > 1$, whereas $\sin\theta$ is limited by the maximum value $+1$. For this case we observe from inspection of Eq. (12–16) (with limits appropriate to the circular aperture) that the integrand does not vary significantly over the range of integration and that the corresponding intensity is independent of θ. Thus a circular aperture which is extremely small compared with a wavelength radiates a spherical wave whose intensity is the same in all directions. This is an obvious result, since it simply restates the Huygens-Fresnel assumption that a single point on a wavefront acts as the source of a uniform spherical wave.

When the diameter of the aperture is reasonably large compared with a wavelength, the Airy disk is surrounded by a large number of annular rings. The successive rings diminish rapidly in intensity. It can be shown that the Airy disk accounts for approximately 85% of the energy flux in the entire pattern.

We can think of the energy flux of the original plane wave, which was highly concentrated about the direction $\theta = 0$, as having been dispersed through a range of angles between $\theta = 0$ and the radius of the first dark ring. For $a \gg \lambda$ this is given by $\sin\theta_1 \doteq \theta_1 = 1.22(\lambda/a)$. Thus, roughly speaking, the aperture produces a "conical beam" of semiangle $1.22\lambda/a$. The intersection of this cone with the observing screen is the Airy disk.

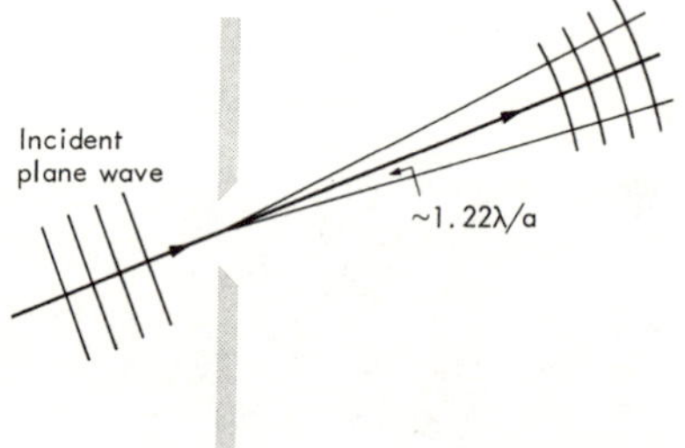

Fig. 12–19. Cross-sectional representation of the diffracted wave at large distances from a circular aperture.

When the case of oblique incidence is considered (and $a \gg \lambda$) the same qualitative result is obtained except that the axis of the "cone" turns out to be parallel to the direction of the incident plane wave. Near the axis of the cone the diffracted wave has nearly uniform intensity and resembles a portion of a spherical wave emanating from the center of the aperture (see Fig. 12–19). Almost complete destructive interference takes place at points which lie outside the diffraction cone. Clear interference bands in a double pinhole experiment will be obtained only where the Airy disks overlap. Since the center of each Airy disk is determined by a ray traced according to geometrical optics, maximum overlap can be arranged by using lenses to focus these rays at a common point on the observing screen. Interference also takes place outside the overlap region. However, there will be no bands at those locations where either signal vanishes, and the fringe contrast will be poor unless the two signals are of approximately equal amplitude.

12-13 ACOUSTIC RADIATION FROM A CIRCULAR PISTON

Every diffraction problem using the Huygens-Fresnel principle is based on the calculation of the radiation pattern from some distribution of coherently radiating sources. The results can be interpreted *mutatis mutandi* as the solution to an actual distribution of sources, if such an arrangement can be realized physically. For example, the diffraction of a plane wave by a circular aperture deals with fictitious point sources distributed over a circular area. But the corresponding array of actual point sources is precisely what is involved in determining the acoustic radiation from a flat circular piston which is executing simple harmonic motion in a direction perpendicular to a face. Let the x-axis be chosen parallel to this direction, with the origin located at the center of the piston, and let the diameter of the piston be a. Suppose that the piston is surrounded by an infinite plane baffle in the plane $x = 0$, so that only one side of the piston is exposed to the region $x > 0$. Each point on the piston then acts as a source of spherical waves which radiate into $x > 0$. The net disturbance at any point of observation P is the superposition of contributions from all the points on the piston face. At distances sufficient for the Fraunhofer approximation to be valid, the results asserted in the last section can be applied. In particular, if $a \ll \lambda$, the radiation is uniform in all directions, and if $a \gg \lambda$, the radiation at large distances is roughly speaking confined to a cone of semiangle $1.22\lambda/a$.

It follows then that for low-frequency sounds a loudspeaker radiates isotropically, but for high-frequency sounds it produces a distinct "beam" along the axis. To avoid directionality a high-frequency speaker should have a small diameter. As another example, we can see from consideration of the wavelengths involved why a tuba can afford a larger bell than a piccolo.

Conversely, if directional beaming of sound *is* desired, as in public address systems, a horn of large radius is required. For qualitative purposes we may think of the end of the horn as a circular aperture whose diameter determines the directionality of the radiation pattern. (The coupling between the speaker diaphragm at the throat of the horn and the outside medium is a more complex matter, however, and goes beyond the context of the present discussion.)

To a person who is unacquainted with diffraction phenomena it might seem that the most efficient way to direct sound at a distant listener would be to shout into a cylindrical tube pointed at the listener, thus "starting the sound

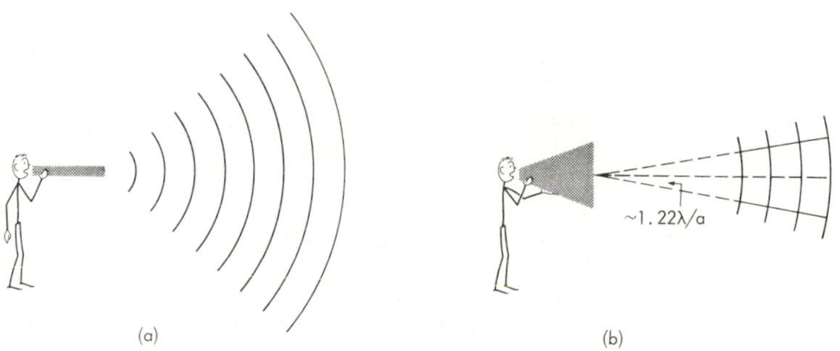

Fig. 12-20. The directionality of a sound beam is enhanced by using a megaphone of large diameter compared with a wavelength.

off in the right direction" (Fig. 12–20a). Instead, of course, a megaphone of large terminal diameter is more efficient, since then at least the shorter-wavelength sounds produce a narrow beam (Fig. 12–20b).

12-14 LIMIT OF RESOLUTION OF IMAGE FORMING INSTRUMENTS

The approximations required to obtain the Fraunhofer diffraction formulas are equivalent to assuming that both the source and the observing screen are at large distances from the aperture screen. As in Section 11–8, lenses can be used either to produce a virtual source at infinity or to make the focal plane of a lens equivalent to an observing screen at infinity. In each of the examples of this section the Fraunhofer condition will be satisfied by some such arrangement of lenses.

The primary purpose of the optical instruments to be described is not to demonstrate diffraction patterns. However, it must be recognized that a lens of finite diameter is equivalent to a lens of infinite extent mounted in front of a circular aperture whose diameter is equal to that of the lens. The diffraction associated with this aperture must be taken into account in a discussion of images formed by the lens. In reality, what we see is not the geometrical image of an object but the diffraction pattern produced by the aperture with the given object as source. When the diameter of the lens is large, the diffraction image of each point on the source is in effect an Airy disk of extremely small radius and hence closely resembles the ideal point predicted on the basis of geometrical considerations alone. But for a lens of given diameter there exists a critical separation of two image points which permits them to be distinguished from each other. For example, when the points are very close compared with the radius of their Airy disks, the latter overlap to such an extent that they look like a single disk, and the two points are said to be *unresolved*. In the following illustrations, Fraunhofer diffraction theory is used to calculate the limitation on the fineness of detail which can be perceived with a given optical instrument.

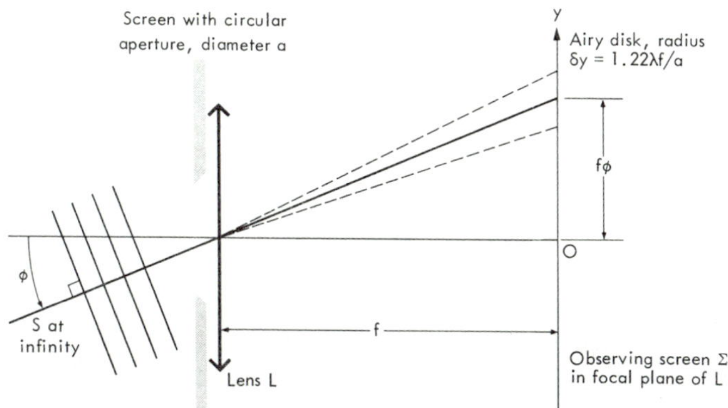

Fig. 12–21. Aperture limitation causes a plane wave to image as an Airy disk.

Consider first an astronomical telescope receiving light from a distant star in the form of plane waves whose normal makes a small angle ϕ with the axis (Fig. 12–21). The center of the diffraction image lies at $y = f\phi$, and the radius of the Airy disk is $\delta y = 1.22\lambda f/a$, where a is the diameter of the objective and f its focal length. A second star at an angle $\phi + \Delta\phi$ forms an Airy disk of the same size but its center is displaced by the amount $\Delta y = f\,\Delta\phi$. If $\Delta y = 2\,\delta y$, the Airy disks are tangent* and the pattern can still be recognized as due to two distinct point sources (Fig. 12–22a). Since different observers do not agree on the exact condition at which recognition is lost, it is convenient to adopt a convention suggested by Lord Rayleigh: The images are said to be *just resolved* when the center of one of the Airy disks coincides with the edge of the other (Fig. 12–22b). To write this algebraically, let $(\Delta y)_{\text{min}}$, known as the *limit of resolution*, be the distance between the centers of a pair of images which are just resolved. The Rayleigh criterion is then

$$(\Delta y)_{\text{min}} = \delta y. \tag{12–23}$$

Substituting the expression for the radius of the Airy disks, we have

$$(\Delta y)_{\text{min}} = 1.22\lambda f/a. \tag{12–24}$$

This gives the minimum distance between the images of two stars which can be distinguished from one another using the given telescope objective. If all available detail is to be recorded photographically, the grain diameter of the film should be at least as small as $(\Delta y)_{\text{min}}$. On the other hand, the use of a much smaller grain size will not improve the resolution even if the plate is later to be enlarged.

---

* By definition the *edge* of the Airy disk is taken to be the center of the first dark ring.

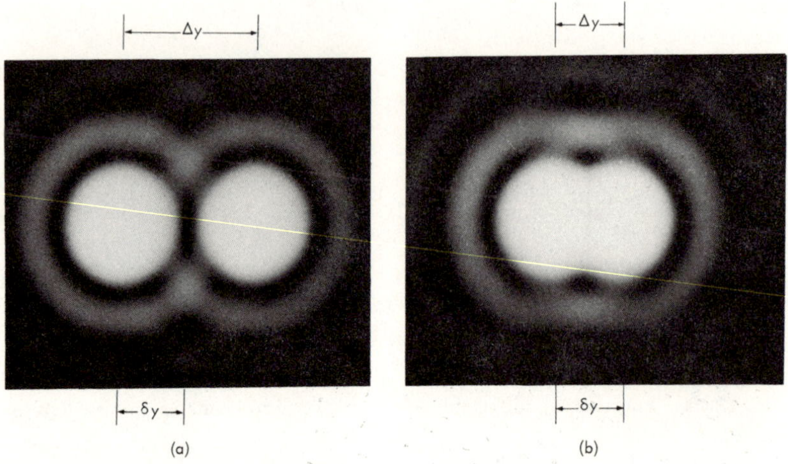

Fig. 12-22. (a) Images of two point sources with first dark rings tangent. (b) Images of two point sources which are *just resolved* according to the Rayleigh criterion.

The *angular limit of resolution* of the telescope is $(\Delta\phi)_{\min}$, the angular separation of two point sources which are just resolved. From Eq. (12-24) we obtain

$$(\Delta\phi)_{\min} = (\Delta y)_{\min}/f = 1.22\lambda/a. \tag{12-25}$$

The human eye focused for infinity is another example to which Fig. 12-21 is applicable. In this case Σ represents the retina, and the aperture is the pupil. In broad daylight an average person has a pupil diameter $a \sim 2.5$ mm. Using a wavelength of $\lambda = 5550$ Å, the value calculated from Eq. (12-25) is

$$(\Delta\phi)_{\min} = 2.7 \times 10^{-4} \text{ rad}$$

or a little under one minute of arc. This is the same order of magnitude as the experimentally determined minimum angle between two distant point sources which an observer can tell apart. Since the focal length of the eye lens is about 2 cm, $(\Delta y)_{\min}$, the linear distance on the retina between images which are just resolved, is about 5.4×10^{-4} cm. The distance between adjacent rods or cones varies in the range from 1.5×10^{-4} cm to 5×10^{-4} cm. We see therefore that the two images actuate different receptors. It is interesting to note, however, that nature does not provide an extravagant excess of receptors.

Figure 12-23 illustrates the use of a simple magnifier L_1 placed directly in front of the eye. The object to be examined is located in the focal plane of L_1 and the eye is focused for infinity. A typical point on the source is indicated by S at an angle ϕ with the axis. Because this gives rise to plane waves at the pupil, the conditions of the Fraunhofer approximation are satisfied. The linear distance of separation of two points S and S' that will be just resolved is

$$\Delta y_S = f_1(\Delta\phi)_{\min} = 1.22\lambda f_1/a_{\text{pupil}}. \tag{12–26}$$

This number gives an estimate of the fineness of detail which can be observed in an object of complex structure with the aid of a magnifier. There is, however, an important limitation on the context of Eq. (12–26). It must be assumed that the point sources S and S' are relatively *incoherent*, and hence the formula applies in a strict sense only to *self-luminous* objects. When an object to be viewed is illuminated by an external source, at least a partial correlation is to be expected in the radiation scattered from different points of the object.

The extreme case of complete coherence was first investigated by Abbé (1873).* Abbé's theory leads to some fascinating insights into the formation of images by a microscope, showing that in some cases "detail" which is not present in the object can be observed in an image. Abbé devised a series of simple experiments which give dramatic demonstration of some of these remarkable conclusions from the theory. The Nobel Prize was awarded in 1953 to Fritz Zernike for development of the *phase-contrast microscope*, a device which applies concepts of the Abbé theory to show up small *transparent* objects in the field of the microscope.

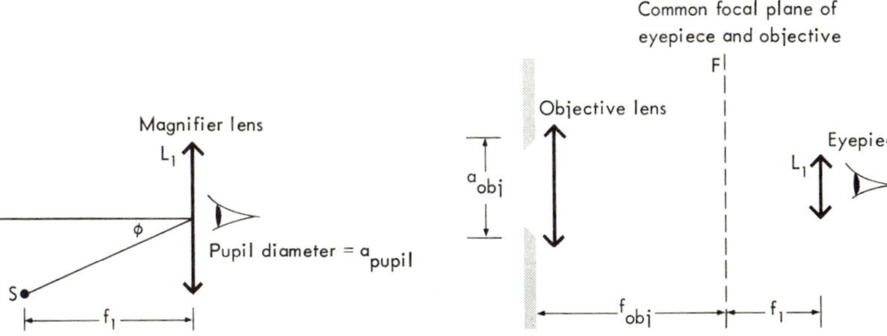

Fig. 12–23. Viewing with a simple magnifier.

Fig. 12–24. Eyepiece used to examine the image in the focal plane of a telescope.

Figure 12–24 shows an eyepiece being used to examine an image in the focal plane of the objective of a telescope. Two conditions must be satisfied

---

* M. Born and E. Wolf, *Principles of Optics*. New York: Pergamon Press, 1959, Section 8.6(b). R. W. Ditchburn, *Light*. New York: Interscience Publishers, 1952. A knowledge of Fourier series is required to follow the complete discussion of this subject.

for the images of a pair of point sources to be resolved:

i) The images in the focal plane F must be resolved.
ii) The linear distance of separation of the centers of the images in F must be greater than the value specified by Eq. (12–26).

When working with a given objective, the choice of an eyepiece with too small a magnification will result in a failure of the eyepiece to resolve images which are resolved by the objective. Increasing the magnification of the eyepiece will improve matters up to the point at which the eyepiece can resolve points whose distance of separation is the same as that of images which are just resolved in the focal plane of the objective. The latter distance is given from Eq. (12–24) by

$$(\Delta y)_{\min} = 1.22\lambda f_{\text{obj}}/a_{\text{obj}}. \tag{12–27}$$

The desired focal length f_1 of the eyepiece is therefore obtained by equating the right-hand sides of Eqs. (12–26) and (12–27). Solving for f_1 we obtain

$$f_1 = f_{\text{obj}}(a_{\text{pupil}}/a_{\text{obj}}). \tag{12–28}$$

The corresponding magnification is referred to as the *optimum magnification*:

$$f_{\text{obj}}/f_1 = a_{\text{obj}}/a_{\text{pupil}}.$$

We see that the optimum magnification of a telescope is determined solely by the ratio of the diameter of the objective to the diameter of the pupil of the eye. With eyepiece focal lengths shorter than that given by Eq. (12–28), larger images are obtained, but the size of the Airy disks is magnified in the same proportion as the linear separation of the centers and no improvement in resolution is obtained. Such a situation is referred to as "empty magnification."

*12–15 VALIDITY OF THE HUYGENS-FRESNEL PRINCIPLE

Introduction of the Huygens-Fresnel principle as a separate postulate is a departure from the ideal program of explaining all the phenomena of acoustics as a deduction from mechanics and all the phenomena of electromagnetic waves as a deduction from Maxwell's equations. Historically, Fresnel's application of this principle to optical diffraction patterns (1818) was instrumental in establishing that light might fruitfully be considered a wave phenomenon, and long preceded the postulation of the electromagnetic character of light by Maxwell in 1864. As an intuitive principle, Fresnel's postulate has shown itself to be immensely practical in interpreting a great wealth of experimentally observed diffraction phenomena. Nevertheless it would be an unwarranted conclusion to suppose that the principle is of unlimited applicability. The question arises whether the principle itself, apart from whatever mathematical approximations are introduced for convenience in particular calculations, is valid for all types of wave and for all ranges of the parameters involved. When the problem is viewed from the point of view of more comprehensive theory, it can be seen immediately that our statement of the Huygens-Fresnel principle is somewhat incomplete, since it has given only a vague specification of certain details of the problem. For example, in summing the contributions from various portions of the aperture we have used scalar addition. No mention was made of the fact that the variables **E** and **H** in an electromagnetic wave are vector quantities. Thus any possibility of describing the diffraction of different types of polarized light is ruled out by our tacit assumptions. As another example, we did not state whether the diffracting screen is made of a perfectly absorbing material or of a perfect conductor. If there is a distinction between these two cases, our naive formulation is powerless to describe it. The incompleteness of the Huygens-Fresnel principle is both a weakness and a virtue. The principle "cuts many corners" and provides a simple framework of interpretation for diffraction patterns in a variety of contexts. The close agreement with experiment shows that whatever details have been omitted from the discussion are evidently unimportant to the extent that they effect those features of the pattern which we customarily observe.

Starting with the scalar wave equation in three dimensions, Kirchhoff in 1883 deduced an exact formulation (Kirchhoff's theorem) which in many respects embodies the basic ideas of the Huygens-Fresnel principle.* When a "short-wavelength" approximation is made, a form is obtained which more closely resembles the original statement of the principle.

There are several respects in which a study of the Kirchhoff formulas sharpens our insight into the Huygens-Fresnel principle. These will be mentioned here only briefly, for they are not important in the applications made in this text.

1) *The secondary sources (points on a given initial wavefront) radiate spherical waves which do not have uniform intensity in all directions.* If the point of observation P lies in a direction which makes an angle χ with the normal to the original wavefront at the point being considered, the contribution $d\psi$ to the disturbance at P is dependent on χ through a multiplicative factor $f(\chi)$ called the *obliquity factor*. In our applications above we tacitly assumed

---

* See M. Born and E. Wolf, *Principles of Optics*. New York: Pergamon Press, 1959, Section 8.3. A justification is given in Section 8.4 for treating optical diffraction problems by means of a scalar variable. See also J. M. Pearson, *A Theory of Waves*. Boston: Allyn and Bacon, 1966, Chapter 6. An excellent critique of diffraction theory will be found in C. F. Meyer, *The Diffraction of Light, X-rays and Material Particles*. Chicago: University of Chicago Press, 1934, Chapter 7, Sections 25–31.

that $f(\chi)$ varies only slightly from point to point across the aperture. The factor can therefore be moved out of the integrand of the diffraction integral and absorbed in the constant of proportionality I_0. Fresnel was aware of the desirability of including an obliquity factor, but was unable to do more than guess at a suitable form. The Kirchhoff formulation removes the arbitrariness of this assignment and shows that the obliquity factor has the form

$$f(\chi) = \cos^2 (\chi/2).$$

Thus $f(\chi)$ has a maximum in the forward direction $\chi = 0$ and is zero in the backward direction $\chi = \pi$. This feature incidentally removes a paradox associated with Huygens' construction. If one constructs the spherical wavelets emanating from a given wavefront, the envelope on the forward side represents the wavefront as it has progressed at a later time. But there is also an envelope on the back side of the original wavefront, apparently representing a wave propagating in the backward direction. The Kirchhoff formulation dispenses with the back wave automatically, for, although there is in principle an envelope in this position, in fact the amplitude of each of the contributing wavelets is zero in this direction. Earlier versions of Huygens' principle required *ad hoc* assumptions to rule out the back wave. Fresnel, for example, assumed that $f(\chi) = 0$ for $\pi/2 \leq |\chi| \leq \pi$ so that the secondary sources made no contribution at all on the back side of the initial wavefront. The first derivation of the correct form of the obliquity factor was given by Stokes (1849) whose theory is a limited version of the treatment later made more general by Kirchhoff.

2) *Even in the case of plane waves incident it is not strictly correct to assume that the source strength of the secondary sources is uniform across the aperture.* In principle, the complex amplitude **B** might be a function of position, and justification would be required before moving it out of the integrand in going from Eq. (12–2) to Eq. (12–3). The Kirchhoff formulation shows that **B** depends on the field which *actually exists* within the aperture and is not determined solely by the field of the incident wave, i.e., the field that would exist in the aperture if the diffraction screen were not present. The field within the aperture depends on the type of boundary condition which must be satisfied (e.g. **E** perpendicular to a perfect conductor) and hence is not universal to all diffraction problems having the same geometry. Significant departure from the undisturbed field occurs only within a distance of a few wavelengths of the edge of the aperture. Consequently, if the total dimensions of the aperture are large compared with a wavelength, the assumption that **B** is proportional to the complex amplitude of the incident wave alone is a good one. This condition is usually satisfied in optics but is not so often satisfied in the microwave region of the electromagnetic spectrum or in acoustical diffraction problems. The Huygens-Fresnel principle may still be useful to

indicate qualitatively the type of diffraction pattern to be expected in these contexts, but a more elaborate theoretical treatment, or direct appeal to experiment, may be required to determine how closely the facts agree with the simplified theory.

3) *The secondary sources are $\pi/2$ out of phase with the disturbance in the incident wave.* Since we have been concerned only with a determination of the relative intensities in the diffraction pattern, it has not been necessary to relate **B**, the complex amplitude per unit area of the secondary sources, to **A**, the amplitude of the disturbance in the primary plane wave. The Kirchhoff formulation shows that $\mathbf{B} = i\mathbf{A}/\lambda$ and implies therefore that the arguments of the complex numbers **B** and **A** differ by a phase angle of $\pi/2$.

PROBLEMS

12–1. Let θ be the angle in Fig. 12–2 which corresponds to the *first* minimum, under the conditions of the Fraunhofer approximation, in the radiation pattern from a coherent line source. For this case what is the phase difference at P between the contributions from O, the center of the source, and C, the lower end? Compare the contributions from other pairs of corresponding points, i.e., other pairs in which the first point lies on the upper half of the source a certain distance above O and the second lies on the lower half the same distance above C. From this information can you predict what the resultant of all the contributions should be?

Devise a similar interpretation of the second minimum in the Fraunhofer radiation pattern.

12–2. Consider the average power flux out of portions of a sphere of large radius surrounding the coherent line source of Fig. 12–2. a) Show that this quantity is given by

$$P(\beta) = \frac{8\pi r^2 I_0}{ka} \int_0^{\beta} \left(\frac{\sin t}{t}\right)^2 dt$$

for the zone lying between $-\theta$ and $+\theta$ where $\beta = (ka/2) \sin \theta$. b) By suitably integrating by parts show that the average power flux through the first n lobes of the Fraunhofer pattern is given by

$$P(n\pi) = \frac{8\pi r^2 I_0}{ka} \int_0^{2\pi n} \frac{\sin t}{t}\, dt\,.$$

c) Consider a case in which ka is sufficiently large so that the average power flux out of the whole sphere is given by $P(\infty)$. Use numerical values obtained from a table of $\mathrm{Si}(x) = \int_0^x [(\sin u)/u]\, du$ to calculate the quantity $P(\pi)/P(\infty)$ which represents the relative power flux through the principal maximum of the Fraunhofer pattern. (Tables of the sine integral function $\mathrm{Si}(x)$ are to be found in M. Abramowitz and Irene Stegun, *Handbook of Mathematical Functions.* New York: Dover Publications, 1965.)

12–3. a) Carry out the integration in Eq. (12–15) and find the real and imaginary parts of the function $\mathbf{Z}(y)$ which defines the vibration curve for the Fraunhofer approximation. Eliminate the parameter y to obtain an explicit equation for the locus. Use standard techniques of analytic geometry to determine the location of the center C and the radius of the circle. b) Given the results of (a), locate the points $\mathbf{Z}_1$ and $\mathbf{Z}_2$ corresponding to $y = -a/2$ and $y = +a/2$ respectively. Show that the angle Z_1CZ_2 is $2\beta = ka \sin \theta$. Use the geometry of the figure to deduce $|Z_1Z_2|^2$, the square of the length of the chord joining $\mathbf{Z}_1$ and $\mathbf{Z}_2$, and show that this result agrees with Eq. (12–8).

12–4. Sketch a polar plot of intensity vs. angular position which corresponds to the Fraunhofer diffraction of a normally incident acoustic plane wave by an extremely narrow slit which is three wavelengths in length.

12–5. Suppose that the edges of a laboratory slit are opened symmetrically about the center and are driven at constant velocity. A monochromatic plane wave is incident on the slit and the intensity variations at a fixed point of observation are recorded by means of a photocell. Describe these intensity variations as a function of time.

12–6. Sketch a polar plot for the same conditions as those stated in Problem 12–4 except that the angle of incidence of the plane wave is 45°. (Assume that the normal to the plane wave lies in a plane perpendicular to the long dimension of the slit.)

12–7. It has been assumed in the examples of optical diffraction experiments in this chapter that the aperture screen is illuminated by means of a plane wave produced either directly by a laser, or by placing a point source at the focal point of a lens. This has come into the derivation through the assumption that there is no significant phase difference among the secondary sources distributed across the aperture. It is commonly said that the waves arriving from a point source which is sufficiently far away can be considered plane waves. Consider a point source on the axis of a rectangular aperture. How far away would this source have to be located for one to consider that the waves arriving at the aperture are effectively plane waves?

12–8. A double pinhole card contains two circular apertures of diameter 0.2 mm having a distance between centers of 0.7 mm. The card is placed directly in front of the eye and the eye is focused in the plane of a distant point source. This focusing action causes the Airy disks associated with the individual pinholes to coincide, and double-source interference fringes are seen across this region. Approximately how many dark bands will occur within the central disk?

12–9. Consider sound waves generated in water by means of the sinusoidal vibration of a circular piston face of diameter 2 cm. How high must the frequency be to obtain plane waves collimated to within 1°?

12–10. Let the "beam angle" in the radiation of sound by a loudspeaker be defined as the angle subtended by the first minimum in the Fraunhofer pattern of a normally incident plane wave diffracted by a circular aperture whose diameter is the same as the outside diameter of the loudspeaker cone. For a 10-in speaker sketch the beam angles for middle C (256 cps) and successively higher octaves.

12–11. Assume that the limiting aperture of an optical system is rectangular rather than circular. For example, suppose that objects are to be viewed by an eye placed directly behind a laboratory slit. Consider two point objects which are aligned parallel to the narrow dimension a of the slit. Assume that the images of two objects are unresolved if there is any overlapping of the central lobes in their Fraunhofer patterns. a) Show that the angular limit of resolution is given by

$$(\Delta\theta)_{\text{min}} = 2\lambda/a \cos \theta,$$

where θ is the angle which the objects make with the axis normal to the slit at its center. b) Consider a horizontal string of light bulbs spaced 1 m apart along the railing of a bridge. These are viewed from a distance of 1 km through a vertical slit. Apply the criterion of (a) to find the slit width at which resolution would be lost. (Assume that $\lambda = 5500$ Å.)

12–12. What is the angular limit of resolution of a 12-in diameter telescopic objective lens? What linear distance does this correspond to in the focal plane of the objective if the focal length is 15 ft? (Use $\lambda = 5500$ Å.) What is the optimum magnification and what is the focal length of the eyepiece that would be needed to attain this?

13 Fresnel diffraction

13-1 INTRODUCTION

We saw in the last chapter that a typical diffraction problem leads to a diffraction integral of the form

$$\Phi = \int e^{-ikr(y)}\, dy,$$

where y is a variable that ranges over the openings in the aperture screen. Under the *Fraunhofer* approximation $r(y)$ is replaced by a *linear* function in y and the integration is simple to perform. In this chapter we shall consider a more sensitive approximation, the *Fresnel* approximation, in which $r(y)$ is expanded as a *quadratic* function of y. The resulting integral cannot be integrated in terms of elementary functions. The basic form of the integral is of sufficient importance, however, to compute and tabulate numerical values. Since these tables are readily available, we may consider the diffraction problem solved in as real a sense as if it were expressed in terms of the sine or exponential functions for which we also rely upon tabulated values.

The Fresnel formulas for diffraction by a single slit include the Fraunhofer formula as a special case when the slit is sufficiently narrow so that the phase variations associated with the quadratic term in $r(y)$ are negligible. For wider slits the Fresnel and Fraunhofer formulas agree in the "wings" of the pattern, but the Fresnel formula begins to show some interesting changes in structure at the center of the pattern. The portions of the pattern where the Fresnel approximation is important lie within or very close to those regions which would be illuminated if diffraction did not occur. That is, if light propagated according to geometrical rays, the observing screen would be divided into regions of shadow and light; the Fresnel formulas give an adequate description of both regions, whereas the Fraunhofer approximation is adequate only at positions well within a shadow region. The Fresnel approximation therefore permits the study of a situation which cannot be handled by the Fraunhofer approximation, namely, the *transition* from a shadow region to a fully illuminated region. It is particularly interesting to note how the abrupt transition predicted by ray theory can be obtained from the Fresnel analysis in the limit of very short wavelengths.

The relation between the diffraction pattern of a given aperture and the radiation pattern of a continuous distribution of point sources remains the same under the Fresnel approximation as it was under the Fraunhofer approximation. We therefore pass freely from the calculation of the radiation pattern of a source distribution to its interpretation in the context of a diffraction problem. The subtle questions concerning the applicabiiity of the Huygens-Fresnel principle (Section 12–15) remain the same as before. We shall be content to note that the main features of the patterns derived in this chapter are observed in the corresponding diffraction experiments.

13-2 FRESNEL APPROXIMATION FOR THE RADIATION PATTERN OF A LINEAR DISTRIBUTION OF COHERENT SOURCES

The following is a brief recapitulation of the derivation carried out in Chapter 12 prior to the introduction of the Fraunhofer approximation. If the line segment CD in Fig. 13–1 represents a continuous distribution of point sources of strength B per unit length, the contribution to the disturbance at the point of observation P from the strip dy is

$$d\psi = \mathcal{R}e\left\{\frac{Be^{i[\omega t - kr(y)]}}{r(y)}\,dy\right\}.$$

We make the approximation $r(y) \doteq R$ in the amplitude factor and sum the contributions from

$$y_1 = -a/2$$

to

$$y_2 = a/2.$$

This yields

$$\psi = \mathcal{R}e\left\{\frac{Be^{i\omega t}}{R}\,\mathbf{\Phi}\right\}, \tag{13–1}$$

where

$$\mathbf{\Phi} = \int_{y_1}^{y_2} e^{-ikr(y)}\,dy. \tag{13–2}$$

A Maclaurin series expansion of $r(y)$ about $y = 0$ yields, as in Eq. (12–5),

$$r(y) \doteq R - y\sin\theta + \frac{y^2\cos^2\theta}{2R}.$$

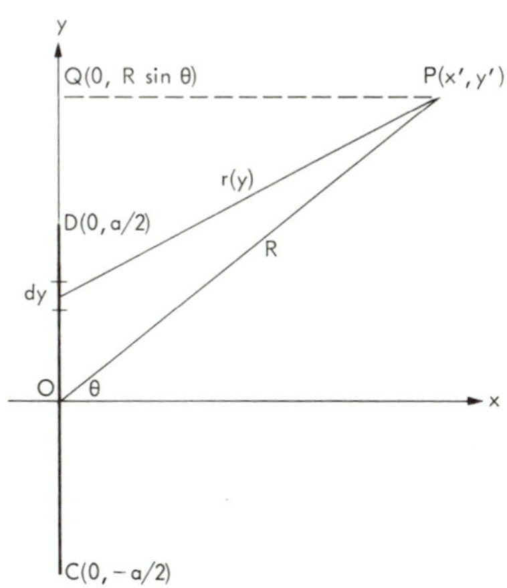

Fig. 13–1. Geometry associated with a continuous distribution of coherent sources along line segment *CD*.

It will simplify the following development if we decide at this point to examine the radiation pattern only in the region where θ is relatively small. In the typical experimental setup to which our results will be applied, the intensity is negligible except near $\theta = 0$. It is therefore convenient to replace $\cos^2 \theta$ by unity in the expansion for $r(y)$. (This replacement is not essential. Retaining $\cos^2 \theta$ adds a few extra terms to the algebra and complicates the discussion somewhat, but does not lead to any real difficulties.)

In anticipation of the reduction of the diffraction integral to a convenient form, we further transform the expression for $r(y)$ by completing the square in y:

$$r(y) = R - (R/2) \sin^2 \theta + (1/2R)(y - y_0)^2, \tag{13–3}$$

where $y_0 = R \sin \theta$.

Before proceeding further with the calculations it is appropriate to make a few general qualitative observations. The form of the diffraction integral as reduced so far is

$$\int_{y_1}^{y_2} e^{-i\phi(y)}\, dy,$$

where $\phi(y) = (k/2R)(y - y_0)^2$. The integrand is an oscillating function; consequently there is a tendency for contributions from some values of y to cancel others. The most important contributions to the integral may be expected to come from those regions where the oscillations are slow, i.e. from those values of y for which the phase $\phi(y)$ is changing least rapidly. The rate of change of ϕ is

$$\phi'(y) = (k/R)(y - y_0).$$

This shows that ϕ varies slowly in the vicinity of $y = y_0$. We therefore expect the most important contributions to the integral to result from the neighborhood of y_0, provided that y_0 lies within the range of integration. To interpret y_0 geometrically we note that $y_0 = R \sin \theta$ is the coordinate of Q, the foot of the perpendicular erected from P to the y-axis. If $y_1 < y_0 < y_2$, then, in the context of a diffraction problem, P is within the geometrically illuminated region. In this region the diffraction integral takes on large values associated with the stationary property of $\phi(y)$ about y_0. This is essentially the reason that the formulas developed below are capable, as the Fraunhofer formula is not, of accounting for a marked increase in intensity as the point P moves into the geometrically illuminated region. We also note in passing that since $\mathbf{\Phi}$ depends mainly on contributions from the portion of the slit around the foot of the perpendicular, the diffraction integral becomes indifferent to conditions which exist far away from this point. Thus, when the slit is very wide and P is opposite a point well within the slit (not near one of the edges), the intensity at P is independent of the width of the slit. This means that the wave arriving at such a point is simply a continuation of the incident wave.

When we collect the information from the last few steps and make a simple shift of origin in the diffraction integral, we have

$$\mathbf{\Phi} = e^{-ikR[1-(1/2)\sin^2\theta]} \int_{y_1-y_0}^{y_2-y_0} e^{-i(k/2R)y^2}\, dy. \tag{13–4}$$

It is now convenient to introduce a change of variable which will reduce the diffraction integral to the standard form in terms of which it is tabulated. This consists in the replacement of $ky^2/2R$ by $\pi u^2/2$. That is, a dimensionless variable of integration u is defined by

$$u = y/\sqrt{\lambda R/2}\,. \tag{13–5}$$

We may think of u as measuring distance along the source in terms of the unit $\sqrt{\lambda R/2}$. This unit depends on both the wavelength and the distance of the point of observation from the source. As we shall see later, the source is to be considered "long" or "short" depending on its length measured in these units. The relation of the source dimensions to wavelength alone is not a sufficient criterion.

Under the indicated change of variables the limits of integration become

$$\begin{aligned} u_1 &= \sqrt{\frac{2}{\lambda R}}\,(y_1 - y_0) = \sqrt{\frac{2}{\lambda R}}\left(-\frac{a}{2} - R\sin\theta\right), \\ u_2 &= \sqrt{\frac{2}{\lambda R}}\,(y_2 - y_0) = \sqrt{\frac{2}{\lambda R}}\left(\frac{a}{2} - R\sin\theta\right). \end{aligned} \tag{13–6}$$

The values of u_1 and u_2 are functions of θ, through $y_0 = R\sin\theta$. The value of u_1 is the directed distance QC measured in units of $\sqrt{\lambda R/2}$; u_2 is the directed distance QD measured in the same units. When we make the change of variables we obtain the reduced form of the diffraction integral:

$$\mathbf{\Phi} = \sqrt{\frac{\lambda R}{2}}\, e^{-ikR[1-(1/2)\sin^2\theta]} \int_{u_1}^{u_2} e^{-i\pi u^2/2}\, du. \tag{13–7}$$

Since $\mathbf{\Phi}$ determines the complex amplitude of the disturbance at P (cf. Eq. 13–1), the angular dependence of the radiation pattern is given by $|\mathbf{\Phi}|^2$. Thus the factor

$$e^{-ikR[1-(1/2)\sin^2\theta]}$$

is of significance only as far as the resultant phase at P is concerned, but does not enter into the intensity distribution. Introducing a constant of proportionality $I_0/2$ we can write the angular distribution of intensity as

$$I(\theta) = (I_0/2)\left|\int_{u_1}^{u_2} e^{-i\pi u^2/2}\, du\right|^2. \tag{13–8}$$

13-3 THE FRESNEL INTEGRALS AND THE CORNU SPIRAL

Calculation of numerical values in Eq. (13–8) can be based on tabulated functions,* known as the *Fresnel integrals*, which are defined by

$$C(u) = \int_0^u \cos(\pi u^2/2)\, du,$$
$$S(u) = \int_0^u \sin(\pi u^2/2)\, du. \tag{13–9}$$

In terms of these functions Eq. (13–8) can be written

$$I(\theta) = (I_0/2)\{[C(u_2) - C(u_1)]^2 + [S(u_2) - S(u_1)]^2\}. \tag{13–10}$$

Radiation patterns for a variety of problems can be computed from Eq. (13–10). However, such a strictly analytical approach does not give much of a feeling for the general features to be expected in a Fresnel pattern. For this reason it is profitable to interpret Eq. (13–8) in terms of a vibration curve of the type considered in Eq. (12–10). If we take

$$\mathbf{Z}(u) = \int_0^u e^{i\pi u^2/2}\, du, \tag{13–11}$$

then Eq. (13–8) can be written

$$I(\theta) = (I_0/2)|\mathbf{Z}_{21}|^2, \tag{13–12}$$

and the intensity becomes associated with the square of the length of the chord joining the points $\mathbf{Z}_1$ and $\mathbf{Z}_2$ on the vibration curve defined by $\mathbf{Z}(u)$. If we write $\mathbf{Z} = X + iY$, we observe from Eqs. (13–11) and (13–9) that the vibration curve is given parametrically by the equations

$$X = C(u), \qquad Y = S(u). \tag{13–13}$$

An accurate plot can therefore be obtained from the tabulated values of the Fresnel integrals. The resulting curve, known as the *Cornu spiral*, is shown in Fig. 13–2.

Certain features of the Cornu spiral can be simply deduced from the general considerations of Section 12–5. Recall that the parameter u is the distance measured along the curve from the origin to the general point $\mathbf{Z}(u)$. The curvature is given by Eq. (12–14), with $\phi(u) = \pi u^2/2$:

$$K = \phi'(u) = \pi u. \tag{13–14}$$

* M. Abramowitz and Irene A. Stegun, *Handbook of Mathematical Functions*. New York: Dover Publications, 1965. Jahnke, Emde and Lösch, *Tables of Higher Functions*. New York: McGraw-Hill, 1960.

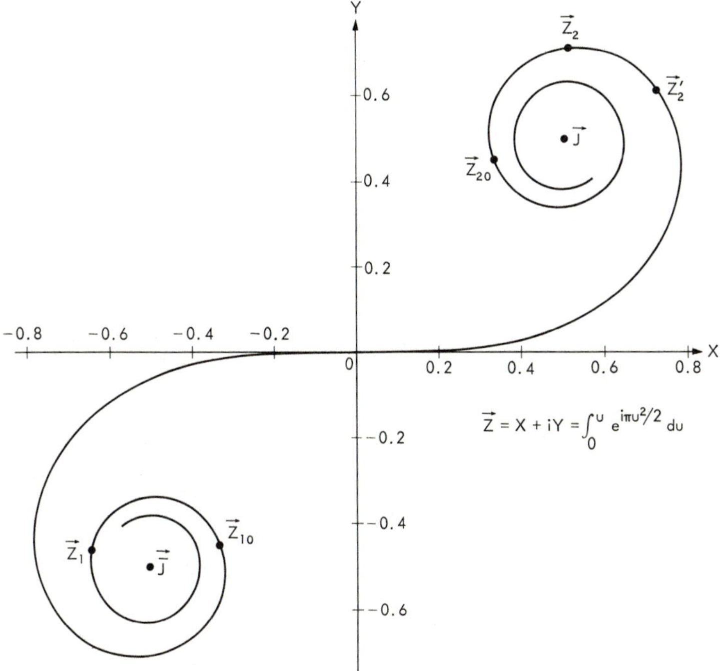

Fig. 13–2. Cornu spiral.

The radius of curvature therefore decreases as the point moves out from the origin along the curve. The inclination of the tangent line is given by $\phi(u)$ itself. Horizontal tangents occur where $\phi = \pi u^2/2 = m\pi$, or $u^2 = 0, 2, 4, \ldots,$ and vertical tangents where $\phi = (m + \frac{1}{2})\pi$, or $u^2 = 1, 3, 5, \ldots$ The spiral nature of the curve is therefore evident, and the values of u can be marked off at each quarter-turn of the spiral.

Inspection of Eq. (13–9) shows that both $C(u)$ and $S(u)$ are *odd* functions of u and hence that $\mathbf{Z}(-u) = -\mathbf{Z}(u)$. This implies that the Cornu spiral is symmetric with respect to the origin. Geometrical properties of the third quadrant portion of the spiral ($u < 0$) can be inferred from a study of the first quadrant portion ($u > 0$).

By analytical techniques which are beyond the level of this text it can be shown that when u is large and positive, approximate values of the Fresnel integrals are given by the *asymptotic forms*

$$\begin{aligned} C(u) &\approx \tfrac{1}{2} + (1/\pi u) \sin(\pi u^2/2) \\ S(u) &\approx \tfrac{1}{2} - (1/\pi u) \cos(\pi u^2/2). \end{aligned} \tag{13–15}$$

Thus, as u tends to $+\infty$ the coordinates of the points $\mathbf{Z}(u)$ differ by only a small amount from the coordinates of the *asymptotic point* $\mathbf{J} = (1 + i)/2$. The difference vector, $\mathbf{Z} - \mathbf{J} = (1/\pi u)[\sin(\pi u^2/2) - i\cos(\pi u^2/2)]$, has slope $-\cot(\pi u^2/2)$ and is therefore perpendicular to the tangent to the Cornu spiral at the point $\mathbf{Z}$. (These remarks apply only to the case $u > 0$ and $u \gg 1$.) The length of the difference vector is $|\mathbf{Z} - \mathbf{J}| = 1/\pi u$, which from Eq. (13–14) is the radius of curvature at $\mathbf{Z}$. The point $\mathbf{J}$ therefore becomes identified as a *fixed* center of curvature for all portions of the curve where u is large. Consequently we can speak of the eye of the spiral descriptively as "a circle of gradually decreasing radius."

The asymptotic behavior for $u < 0$ and $u \gg 1$ can be inferred from the above mentioned symmetry with respect to the origin. The asymptotic point $\bar{\mathbf{J}} = -\mathbf{J}$ has coordinates $(-\frac{1}{2}, -\frac{1}{2})$.

13-4 THE FRESNEL DIFFRACTION PATTERN OF A SINGLE SLIT

The angular distribution of intensity in the Fresnel diffraction pattern of a single slit can be studied by following the migration of the points $\mathbf{Z}_1 = \mathbf{Z}(u_1)$, $\mathbf{Z}_2 = \mathbf{Z}(u_2)$ along the Cornu spiral as the point of observation is varied. The value of $I(\theta)$ is proportional to the square of the length of the chord joining $\mathbf{Z}_1$ and $\mathbf{Z}_2$. The values of u_1 and u_2 are given by Eq. (13–6). The *physical* significance of u_1 is that it represents the directed distance from Q, the foot of the perpendicular dropped from the point of observation P onto the aperture screen, to C, the lower edge of the slit. The *geometrical* significance of u_1 is that it represents the distance along the Cornu spiral from the origin to the point $\mathbf{Z}_1$. Consequently, as P moves through a certain distance on the observing screen, $\mathbf{Z}_1$ moves through the same distance (measured in units of $\sqrt{\lambda R/2}$) along the spiral. Similar remarks apply to $\mathbf{Z}_2$ in relation to the upper edge of the slit. From Eq. (13–6) the arc length from $\mathbf{Z}_1$ to $\mathbf{Z}_2$ is

$$u_2 - u_1 = a\sqrt{2/\lambda R}\,. \tag{13–16}$$

Since the arc length is independent of θ, we can study the pattern by sliding an arc of fixed length along the spiral. Starting at $\theta = 0$, $\mathbf{Z}_1$ and $\mathbf{Z}_2$ are symmetrically located with respect to the origin, in the third and first quadrants respectively. (See the points labeled $\mathbf{Z}_{10}$ and $\mathbf{Z}_{20}$ in Fig. 13–2.) As θ is increased, $\mathbf{Z}_1$ remains in the third quadrant and moves toward the asymptotic point $\bar{\mathbf{J}}$; $\mathbf{Z}_2$ moves along the spiral toward the origin and eventually crosses over into the third quadrant. As a specific example consider the points $\mathbf{Z}_1$ and $\mathbf{Z}_2$ in Fig. 13–2. It is clear that $|\mathbf{Z}_2 - \mathbf{Z}_1| > |\mathbf{Z}_{20} - \mathbf{Z}_{10}|$, and hence $I(\theta) > I(0)$. This is in contrast with the Fraunhofer case for which maximum intensity occurs at $\theta = 0$. (By symmetry considerations $\theta = 0$ must be a relative extremum; it is a relative minimum in the present case.) It is also clear that $I(\theta)$ is approaching a relative maximum which will be attained when $\mathbf{Z}_1$ and $\mathbf{Z}_2$ move slightly beyond the positions indicated.

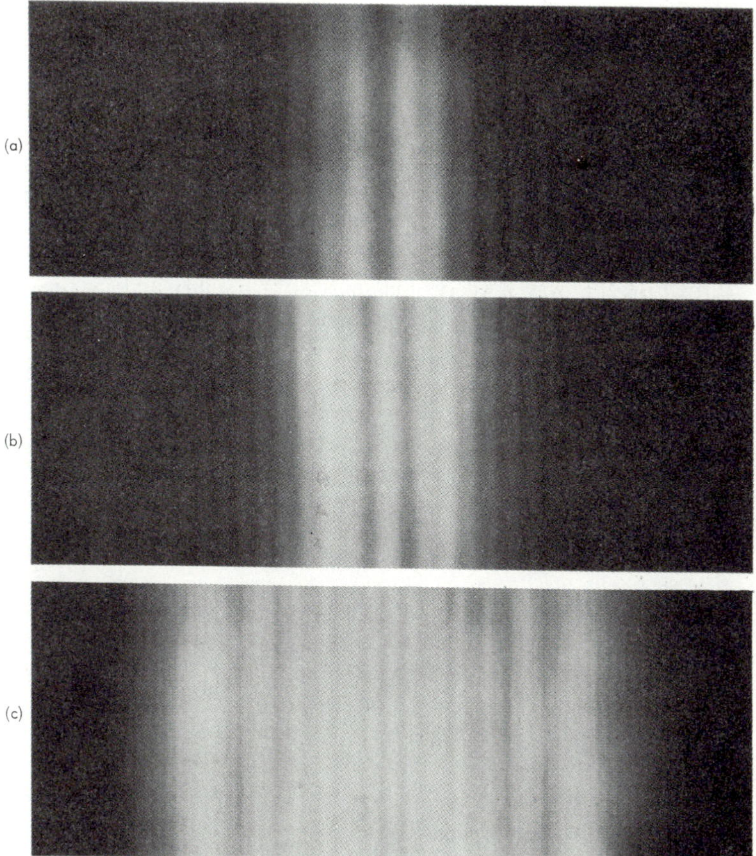

Fig. 13-3. Fresnel single-slit diffraction patterns. (Exposures at a fixed location from progressively wider slits.)

Suppose that the slit width is such that $\mathbf{Z}_{10}$ and $\mathbf{Z}_{20}$ are located several turns within their respective spiral eyes. Then as θ increases the distance $|\mathbf{Z}_{21}|$ goes through an intricate sequence of relative maxima and minima as the points approach and recede from each other in moving around the coils of the spiral. Although the minima are not places where $I(\theta) = 0$, the pattern is perceived as a series of light and dark bands, known as *internal fringes.* To have a rough criterion for when fringes of this type will occur, recall that the first horizontal tangent on the spiral eye is given by $u^2 = 2$. From Eq. (13-6) we find that the value of u_2 at $\theta = 0$ is $u_{20} = a/\sqrt{2\lambda R}$. Thus if $a^2 > 4\lambda R$, the point $\mathbf{Z}_{20}$ lies "inside" the spiral eye and internal fringes are expected. Photographs are shown in Fig. 13-3 of the diffraction patterns obtained when this condition is satisfied.

The crossing of $\mathbf{Z}_2$ from the first to the third quadrants occurs when $u_2 = 0$, or $y_2 = y_0$. The point of observation is then perpendicularly opposite the upper edge of the slit. In other words, P is moving into what would be a shadow region according to geometrical optics. As $\mathbf{Z}_2$ follows $\mathbf{Z}_1$ around the eye of the spiral, the distance between the two points becomes much shorter than was the case with $\mathbf{Z}_2$ in the first quadrant. A relative minimum will be reached when the arc of fixed length fits around one turn of the spiral. This will be followed by a relative maximum when the arc moves into the smaller turns of the spiral and corresponds to one and one-half turns. In this way a succession of relative minima and maxima is generated. The corresponding fringes are referred to as *external* fringes.

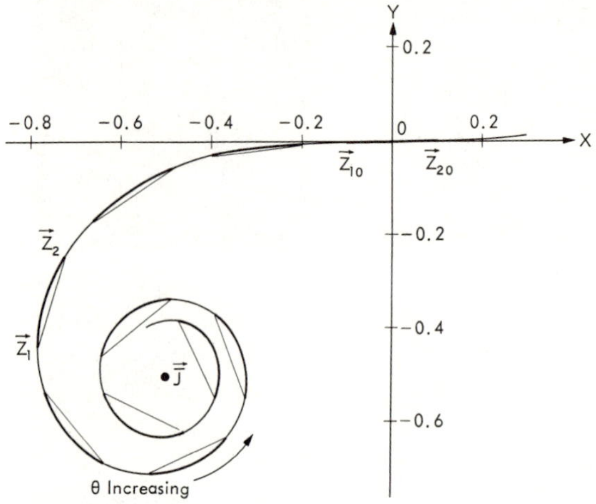

Fig. 13-4. Cornu spiral applied to the case of a narrow slit.

We remarked in connection with Eq. (12–5) that the Fraunhofer approximation is a special case of the Fresnel approximation when the condition $a^2 \ll \lambda R$ is satisfied. Comparison with Eq. (13–16) shows that this condition is equivalent to the assumption that the width of the slit measured in units of $\sqrt{\lambda R/2}$ is a small number. This implies that the arc which slides on the Cornu spiral is extremely short. The criterion for internal fringes is not satisfied. The sequence indicated in Fig. 13–4 shows that $|\mathbf{Z}_{21}|$ remains very nearly constant until the arc has moved into the eye of the spiral where the radius of curvature is small. The inner coils of the spiral are adequately described by the asymptotic formulas, Eqs. (13–15), and these can be used to determine the angular positions at which the bright and dark fringes of external type appear. It is left as an exercise (Problem 13–4) to show that the results are identical with those obtained from the Fraunhofer approximation. The "circles of gradually decreasing radius" are related to the succession of Fraunhofer vibration curves shown in Fig. 12–7.

13-5 FRESNEL DIFFRACTION BY A WIDE SLIT

In this section we will consider the limiting case of Fresnel diffraction by a slit which is assumed to be extremely wide in units of $\sqrt{\lambda R/2}$. The condition is opposite to that which leads to the Fraunhofer approximation. The graphical significance of $a \gg \sqrt{\lambda R/2}$ is that the arc which slides on the Cornu spiral is extremely long. When $\theta = 0$ the two points $\mathbf{Z}_{10}$ and $\mathbf{Z}_{20}$ can be considered to be effectively coincident with the asymptotic points $\bar{\mathbf{J}}$ and $\mathbf{J}$ respectively. The intensity in the direction $\theta = 0$ is then given by Eq. (13–12) as

$$I(0) = (I_0/2)|\mathbf{J} - \bar{\mathbf{J}}|^2 = (I_0/2)[(\tfrac{1}{2} + \tfrac{1}{2})^2 + (\tfrac{1}{2} + \tfrac{1}{2})^2] = I_0.$$

The interpretation of I_0 afforded by this equation is the reason for introducing $(I_0/2)$ as the constant of proportionality in Eq. (13–12). Additional physical meaning is given to I_0 when it is realized that the intensity at a point opposite the center of an extremely wide slit must be the same as that of the incident wave. Thus I_0 can be interpreted in the general application of Eq. (13–12) as the intensity of the plane wave incident on the slit.

Since $I(\theta)$ is symmetrical about $\theta = 0$, it is sufficient to study the pattern for θ positive. As θ increases from $\theta = 0$ the point $\mathbf{Z}_2$ moves outward from the first quadrant eye of the spiral, but $\mathbf{Z}_1$ remains effectively coincident with the asymptotic point $\bar{\mathbf{J}}$. Provided the quantity $u_2 = \sqrt{2/\lambda R}[a/2 - R \sin \theta]$ is large, the coordinates of $\mathbf{Z}_2$ are still close to those of $\mathbf{J}$ and the intensity remains close to I_0. It is only as $\mathbf{Z}_2$ moves significantly far away from the center of the spiral that the chord $\bar{J}Z_2$ varies significantly in length. We can then see that the intensity goes through a series of oscillations as θ increases. The contrast between successive maxima and minima becomes greater as $\mathbf{Z}_2$ reaches the outer turns of the spiral, and an absolute maximum is achieved when $\mathbf{Z}_2$ is at the position marked $\mathbf{Z}_2'$ on Fig. 13–2. (See Fig. 13–5 for a plot of I vs. u_2 and Fig. 13–6 for a photograph of the corresponding fringe pattern.)

When the point of observation is at the exact edge of the geometrical shadow, $R \sin \theta = a/2$ and $u_2 = 0$, so that $\mathbf{Z}_2 = 0$. The intensity is then

$$I = (I_0/2)|\bar{\mathbf{J}}|^2 = (I_0/2)[(\tfrac{1}{2})^2 + (\tfrac{1}{2})^2] = I_0/4.$$

From this point on, the chord length $\bar{J}Z_2$ decreases monotonically as $\mathbf{Z}_2$ moves in around the spiral toward $\bar{\mathbf{J}}$. There are thus no oscillations in intensity within the geometrical shadow region.*

---

* Since $\mathbf{Z}_1$ does not coincide precisely with $\bar{\mathbf{J}}$, Fraunhofer-like external fringes do in principle occur when $\mathbf{Z}_1$ and $\mathbf{Z}_2$ are both far inside the spiral. However, this does not take place until the value of I has become generally insignificant in comparison with I_0.

The intensity variations adjacent to the edge at $y = a/2$ are clearly indifferent to the value of y at the other edge of the slit provided that it is sufficiently large so that the coordinates of $\mathbf{Z}_1$ are close to those of $\bar{\mathbf{J}}$. We may therefore think of the lower edge of the slit as being infinitely far away. Although it is true that a number of the approximations which have been made in the Fresnel calculations would not be valid at all points across an infinite slit, we noted previously that the important contributions to the diffraction integral come from the vicinity of Q, the foot of the perpendicular from P. It is therefore sufficient that the approximations be valid in this region. Large errors can be tolerated in connection with portions of the slit which make only minor contributions. The portion of the pattern surrounding an individual edge of the wide slit is often referred to as a *straight-edge diffraction* pattern.

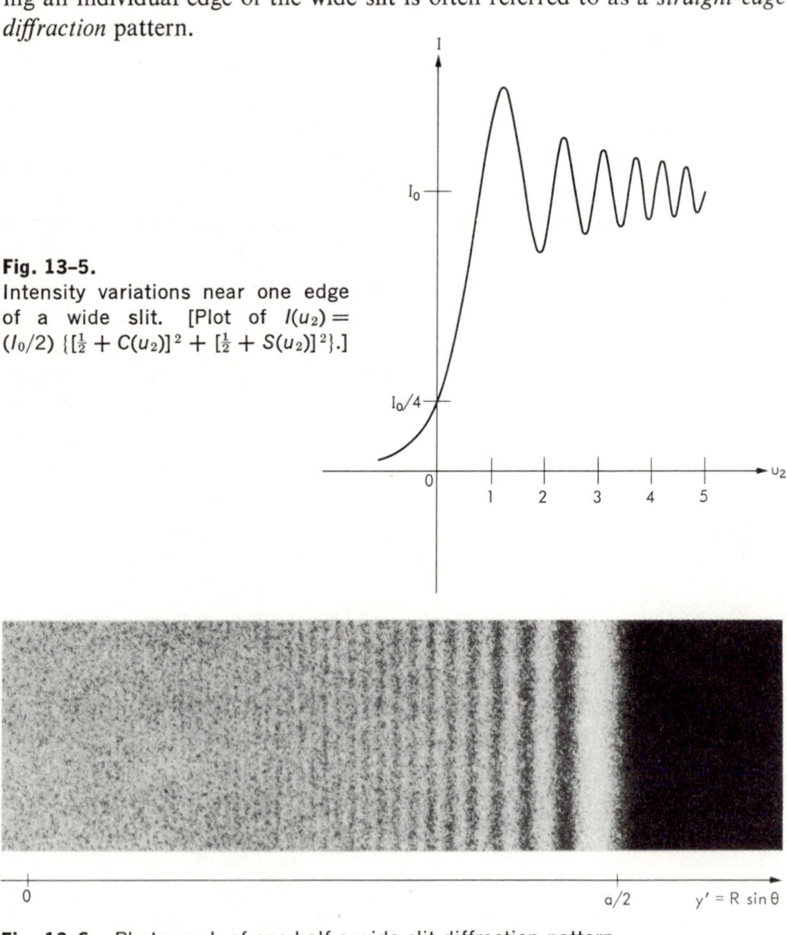

Fig. 13-5. Intensity variations near one edge of a wide slit. [Plot of $I(u_2) = (I_0/2)\{[\frac{1}{2} + C(u_2)]^2 + [\frac{1}{2} + S(u_2)]^2\}$.]

Fig. 13-6. Photograph of one-half a wide-slit diffraction pattern.

The scale of a straight-edge diffraction pattern depends on both the wavelength and the distance of the observing screen from the diffracting edge. On the other hand, the values of u which correspond to different features in the pattern are independent of λ and R. For example, we can estimate the value of u that corresponds to the highest intensity in the entire pattern. The point $\mathbf{Z}_2'$ in Fig. 13–2 lies somewhere between the horizontal tangent where $u^2 = 2$ and the vertical tangent where $u^2 = 1$, say at $u_2 = \sqrt{\frac{3}{2}}$. From Eq. (13–6) this implies that the y-coordinate of the point on the observing screen where this maximum occurs is given by

$$y' = R \sin\theta = a/2 - \sqrt{\lambda R/2}\, u_2 = a/2 - \tfrac{1}{2}\sqrt{3\lambda R}. \qquad (13\text{–}17)$$

As the wavelength is decreased, this maximum, and all the others as well, moves in closer toward the shadow boundary at $y' = a/2$. For extremely short wavelengths the detail in the structure becomes fine, and is confined to a narrow interval. There thus appears to be a discontinuous transition from a uniform value I_0 when $y' < a/2$ to $I = 0$ when $y' > a/2$. This illustrates the fact that the geometrical ray model of light propagation is incorporated in the wave theory as the limiting case obtained by letting the frequency tend to infinity.

A simple demonstration experiment can be used to show the dependence of the scale of the pattern on R, the distance of the observing screen from the diffracting slit. Let a plane wave of wavelength λ be normally incident on a slit of width a. The diffraction pattern is observed with a microscope which focuses on a plane at a distance R from the slit. When the microscope is set far away from the slit, so that $\lambda R \gg a^2$, the Fraunhofer pattern is observed. As the microscope is moved closer to the slit, changes occur near the center of the pattern in accordance with the Fresnel analysis when $\lambda R \sim a^2$. Eventually, when $\lambda R \ll a^2$, independent straight-edge patterns are observed adjacent to the shadow boundaries of the two edges of the slit. Finally the band separation in the straight-edge pattern becomes extremely narrow and one ends with the microscope focused on the slit edges.

PROBLEMS

13–1. a) Use the general techniques of analytic geometry to find the center of curvature at an arbitrary point on the Cornu spiral. b) Consider a point on the curve for which the asymptotic forms of Eq. (13–15) are a valid approximation. Show that at such a point the center of curvature is the asymptotic point **J**.

13–2. a) Estimate the value of u corresponding to the point $\mathbf{Z}_{20}$ in Fig. 13–2. Consider a slit of fixed width $a = 2$ mm and a wavelength $\lambda = 5000$ Å. At what distance should the observing screen be placed to have the slit edges correspond to $\mathbf{Z}_{10}$ and $\mathbf{Z}_{20}$ when the point of observation is on the axis $\theta = 0$? Sketch $I(\theta)$ for this case. b) If the observing screen is moved closer to the slit, at approximately what position will the center of the pattern be a strong relative maximum? Sketch $I(\theta)$ for this case.

13–3. Let a plane wave of sound be normally incident on a rectangular slit. Assume that the narrow dimension of the slit is sufficiently small so that the diffraction problem can be treated by application of the formulas for the radiation from a coherent linear source. Take the long dimension of the slit to be 10 wavelengths. Consider a microphone which can be moved to different positions along the axis $\theta = 0$ (the perpendicular to the slit at its midpoint). According to the Cornu spiral analysis there should be relative maxima and minima as the microphone is moved along the axis. Approximately where are some of these expected to be located? In what frequency range might this be a feasible experiment? (Consider the limited ability of the microphone to resolve the maxima and minima and the requirement that the dimensions of the slit be reasonable.)

13–4. *The Fraunhofer approximation as a limiting case of the Fresnel.* When $R \gg a$ it can be seen from Eq. (13–6) that except for θ very close to zero the values of u_1 and u_2 have the same sign. Furthermore, with $R \gg \lambda$, the values of u_1 and u_2 are large in magnitude and the corresponding points $\mathbf{Z}_1$ and $\mathbf{Z}_2$ are far inside the same eye of the spiral. This enables the asymptotic form, Eq. (13–15), to be used in calculating the length of the chord Z_1Z_2. a) Show that the resulting angular distribution of intensity is of the same form as that obtained in Eq. (12–8) for the Fraunhofer approximation. (An additional standard approximation will be required to simplify the Fresnel results.) b) Suppose that the criterion $a^2 \ll \lambda R$ is not satisfied. Why would we then not be willing to say that the calculations in (a) show that essentially the same pattern is predicted by the Fresnel and Fraunhofer formulas?

13–5. Consider a point of observation lying in the bisecting plane of a coherent linear source at a distance of 50 wavelengths. Let a, the length of the source, increase, while maintaining a constant value of B, the source strength per unit length. When a is small how does the intensity at the point of observation vary as a function of a? (Can you think of a physical reason for this?) At approximately what value of a do you expect a departure from this initial trend? Estimate a value of a beyond which the variations in intensity will be slight.

13–6. As indicated in Fig. 13–5 the relative maxima in a straight-edge diffraction pattern are characterized by fixed values of u_2. Show that when $x' \gg y'$ the locus $y'(x')$, along which u_2 maintains a fixed value, is a parabola. Sketch the loci of a few of these intensity maxima.

13–7. Show that within the shadow region of a straight-edge diffraction pattern the intensity falls off asymptotically as the inverse square of the distance to the geometrical shadow boundary.

***13–8.** In deriving the Fresnel radiation pattern for a coherent line source we assumed that all source points were in phase. In application to a diffraction problem this corresponds to having a normally incident plane wave. It is clear by inspection that the basic structure of the calculations will not be changed if we allow the phase of the source points to vary as a *quadratic* function of y. Thus consider $\mathbf{B} = |\mathbf{B}|e^{-i\psi}$, where $|\mathbf{B}|$ is independent of y and $\psi(y)$ is quadratic in y. The factor $e^{-i\psi}$ must now be included in the diffraction integral of Eqs. (13–1, 2). Let a point source be located in the third quadrant of Fig. 13–1, the coordinates being $(-R' \cos \phi, \ -R' \sin \phi)$ where ϕ is a relatively small angle. a) Expand the phase of the wave arriving at a general point along CD as a quadratic in y. Show that this will result in a diffraction integral of the same form as previously calculated, where R is replaced by $\rho = (1/R + 1/R')^{-1}$ and $y_0 = \rho\,(\sin\theta - \sin\phi)$. b) What changes are required for a case in which the incident wave is a *converging* spherical wave? c) Suppose that the object of an experiment is to study a diffraction pattern that is as close to the extreme case of the Fraunhofer approximation as possible. With the aperture and observing screens at fixed locations, are conditions made better or worse by having the point source at a finite distance? How can a point source and a single lens be used to satisfy the Fraunhofer conditions ideally at the observing screen?

***13–9.** *The Fresnel approximation applied to a rectangular distribution of coherent point sources.* a) Carry out a development parallel to that of Section 12–8, but retain terms quadratic in y and z. [Terms of the type $y^2y'^2$ are to be neglected; this is comparable to the replacement of $\cos^2\theta$ by unity in the development following Eq. (13–2).] Show that the diffraction integral can be written as the product of separate Fresnel integrals in y and z and indicate how the intensity distribution can be obtained. b) Consider the case of an on-axis point of observation ($y' = 0 = z'$) and a rectangular aperture which is extremely wide in both dimensions. Keep track of all numerical factors in the calculation of ψ. Since the wave at the point of observation is expected to be merely the incident plane wave propagated through an additional distance, a relation can be deduced between $\mathbf{B}$, the source strength per unit area, and $\mathbf{A}$, the amplitude of the incident plane wave (cf. the asserted result in item (3) of Section 12–15).

***13–10.** Discuss how the diffraction of a normally incident plane wave by an opaque strip can be considered as an application of the Cornu spiral. If $\mathbf{Z}_1$ and $\mathbf{Z}_2$ correspond to the edges of the strip and $\mathbf{J} - \bar{\mathbf{J}} = 1 + i$, show that the intensity is proportional to $|(\mathbf{J} - \bar{\mathbf{J}}) - \mathbf{Z}_{21}|^2$. Find a condition in which the center of the "shadow" should be a strong relative maximum and give reasonable specifications for an optics experiment to demonstrate the effect.

14 The double slit; multiple-slit arrays; diffraction gratings

14-1 INTRODUCTION

In the discussion of Young's experiment in Chapter 11 we treated the individual pinholes as if they were point sources each of which gives rise to a uniform spherical wave beyond the aperture screen. In Chapter 12 we discussed the diffraction which occurs at a single aperture of realistic dimensions. We will now combine these two ideas to give a comprehensive derivation of the pattern to be expected when a plane wave is incident on an aperture screen containing two apertures of finite dimensions. The phenomenon is not readily classed as being either "interference" or "diffraction," since both of these effects play an important role. Diffraction is important in that the wave spreads out to a certain extent after it passes through each of the apertures. Interference is important in that destructive and constructive relations exist between the contributions from the individual apertures. Indeed, as we saw in Chapter 12, the distinction between "interference" and "diffraction" is a tenuous one, since the "diffraction pattern" of an individual aperture is the result of interference effects among the various contributions from within the aperture.

A second idea which will be pursued in this chapter is the discovery of the "magic" that is accomplished by the use of many slits in a diffraction grating. The interference maxima in Young's experiment are at positions given by the same formula ($n\lambda = d \sin \theta$) which locates the spectral images produced by a diffraction grating. It is quite clear, however, that a double slit is not suitable for purposes of spectral analysis. A diffraction grating produces a sharp concentration of energy into spectral lines, whereas the illumination in a double-slit pattern is spread out in fringes which are relatively broad. We shall seek to understand how this sharpening is brought about. The effect is conveniently studied by examining the sequence of patterns produced by N-tuple slits starting with the double slit ($N = 2$) and ranging to the diffraction grating where N is large.

As we have stated previously, the Huygens-Fresnel principle relates the diffraction pattern produced by a plane wave incident on an aperture screen to a corresponding radiation pattern by a continuous distribution of point sources. In the following sections the derivations are carried out in terms of abstract source distributions. Without further comment the results are then interpreted in the context of diffraction experiments. We shall also confine the derivations to arrays of point sources along a straight line, since the extension of these results to include rectangular or circular distributions is similar to that performed for the single slit in Chapter 12.

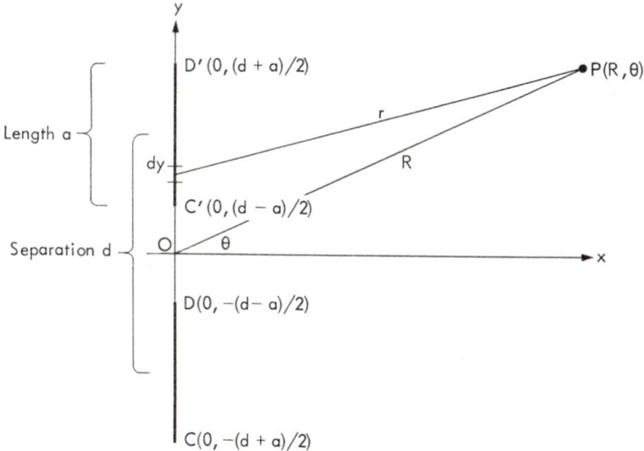

Fig. 14–1. Point sources uniformly distributed along a pair of line segments *CD* and *C′D′*.

14–2 THE DOUBLE SLIT

The arrangement of radiating sources which is the prototype for a Huygens-Fresnel derivation of the diffraction by a pair of slits of width a is indicated in Fig. 14–1. Let B designate the source strength per unit length. For application to the case of normal incidence in the diffraction problem we assume that B is real and independent of y. The contribution from the element dy to the disturbance at P has the form

$$d\psi = \mathcal{R}e\left\{B\,dy\,\frac{e^{i(\omega t-kr)}}{r}\right\}.$$

Making the Fraunhofer approximation we replace r by $(R - y\sin\theta)$ in the phase factor and by R in the amplitude factor. The simplified expression for the contribution from dy is then

$$d\psi = \mathcal{R}e\left\{\frac{B}{R}\,e^{i(\omega t-kR)}e^{iky\sin\theta}\,dy\right\}.$$

The total disturbance at P is obtained by integrating over those portions of the y-axis where the sources lie. Note incidentally that prior to the substitution of the actual limits for y the derivation has been general and can be adapted to a variety of source arrays. Thus we obtain

$$\psi = \mathcal{R}e\left\{\frac{Be^{i(\omega t - kR)}}{R}\mathbf{\Phi}\right\}, \tag{14-1}$$

where

$$\mathbf{\Phi} = \int_{-(d+a)/2}^{-(d-a)/2} e^{iky\sin\theta}\, dy + \int_{(d-a)/2}^{(d+a)/2} e^{iky\sin\theta}\, dy. \tag{14-2}$$

Through the change of variables $u = y + d/2$ in the first integral and $u = y - d/2$ in the second, this reduces to

$$\begin{aligned}\mathbf{\Phi} &= [e^{ik(d/2)\sin\theta} + e^{-ik(d/2)\sin\theta}]\int_{-a/2}^{a/2} e^{iku\sin\theta}\, du \\ &= 2\cos\,[(kd/2)\sin\theta]\int_{-a/2}^{a/2} e^{iku\sin\theta}\, du.\end{aligned}$$

The remaining integral is exactly the same as the one appearing in Eq. (12–6). As before, we define $\beta = (ka/2)\sin\theta$. After also setting $(kd/2)\sin\theta = \gamma$, we obtain

$$\mathbf{\Phi} = 2a\cos\gamma\,\frac{\sin\beta}{\beta}\cdot \tag{14-3}$$

Returning to Eq. (14–1), we now have

$$\psi = \frac{2aB}{R}\cos\gamma\left(\frac{\sin\beta}{\beta}\right)\cos\,(\omega t - kR). \tag{14-4}$$

Comparison with Eq. (12–7) shows that the disturbance at P due to the pair of segments differs by the factor $2\cos\gamma$ from the disturbance produced by a single one. The angle θ appears in the amplitude of ψ through the two parameters β and γ. Since the intensity varies as the square of the amplitude, we write

$$I(\theta) = 4I_0\cos^2\gamma\left(\frac{\sin\beta}{\beta}\right)^2, \tag{14-5}$$

where I_0 is the intensity along $\theta = 0$ associated with a *single* line segment. The two special cases which have been studied previously can be recovered from this formula:

i) *Two point sources:* Since $\beta = (ka/2)\sin\theta$, if we assume that each line segment is extremely short in comparison with a wavelength ($ka \ll 1$), then $(\sin\beta)/\beta \doteq 1$ and Eq. (14–5) becomes identical with the radiation pattern from a pair of point sources which are in phase, namely, Eq. (11–5) with $(\Delta\phi)_0 = 0$.

ii) *The single line source.* The effect of setting $d = 0$ is to bring the centers of the two line sources into coincidence, resulting in a single line source of double strength. Thus with $d = 0$, $\gamma = (kd/2)\sin\theta = 0$, and Eq. (14–5) becomes $I = 4I_0[(\sin\beta)/\beta]^2$ which is the same as Eq. (12–8) except for the factor of 4 which comes from doubling the source strength.

To obtain an idea of the plot of Eq. (14–5) for more general values of a and d, consider values of θ which are sufficiently small so that $\sin\theta$ can be replaced by θ. It is then suitable to use the approximations $\gamma \doteq kd\,\theta/2$ and $\beta \doteq ka\,\theta/2$. In all applications to diffraction we have $d > a$ (the line sources do not overlap) and therefore, as a function of θ, the factor $\cos^2(kd\,\theta/2)$ oscillates more rapidly than the factor $\sin^2(ka\,\theta/2)$. We may thus consider the factor $I_e = 4I_0[(\sin\beta)/\beta]^2$ as an envelope curve within which the oscillations of $\cos^2\gamma$ are inscribed (Fig. 14–2). The regular double-point source pattern is modulated by a more gradual intensity variation which depends on the width of the individual source. A double slit therefore produces what looks like a set of normal Young's interference fringes except that the intensity fades out at those positions where the Fraunhofer pattern associated with the single slit has its minima (see Fig. 14–5b). The intensity also drops off at large values of θ due to the envelope factor $1/\beta^2$. If a zero of the Fraunhofer pattern should happen to coincide with the center of a Young's fringe, this particular fringe is suppressed and there is said to be a *missing order.* [The maxima of the Young's pattern occur at $\gamma = (kd/2)\sin\theta = n\pi$, or equivalently, $n\lambda = d\sin\theta$, $n = 0, \pm 1, \pm 2, \ldots$ The integer n may be used to designate the *order* of a fringe.]

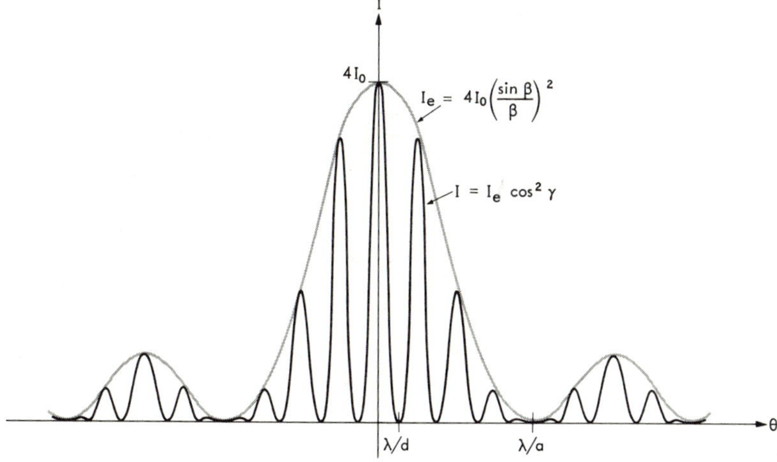

Fig. 14–2. Double-source pattern inscribed within Fraunhofer envelope.

It should be clear that the criterion for complete destructive interference by a single slit is a *sufficient* condition for vanishing of the double-slit pattern; if the contributions summed over each of the slits separately yield zero resultant at a given point of observation, then the combination obviously must be zero. There is likewise a simple reason that the criterion for a zero in the double point-source pattern is a *sufficient* condition for a·zero in the double-slit pattern. Equation (12–7) shows that the resultant phase of the signal obtained from a single slit is the same as the phase along a ray from the *center* of the slit. Therefore, if the distances from P to the two slit centers differ by an odd number of half-wavelengths, the two signals cancel each other.

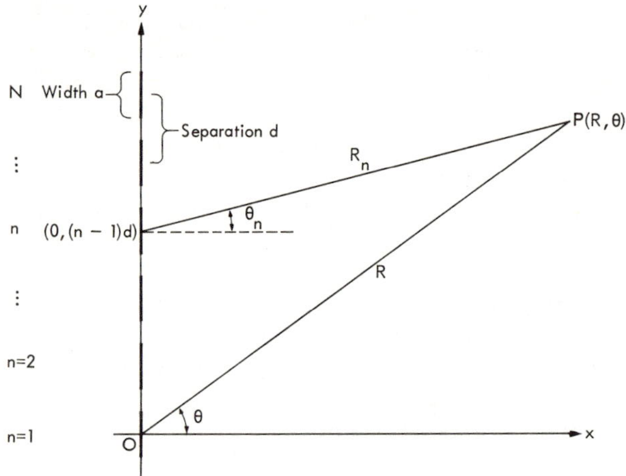

Fig. 14–3. Segmented distribution of coherent sources.

14–3 MULTIPLE-SLIT ARRAYS

In this section we aim to deduce the nature of the diffraction pattern obtained from an aperture screen with three or more equally spaced slits of equal width. The equivalent segmented linear distribution of coherent sources is indicated in Fig. 14–3. The disturbance at P can be obtained by substituting appropriate limits into a sum of integrals as in Eq. (14–2). Equivalently, we can make use of Eq. (12–7) in which the integration over the width of an individual segment has already been carried out. Thus the nth segment contributes a signal

$$\psi_n = \mathcal{R}e \left\{ \frac{aB}{R_n} \left(\frac{\sin \beta_n}{\beta_n} \right) e^{i(\omega t - kR_n)} \right\}, \tag{14–6a}$$

where $\beta_n = (ka/2) \sin \theta_n$. We must assume that the sources are sufficiently close to the origin so that the Fraunhofer approximation applies not only across each segment but also across the entire array of sources. (The scale of

Fig. 14–3 is greatly exaggerated.) Thus we replace R_n by R and β_n by $\beta = (ka/2)\sin\theta$ in the amplitude factors of Eq. (14–6a) and take $R_n = R - (n-1)d\sin\theta$ in the phase factors. If there are N segments altogether, the total disturbance at P is thus given by

$$\psi = \sum_{n=1}^{N} \psi_n = \mathcal{R}e\left\{\frac{aB}{R}\frac{\sin\beta}{\beta}e^{i(\omega t - kR)}\sum_{n=1}^{N} e^{i(n-1)kd\sin\theta}\right\}. \quad (14\text{–}6\text{b})$$

The sum called for is a *geometric series* of ratio $e^{ikd\sin\theta}$ and can be transformed as follows:

$$\begin{aligned}\sum_{n=1}^{N} e^{i(n-1)kd\sin\theta} &= \frac{1 - e^{iNkd\sin\theta}}{1 - e^{ikd\sin\theta}}\\ &= \frac{e^{-i(Nkd/2)\sin\theta} - e^{i(Nkd/2)\sin\theta}}{e^{-i(kd/2)\sin\theta} - e^{i(kd/2)\sin\theta}}\cdot\frac{e^{i(Nkd/2)\sin\theta}}{e^{i(kd/2)\sin\theta}}\\ &= \frac{\sin\left(\dfrac{Nkd}{2}\sin\theta\right)}{\sin\left(\dfrac{kd}{2}\sin\theta\right)}\cdot e^{i(N-1)(kd/2)\sin\theta}.\end{aligned}$$

Thus with $\gamma = (kd/2)\sin\theta$ and $\overline{R} = R - (N-1)(d/2)\sin\theta$, we have

$$\psi = \mathcal{R}e\left\{\frac{aB}{R}\left(\frac{\sin\beta}{\beta}\right)\left(\frac{\sin N\gamma}{\sin\gamma}\right)e^{i(\omega t - k\overline{R})}\right\}. \quad (14\text{–}7)$$

In passing we note that the net phase is the same as that of a ray which has traveled a distance $\overline{R}$ and can be associated with the midpoint of the source array. The important effect of the summation over the total number of segments has been the introduction of the factor $(\sin N\gamma)/\sin\gamma$. The other factors in the amplitude of the sinusoidal function in Eq. (14–7) are the same as those for a single segment. Thus the intensity distribution has an angular dependence given by

$$I(\theta) = I_0\left(\frac{\sin\beta}{\beta}\right)^2\left(\frac{\sin N\gamma}{\sin\gamma}\right)^2, \quad (14\text{–}8)$$

where I_0 is the intensity along $\theta = 0$ for a single segment. As an immediate check note that with $N = 2$, $\sin 2\gamma = 2\sin\gamma\cos\gamma$, and the formula reduces to Eq. (14–5). The correct reduction when $N = 1$ is obvious.

One simple fact which can be read from Eq. (14–8) is the intensity in the direction $\theta = 0$. Here $\beta = 0$ and $(\sin\beta)/\beta = 1$. Also $\gamma = 0$ and, by applying L'Hôpital's rule, we have

$$\lim_{\gamma\to 0}\frac{\sin N\gamma}{\sin\gamma} = \lim_{\gamma\to 0}\frac{N\cos N\gamma}{\cos\gamma} = N.$$

Hence $I(0) = N^2 I_0$. The meaning of this is to be found in the nature of the Fraunhofer approximation. The latter consists of representing the distance $R(y)$ from P to the point $(0, y)$ by the Maclaurin series approximation $R(y) \doteq R(0) + yR'(0)$. But, when $\theta = 0$, the point P lies on the x-axis, and the distance from P to the origin is a relative minimum. Hence for this case $R'(0) = 0$. This means that "to first order" the distances from P to all points on the source are the same. Hence when P lies on the x-axis the contributions from the N line segments are all of the same phase and the resultant has an amplitude N times that from a single segment.

The presence of the factor $(\sin^2 \beta)/\beta^2$ in Eq. (14–8) shows that the pattern determined by the remaining factors is modulated by the Fraunhofer envelope associated with the individual line segment. Hence all multiple-slit patterns "fade out" at those positions where the single-slit pattern goes to zero.

To study the remaining angular dependence of the pattern, consider the function

$$J_N(\gamma) = \frac{1}{N^2}\left(\frac{\sin N\gamma}{\sin \gamma}\right)^2. \tag{14–9}$$

The factor of N^2 has been included so that this function represents the intensity distribution relative to *unit* intensity in the direction $\theta = 0$. The intensity distribution is proportional to J_N alone if the segments are short compared with a wavelength and $(\sin \beta)/\beta \sim 1$ over the entire pattern. This implies that J_N is essentially the radiation pattern which would be obtained from N uniformly spaced *point* sources. It can be verified as an exercise that direct summation of the signals arriving from such an array of point sources yields the function $J_N(\gamma)$ for the angular distribution function of intensity.

Although it is ultimately J_N vs. θ which is of interest, a plot of J_N vs. γ gives the main features of the pattern, if we bear in mind that γ is related to θ through the equation $\gamma = (kd/2) \sin \theta$. For $N = 2$, $J_N(\gamma) = \cos^2 \gamma$, which gives the Young's double-slit pattern studied previously. For the triple slit we can make use of the identity $\sin 3\gamma = 3 \sin \gamma - 4 \sin^3 \gamma$. Thus

$$J_3(\gamma) = [1 - \tfrac{4}{3} \sin^2 \gamma]^2 = [\tfrac{1}{3} + \tfrac{2}{3} \cos 2\gamma]^2.$$

This curve is plotted in Fig. (14–4). The curve has *principal* maxima at $\gamma = n\pi$ or $n\lambda = d \sin \theta$, which are the same locations as the maxima of the double-slit pattern. In addition there are *secondary* maxima at $\gamma = (n + \frac{1}{2})\pi$. The intensity at one of these "pips" is $J_3(\pi/2) = \frac{1}{9}$. The effect of adding the third slit has been to cause a narrowing of the band around the $n\lambda = d \sin \theta$ positions and to add a smaller secondary maximum between each of the principal maxima. In Fig. 14–5(c) a photograph of the diffraction pattern from a triple slit is shown. (Note that the slits used were not of negligible width and that the modulation due to the single-slit Fraunhofer envelope is evident.)

The changes which take place in going from $N = 2$ to $N = 3$ are indicative of what continues to happen as N is increased. The general features of the graph of $J_N(\gamma)$ can be seen from the following properties:

a) $J_N(\gamma + \pi) = J_N(\gamma)$. Since $J_N(\gamma)$ is periodic, we can confine our attention to the range $0 \leq \gamma < \pi$.

b) $J_N(0) = J_N(n\pi) = 1$. Thus the intensity is unity at all positions where $\gamma = n\pi$. In terms of the angle θ this condition is $n\lambda = d \sin \theta$. The location of these *principal maxima* is the same for all N.

c) $J_N(n\pi/N) = 0$, $n = 1, 2, \ldots, (N - 1)$. The zeros of J_N occur where $\sin(N\gamma) = 0$, or $N\gamma = n\pi$. Thus J_N vanishes exactly $N - 1$ times in the range $0 < \gamma < \pi$.

d) Since $J_N(\gamma) \geq 0$ there must be at least one relative maximum between each pair of zeros of the function. Further study shows that only one such maximum is attained. Thus there are exactly $N - 2$ subsidiary maxima between adjacent principal maxima. For example, the quintuple slit shows three faint bands between the bright bands at the $n\lambda = d \sin \theta$ positions (Fig. 14–5e).

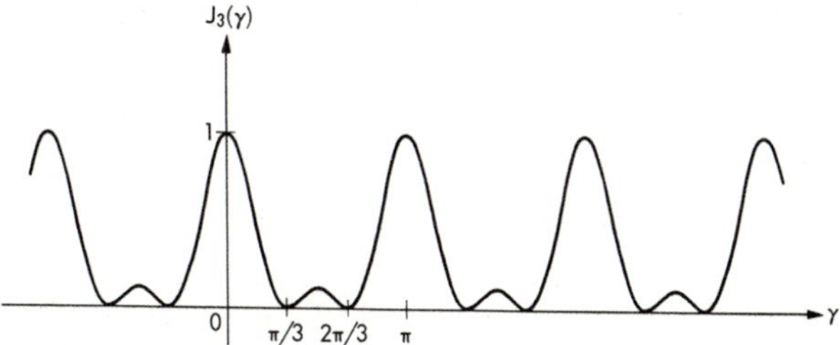

Fig. 14–4. Triple-slit interference pattern.

14–4 THE DIFFRACTION GRATING

The features of the pattern produced by a diffraction grating are obtained by extending the discussion of the preceding section to include special statements which apply when the number of line segments in the source array is extremely large.

e) For large N the oscillations of $\sin^2 N\gamma$ are quite rapid compared to those of $\sin^2 \gamma$. We therefore write

$$J_N(\gamma) = J_e(\gamma) \sin^2 N\gamma,$$

Fig. 14–5. (a) Fraunhofer diffraction pattern of a single slit. (b), (c), (d), (e) Interference patterns of 2, 3, 4, and 5 slits, respectively. The slit systems are indicated at the left. (From F. W. Sears, *Optics*, 3rd ed. Reading, Mass., Addison-Wesley, 1949.)

where $J_e(\gamma) = (\csc^2 \gamma)/N^2$ acts as an envelope to the curve $\sin^2 N\gamma$ (Fig. 14-6). The behavior of this envelope points out most clearly how the broad interference band of the double-slit pattern is transformed into the sharp spectral line of the diffraction grating. Due to the factor of $1/N^2$, the envelope (below which the entire pattern must lie) is negligible when N is large except in the immediate vicinity of the positions $\gamma = n\pi$, $n = 0, \pm 1, \pm 2, \ldots$

f) Since the pattern for large N is negligible except near the locations of the principal maxima, and since the behavior near $\gamma = 0$ is repeated identically at $\gamma = \pi, 2\pi$, etc., we can study the structure of the important part of the pattern by specializing to the immediate vicinity of $\gamma = 0$. We may then take $\sin \gamma \doteq \gamma$. However, since N is large, a similar replacement for $\sin (N\gamma)$ is not allowed. Thus $J_N(\gamma) = (\sin^2 N\gamma)/(N\gamma)^2$ is the function which describes what we may refer to as the detailed structure of a *spectral line*. With the substitution $N\gamma = \beta'$, this function is seen to be of exactly the same *form* as the Fraunhofer pattern of a single slit:

$$J_N(\gamma) = \left(\frac{\sin \beta'}{\beta'}\right)^2. \tag{14-10}$$

The width a' of the associated single slit is found from the equivalence

$$\beta' = \frac{ka'}{2} \sin \theta = N\gamma = \frac{Nkd}{2} \sin \theta.$$

Thus $a' = Nd$. But the *total width* of the grating is $(N - 1)d + a \doteq Nd$. The conclusion is therefore that each spectral line has a Fraunhofer-like

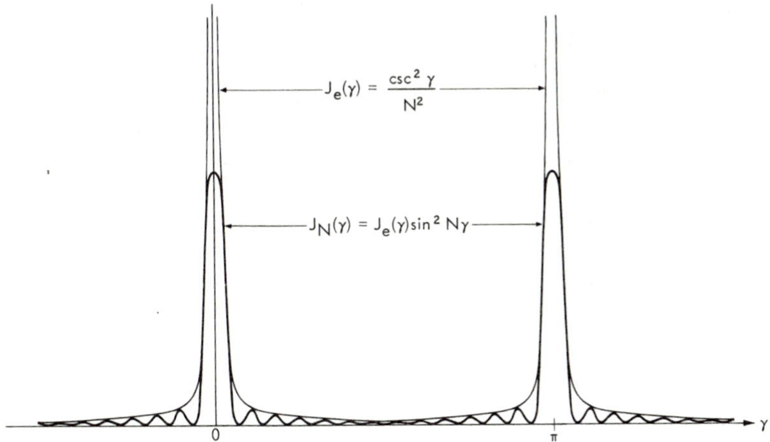

Fig. 14-6. Interference pattern for N large.

structure on a scale determined by the entire width of the grating. This result is easily comprehended if we imagine that the grating is part of an infinite grating which has been blocked off by a rectangular aperture of width a'. As the perfect plane waves produced by the infinite grating pass through this aperture, they are subject to ordinary single-slit diffraction.

When a wide grating is used, the structure associated with the factor $\sin^2 N\gamma$ is too fine to be resolved. However, the effect can be demonstrated beautifully by blocking off the grating by a variable slit parallel to the rulings. When the slit is sufficiently narrow, each spectral line is seen to have the banded Fraunhofer pattern. As the slit is narrowed further, the pattern fans out and the bands around *each* line appear to flee from the line center. The pattern of the zero-order line is exactly what would be obtained if the slit were kept in place and the grating removed. The pattern around each colored spectral line is identical with that of the zero-order line.

Since 91% of the energy is contained in the central maximum of a single-slit pattern, we can define an effective width of the spectral line. A principal maximum centered about $\gamma = n\pi$ terminates at the adjacent zero of $\sin N\gamma$. This implies a change in γ by the amount $\delta\gamma = \pi/N$. The corresponding change in θ is obtained from

$$\delta\gamma = \delta\left[\frac{kd}{2}\sin\theta\right] = \left[\frac{\pi d}{\lambda}\cos\theta\right]\delta\theta.$$

Thus the angular separation from the center of a line to its effective edge is

$$\delta\theta = \lambda/Nd\cos\theta. \tag{14-11}$$

(Note that the factor $Nd\cos\theta$ can be interpreted as the projection of the width of the grating perpendicular to the line of sight.)

In the preceding discussion we have assumed that the incident wave is perfectly monochromatic. The line width which we have derived is that due to *instrumental broadening*. This effect will be superimposed on any line widths which are associated with a distribution of frequencies in the incident wave.

The occurrence of instrumental broadening raises the question of whether two distinct spectral lines (each of which is assumed to be associated with a single frequency) will be resolved in the spectrum produced by a grating. The location of the centers of these lines in the nth order is given by $n\lambda = d\sin\theta$. If the wavelengths differ by $\Delta\lambda$, the corresponding difference in angular position is

$$\Delta\theta = (n\,\Delta\lambda)/d\cos\theta. \tag{14-12}$$

This formula shows that the angular separation of centers is greater in the higher-order spectra. The angular separation per unit wavelength,

$\Delta\theta/\Delta\lambda = n/d\cos\theta$, is known as the *dispersion* in the nth order spectrum. When the wavelength difference is sufficiently small, the distance between line centers is smaller than the instrumental width of each line, and the lines are unresolved. In Section 12–14 we discussed the Rayleigh criterion for an *image-forming* instrument (as opposed to the grating, which is an *analyzing* instrument). By similar reasoning, we consider two spectral lines to be *just resolved* if the center of one line falls at the first minimum in the diffraction pattern of the other. Thus, from Eq. (14–11), $\Delta\theta_{\min} = \delta\theta = \lambda/Nd\cos\theta$. If $\Delta\lambda_{\min}$ is the minimum difference in wavelengths which can be resolved (the so-called *limit of resolution*), Eq. (14–12) shows that $\Delta\theta_{\min} = n(\Delta\lambda)_{\min}/d\cos\theta$. Eliminating between these two equations we find $(\Delta\lambda)_{\min} = \lambda/nN$. This result is often expressed in terms of the *resolving power*, $\lambda/(\Delta\lambda)_{\min}$:

$$\lambda/(\Delta\lambda)_{\min} = nN. \tag{14–13}$$

From this formula the following conclusions can be drawn:

a) Within any given order the minimum resolvable difference in wavelengths is a fixed percentage of the wavelength. For example, a grating of 100 slits can resolve no more than 1% wavelength difference in any portion of the first-order spectrum.

b) The resolving power is greater in higher-order spectra. For example, it may happen that the sodium doublet is not resolved in the first two orders, but is resolved in the third and all higher orders.

c) The resolving power depends on the parameters of the grating only through N, the total number of slits. (In case part of the grating is blocked off, this refers, of course, to the actual number of slits used.)

In view of the fact that the width of an individual spectral line varies inversely as the total width of the grating, it might seem that the resolving power of a *fixed* number of slits would be improved by distributing the slits over a wider area. Thus according to Eq. (14–11), each spectral line is made sharper if we keep N fixed and increase d, the spacing between slits. That is, the *coarser* the grating, the *narrower* the spectral lines. And doesn't a narrowing of the spectral lines improve the resolution? The catch is, as shown by Eq. (14–12), that the angular separation between the centers of two lines decreases as d increases. Thus although the lines are made sharper by the process, they are also moved closer together and the resolving power remains the same. Although coarse and fine gratings with the same number of lines have the same resolving power, the latter has several advantages over the former. Being of lesser width, the fine grating requires smaller lenses to make effective use of the whole grating. Also the fine grating does not require as fine a grain in the photographic emulsion or as high-powered an eyepiece in visual work, if all available detail is to be recorded.

The desire to produce gratings of high resolving power has led to the development of ruling engines of fantastic precision.*

The pattern we have been describing in this section is that arising from the factor $J_N(\gamma)$ in the total expression for the angular distribution, $I(\theta)$. Under usual circumstances the intensity modulation due to the Fraunhofer envelope, $(\sin \beta)^2/\beta^2$, is sufficiently gradual so that it need not come into the discussion of the structure of an individual spectral line. It is, however, of importance in determining the relative intensities of the images of a line of given wavelength as they appear in the different spectral orders. For light of wavelength λ, the nth order line appears in the position $\sin \theta = n\lambda/d$. The corresponding value of β is $\beta = (\pi a/\lambda) \sin \theta = n\pi a/d$; hence the relative intensities are affected by the factor $[\sin^2 (n\pi a/d)]/[n\pi a/d]^2$. Thus the intensity of this particular spectral line as seen in the various orders falls off generally as n^2. Furthermore, if the slit width a and the slit separation d are commensurable, there will exist values of n for which the sine function vanishes and the corresponding spectral order is suppressed. For example, if $d = 2a$, none of the spectra of *even* order appear.

There are two types of grating for which the above derivation based upon equally-spaced coherent linear sources is applicable:

a) a *transmission* grating with perfectly transparent openings alternating with perfectly opaque strips;

b) a *reflection* grating with perfectly reflecting strips alternating with perfectly absorbing strips.

It can be shown more generally that any periodic structure gives rise to a pattern similar to the above except that the factor $(\sin^2 \beta)/\beta^2$, which was associated with the diffraction pattern of a rectangular slit, becomes replaced by a factor whose form is determined by the diffraction pattern of an individual element of the periodic structure. This principle has led to experimentation on the groove form in ruled gratings to obtain a more favorable intensity distribution among the various orders than that governed by $(\sin^2 \beta)/\beta^2$. In particular, it has been possible to concentrate most of the energy into a single higher-order spectrum.

PROBLEMS

14–1. It is obvious that a "double slit" for which the slit width a is equal to the distance between centers d is in reality just a single slit of width $2a$. Show that this result follows from the double-slit formula Eq. (14–5).

---

* For a description of the precision required of a ruling engine and the ingenious methods of conquering inherent construction problems, see A. G. Ingalls, *Scientific American*, **186,** 45 (1952).

14–2. Sketch a graph of the angular distribution of intensity for a double slit for which the distance between slit centers is ten times the slit width.

14–3. *Effect of finite slit width in a Lloyd's-mirror experiment.* Consider a reflecting surface set perpendicular to an aperture screen containing a slit of width a (Fig. 14–7). Let the center of the slit be a distance $d/2$ above the reflecting surface. a) Use the Fraunhofer approximation to calculate the diffraction pattern at large distances due to a plane wave normally incident on the aperture screen. b) Sketch the angular distribution of intensity for the case $d = 5a/2$ and describe what modifications have been introduced in the Lloyd's-mirror pattern due to the finite width of the slit. c) Sketch the angular distribution of intensity for the case $a = d$. Find to within an accuracy of 1% the value of $\beta = (ka/2) \sin \theta$ at which the highest maximum occurs.

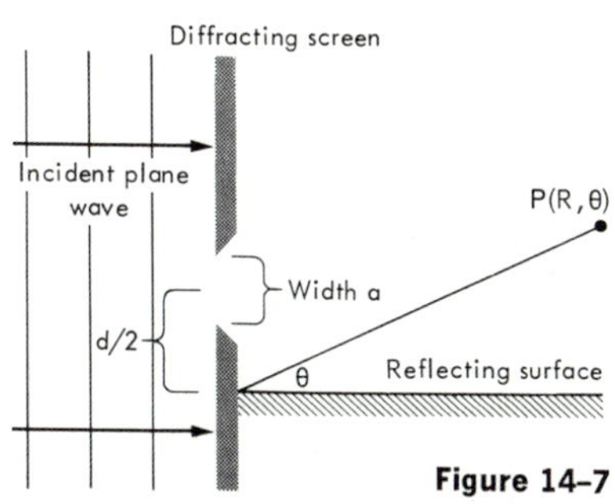

Figure 14–7

14–4. In the radiation pattern of the sextuple source (six coherent point sources equally spaced along a straight line), there are five zeros occurring in the range $0 < \gamma < \pi$ where $\gamma = (kd/2) \sin \theta$. How many of these can you account for in terms of cancellation of contributions between pairs of points?

14–5. Consider an array of N coherent point sources equally spaced along a straight line, the distance between each pair being d. Assume that the sources are of equal strength and in phase. Use the Fraunhofer approximation and sum the contributions from each source to determine the net signal at a distant point of observation. (This method of direct summation is equivalent to that used in Chapter 11 for the double source.) Show that the angular distribution of intensity is the same as that obtained from Eq. (14–8) by taking the limit as the length of the individual line segments tends to zero.

14–6. Five radio antennas are equally spaced along a straight line. Assume for the purposes of this problem that each is the source of spherical waves of uniform intensity. Let the distance between adjacent antennas be one wavelength. Assume that the waves radiated by the various antennas are in phase and of equal strength. Sketch a polar plot of the intensity distribution as a function of angular position of the point of observation.

14–7. a) Sketch the angular intensity distribution for diffraction of a normally incident plane wave by a double slit having a slit width equal to one-half the distance between centers. b) Consider a grating having a slit width equal to one-half the distance between slit centers. Plot a bar graph indicating the relative intensities of any particular spectral line as it appears in the various orders.

14–8. Discuss the general appearance of the diffraction pattern obtained using a transmission grating and a) a point source, b) a narrow incoherent line source (source slit) parallel to the grating slits, c) a narrow incoherent line source not parallel to the grating slits, d) an incoherent line source parallel to the grating slits but of finite breadth.

14–9. What resolving power is required to resolve the sodium doublet? b) What information is required about the grating to predict the smallest-order spectrum in which the doublet will be resolved? c) Show that a knowledge of the total width of the grating is sufficient to obtain a lower bound for the *angle* at which a resolved doublet will be obtained. d) Calculate the order number and the angle of the first doublet resolved by a one-inch grating of 200 lines.

***14–10.** a) Examine the derivation of the multiple-slit diffraction pattern under the Fraunhofer approximation and consider what changes will be required for the case of an obliquely incident plane wave. b) Suppose that the spectral lines associated with a particular wavelength are to be observed by means of an arrangement similar to that in Fig. 11–20 where the aperture screen is replaced by a transmission grating. Describe qualitatively how the pattern will change if the grating is rotated about an axis parallel to its slits. c) The wavelength is to be measured by means of the arrangement described in (b). The telescope will be used to determine the angle $\Delta\theta$ between the blaze (the zero-order image) and the first-order image on one side. The wavelength is then calculated from the formula $\lambda = d \sin \Delta\theta$. Suppose that the grating has been carelessly set so that there is an angle ϕ between its plane and the plane of the incident wave. Determine to first order in ϕ the percentage error this will cause in the wavelength calculated as above. Calculate a numerical value if $\Delta\theta = 20°$, $\phi = 10°$. d) The error discussed in (c) can be reduced to second order in ϕ by making additional use of the first-order line on the other side of the blaze. The experimenter has two possible procedures: take the average of λ calculated on both sides, or, average the magnitudes of the angles measured on both sides and compute λ from the formula $\lambda = d \sin \overline{\Delta\theta}$. Which of these is the more accurate procedure? Estimate the percentage error for the data given in (c).

***14–11.** *Lecture demonstration of the single slit diffraction pattern.* Prior to the development of lasers it was difficult to project diffraction patterns onto a screen with sufficient intensity to be seen by a large group. The following method devised by R. W. Pohl can be used to demonstrate the single-slit diffraction pattern.

Consider a coarse grating (i.e. a multiple-slit array) for which the individual slit is of suitable width to produce a projected diffraction pattern of convenient size. Recall that the function graphed in Fig. 14–6 must be multiplied by the factor $[(\sin\beta)/\beta]^2$ associated with diffraction by a single slit, to obtain the diffraction pattern of the grating. If the number of slits is large, the fine structure in between the principal maxima of the multiple-slit pattern will be unresolved. The net pattern can then be thought of as the principal maxima inscribed under the $[(\sin\beta)/\beta]^2$ envelope. Suppose now that the source slit is opened so that the projected pattern consists of the noncoherent superposition of the independent patterns associated with different points across the source slit. When the slit is sufficiently broad the structure associated with the principal maxima will be lost and we are left with an intensity distribution governed by the Fraunhofer envelope. This pattern is, of course, somewhat smeared out, but if the parameters are chosen appropriately, a source slit which is "broad" in the sense of the grating pattern can still be considered a point source as far as the individual slit diffraction pattern is concerned.

Suppose the collimating lens (between the source slit and the grating) has a focal length of 20 cm and the projecting lens (between the grating and the screen) has a focal length of 1 m. What grating slit width will produce a central maximum of total width 10 cm? (Use $\lambda = 5500$ Å.) Pick a reasonable value for the number of lines per inch of a coarse grating which would achieve the desired results. How wide should the source slit be?

***14–12.** *Repetition of identical radiating elements.* Assume that the angular distribution of intensity is known to be $f(\theta)$ for some particular arrangement of sources. This arrangement of sources, which will be referred to as a radiating *element*, may consist of a number of point sources or coherent line segments, or some combination thereof. Suppose that an array is formed by placing N identical elements along a straight line. Let D be the distance of separation of corresponding points in adjacent elements. The Fraunhofer approximation is assumed to hold over the entire array. The angle θ is measured from a reference axis perpendicular to the line along which the elements are distributed. a) Show that the radiation pattern of the array is given by

$$I(\theta) = f(\theta)\left(\frac{\sin N\Gamma}{\sin \Gamma}\right)^2,$$

where $\Gamma = (kD/2)\sin\theta$. Show that this checks with Eq. (14–8) which refers to an array of coherent line segments. b) Check the preceding formula by considering a grating of $2N$ slits as an array of N elements each of which is a double slit. c) Consider a transmission grating which consists of N pairs of double slits, each slit having a separation d, the distance between pairs being $D = 20\,d$. Indicate how such a grating can be used in conjunction with a broad source to project a double-slit pattern suitable for viewing by a large audience (cf. Problem 14–11).

***14–13.** *A simple model to explain the appearance of "ghosts" in the spectrum produced by an imperfectly ruled diffraction grating.* If there is a periodic imperfection in the ruling of a grating, the actual grating acts in a crude sense like two independent gratings, one of which has the spacing of the ideal rulings, the other of which has the spacing associated with the repetitive error. Since the latter spacing is larger than the former, it results in a more closely spaced series of spectral lines. These *ghosts* are usually much fainter than the corresponding lines which occur at the "ideal" positions. However, the strongest of them can sometimes be seen in positions which are quite unexpected to the uninitiated observer.

As a simple example, consider a grating in which every qth slit is missing. Let there be $p(q-1)$ slits altogether, i.e. an array of p elements each of which consists of $q-1$ slits. Let the distance of separation between slits be d. Apply the formula deduced in Problem 14–12 to determine the angular distribution of intensity. (Assume that the individual slit width is small enough so that the factor $(\sin\beta)/\beta$ can be set equal to unity.) Show that the maxima at the locations given by $n\lambda = d\sin\theta$ have intensity $I = p^2(q-1)^2 I_0$ where I_0 is the intensity associated with a single slit and that the maxima at $n\lambda = qd\sin\theta$ have intensity $I = p^2 I_0$.

15 Waves confined to a limited region

15-1 INTRODUCTION

The determination of the transverse motion of a string segment rigidly clamped at both ends is illustrative of the kind of situation which will be discussed in this chapter. When the condition of rigid clamping was considered in Chapter 3, it was applied at only one end, and the string was taken to be semi-infinite. It was then possible to speak simply of an incident wave being reflected and proceeding off to infinity. The situation is more complicated when there are two fixed ends. Due to reflection at both ends the disturbance remains confined to the given limited segment of the string. For a short pulse, the reflections at the two ends of the string can be handled independently, and the future motion can be obtained by following through the successive reflections of the pulse. A more formal mathematical approach is not necessary in this simple situation. However, if the width of the pulse is greater than the length of the string, there is never any time at which the wave is a pure running wave propagating in a single direction. That is, the shape of the string results at all times from an overlapping of plus and minus component waves. It is then less convenient to visualize the problem in terms of multiple reflections, and the techniques developed in this chapter are more suitable.

From the mathematical point of view, by requiring the string to be fixed at *two* points we add a further restriction on the class of acceptable functions over and above what we have found for the string fixed at one point. The added condition turns out to imply that the general solution is *periodic*. A way of representing the general solution is therefore provided by *Fourier's theorem*, which allows a periodic function to be expanded in a trigonometric series. The individual terms in this series describe possible motions of a special sort for the string with fixed ends. These special motions are referred to as *normal modes of vibration*. Though interesting in their own right, the overwhelming importance of the functions describing the normal modes is, as was noted above, that the general solution is a superposition of normal-mode solutions. The techniques for obtaining this so called *normal-modes expansion* are the main subject of the present chapter.

At first glance this chapter will seem to have a heavy concentration of formal mathematics. There is also a great deal of "name-calling," i.e., the introduction of terminology associated with certain of the mathematical operations. These two aspects of the material are complementary, since the language helps one to cultivate an intuitive feeling for the meaning of the formal representations. The study of specific examples does more than merely illustrate the procedures, for it enables us to build up a lore in terms of which we can understand how the individual normal modes contribute to the general solution.

The boundary condition for a *free* end can easily be substituted at either end in place of the condition of a *rigid* boundary. A more significant generalization is to consider the waves generated on a limited string segment if one of the ends is *driven* with a specified motion. In this context the frequencies associated with the normal-modes solutions turn out to be resonance frequencies of the system.

It goes without saying that the various problems discussed here for transverse waves on a string can be translated into other wave contexts. For example, the problem of the string with two fixed ends is isomorphic to that of an acoustic plane wave confined to a tube with rigidly reflecting ends; driving one end of the string is equivalent to replacing one end of the tube with a vibrating piston. Although we shall consider only the case of one-dimensional waves and the two-dimensional waves corresponding to the transverse vibrations of a rectangular membrane, the mathematical techniques developed are quite general. For example, given the three-dimensional wave equation, the present techniques can be applied to a discussion of the waves confined by a cavity with reflecting walls. Applications of this problem are found in architectural acoustics (the determination of room resonances) and in electromagnetism (the generation of microwaves by a cavity resonator).

The basic ideas of the normal-modes expansion apply not only to the wave equation but also to partial differential equations of a more general type. For simplicity, this chapter only considers examples in which the normal-modes functions are *sinusoidal* and the series expansions are Fourier-series representations. Other problems give rise to representation of the general solution as a sum of functions which are other than sinusoidal. Each of these functions by itself is a special kind of solution to the partial differential equation and the prescribed boundary conditions. (In Appendix VIII an example is considered in which these fundamental solutions are Bessel functions.) Generically these special solutions are referred to as *eigenfunctions* and the analog of the normal-modes expansion is the representation of the general solution as a superposition of eigenfunctions.

A customary method of solving the Schrödinger equation of quantum mechanics is expansion in a series of eigenfunctions. It is unfortunate that the student often first encounters this technique at the same time that he is transplanting his philosophical roots from classical to modern soil. Unfamiliarity with the mathematical methods may make the quantum mechanics seem unnecessarily strange. In a certain sense then, this chapter is a preparation for the study of quantum mechanics. The motion of a string is easily visualized, and is purely classical physics. There is therefore every opportunity to comprehend what the sophisticated mathematics is doing for us. Once confidence has been gained in these techniques, the student can concentrate on conceptual aspects which can more properly be said to be peculiar to quantum mechanics.

15-2 TRANSVERSE WAVES ON A STRING SEGMENT WITH FIXED ENDS

Let a string of length l be rigidly fixed at the ends $x = 0$ and $x = l$. The function $y(x, t)$ which specifies transverse displacements of the string must then satisfy the wave equation

$$\frac{\partial^2 y}{\partial x^2} = \frac{1}{c^2}\frac{\partial^2 y}{\partial t^2} \tag{15–1}$$

and the boundary conditions

$$y(0, t) = 0, \tag{15–2}$$

$$y(l, t) = 0. \tag{15–3}$$

Equations (15–1), (15–2), and (15–3) jointly constitute a mathematical statement of the boundary-value problem which will be considered throughout most of this chapter.

An important property of the general solution to this problem is that any function satisfying all three of these conditions is necessarily periodic in time. Before giving a formal proof of this proposition, we present an argument which leads simply to this conclusion and which assigns a numerical value to the period. Consider a short pulse which starts out near the center of the string and travels in the positive x-direction. This pulse is reflected with inversion at the fixed end $x = l$ and is *reinverted* upon reflection at $x = 0$. When the pulse returns to the original position, traveling in the same direction that it was initially, the string is in identically the same condition as it was initially, and the process repeats itself. The period T is therefore the length of time required for the pulse to travel twice the length of the string: $T = 2l/c$. If the disturbance on the string is the superposition of any number of pulses traveling in either direction, they all repeat themselves in this interval and the entire motion is periodic.

From a more formal point of view, let us consider the successive restrictions on the class of functions which meet each of the imposed conditions. The wave equation itself requires that the solution be of the form

$$y(x, t) = f(t - x/c) + g(t + x/c). \tag{15–4}$$

As was seen in Section 3–1, the boundary condition $y(0, t) = 0$ further requires that $g(t) = -f(t)$, or, as in Eq. (3–4), that

$$y(x, t) = f(t - x/c) - f(t + x/c). \tag{15–5}$$

The additional boundary condition $y(l, t) = 0$ therefore implies that

$$0 = f(t - l/c) - f(t + l/c),$$

or

$$f(t + l/c) = f(t - l/c). \tag{15–6}$$

Letting $u = t - l/c$, Eq. (15–6) can be written

$$f(u + 2l/c) = f(u). \tag{15–7}$$

Thus to meet all three of the imposed conditions, the function $f(t)$ is required to be periodic with period $T = 2l/c$. Inspection of Eq. (15–5) then shows that $y(x, t)$ is periodic with this same period.

15–3 SINUSOIDAL SOLUTIONS

Since it was established in the last section that the general motion of the string with two fixed ends is periodic, it is natural to try the simplest type of periodic function, namely the sinusoidal. That is, we ask whether the conditions of Eqs. (15–1), (15–2), and (15–3) admit a solution in which the particles of the string execute simple harmonic motion all of the same frequency. It is obvious that if such solutions are possible at all, the frequency cannot be chosen arbitrarily. A sinusoidal function of angular frequency ω' has *minimum* period $2\pi/\omega'$. From Section 15–2, the quantity $T = 2l/c$ must be *a* period of the motion. If it is not the minimum period, it must be some integral multiple of it. This requirement is met by having $2l/c = n(2\pi/\omega')$, where n is any integer. Hence sinusoidal motion is only possible at the angular frequencies given by

$$\omega' = n(\pi c/l), \qquad n = 1, 2, 3, \ldots \tag{15–8}$$

The question can be tackled more directly by inquiring whether $f(t)$ in Eq. (15–5) can be of the form

$$f(t) = \mathcal{R}e\,\{\mathbf{A}e^{i\omega' t}\}.$$

If we substitute this form into Eq. (15–5) and take $k' = \omega'/c$, the implied form of $y(x, t)$ becomes

$$\begin{aligned} y(x, t) &= \mathcal{R}e\,\{\mathbf{A}e^{i\omega' t}[e^{-ik'x} - e^{ik'x}]\} \\ &= \mathcal{R}e\,\{\mathbf{A}e^{i\omega' t}[-2i \sin k'x]\} \\ &= -2 \sin k'x\, \mathcal{R}e\,\{i\mathbf{A}e^{i\omega' t}\}. \end{aligned}$$

If we now set $\mathbf{A} = |\mathbf{A}|e^{i\phi}$ and use the identity

$$\mathcal{R}e\,\{ie^{i\theta}\} = \mathcal{R}e\,\{i(\cos\theta + i\sin\theta)\} = -\sin\theta,$$

the implied form of $y(x, t)$ becomes

$$y(x, t) = 2|\mathbf{A}| \sin k'x \sin(\omega' t + \phi). \tag{15–9}$$

This result is based on Eq. (15–5) and hence satisfies only the wave equation and the boundary condition at the end $x = 0$. Since $y(l, t)$ must vanish for *all* t, applying Eq. (15–3) to Eq. (15–9) implies that $|\mathbf{A}| \sin k'l = 0$. The case $|\mathbf{A}| = 0$ is a trivial solution in which no motion takes place. The criterion for a nontrivial solution is thus

$$\sin k'l = 0, \qquad \text{or} \qquad k'l = n\pi, \qquad n = 1, 2, 3, \ldots$$

(The solution corresponding to $n = 0$ is also clearly trivial.) Solving for $\omega' = k'c$ again yields Eq. (15–8).

We see therefore that sinusoidal motion is possible, but only if the frequency is chosen from a *select set of values.* These special frequencies which are characteristic of a given string (of given length, tension, and mass per unit length) are referred to as *eigenfrequencies.**

The nth eigenfrequency will be designated with a subscript n:

$$\omega_n = n(\pi c/l) = n\omega_1. \tag{15–10}$$

The associated *eigenvalues of k* are

$$k_n = n(\pi/l) = nk_1, \tag{15–11}$$

and the *eigenfunction* describing this special motion of the string is of the form

$$y(x, t) = C \sin k_n x \sin(\omega_n t + \phi). \tag{15–12}$$

(The arbitrary constant $2|\mathbf{A}|$ in Eq. (15–9) has been designated by C in Eq. (15–12).) The corresponding motion of the string is referred to as the nth *normal mode of vibration.* The mode of lowest frequency is referred to as the *fundamental.* Since the modes with $n > 1$ have frequencies that are

* The German word *eigen* means special, characteristic.

integral multiples of the fundamental frequency, the musical tones associated with the various modes are consonant with one another. In such a case the nth mode is also referred to as the nth *harmonic*.

The motion of the string described by a function of the form given in Eq. (15–12) has already been discussed in connection with Eq. (3–5). The context there was that of the standing wave produced when a sinusoidal wave of *arbitrary* frequency is reflected at the fixed end of a semi-infinite string. In the present context the string is of finite length. Equation (15–10), known as the *eigenfrequency* condition, must be satisfied to assure that both ends of the string are nodes of y in the standing wave pattern.

If the standing wave is interpreted as a superposition of equal sinusoidal waveforms propagating in opposite directions, the wavelength associated with each of these forms in the nth mode is

$$\lambda_n = 2\pi/k_n = 2\pi(l/n\pi) = 2l/n.$$

Thus the nth mode can be characterized by the fact that the string is divided into n half-wavelength intervals.

15–4 SOLUTIONS OF PRODUCT FORM

We observe that the eigenfunction given by Eq. (15–12) is a product of two factors, one of which is a function of x only and the other a function of t only. We shall now show that this is a unique feature of the normal-mode solutions. That is, a solution will be said to be of *product form* if it can be written

$$y(x, t) = X(x)T(t), \tag{15–13}$$

where X is a function of x only and T is a function of t only. What will be shown is that the *only* functions which satisfy Eq. (15–13) in addition to Eqs. (15–1), (15–2), and (15–3) are the sinusoidal functions of Eq. (15–12). If Eq. (15–13) is assumed, then

$$\frac{\partial^2 y}{\partial x^2} = X''(x)T(t) \qquad \text{and} \qquad \frac{\partial^2 y}{\partial t^2} = X(x)T''(t).$$

(According to standard convention a prime denotes the derivative of a function of a single variable with respect to its argument.) Equation (15–1) then implies that

$$X''(x)T(t) = (1/c^2)X(x)T''(t).$$

Dividing both sides by $X(x)T(t)$, we obtain

$$\frac{X''(x)}{X(x)} = \frac{T''(t)}{c^2T(t)}\cdot \tag{15–14}$$

The left-hand side of this equation is a function of x only and the right-hand side is a function of t only. The equation is to be satisfied for all values of x and t. Considering any fixed value of t, we see that the value of the right-hand side is a fixed number which does not depend on x. It is clear therefore that the left-hand side must have the same value for all x. The only way this condition can be met is by the existence of some number b, not a function of x or t, such that

$$X''(x)/X(x) = b \tag{15–15}$$

and

$$T''(t)/c^2T(t) = b. \tag{15–16}$$

The arguments just given have enabled us to *separate* Eq. (15–14) into two equations, one involving x only (Eq. 15–15), the other involving t only (Eq. 15–16). The constant b appears as a mathematical artifice and is referred to as a *separation constant.* (Whatever physical significance b may have emerges from the succeeding discussion.) The general solution to Eq. (15–15) is immediately recognized:

$$X(x) = c_1e^{\sqrt{b}x} + c_2e^{-\sqrt{b}x}, \tag{15–17}$$

where c_1 and c_2 are arbitrary (and possibly complex) constants. So far, we have not considered the requirement that the presumed solution of product form should satisfy the boundary conditions. Equation (15–2) in conjunction with Eq. (15–13) implies $X(0) = 0$. Applying this to Eq. (15–17) we have $c_1 + c_2 = 0$, or $c_2 = -c_1$, and hence

$$X(x) = c_1[e^{\sqrt{b}x} - e^{-\sqrt{b}x}]. \tag{15–18}$$

The boundary condition Eq. (15–3) reduces for functions of the form given by Eq. (15–13) to the requirement $X(l) = 0$. Applying this to Eq. (15–18), we have

$$0 = c_1[e^{\sqrt{b}l} - e^{-\sqrt{b}l}]. \tag{15–19}$$

The case $c_1 = 0$ is a trivial solution. The only possibility for a nontrivial solution is therefore

$$e^{\sqrt{b}l} = e^{-\sqrt{b}l} \qquad \text{or} \qquad e^{2\sqrt{b}l} = 1. \tag{15–20}$$

This equation has a solution for either $l = 0$ or $b = 0$. The former is obviously a trivial solution. The latter is also trivial, for it can be seen from Eq. (15–18) that this would imply $X = 0$ for all x, and hence $y(x, t) = 0$ for all x and t. Nontrivial solutions to Eq. (15–20) are obtained only if we allow for the possibility that $2\sqrt{b}\, l$ is a complex number. The totality of roots to the equation $e^z = 1$ are given by $z = 2\pi ni$, $n = 0, \pm 1, \pm 2, \ldots$ Thus Eq. (15–20) implies $2\sqrt{b}\, l = 2\pi ni$. The case $n = 0$ leads again to the trivial result $b = 0$; hence the nontrivial possibilities are*

$$\sqrt{b} = (n\pi/l)i, \qquad n = 1, 2, 3\ldots \tag{15–21}$$

The result of imposing the boundary conditions at $x = 0$ and $x = l$ has been to show that the separation constant b introduced in Eqs. (15–15) and (15–16) must be a negative real number of the form

$$b = -(n\pi/l)^2, \qquad n = 1, 2, 3, \ldots \tag{15–22}$$

When Eq. (15–21) is substituted into Eq. (15–18), we obtain

$$X(x) = c_1[e^{ik_nx} - e^{-ik_nx}] = c_3 \sin k_n x, \tag{15–23}$$

where we have let $n\pi/l = k_n$, and $c_3 = 2ic_1$ is a new arbitrary constant. Knowing that the separation constant b is restricted to a set of eigenvalues prescribed by Eq. (15–22), we see that the solution for $T(t)$ now proceeds directly. Equation (15–16) implies that

$$T''(t) = -k_n^2c^2T(t) = -\omega_n^2T(t),$$

where we set $\omega_n = k_nc$. The general solution to this equation is

$$T(t) = c_4 \sin(\omega_n t + \phi), \tag{15–24}$$

where c_4 and ϕ are arbitrary constants. Combining Eqs. (15–23) and (15–24), we have

$$y(x, t) = X(x)T(t) = C \sin k_n x \sin(\omega_n t + \phi). \tag{15–25}$$

In this equation ϕ and $C = c_3c_4$ are arbitrary constants, and

$$\omega_n = k_nc = n(\pi c/l), \qquad n = 1, 2, 3, \ldots$$

Equation (15–25) has been deduced as the necessary form of any function which satisfies the wave equation, the boundary conditions, and Eq. (15–13). Comparison with Eq. (15–12) shows that these functions are identical with the eigenfunctions which were demonstrated in the previous section to be the only solutions of sinusoidal form. We have arrived at the same set of eigenfunctions by asking the question in two different ways, namely, "What are the possible sinusoidal solutions?" and "What are the possible solutions of product form?" Since the latter can be extended to problems involving more than one space dimension, it eventually comes to be the preferred way of defining and deducing the eigenfunctions. The technique is often referred to as the *method of separation of variables.*

---

* Since Eq. (15–17) involves $\sqrt{b}$ and $-\sqrt{b}$, it would be *redundant* to include $n = -1, -2, -3, \ldots$ in Eq. (15–21).

15-5 LINEAR COMBINATION OF NORMAL-MODE SOLUTIONS

It is easy to see that linearity of superposition applies to solutions of the boundary-value problem specified by Eqs. (15–1), (15–2), and (15–3). Since the wave equation is linear, any linear combination of solutions is also a solution. Furthermore, if each term in the sum vanishes at $x = 0$ and $x = l$, the same is true for the combination. In particular, the normal-mode eigenfunctions of Eq. (15–25) are solutions to the boundary-value problem. Therefore if C_n and ϕ_n are assigned arbitrary values for each n, the linear combination

$$y(x, t) = \sum_n C_n \sin k_n x \sin (\omega_n t + \phi_n) \tag{15–26}$$

is assuredly a function which satisfies Eqs. (15–1), (15–2), and (15–3).

If all but one of the values of C_n are zero, this sum of course reduces to a single normal-mode solution. Borrowing language from quantum mechanics, we shall refer to such a situation as a *pure state*. If $C_n \neq 0$ for two or more values of n, the solution represents a *mixed state*.

Consider a string which is vibrating in a pure state of the second harmonic, described by a function of the form

$$y_2(x, t) = C_2 \sin 2k_1 x \sin (2\omega_1 t + \phi_2).$$

The associated waveform is a standing wave with nodes at the center and the two ends. Suppose that while the string is vibrating in this way, it is given a push at the center. It seems likely that this will not change the amplitude or phase of the second harmonic, but that it will excite modes of vibration which have an antinode at the center, e.g. the fundamental, given by a function of the form

$$y_1(x, t) = C_1 \sin k_1 x \sin (\omega_1 t + \phi_1).$$

(In a demonstration experiment it looks convincingly as if the second harmonic continues unaltered except for the addition of the slower rise and fall of the fundamental.) In the resulting mixed state there are no nodes in $0 < x < l$. The sum $y_1 + y_2$ cannot be expressed in product form as in Eq. (15–13), since that was shown to be a unique characteristic of the pure states.

So far Eq. (15–26) states only that any sum of normal-mode solutions is a function which satisfies the conditions of the boundary-value problem. The converse is a much more powerful statement referring to the *necessity* of this form of the general solution: *every* solution to Eqs. (15–1), (15–2), and (15–3) can be written as a linear combination of normal-mode solutions. The proof of this proposition is a relatively simple application of *Fourier's theorem*. It is beyond the scope of this text to prove Fourier's theorem itself. Indeed, an adequate discussion of Fourier's theorem for a general case can occupy the better part of a semester course. An assertion of the theorem will be made in a limited version, sufficiently general for our present purposes. If this assertion is accepted, no prior knowledge of the subject is required. Whatever additional techniques are needed to deal with applications of the theorem will be derived.

Statement of Fourier's theorem. Given a twice differentiable function $f(t)$ which is periodic with period $T = 2\pi/\omega$, the function can be represented by a uniformly and absolutely convergent series of the form

$$\begin{aligned} f(t) &= A_0 + A_1 \cos \omega t + A_2 \cos 2\omega t + A_3 \cos 3\omega t + \cdots \\ &\quad + B_1 \sin \omega t + B_2 \sin 2\omega t + B_3 \sin 3\omega t + \cdots \\ &= A_0 + \sum_{n=1}^{\infty} \{A_n \cos \omega_n t + B_n \sin \omega_n t\}, \end{aligned} \tag{15-27}$$

where $\omega_n \equiv n\omega$ and A_n and B_n are referred to as the *coefficients* in the Fourier-series representation of $f(t)$. Their values are uniquely determined by the given function $f(t)$. The series can be differentiated term by term, and the derivative series converges to the value of $f'(t)$.

The desired proof of the *necessity* of the normal-modes form of expansion now follows. We have already seen in Eqs. (15–5) and (15–7) that the general solution of the boundary-value problem has the form

$$y(x, t) = f(t - x/c) - f(t + x/c),$$

where $f(t)$ is periodic with period $T = 2l/c$. If we let $\omega = 2\pi/T = \pi c/l$, then $f(t)$ can be expanded as in Eq. (15–27). This enables us to write $y(x, t)$ as

$$\begin{aligned} y(x, t) &= f(t - x/c) - f(t + x/c) \\ &= \left\{A_0 + \sum_{n=1}^{\infty} [A_n \cos(\omega_n t - k_n x) + B_n \sin(\omega_n t - k_n x)]\right\} \\ &\quad - \left\{A_0 + \sum_{n=1}^{\infty} [A_n \cos(\omega_n t + k_n x) + B_n \sin(\omega_n t + k_n x)]\right\}. \end{aligned}$$

Using trigonometric identities to combine the terms in A_n and also in B_n, we have

$$y(x, t) = \sum_{n=1}^{\infty} \{2A_n \sin \omega_n t \sin k_n x - 2B_n \cos \omega_n t \sin k_n x\},$$

or, setting $a_n = 2A_n$ and $b_n = -2B_n$, we have

$$y(x, t) = \sum_{n=1}^{\infty} \{\sin k_n x [a_n \sin \omega_n t + b_n \cos \omega_n t]\}. \tag{15-28}$$

But $a_n \sin \omega_n t + b_n \cos \omega_n t$ is the same as $C_n \sin(\omega_n t + \phi_n)$, where

$$C_n = \sqrt{a_n^2 + b_n^2}$$

and

$$\tan \phi_n = b_n/a_n.$$

Thus Eq. (15–28) can also be written in the form

$$y(x, t) = \sum_{n=1}^{\infty} C_n \sin k_n x \sin(\omega_n t + \phi_n). \tag{15–29}$$

Comparison of Eq. (15–29) with Eq. (15–25) then shows that the general solution to the boundary-value problem has been written as a sum of terms each one of which has the form of a normal-mode eigenfunction.

Since it is usually more convenient to work with a_n and b_n in place of C_n and ϕ_n, Eq. (15–28) will be taken as the standard form of the *normal-modes expansion*. In the next section we will discuss the procedure which leads to an evaluation of the *expansion coefficients* a_n and b_n.

15-6 DETERMINATION OF THE COEFFICIENTS IN A NORMAL-MODES EXPANSION

As written in Eq. (15–28), the coefficients a_n and b_n are arbitrary constants appearing in the general solution to the boundary-value problem of Eqs. (15–1), (15–2), and (15–3). In keeping with the fact that we began our analysis of the string as a mechanical system subject to Newton's laws of motion, it should come as no surprise that these constants depend on the initial positions and velocities of the particles of the system.

Consider, for example, what can be learned if the shape of the string at $t = 0$ is specified by the given function $y(x, 0)$. From Eq. (15–28), we have

$$y(x, 0) = \sum_{n=1}^{\infty} b_n \sin k_n x. \tag{15–30}$$

Select an arbitrary integer m, multiply both sides of Eq. (15–30) by $\sin k_m x$, and integrate from $x = 0$ to $x = l$. Since the series on the right-hand side of Eq. (15–30) is uniformly convergent and remains so after multiplication by $\sin k_m x$, we are permitted to integrate term by term:

$$\int_0^l y(x, 0) \sin k_m x \, dx = \sum_{n=1}^{\infty} \left\{ b_n \int_0^l \sin k_m x \sin k_n x \, dx \right\} \cdot \tag{15–31}$$

Upon changing variables to $u = k_1 x = \pi x/l$ and remembering that

$k_m = mk_1$, we obtain

$$\int_0^l \sin k_m x \sin k_n x \, dx = \frac{l}{\pi} \int_0^\pi (\sin mu)(\sin nu) \, du$$

$$= \frac{l}{2\pi} \int_0^\pi \{\cos (m - n)u - \cos (m + n)u\} \, du$$

$$= \frac{l}{2\pi} \left[\frac{\sin (m - n)u}{m - n} - \frac{\sin (m + n)u}{m + n} \right] \Bigg|_0^\pi = 0.$$

The integration is only legitimate as performed if neither $m - n$ nor $m + n$ is zero. The latter is not possible since m and n are ≥ 1 in the present context. Thus we have shown that

$$\int_0^l (\sin k_m x)(\sin k_n x) \, dx = 0, \tag{15–32}$$

provided $m \neq n$. This result is well worth remembering, as it is frequently used in applications of trigonometric series. We refer to it by saying the functions $\sin k_m x$ and $\sin k_n x$ are *orthogonal* over the interval $(0, l)$ when m and n are different integers. If $m = n$ the integral $\int_0^\pi [\cos (m - n)u] \, du$ has a constant integrand and yields the value π. Thus

$$\int_0^l \sin^2 k_m x \, dx = \frac{l}{2} \cdot \tag{15–33}$$

Applying Eqs. (15–32) and (15–33) to Eq. (15–31) we see that all terms in the sum on the right-hand side vanish except that one for which $n = m$. Consequently, the series reduces to $b_m l/2$ and, solving for b_m, we have

$$b_m = \frac{2}{l} \int_0^l y(x, 0) \sin k_m x \, dx. \tag{15–34}$$

This equation is an explicit formula which enables the value of b_m to be determined from the given $y(x, 0)$ for an arbitrary integer m. Hence the entire set of series coefficients b_n is determined.

Similarly, suppose that we are given the initial velocity-distribution function $\dot{y}(x, 0)$. Equation (15–28) can be differentiated term by term to yield

$$\dot{y}(x, t) = \sum_{n=1}^{\infty} \{\sin k_n x [a_n \omega_n \cos \omega_n t - b_n \omega_n \sin \omega_n t]\}, \tag{15–35}$$

and therefore

$$\dot{y}(x, 0) = \sum_{n=1}^{\infty} a_n \omega_n \sin k_n x. \tag{15–36}$$

But this is an equation of the same type as Eq. (15–30) with $a_n\omega_n$ in place of b_n, and $\dot{y}(x, 0)$ in place of $y(x, 0)$. Therefore the same techniques as those used above will yield the result

$$a_n\omega_n = \frac{2}{l}\int_0^l \dot{y}(x, 0) \sin k_n x \, dx, \qquad n = 1, 2, 3 \ldots \qquad (15\text{–}37)$$

This is an equation which enables the set of expansion coefficients a_n to be determined from the given initial-velocity profile $\dot{y}(x, 0)$.

15–7 INDEPENDENCE OF THE ENERGY CONTRIBUTIONS FROM DIFFERENT MODES

Let

$$y(x, t) = \sum_{n=1}^{\infty} y_n(x, t)$$

be the normal-modes expansion of an arbitrary solution to the boundary-value problem of Eqs. (15–1), (15–2), and (15–3). The instantaneous kinetic energy of the string is

$$\text{KE} = \frac{\sigma}{2}\int_0^l \dot{y}^2 \, dx.$$

However,

$$\dot{y} = \sum_{n=1}^{\infty} \dot{y}_n, \qquad \text{and} \qquad \dot{y}^2 = \left(\sum_{m=1}^{\infty} \dot{y}_m\right)\left(\sum_{n=1}^{\infty} \dot{y}_n\right) = \sum_{m=1}^{\infty}\sum_{n=1}^{\infty} \dot{y}_m\dot{y}_n.$$

Therefore the kinetic energy can be written as the double sum

$$\text{KE} = \frac{\sigma}{2}\sum_{m=1}^{\infty}\sum_{n=1}^{\infty}\int_0^l \dot{y}_m\dot{y}_n \, dx. \qquad (15\text{–}38)$$

Now from Eq. (15–29) we have

$$y_n = C_n \sin k_n x \sin(\omega_n t + \phi_n)$$

and hence,

$$\dot{y}_n = C_n\omega_n \sin k_n x \cos(\omega_n t + \phi_n). \qquad (15\text{–}39)$$

As far as dependence on x is concerned, the function $\dot{y}_n$ is proportional to $\sin k_n x$. The orthogonality of the sine functions for different values of n, Eq. (15–32), implies that all the terms in Eq. (15–38) vanish except those for which $m = n$. The sum then reduces to

$$\text{KE} = \sum_{n=1}^{\infty}\left\{\frac{\sigma}{2}\int_0^l \dot{y}_n^2 \, dx\right\}\cdot \qquad (15\text{–}40)$$

The significant thing which has occurred is the dropping out of *cross terms* depending jointly on two modes. The quantity

$$(\text{KE})_n \equiv \frac{\sigma}{2} \int_0^l \dot{y}_n^2 \, dx$$

is the kinetic energy associated with the nth mode. It is the kinetic energy that would be assigned to the string if its state of motion were described by the function y_n alone. Equation (15–40) shows that the kinetic energy in a mixed state is the simple sum of the kinetic energies associated with the contributing modes:

$$\text{KE} = \sum_{n=1}^{\infty} (\text{KE})_n.$$

No interference effects between different modes need to be considered in calculating the kinetic energy of the string as a whole. (It should be emphasized that this proposition does not refer to energy *density*, but to the integral of the latter over the length of the string.)

The kinetic energy associated with the nth mode can be calculated explicitly from Eq. (15–39):

$$\begin{aligned}(\text{KE})_n &= \frac{\sigma}{2} \int_0^l \dot{y}_n^2 \, dx = \frac{\sigma}{2} (C_n\omega_n)^2 \cos^2 (\omega_n t + \phi_n) \int_0^l \sin^2 k_n x \, dx \\ &= \frac{\sigma l}{4} (C_n\omega_n)^2 \cos^2 (\omega_n t + \phi_n). \qquad (15\text{–}41)\end{aligned}$$

In similar manner it can be shown that the instantaneous potential energy of the string is the sum of independent contributions from the participating modes:

$$\text{PE} = \frac{T}{2} \int_0^l \left(\frac{\partial y}{\partial x}\right)^2 dx = \sum_{n=1}^{\infty} (\text{PE})_n,$$

where

$$(\text{PE})_n = \frac{T}{2} \int_0^l \left(\frac{\partial y_n}{\partial x}\right)^2 dx = \frac{\sigma l}{4} (C_n\omega_n)^2 \sin^2 (\omega_n t + \phi_n). \qquad (15\text{–}42)$$

It can be seen from Eqs. (15–41) and (15–42) that the sum of the instantaneous kinetic- and potential-energy contributions from the nth mode is a constant:

$$\mathcal{E}_n = (\text{KE})_n + (\text{PE})_n = (\sigma l/4)(C_n\omega_n)^2.$$

It is convenient to express this total energy contribution from the nth mode in terms of the expansion coefficients a_n and b_n [cf. Eq. (15–28)]:

$$\mathcal{E}_n = \frac{\sigma l \omega_n^2}{4} (a_n^2 + b_n^2). \qquad (15\text{–}43)$$

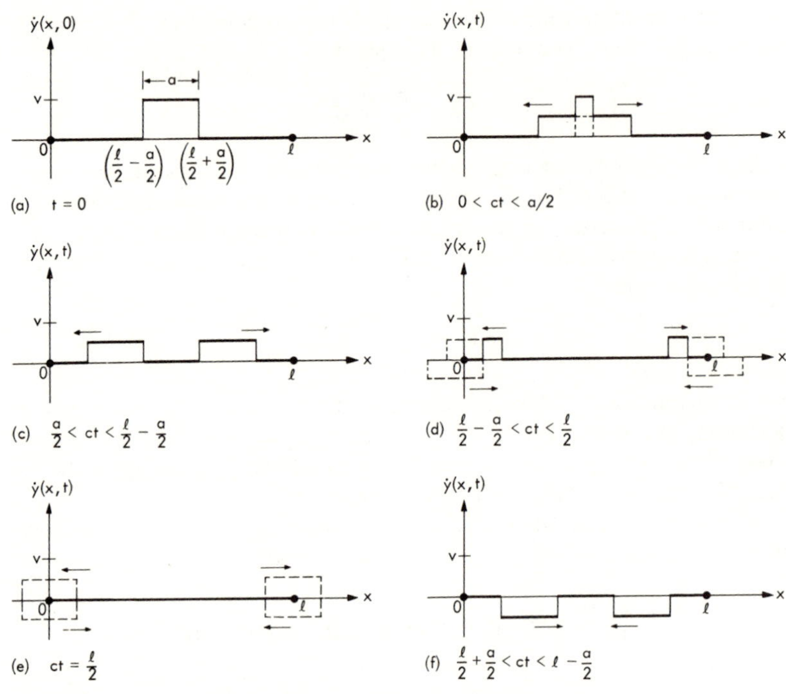

Fig. 15–1. Pulse with rectangular velocity profile reflected at two fixed ends of a string segment.

15–8 NORMAL-MODES EXPANSION OF A RECTANGULAR PULSE

As a specific application of the formulas obtained in the preceding sections, consider a string which satisfies the initial conditions

$$y(x, 0) = 0 \tag{15–44}$$

and

$$\dot{y}(x, 0) = \begin{cases} 0, & |x - l/2| \geq a/2 \\ v, & |x - l/2| < a/2. \end{cases} \tag{15–45}$$

This may be thought of as a string which has been at rest and which at $t = 0$ is struck from below by a flat mallet. The mallet, which is a units long, hits the string symmetrically about the midpoint and imparts an initial velocity v to the particles immediately above it. Since the initial-velocity profile has the shape indicated in Fig. 15–1(a), this will be referred to as a *rectangular pulse*.

Before we proceed with the normal-modes analysis, let us observe that this is a problem which we can solve easily by the graphical techniques of

Fig. 3–1, taking explicit account of the reflections at both ends. The velocity profiles* at various times $t > 0$ are plotted in Fig. 15–1. This relatively simple method of plotting waveforms is to be contrasted with the procedure of the normal-modes analysis. The latter represents the situation of Fig. 15–1(c), for example, by means of an infinite trigonometric series and requires the summation of a large number of terms to obtain an accurate calculation of the velocity profile. It would hardly be suspected from an inspection of the series that the sum has the shape of two rectangles.

It can be seen therefore that the normal-modes analysis does not have the advantage of computational simplicity in a case such as the present one. One may inquire what object there is in studying the normal-modes representation. A preliminary idea of the ways in which the normal-modes concept turns out to be fruitful is conveyed by the following three observations:

i) The normal-modes representation dispenses with *piecewise definitions.* That is, one single equation, having the form of Eq. (15–35), represents $\dot{y}(x, t)$ for *all* x and t. This equation replaces what must otherwise be expressed by a series of statements of the form:

When $0 \leq ct \leq a/2$,

$$\dot{y}(x,t) = \begin{cases} 0, & 0 \leq x \leq \frac{l}{2} - \frac{a}{2} - ct; \\ \frac{v}{2}, & \frac{l}{2} - \frac{a}{2} - ct < x \leq \frac{l}{2} - \frac{a}{2} + ct; \\ v, & \frac{l}{2} - \frac{a}{2} + ct < x < \frac{l}{2} + \frac{a}{2} - ct; \\ \frac{v}{2}, & \frac{l}{2} + \frac{a}{2} - ct \leq x < \frac{l}{2} + \frac{a}{2} + ct; \\ 0, & \frac{l}{2} + \frac{a}{2} + ct \leq x \leq l. \end{cases} \tag{15–46a}$$

When $a/2 < ct \leq l/2 - a/2$,

$$\dot{y}(x,t) = \begin{cases} 0, & 0 \leq x \leq \frac{l}{2} - \frac{a}{2} - ct; \\ \frac{v}{2}, & \frac{l}{2} - \frac{a}{2} - ct < x \leq \frac{l}{2} + \frac{a}{2} - ct; \\ 0, & \frac{l}{2} + \frac{a}{2} - ct < x < \frac{l}{2} - \frac{a}{2} + ct; \\ & \text{etc.} \end{cases} \tag{15–46b}$$

* As a matter of convenience in the illustrative examples of this chapter, we will confine our attention to the determination of the velocity-distribution function $\dot{y}(x, t)$. It should be clear that the wave profile, $y(x, t)$, can be obtained by integration of $\dot{y}(x, t)$.

ii) The harmonic content of a particular motion of the string is significant in numerous practical applications. For example, the sound wave generated in air by the vibration of a string is a periodic disturbance whose harmonic content is related to the harmonic content of the motion of the string. Since the ear is an instrument that informs us of the harmonic content of a sound wave, we are interested in knowing how this is affected by the use of different methods of initiating the disturbance on the string. As a second example, a property of a real string not described by the idealized wave equation is dissipation of energy. If the attenuation of the higher frequency modes is more drastic than that of the lower frequency modes, after a certain length of time the disturbance is not given exactly by the rectangles of Fig. 15–1, but is more adequately described by the sum of the first few terms in the harmonic series. It is obvious also that the example of a wave on a string is a prototype for a number of other wave situations, e.g., waves on a transmission line. Detecting instruments often "filter out" certain ranges of frequencies from an incident wave. In such contexts it is important to be able to characterize the wave in terms of its harmonic content.

iii) The device of a trigonometric-series representation is necessary to discuss the propagation of a nonsinusoidal disturbance in a *dispersive* medium. Consider a pulse in deep water, for example. Since the velocity is dependent on wavelength, there is no wave equation with a single velocity c which describes the propagation of an arbitrary disturbance. The required procedure is an analysis of the initial waveform for its harmonic content. Each component wave is then propagated at its own characteristic velocity. When these waves are added together at successive later times, the progressive modifications in waveform are obtained.

We turn then to the case of the string struck by a mallet to learn how the frequency distribution is affected by the parameters at our disposal. Since we are given $y(x, 0) = 0$, Eq. (15–34) shows that the coefficients b_m are all zero. From Eqs. (15–45) and (15–37) we determine the coefficients a_n by

$$a_n\omega_n = \frac{2v}{l}\int_{(l/2)-(a/2)}^{(l/2)+(a/2)} \sin k_n x\, dx = \frac{4v}{k_n l}\sin\left(\frac{k_n l}{2}\right)\sin\left(\frac{k_n a}{2}\right).$$

But, since $k_n = nk_1 = n\pi/l$, this can be written

$$a_n\omega_n = \frac{4v}{n\pi}\sin\left(\frac{n\pi}{2}\right)\sin\left(\frac{n\pi a}{2l}\right). \tag{15–47}$$

Inasmuch as $\sin(n\pi/2)$ vanishes for n even, Eq. (15–47) implies that a_n is zero for even n. This means that no even harmonics appear in the trigonometric series, a result which is significant in connection with the timbre of the corresponding musical note. Finally, from Eq. (15–35), the velocity profile is given for all $t \geq 0$ by

$$\dot{y}(x, t) = \sum_{n \text{ odd}} a_n\omega_n \sin k_n x \cos \omega_n t, \tag{15–48}$$

where $a_n\omega_n$ is given by Eq. (15–47). Equation (15–48) is the single equation that replaces the piecewise definition of Eq. (15–46).

The pulse considered in this section was chosen to provide algebraically as simple an illustration as possible of the calculation of a normal-modes expansion. Unfortunately, it happens to violate several mathematical conditions. First, as can be seen from Fig. 15–1, $\dot{y}(x, t)$ is a discontinuous function of t at various times corresponding to the arrival of the rectangle edges or their reflections. Thus $\partial^2 y/\partial t^2$ fails to exist at certain times and the wave equation does not apply. We can avoid this matter of principle by smoothing off the edges of the rectangle function in Eq. (15–45). A curve joining the values $\dot{y} = 0$ and $\dot{y} = v$ can be added in the vicinity of $x = (l/2) \pm (a/2)$. This can be done in such a way that the required higher derivatives exist and yet the width of the transition interval is as small as we want to make it. The calculation of the normal-modes expansion of such a function is not as simple as it was for the fictitious rectangular pulse, although the complication is of a purely algebraic nature. The result of the extra labor does not alter in any significant way the conclusions which have been drawn from this as an illustrative example.

Second, Fourier's theorem as stated above calls for the function $f(t)$ to be twice differentiable. This requirement is as strong as actually needed from the point of view of the physics, since $y(x, t) = f(t - x/c) - f(t + x/c)$ must satisfy the wave equation. However, since we have applied the results to a rectangle function, it is pertinent to note that Fourier's theorem can be extended to include this type of function.*

15-9 ENERGY SPECTRUM OF THE RECTANGULAR PULSE

From Eq. (15–43), the energy contribution from the nth mode reduces when $b_n = 0$ to

$$\mathcal{E}_n = \frac{\sigma l}{4} (a_n\omega_n)^2. \tag{15–49}$$

For the rectangular velocity profile studied in the last section, the value of a_n is zero when n is even. When n is odd, $\sin (n\pi/2) = \pm 1$. Hence, substituting from Eq. (15–47) in Eq. (15–49), we have

$$\mathcal{E}_n = \frac{4\sigma l v^2}{n^2\pi^2} \sin^2 \frac{n\pi a}{2l} \qquad \text{for } n \text{ odd}. \tag{15–50}$$

---

* R. Courant, *Differential and Integral Calculus*. London: Blackie & Sons, Ltd., 1937, Vol. I, Chapter IX.

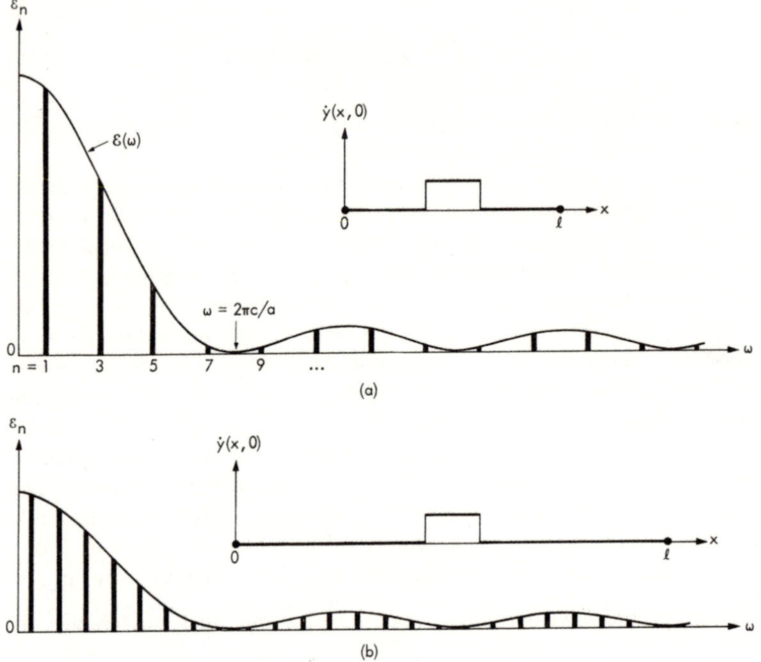

Fig. 15-2. (a) Energy distribution among the normal modes for the rectangular pulse defined by Eqs. (15-44) and (15-45). (b) All conditions the same as in (a) except that the length of the string has been doubled.

To study the energy distribution among the modes as a function of mode frequency ω_n, consider the replacement

$$n = \omega_n/\omega_1 = \omega_n l/\pi c. \tag{15-51}$$

Then

$$\mathcal{E}_n = \frac{4\sigma v^2 c^2}{\omega_n^2 l} \sin^2 \frac{\omega_n a}{2c}. \tag{15-52}$$

The bar graph of $\mathcal{E}_n$ vs. ω_n shown in Fig. 15-2(a) is obtained by inscribing equally spaced vertical lines under the *envelope curve*

$$\mathcal{E}(\omega) = \frac{4\sigma v^2 c^2}{\omega^2 l} \sin^2 \frac{\omega a}{2c}. \tag{15-53}$$

This curve has a shape which is familiar from the study of the Fraunhofer radiation pattern from a coherent line source (cf. Section 12-4). The first zero on the envelope curve occurs at $\omega a/2c = \pi$. Since the subsidiary maxima are much smaller than the principal maximum, the modes which make the most important contributions are those for which

$$\omega_n < 2\pi c/a. \tag{15–54}$$

This frequency criterion depends only on a, the width of the initial rectangular pulse, and is independent of l, the length of the string on which this pulse is traveling. Thus if it can be said of a pulse that the important frequencies are those below 1000 cps, the statement applies to a pulse of the same size on longer or shorter strings, provided that the wave propagation velocity c remains the same.

Inspection of Eq. (15–53) shows in more general form that l has no effect on the horizontal scale of the envelope curve. On the other hand, the spacing between bars is, from Eq. (15–51),

$$\omega_{n+2} - \omega_n = 2\pi c/l,$$

which decreases as l increases. Keeping a pulse of fixed shape and imbedding it in successively longer strings therefore crowds more bars under the principal maximum of the envelope curve (Fig. 15–2b), but does not change the relative importance of different general frequency ranges. The reason for the change in vertical scale in going from Fig. 15–2(a) to Fig. 15–2(b) is the factor of $1/l$ in Eq. (15–53). It is obvious that some such decrease must occur as l increases, to keep the total energy finite.

The way in which the horizontal scale of the envelope curve $\mathcal{E}(\omega)$ is controlled by a, the spatial width of the pulse, is typical of the relation which exists generally between the spatial extent of a pulse and the spread of frequencies which make important contributions to the normal-modes representation of the pulse. From Eq. (15–54), the range of "important" frequencies, or the *spectral width of the pulse*, can be considered to be on the order of $\Delta\omega \sim 2\pi c/a$. If we designate the spatial width of the pulse by $\Delta x = a$, these two widths bear an inverse relation to each other:

$$(\Delta x)(\Delta\omega) \sim 2\pi c. \tag{15–55}$$

Figure 15–3 indicates schematically the way in which a short sharp pulse has a broad energy spectrum, whereas a broad pulse is composed mainly of the very low frequencies.

The two pulses drawn in Fig. 15–3 are adjusted to represent the same total energy. This is accomplished by noting that the *initial* energy is entirely kinetic and is distributed over a length a of the string with kinetic energy

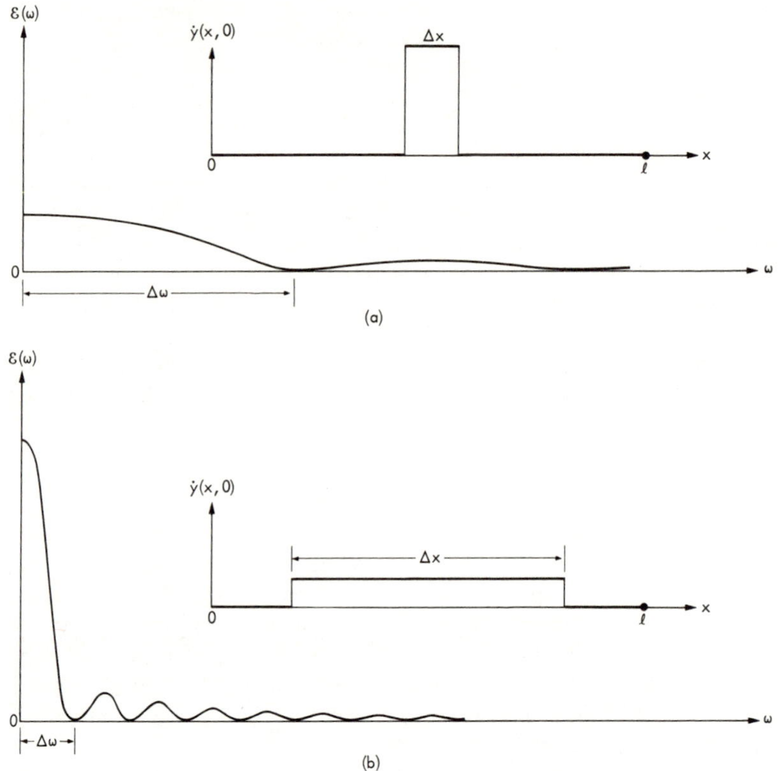

Fig. 15-3. Spectrum envelope curves $\mathcal{E}(\omega)$ for short and long rectangular pulses of the same total energy.

per unit length given by $\frac{1}{2}\sigma v^2$. Therefore the total energy is

$$\mathcal{E} = \tfrac{1}{2}\sigma a v^2. \tag{15-56}$$

Rectangular pulses of different widths a therefore carry the same total energy on a given string if v is taken as inversely proportional to $\sqrt{a}$.

An interesting mathematical identity can be obtained by applying the law of conservation of energy. On one hand, the energy is given by the sum of the contributions from all the participating modes, which from Eq. (15-50) can be written

$$\mathcal{E} = \sum_{n=1}^{\infty} \mathcal{E}_n = \sum_{n\,\text{odd}} \frac{4\sigma l v^2}{n^2\pi^2} \sin^2 \frac{n\pi a}{2l}\cdot \tag{15-57}$$

On the other hand, $\mathcal{E}$ is given by Eq. (15-56). Equating these two expressions and simplifying, we see that the following identity is implied:

$$\sum_{n\,\text{odd}} \frac{1}{n^2} \sin^2 \frac{n\pi a}{2l} = \frac{\pi^2 a}{8l}. \tag{15–58}$$

Although we have used physical language in deriving this formula, the steps can be justified on a purely mathematical basis, and the equation is a general identity. The special case $a = l/2$ is often quoted in textbooks on Fourier series:

$$\sum_{n\,\text{odd}} \frac{1}{n^2} = \frac{\pi^2}{8}.$$

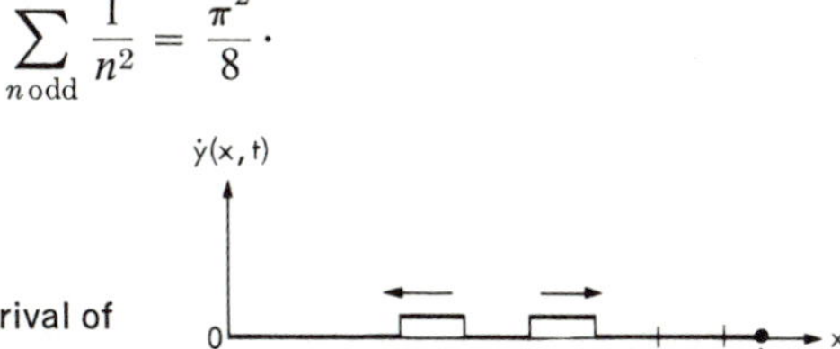

Fig. 15–4. The velocity profile prior to the arrival of any disturbance in (x_1, x_2).

15–10 A TOO LITERAL INTERPRETATION OF THE NORMAL-MODES EXPANSION

The normal-modes expansion of Eq. (15–48) is an exact analytical representation of the situation described graphically in Fig. 15–1. Suppose that we consider an interval (x_1, x_2) in Fig. 15–4 throughout a time interval before the first pulse arrives, i.e. $0 < ct < ct_1 = x_1 - [(l/2) + (a/2)]$. During this time interval, absolutely nothing happens in (x_1, x_2). Yet we are allowed to picture the disturbance in this interval as the superposition of the sinusoidal standing wave functions of the normal-modes expansion. This is indicated schematically in Fig. 15–5. Since the sum of the series is zero for all points in the interval, it is implied that the individual sinusoidal functions contrive to cancel each other completely. As time progresses, each standing wave moves up and down, but as long as $t < t_1$, they manage to do this in such a way that the exact cancellation continues. In a crude sense, just as many are "up" as "down" at any given time. If some switch from "up" to "down," an equal number switch in the opposite direction. That the terms in the infinite trigonometric series can manage to be this clever may seem astonishing, but it is a fact which was established in the proof of the normal-modes expansion as an identity. For $t > t_1$ the sinusoidal functions in (x_1, x_2) cease their

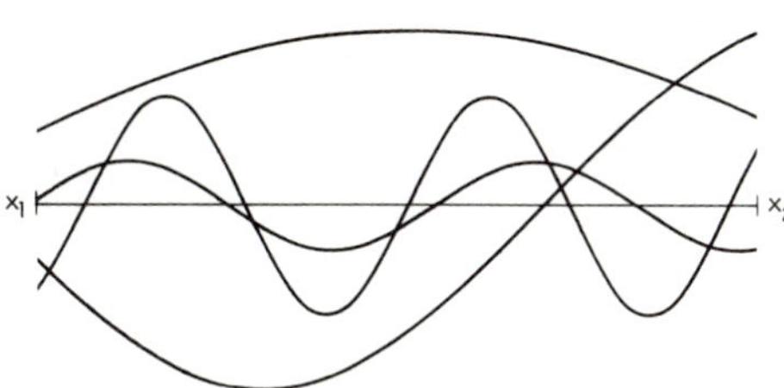

Fig. 15–5. Individual plot of a few normal-mode functions at a given time.

connivance to produce exact cancellation. Starting at the left end of the interval, we find that a number of the values are in the same direction and that they superimpose to produce the constant value v within the rectangle.

There is a strong temptation to feel that since the component waves are "present" in the interval (x_1, x_2) even when $t < t_1$, it should be possible to detect their presence and thus learn about the approaching pulse before it actually arrives. To emphasize the paradoxical nature of this interpretation, let us switch the context to an analogous problem in which it will be clear that something is wrong. Suppose that a supernova located at $x = l/2$ explodes at $t = 0$ and gives rise to a light pulse of rectangular shape. The walls of our laboratory are located at x_1 and x_2, many light years away from the supernova. To give this problem the periodicity required for the analysis in terms of Fourier series, we can imagine that there are vast mirrors at the edges of the universe, $x = 0$ and $x = l$. If the supernova explodes during our lifetime, it is clear that these mirrors should in no way influence what we will perceive. They do, however, allow us to conceive of the problem within the mathematical framework of periodic functions.* A very definite spectral distribution of energies is associated with the explosion pulse. Each sinusoidal component wave extends from $x = 0$ to $x = l$. If we were to point a spectroscope at the supernova at some time $t < t_1$ and obtain the spectrum of these waves, we should be able to learn something about the explosion before the pulse itself has arrived!

The fallacy lies in the assumption that an analyzing instrument at any given time simply registers, according to their respective frequencies, the component waves of which the total disturbance is composed. Clearly there must be an actual energy flux into the instrument during the time that it operates if there is to be any response. As we have seen before, due account must be taken of *interference* effects when energy flux is to be determined. The energy flux can only be calculated from the *resultant* of the components (which in this case is zero) and not from the sum of the fluxes associated with each of the components independently. This statement seems to conflict with Section 15–7 in which it was asserted that the normal modes can be considered independently in the calculation of total energy. It was stated there, and it must be emphasized again, that this refers to the energy of motion of the *whole* string. In Eq. (15–38), for example, the range of integration is from $x = 0$ to $x = l$. An attempt to prove this theorem for a limited length of the string, say from x_1 to x_2, would fail because the orthogonality of the sine functions (Eq. 15–32) applies in general only over an interval of length l.

In summary, it is permissible to think of the component waves as being "present" in the interval (x_1, x_2) only in the sense that their superposition gives the exact value of the disturbance in this interval at all times. Any

---

* The device of the mirrors is unnecessary if the techniques of Fourier *analysis* are employed. See Section 15–18.

further attempt to attribute "physical reality" to the component waves is thwarted by the occurrence of interference effects. We are released from this restriction only if we consider the entire string.

15-11 NORMAL-MODES EXPANSION OF A SINUSOIDAL WAVETRAIN OF LIMITED EXTENT

Suppose that the initial conditions for a string with two fixed ends are $y(x, 0) = 0$, and for some chosen integer M,

$$\dot{y}(x, 0) = \begin{cases} v \sin(M\pi x/a), & 0 \le x \le a \\ 0, & a < x \le l. \end{cases} \tag{15–59}$$

As indicated in Fig. 15–6, the initial velocity profile consists of M half-cycles of a sine curve of wavelength $\Lambda = 2a/M$. The associated wave number and angular frequency will be designated respectively as

$$K = \frac{2\pi}{\Lambda} = \frac{M\pi}{a} \quad \text{and} \quad \Omega = Kc = \frac{M\pi c}{a}.$$

If the wave were a pure sine curve, not cut off at $x = a$, Ω would be expected to be the only frequency contained. It will be interesting to observe how this situation is modified in the case of the limited wavetrain and how it is approximately recaptured in the case of a "nearly sinusoidal wave," when M is extremely large.

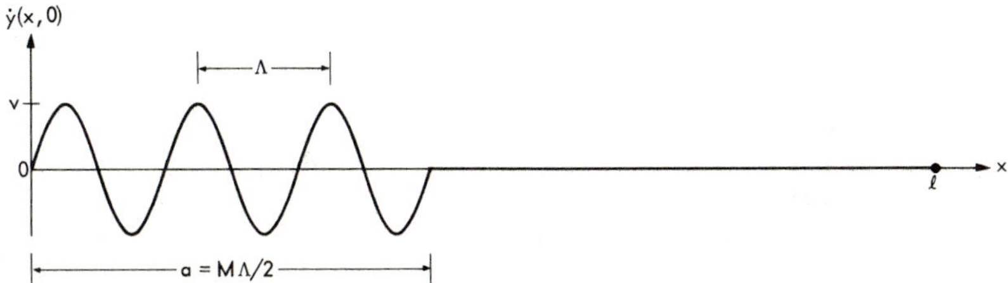

Fig. 15–6. Initial velocity profile consisting of a portion of a sinusoidal wave.

From Eq. (15–37) the expansion coefficients a_n are given by

$$\begin{aligned} a_n\omega_n &= \frac{2}{l}\int_0^l \dot{y}(x, 0)\sin k_n x\, dx = \frac{2v}{l}\int_0^a \sin Kx \sin k_n x\, dx \\ &= \frac{v}{l}\int_0^a \{\cos(K - k_n)x - \cos(K + k_n)x\}\, dx \\ &= \frac{v}{l}\left\{\frac{\sin(K - k_n)a}{K - k_n} - \frac{\sin(K + k_n)a}{K + k_n}\right\}. \end{aligned} \tag{15–60}$$

Inspection of this result reveals that the most important modes are those which make the denominator $K - k_n$ small, i.e. those for which $k_n \sim K$. In other words, the standing waves which contribute most heavily to the waveform of Fig. 15–6 are those whose wavelengths are most nearly equal to Λ. Since the denominator $K + k_n$ does not come close to vanishing, the way $K - k_n$ does, the contribution from the second term on the right-hand side of Eq. (15–60) is never as important as the "resonant" contributions from the first term. To simplify the discussion we will neglect the second term and write

$$a_n\omega_n \doteq \frac{v}{l}\frac{\sin(K - k_n)a}{K - k_n}. \tag{15–61}$$

Since the initial condition $y(x, 0) = 0$ implies $b_n = 0$, the energy contribution from the nth mode is given by Eq. (15–43) as

$$\begin{aligned} \mathcal{E}_n &= \frac{\sigma l}{4}(a_n\omega_n)^2 = \frac{\sigma v^2}{4l}\frac{\sin^2(k_n - K)a}{(k_n - K)^2} \\ &= \frac{\sigma c^2 v^2}{4l}\frac{\sin^2[(\omega_n - \Omega)a/c]}{(\omega_n - \Omega)^2}. \end{aligned} \tag{15–62}$$

Exactly as in Section 15–9, the bar graph of $\mathcal{E}_n$ versus ω_n can be discussed in terms of an envelope curve

$$\mathcal{E}(\omega) = \frac{\sigma c^2 v^2}{4l}\frac{\sin^2[(\omega - \Omega)a/c]}{(\omega - \Omega)^2}. \tag{15–63}$$

The horizontal scale of $\mathcal{E}(\omega)$ is governed solely by the parameter a, the length of the sinusoidal wavetrain, and is independent of l, the length of the string. The spacing of the bars under the envelope is $\omega_{n+1} - \omega_n = \pi c/l$ and, as before, is inversely proportional to l.

The envelope curve is of the familiar $(\sin^2\beta)/\beta^2$ shape where

$$\beta = (\omega - \Omega)a/c.$$

This means that the center, $\beta = 0$, corresponds to $\omega = \Omega$, which is once again the observation that the most important terms in the normal-modes expansion of a limited wavetrain are those whose frequencies are close to the frequency of the original sinusoid.

The graph of $\mathcal{E}(\omega)$ is sketched in Fig. 15–7. The frequencies which lie within the central maximum are those for which $|\beta| < \pi$. Hence the important frequencies are those in the range $\Omega - \pi c/a < \omega < \Omega + \pi c/a$, and the width of the range is $\Delta\omega = 2\pi c/a$. This example illustrates the same inverse relation between spectral width $\Delta\omega$ and spatial width $\Delta x = a$ as was seen before in Eq. (15–55). Narrow confinement in space implies a wide spectral distribution. Conversely, a narrow spectral distribution implies a broad spatial distribution.

It is revealing to express $\Delta\omega$ as a fraction of the center frequency Ω:

$$\frac{\Delta\omega}{\Omega} = \left(\frac{2\pi c}{a}\right)\left(\frac{a}{M\pi c}\right) = \frac{2}{M}.$$

The *relative* frequency spread thus depends solely on the number of cycles being represented. When M is extremely large, the wavetrain is nearly sinusoidal and the curve in Fig. 15–7 becomes a sharp and narrow spike.

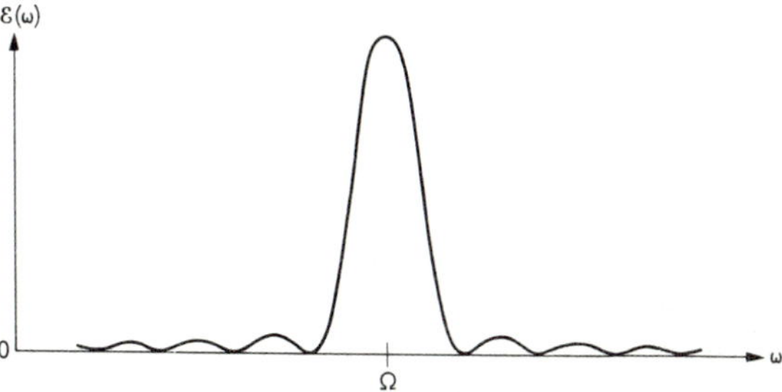

Fig. 15–7. Spectral envelope curve $\mathcal{E}(\omega)$ for a limited sinusoidal wavetrain.

A connection between the present calculations and the collision broadening of spectral lines (Section 11–15) should be evident. It is true that we have been talking about disturbances which are periodic due to reflection between rigid boundaries. However, it has been noted that the envelope curve is characteristic of the intrinsic shape of the pulse and that l, the distance between boundaries, determines the spacing between the inscribed vertical lines of the bar graph. When l is extremely large the mode frequencies ω_n are so closely spaced that they resemble a continuum. Under these circumstances the envelope curve takes on the significance of energy distribution within the continuum. (This transition is discussed more fully in Section 15–17.) The function $\mathcal{E}(\omega)$ in Eq. (15–63) is said to describe a *spectral line.* The pressure in a mercury arc governs M, since the latter is proportional to the mean time between collisions. An increase in the pressure implies a decrease in M, and hence, through $\Delta\omega/\Omega = 2/M$, produces a more impure spectral line.

15–12 FORCED MOTION OF A STRING

The material of the next few sections will extend the discussion of transverse waves on a string in a way which is similar to the transition from the discussion of the free vibration of a harmonic oscillator to the discussion of its response to a driving force. Since the analogy will be helpful, we begin with a brief review of the language used in treating this familiar problem in mechanics. The *free* oscillations of a one-dimensional simple harmonic oscillator are described by solutions to the *homogeneous* differential equation $m\ddot{x} + kx = 0$. Arbitrary constants appearing in the general form of solution receive definite values when it is required that given initial values of x and $\dot{x}$ be matched. The *forced* motion of the oscillator is described by a solution to the *inhomogeneous* equation $m\ddot{x} + kx = F(t)$ where $F(t)$ is a specified driving force. The *general* solution to the inhomogeneous equation is the sum of any *particular* solution thereof and a solution to the homogeneous equation containing two arbitrary constants. If $F(t)$ is sinusoidal with angular frequency $\omega \neq \sqrt{k/m}$, it is possible to select from among the particular solutions one which is sinusoidal with the same angular frequency ω as the driving force. This is referred to as the *steady-state solution.* If a general solution is formed by using the steady-state solution as the particular solution and adding it to a solution of the homogeneous equation, the latter is referred to as the *transient solution.**

If the driving frequency ω is close to the natural frequency $\omega_0 = \sqrt{k/m}$, the amplitude of the steady-state solution becomes large. At *resonance*, where $\omega = \omega_0$, a sinusoidal function of frequency ω no lcnger serves as a particular solution. However, a particular solution can be found by suitable techniques. The result does not represent a "steady state," but corresponds to a continual increase in energy on the part of the oscillator.

The preceding review statements are illustrated by the following example:

EXAMPLE 1. Solve $m\ddot{x} + kx = A \cos \omega t$ subject to $x(0) = \dot{x}(0) = 0$.

We try a particular solution of the form

$$x = \mathcal{R}e\,\{\mathbf{B}e^{i\omega t}\}$$

and find that this works provided we take

$$\mathbf{B} = \frac{A}{m(\omega_0^2 - \omega^2)}$$

and provided $\omega \neq \omega_0$. The general solution is

$$x = \mathcal{R}e\,\{\mathbf{a}e^{i\omega_0 t}\} + \frac{A}{m(\omega_0^2 - \omega^2)} \cos \omega t.$$

The condition $x(0) = 0$ implies

$$\mathcal{R}e\,\{\mathbf{a}\} = \frac{-A}{m(\omega_0^2 - \omega^2)},$$

* The terminology is not entirely appropriate to the ideal equation without a dissipative force term, but is borrowed from a study of the equation when such a dissipative term is present. In that case, solutions to the homogeneous equation approach zero as $t \to \infty$, and all particular solutions approach the steady-state solution.

and the condition $\dot{x}(0) = 0$ implies

$$\omega_0 \mathcal{R}e\,\{i\mathbf{a}\} = 0.$$

Combining these results we obtain the complete solution

$$x = \frac{A}{m(\omega_0^2 - \omega^2)}(\cos \omega t - \cos \omega_0 t).$$

The first term is the *steady-state* solution and the second term is the *transient* solution. The solution when $\omega = \omega_0$ can be obtained by applying L'Hôpital's rule to the previous solution as $\omega \to \omega_0$:

$$\lim_{\omega \to \omega_0}\left[\frac{A(\cos \omega t - \cos \omega_0 t)}{m(\omega_0^2 - \omega^2)}\right] = \lim_{\omega \to \omega_0}\left[\frac{-At \sin \omega t}{-2m\omega}\right] = \frac{At \sin \omega_0 t}{2m\omega_0}.$$

It can be shown directly that

$$x = \frac{At \sin \omega_0 t}{2m\omega_0}$$

is a particular solution to $m\ddot{x} + kx = A \cos \omega_0 t$ which satisfies the initial conditions $x(0) = \dot{x}(0) = 0$. Because of the factor of t in this solution, x executes oscillations of continually increasing amplitude.

So far in this chapter we have obtained what should be called *transient solutions* to the wave equation subject to rigid boundary conditions at the endpoints. These solutions refer to the free motion of the string subject to no applied forces. It was found in Section 15–6 that the parameters in the general solution (the coefficients a_n and b_n) become uniquely determined by specification of the initial shape and velocity profiles, $y(x, 0)$ and $\dot{y}(x, 0)$.

The forced motion of a string can occur in several ways. If the string is driven at the ends, then the expression of Newton's second law within the interval $0 < x < l$ is not altered, and the function $y(x, t)$ still satisfies the *homogeneous* wave equation, Eq. (15–1). But the homogeneous boundary conditions $y(0, t) = 0$ and $y(l, t) = 0$ are now changed. Suppose, for example, that the end $x = 0$ remains fixed, but that a variable transverse force $F(t)$ is applied at $x = l$. In applying Eq. (4–25) we note that $F_y = -F(t)$, since F_y stands for the force which the string exerts on the support. The boundary condition at $x = l$ can then be stated

$$T\frac{\partial y}{\partial x}(l, t) = F(t). \tag{15–64}$$

Another way to specify the condition for a string driven at an end is to require a given displacement $\psi(t)$:

$$y(l, t) = \psi(t). \tag{15–65}$$

Equations (15–64) and (15–65) are both examples of *inhomogeneous boundary conditions*.

A completely different method of driving the string is to apply a *distributed* force along the entire length of the string. If the transverse external force that acts at time t on the section of string between x and $x + \Delta x$ is $F(x, t)\,\Delta x$, then it is easily seen that Eq. (1–2) becomes

$$\sigma\,\Delta x\,\frac{\partial^2 y}{\partial t^2} = \Delta(T\sin\theta) + F(x, t)\,\Delta x.$$

By procedures analogous to those of Section 1–2 we find that Newton's second law becomes expressed in the form of the *inhomogeneous* wave equation

$$\sigma\,\frac{\partial^2 y}{\partial t^2} - T\,\frac{\partial^2 y}{\partial x^2} = F(x, t). \tag{15–66}$$

The boundary conditions to supplement this equation can be homogeneous or inhomogeneous depending on the conditions imposed at the ends.

In summary, the forced motion of a string leads to a replacement of at least one of the homogeneous equations (15–1, 15–2, 15–3) by an inhomogeneous condition. It is a boundary condition that becomes inhomogeneous for a string driven at one of the ends, and it is the wave equation itself that becomes inhomogeneous for a distributed force. The next two sections illustrate how to solve a problem of the former type; Section 15–15 considers one of the latter type.

15–13 EIGENFREQUENCIES AS RESONANCE FREQUENCIES OF A STRING DRIVEN SINUSOIDALLY AT ONE END

Suppose that the end of the string at $x = 0$ is held fixed while the other end is given a specified sinusoidal motion of amplitude A and frequency ω. The boundary conditions are

$$\begin{aligned} y(0, t) &= 0, \\ y(l, t) &= \psi(t) = A\sin\omega t. \end{aligned} \tag{15–67}$$

It has been shown previously that the general solution to the homogeneous wave equation subject to the first of these conditions is given by Eq. (15–5):

$$y(x, t) = f(t - x/c) - f(t + x/c).$$

To find a *particular* solution that satisfies the second boundary condition as well, we will assume $f(t)$ to be of the form

$$f(t) = \Re e\,\{\mathbf{B}e^{i\omega t}\}.$$

It then follows that

$$y(x, t) = f(t - x/c) - f(t + x/c) = -2(\sin kx)\,\Re e\,\{i\mathbf{B}e^{i\omega t}\}. \tag{15–68}$$

Setting $y(l, t)$ equal to

$$A \sin \omega t = -A \,\Re e \{ie^{i\omega t}\}$$

and solving for **B** we have

$$\mathbf{B} = \frac{A}{2 \sin kl}.$$

Substituting this into Eq. (15–68) gives the particular solution which will be called the *steady-state solution*:

$$y_s(x, t) = A \frac{\sin kx}{\sin kl} \sin \omega t. \tag{15–69}$$

This solution is not valid if $\sin kl = 0$, for there is then an unallowed operation of division by zero. If we assume that $\sin kl \neq 0$, the steady-state solution represents a standing wave of amplitude $A/\sin kl$. The end $x = 0$ is a node, but $x = l$ is clearly not a node since the motion at this point has been specified. A graph of Eq. (15–69) is given in Fig. 15–8 for the case $kl = 9\pi/4$. In this case the length of the string is $\frac{9}{8}$ of the wavelength associated with the driving frequency ω. The amplitude is $A/\sin kl = A\sqrt{2}$. Nodes occur at $kx = \pi$ and $kx = 2\pi$, that is, $x = 4l/9$ and $x = 8l/9$.

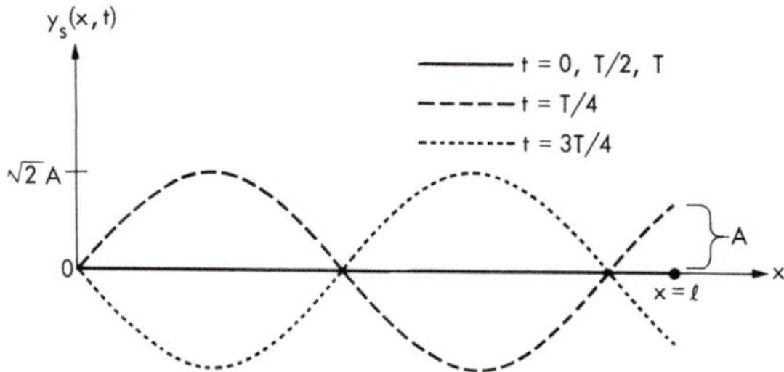

Fig. 15–8. Shape at various times of a string driven sinusoidally at one end. The period of the driving force is $T = 2\pi/\omega$. The graph of Eq. (15–69) is shown at the indicated times for the special case in which $T = 8\ell/9c$.

If the driving frequency is near a value which makes $\sin kl = 0$, the implied amplitude of the steady-state solution becomes large. If the amplitude A of the endpoint is not made correspondingly small, the mathematical solution fails to correspond to physical reality for a number of reasons:

i) Large amplitude motions do not obey the linear wave equation.

ii) *Dissipation* of energy within the string should be taken into account.

iii) Energy losses due to coupling with other systems would have to be taken into account, e.g., energy radiated in the form of sound waves in the air surrounding the string or transmission of energy due to yielding of the support at the "fixed end" $x = 0$.

iv) Strong interaction with the driving mechanism at $x = l$ may prevent the ideally specified motion at this end from being maintained.

The *resonance frequencies*, near which the solution driven by Eq. (15–69) has large amplitude and at which this solution is therefore not valid, are given by

$$kl = n\pi, \qquad n = 1, 2, 3, \ldots$$

The corresponding frequencies $\omega = n(\pi c/l)$ are seen from Eq. (15–10) to be identical with the eigenfrequencies of the string. In contrast to the harmonic oscillator, which has only one resonance frequency, a given string has an infinite number of resonance frequencies. The two systems have the common feature that they show especially sensitive response to any sinusoidal driving force whose frequency is near one of the frequencies at which the system can execute a purely sinusoidal free oscillation.

*15-14 GENERAL SOLUTION FOR THE STRING DRIVEN SINUSOIDALLY AT ONE END

To the steady-state solution of Eq. (15–69) we can add any solution $y_h(x, t)$ to the homogeneous problem defined by Eqs. (15–1), (15–2), and (15–3), and still have a solution to the inhomogeneous problem defined by Eqs. (15–1) and (15–67). Thus

$$y(x, t) = y_s(x, t) + y_h(x, t)$$

certainly satisfies the wave equation, Eq. (15–1), since y_s and y_h are both solutions, and the equation is linear. Also, since $y_s(0, t) = 0 = y_h(0, t)$, we have $y(0, t) = 0$, and since $y_s(l, t) = \psi(t)$ and $y_h(l, t) = 0$, we have $y(l, t) = \psi(t)$. The general form of $y_h(x, t)$, the solution to the homogeneous problem, is given by Eq. (15–28). If we combine this with Eq. (15–69), the general solution to the inhomogeneous problem becomes

$$y(x, t) = A \frac{\sin kx}{\sin kl} \sin \omega t + \sum_{n=1}^{\infty} \sin k_n x \{a_n \sin \omega_n t + b_n \cos \omega_n t\}. \quad (15\text{–}70)$$

The coefficients a_n, b_n are determined by initial conditions. As a specific example consider

$$y(x, 0) = 0 = \dot{y}(x, 0).$$

Then

$$y(x, 0) = 0 = \sum_{n=1}^{\infty} b_n \sin k_n x \qquad \text{or} \qquad b_n = 0.$$

From Eq. (15–70), we have

$$\dot{y}(x, 0) - A\omega \frac{\sin kx}{\sin kl} = \sum_{n=1}^{\infty} a_n \omega_n \sin k_n x.$$

This equation can be solved for the coefficients a_n by making use of the orthogonality of the functions $\sin k_n x$ over the interval $(0, l)$. By comparison with Eqs. (15–36) and (15–37) the solution can be written:

$$a_n \omega_n = \frac{2}{l} \int_0^l \left[\dot{y}(x, 0) - A\omega \frac{\sin kx}{\sin kl}\right] \sin k_n x \, dx.$$

Substituting $\dot{y}(x, 0) = 0$, carrying out the integration, and simplifying, we obtain

$$a_n = (-1)^n \frac{2Ak}{l(k_n^2 - k^2)}. \tag{15–71}$$

Thus

$$y(x, t) = A \frac{\sin kx}{\sin kl} \sin \omega t + \frac{2Ak}{l} \sum_{n=1}^{\infty} \frac{(-1)^n \sin k_n x \sin \omega_n t}{k_n^2 - k^2}. \tag{15–72}$$

This expression represents the motion at all times $t \geq 0$ of a string which is initially in equilibrium and is driven sinusoidally at $x = l$ at a frequency which is not a resonance frequency.

To obtain a solution which is valid at a resonance frequency we can allow k to approach k_m for some particular value of m. There are two terms in Eq. (15–72) which approach infinity. These are the steady-state term, since $\sin kl \to \sin k_m l = 0$, and the mth term in the transient series, because of the denominator $(k_m^2 - k^2)$. These terms are opposite in sign and tend to cancel each other. L'Hôpital's rule can be used to evaluate the limit, and results in

$$\begin{aligned} y(x, t) &\\ = \frac{2Ak_m}{l} &\sum_{n \neq m} (-1)^n \frac{\sin k_n x \sin \omega_n t}{k_n^2 - k_m^2} \\ &+ (-1)^m \frac{A}{l} \left\{x \cos k_m x \sin \omega_m t + ct \sin k_m x \cos \omega_m t - \frac{\sin k_m x \sin \omega_m t}{2k_m}\right\}. \end{aligned} \tag{15–73}$$

This represents the motion of a string which is initially in equilibrium and is driven sinusoidally at $x = l$ at a resonance frequency $\omega = \omega_m$. The solution is well behaved except for the term having t in the coefficient. This shows that as time progresses, the amplitude of motion becomes indefinitely large. The solution obeys the small-amplitudes requirement only for times such that $y \ll \lambda_m$, which requires $Act/l \ll \lambda_m$, or $t \ll (l/A)(\lambda_m/c)$. The ratio λ_m/c is the period of the driving force; hence the solution is valid only for a number of cycles of the driving mechanism which is small compared to the number of times the amplitude of motion at the right-hand end goes into the length of the string.

*15-15 STEADY-STATE SOLUTION FOR A UNIFORMLY DISTRIBUTED SINUSOIDAL FORCE ON A STRING WITH FIXED ENDS

We take $F(x, t) = A \cos \omega t$ in Eq. (15–66), and apply rigid boundary conditions at the two ends of the string:

$$y(0, t) = 0 = y(l, t).$$

A uniformly distributed sinusoidal force of this nature is obtained if, for example, a plane sound wave of frequency ω impinges on the string with its direction of propagation normal to the string. Alternatively, we can imagine that the "string" is a wire carrying an alternating current of frequency ω, placed perpendicular to the lines of force of a uniform magnetic field.

The steady-state solution is defined to be a sinusoidal solution having the same frequency as the driving force. It will therefore have the form

$$y_s(x, t) = \mathcal{R}e\,\{\mathbf{B}(x)e^{i\omega t}\} \tag{15–74}$$

for some complex function $\mathbf{B}(x)$. If Eq. (15–66) is to be satisfied, we have

$$\mathcal{R}e\,\{[-\sigma\omega^2\mathbf{B} - T\mathbf{B}'']e^{i\omega t}\} = \mathcal{R}e\,\{Ae^{i\omega t}\}$$

or

$$\mathbf{B}''(x) + k^2\mathbf{B}(x) = -A/T. \tag{15–75}$$

A particular solution to this inhomogeneous ordinary differential equation is

$$\mathbf{B}(x) = -A/k^2T.$$

The general solution is therefore

$$\mathbf{B}(x) = \mathbf{a}e^{-ikx} + \mathbf{b}e^{ikx} - A/k^2T$$

where $\mathbf{a}$ and $\mathbf{b}$ are arbitrary complex constants. The boundary conditions require $\mathbf{B}(0) = 0 = \mathbf{B}(l)$, which leads to specific values for $\mathbf{a}$ and $\mathbf{b}$:

$$a = \frac{Ae^{ikl/2}}{2k^2T \cos kl/2} = b^*,$$

provided $\cos kl/2 \neq 0$. $\mathbf{B}(x)$ is therefore the *real* function:

$$\mathbf{B}(x) = B_0 \left\{\frac{\cos k(x - l/2)}{\cos kl/2} - 1\right\} \equiv B(x), \tag{15–76}$$

where $B_0 = A/k^2T$. With $B(x)$ so defined, the steady-state solution, Eq. (15–74), can be written

$$y_s(x, t) = B(x) \cos \omega t. \tag{15–77}$$

If $\cos kl/2 = 0$, this solution does not apply. The frequencies at which this occurs are given by $kl/2 = (2n + 1)\pi/2$, or $\omega = (2n + 1)(\pi c/l)$. Since the implied value of $B(x)$ becomes indefinitely large as one of these frequencies is approached, we refer to these as *resonance* frequencies. Comparison with the eigenfrequencies of the string, Eq. (15–10), shows that the resonance frequencies under this mode of excitation are the *odd* eigenfrequencies.

From Eq. (15–77) we see that the profile of the string varies between $+B(x)$ and $-B(x)$ at different times in the cycle. At all times the shape is a scaled version of the graph of $B(x)$. From Eq. (15–76) we see that the latter has the following features:

i) The curve is symmetric about $x = l/2$.

ii) The shape is sinusoidal with wavelength $\lambda = 2\pi c/\omega$, where ω is the given frequency of the driving force; the constant term $-B_0$ shifts the curve downward.

iii) Nodes: by trigonometric identity, $B(x)$ can be written

$$B(x) = \frac{2B_0 \sin kx/2 \sin k(l - x)/2}{\cos kl/2},$$

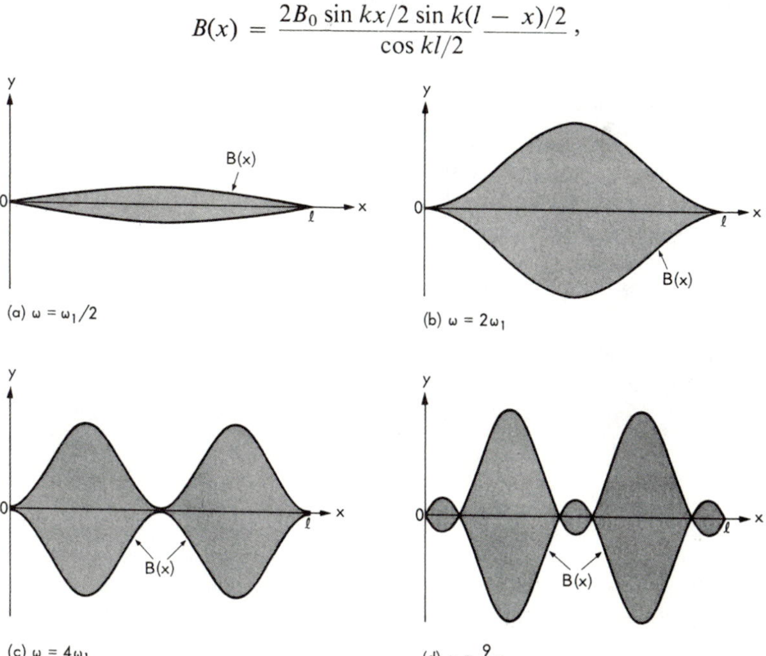

Fig. 15–9. Time exposure blurs of string driven by a uniformly distributed sinusoidal force for various values of the driving frequency ω relative to the fundamental resonance frequency $\omega_1 = \pi c/\ell$.

showing that nodes occur at

$$x = n\lambda, \qquad n = 0, 1, 2, \ldots$$

and at

$$x = l - m\lambda, \qquad m = 0, 1, 2, \ldots$$

iv) Extrema: at

$$x = (l/2) \pm n(\lambda/2), \qquad n = 0, 1, 2, \ldots,$$

$B(x)$ has the extreme values

$$B = B_0[(-1)^n \sec (kl/2) - 1].$$

The even eigenfrequencies of the string provide an interesting special case. If $\omega = 2m(\pi c/l)$ then $\sec (kl/2) = \sec m\pi = (-1)^m$. The extreme values are then either 0 or $-2B_0$, and $B(x) \leq 0$ everywhere.

Several typical graphs are given in Fig. 15–9. The graphs represent a picture of the string taken as a time exposure over an integral number of cycles. The graphs are obtained by filling in the area between $B(x)$ and $-B(x)$.

15-16 NORMAL MODES OF A UNIFORMLY STRETCHED RECTANGULAR MEMBRANE

The equilibrium form of a membrane is taken to be a plane surface lying in the plane $z = 0$. The tension in the membrane at equilibrium is uniform and sufficiently large so that variation in the magnitude of the tension will be considered negligible when a transverse disturbance is present. The tension T is the force per unit length which one part of the membrane exerts on the adjacent part across any line segment lying in the membrane. The mass per unit area, σ, is assumed to be uniform. In a disturbed configuration the transverse displacement, z, will be a function of the three independent variables, x, y, and t. For small-amplitude motion, z satisfies the *two-dimensional wave equation*:*

$$\frac{\partial^2 z}{\partial x^2} + \frac{\partial^2 z}{\partial y^2} = \frac{1}{c^2}\frac{\partial^2 z}{\partial t^2}, \tag{15–78}$$

where $c = \sqrt{T/\sigma}$. To complete the statement of a boundary-value problem we shall consider a rectangular membrane *clamped* at its edges. If the latter are given by the four lines $x = 0$, $x = a$, $y = 0$, $y = b$, the boundary conditions are

$$z(0, y, t) = z(a, y, t) = z(x, 0, t) = z(x, b, t) = 0. \tag{15–79}$$

---

* R. B. Lindsay, *Mechanical Radiation*, New York: McGraw-Hill, 1960.

We seek special solutions of product form

$$z(x, y, t) = X(x)Y(y)T(t). \tag{15–80}$$

The boundary conditions applied to a function of this form can be satisfied if and only if

$$X(0) = X(a) = Y(0) = Y(b) = 0. \tag{15–81}$$

If we insert a function of the form specified by Eq. (15–80) into (15–78) and divide through by XYT, we obtain

$$\frac{X''}{X} + \frac{Y''}{Y} = \frac{1}{c^2}\frac{T''}{T}. \tag{15–82}$$

In this form it is seen that the equation has been successfully separated into various terms each of which refers to only one of the independent variables. For this equation to be satisfied for all x, y and t, each term can be at most a constant. Thus we set

$$X''/X = -k_x^2, \qquad Y''/Y = -k_y^2, \qquad T''/c^2T = -k^2, \tag{15–83}$$

where k_x and k_y can be chosen independently, but to satisfy Eq. (15–82),

$$k^2 = k_x^2 + k_y^2. \tag{15–84}$$

In choosing symbols for these separation constants (in particular the minus signs) we have anticipated the ultimate physical significance of the quantities, though this does not in any way prejudice the generality of the results. The general solutions to the three equations in (15–83) are

$$X = A\sin(k_x x + \alpha), \qquad Y = B\sin(k_y y + \beta), \qquad T = C\sin(\omega t + \gamma), \tag{15–85}$$

where A, B, C, α, β, γ are arbitrary constants and $\omega \equiv kc$. The boundary conditions on X yield the following information:

$$X(0) = 0 = A\sin\alpha, \qquad X(a) = 0 = A\sin(k_x a + \alpha). \tag{15–86}$$

Both these equations are satisfied if $A = 0$, but this is a trivial case implying $z = 0$ for all t. The first equation implies that a nontrivial solution exists only if $\sin\alpha = 0$ or $\alpha = m'\pi$, with $m' = 0, \pm 1, \pm 2, \ldots$ The corresponding function for $X(x)$ is

$$X(x) = A\sin(k_x x + m'\pi) = (-1)^{m'}A\sin k_x x.$$

Since A is an arbitrary constant, we see that the choice $\alpha = 0$ can be made without loss of generality. The second equation in (15–86) then implies that

$A \sin k_x a = 0$. This equation restricts the possible values of the separation constant k_x to one of the set of eigenvalues

$$k_x = m(\pi/a), \qquad m = 1, 2, 3, \ldots \tag{15–87}$$

In like manner, the boundary conditions applied to $Y(y)$ result in the eigenvalue condition $\sin k_y b = 0$, or,

$$k_y = n(\pi/b), \qquad n = 1, 2, 3, \ldots \tag{15–88}$$

The values of ω are now fixed by Eq. (15–84). The conclusion is therefore that any solution which is to be of product form must be sinusoidal in t with a frequency selected from the list

$$\omega_{mn} = \pi c\sqrt{(m/a)^2 + (n/b)^2}, \tag{15–89}$$

where m and n are any two independently chosen integers. Lumping together the product of the three arbitrary constants A, B, and C, we obtain the corresponding eigenfunctions:

$$z = D \sin(m\pi x/a) \sin(n\pi y/b) \sin(\omega_{mn}t + \gamma), \tag{15–90}$$

where D and γ remain arbitrary. The motion described by such a function is referred to as a *normal mode.* A particular mode will be labeled (m, n) according to the choice of integers m and n in Eqs. (15–87) and (15–88). The order in which the integers are stated is important, since the (1, 2) mode is entirely distinct from the (2, 1) mode.

Each normal mode is characterized by a set of nodal lines, which are positions at which the membrane undergoes no displacement throughout the motion.* These are sets of lines parallel to the edges of the plate. For the (m, n) mode, the nodes are given by the equations

$$\begin{aligned} x &= pa/m, \qquad p = 0, 1, \ldots, m - 1, m; \\ y &= qb/n, \qquad q = 0, 1, \ldots, n - 1, n. \end{aligned} \tag{15–91}$$

As an illustration, consider the (2, 2) mode. The nodal lines are

$$\begin{aligned} x &= 0, a/2, a \\ y &= 0, b/2, b. \end{aligned}$$

With the choice $\gamma = 0$ the eigenfunction is

$$z_{22} = D \sin(2\pi x/a) \sin(2\pi y/b) \sin \omega_{22}t.$$

* The well-known demonstration of the Chladni plate is based upon the collection of sand at the nodal lines when a plate, usually under different boundary conditions from those discussed here, is excited in a pure mode.

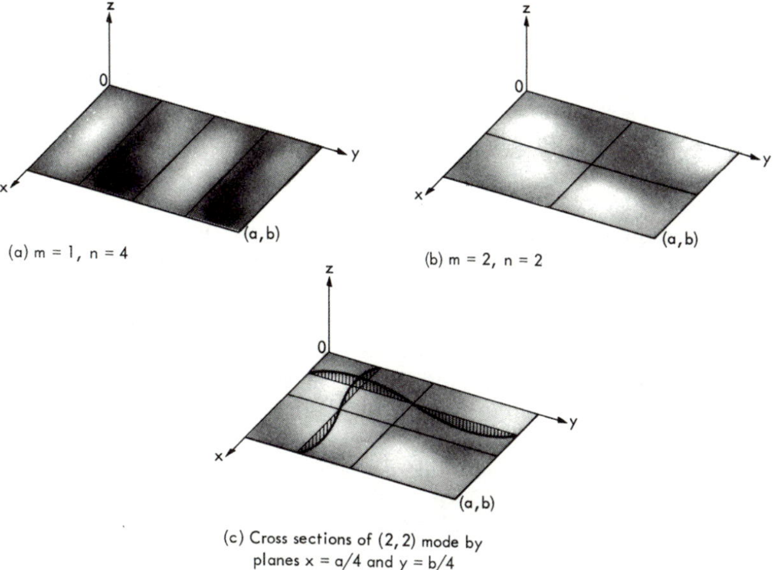

Fig. 15–10. Illustrative normal modes of a clamped rectangular membrane. The surface is depicted at $t = \pi/2\omega_{mn}$ with the choice $\gamma = 0$ in Eq. (15–90). Solid lines are nodal lines.

At $t = 0$ the entire membrane is flat. During the first half-period, $0 < t < \pi/\omega_{22}$, the membrane is divided by the nodal lines into four sectors in which z is alternately positive and negative. (See Fig. 15–10b.) Each sector forms a hummock or a depression within which the displacement is an extremum at the center. The cross section with any plane parallel to the xz- or yz-planes is a sine curve. At $t = \pi/\omega_{22}$ the membrane is again flat, and during the next half-cycle the motion is the mirror image in the $z = 0$ plane of the motion during the first half-cycle.

It is to be noted that the eigenfrequencies do not form a harmonic series. That is, the eigenfrequencies are not all integral multiples of a fundamental frequency. Consider, for example, a membrane with $b = 2a$. Then

$$\omega_{mn} = (\pi c/b)\sqrt{4m^2 + n^2}\,.$$

The (1, 1) mode is the lowest frequency mode. The frequencies of a few other modes, in order of increasing frequency, are

$$\omega_{11} = \frac{\pi c}{b}\sqrt{5}\,, \qquad \omega_{12} = \frac{\pi c}{b}\sqrt{8}\,, \qquad \omega_{13} = \frac{\pi c}{b}\sqrt{13}\,,$$

$$\omega_{21} = \frac{\pi c}{b}\sqrt{17}\,, \qquad \omega_{22} = \frac{\pi c}{b}\sqrt{20}\,.$$

It is seen that these are not all integral multiples of ω_{11}. (There are, of course, subsets that form a harmonic series, for example, $\omega_{mm} = m\omega_{11}$.) This fact reflects upon the musicality of the note produced by a membrane when it is struck in such a way as to excite an arbitrary mixture of modes.

An interesting feature is revealed by the calculation of ω_{14}. We find that ω_{14} and ω_{22} are identical. That is, the (1, 4) mode and the (2, 2) mode happen to correspond to the same frequency of vibration. These modes are physically quite distinct, as is illustrated by comparison of Figs. 15–10(a) and 15–10(b). Although these two modes share the same value of the eigenfrequency, the corresponding eigenfunctions are linearly independent. Such a case is referred to as a situation of *degeneracy* and the corresponding eigenfrequencies are said to be *degenerate eigenfrequencies.* The existence of degeneracy has a bearing upon the relation between *sinusoidal solutions* and *solutions of product form.* In Section 15–4 it was shown that for the one-dimensional wave equation these two ways of seeking special solutions lead to equivalent results. This is still the case in connection with the non-degenerate eigenvalues of the boundary-value problem defined by Eqs. (15–78) and (15–79). Thus with $b = 2a$, the unique form of solution corresponding to $\omega = \omega_{11} = (\pi c/b)\sqrt{5}$ is

$$z = D \sin(2\pi x/b) \sin(\pi y/b) \sin(\omega_{11}t + \gamma).$$

But, corresponding to $\omega = \omega_{22} = \omega_{14} = (\pi c/b)\sqrt{20}$, the general solution is

$$\begin{aligned} z = {} & D_1 \sin(4\pi x/b) \sin(2\pi y/b) \sin(\omega_{22}t + \gamma_1) \\ & + D_2 \sin(2\pi x/b) \sin(4\pi y/b) \sin(\omega_{22}t + \gamma_2). \end{aligned} \tag{15–92}$$

For a general choice of the arbitrary constants D_1, D_2, γ_1, γ_2, the function cannot be written as a product of factors each of which depends exclusively on a single one of the independent variables.

If there are q normal modes whose eigenfrequencies are the same, this eigenfrequency is said to be *q-tuply degenerate.* For the rectangle $b = 2a$ the reader can show that $(\pi c/b)\sqrt{200}$ is a triply degenerate eigenvalue, being the eigenfrequency of the (7, 2), (5, 10) and (1, 14) modes. Likewise, $(\pi c/b)\sqrt{340}$ is quadruply degenerate, being the eigenfrequency of the (9, 4), (7, 12), (6, 14) and (2, 18) modes.

This is as far as the development of the transverse motion of a stretched rectangular membrane will be carried out. We conclude with a few assertions of what can be shown.

i) The most general motion can be written as a superposition of normal-mode solutions:

$$z = \sum_{m,n=1}^{\infty} D_{mn} \sin(m\pi x/a) \sin(n\pi y/b) \sin(\omega_{mn}t + \gamma_{mn}).$$

The constants D_{mn} and γ_{mn} are determined by the initial shape and velocity distributions, $z(x, y, 0)$ and $\dot{z}(x, y, 0)$. Since the eigenfrequencies are not harmonic, the general solution is not periodic as was the case for the string.

ii) The total energy of motion in a mixed state is the sum of the energies associated with the individual modes contained.

iii) The resonance frequencies in forced motion are the eigenfrequencies, or a subset of them, depending on the manner of excitation.

iv) The techniques developed are easily adapted to apply to a rectangular membrane which is *free* along some of its edges. If the membrane is clamped part way along an edge and is free along the rest of the edge, the problem is vastly more complicated. The normal modes of a membrane clamped along a *circular* boundary are derived in Appendix VIII.

15-17 FOURIER INTEGRAL ANALYSIS ON A SEMI-INFINITE STRING

We have learned earlier in this chapter how to represent the future motion of a string of finite length as a trigonometric series determined by the initial conditions. The representation is chiefly useful because it characterizes a given disturbance as a specified mixture of waves of different frequencies. The examples of the rectangular pulse (Section 15–8) and the limited sinusoidal wavetrain (Section 15–11) have both illustrated that the range of frequencies associated with a given waveform is independent of l, the length of the string. If l is increased, keeping a *fixed* initial waveform, the eigenfrequencies become more closely spaced, but the *envelope* of the energy-versus-frequency graph maintains the same form. In the limit as l approaches infinity, the fundamental frequency $\omega_1 = \pi c/l$ approaches zero, and any two adjacent eigenfrequencies become "infinitesimally close to each other":

$$\omega_{n+1} - \omega_n = \omega_1 \to 0.$$

We are then dealing with a continuous, rather than a discrete, distribution of frequencies. The concept of the mode number n is no longer useful and disappears from the formulation by shifting attention directly to the frequencies involved. For example, the maximum on the graph in Fig. 15–7 occurs at a frequency $\Omega = M\pi c/a$. This is invariant of l. On the other hand, the mode number n corresponding to this frequency is $n = \Omega/\omega_1 = Ml/a$, which tends to infinity with l. Crudely speaking, the mode number of all "important" frequencies must be infinity since the fundamental frequency is approaching zero.

To effect a convenient representation which allows consideration of the limit as l approaches infinity, we take an initial pulse which is identically zero outside a specified region. That is, we assume that for some $a < l$ the initial profiles $y(x, 0)$ and $\dot{y}(x, 0)$ vanish in $a < x \leq l$. As a matter of convention let it be understood that $y(x, 0)$ and $\dot{y}(x, 0)$ also vanish for $x > l$. Define two functions

$$\beta(k) = \sqrt{\frac{2}{\pi}} \int_0^\infty y(x, 0) \sin kx \, dx$$
$$\alpha(k) = \frac{1}{kc} \sqrt{\frac{2}{\pi}} \int_0^\infty \dot{y}(x, 0) \sin kx \, dx. \tag{15-93}$$

Since the integrands vanish for $x > a$, we can write the upper limit equivalently as either a, l or ∞. The functions $\alpha(k)$ and $\beta(k)$ are independent of l. In this notation we can rewrite Eqs. (15–34) and (15–37) which determine the coefficients in the eigenfunction expansion of the general solution as:

$$b_n = \frac{\sqrt{2\pi}}{l} \beta(k_n), \qquad a_n = \frac{\sqrt{2\pi}}{l} \alpha(k_n). \tag{15-94}$$

The series expansion itself, Eq. (15–28), can then be written:

$$y(x, t) = \sqrt{\frac{2}{\pi}} \sum_{n=1}^{\infty} \sin k_n x \{\alpha(k_n) \sin \omega_n t + \beta(k_n) \cos \omega_n t\} \left(\frac{\pi}{l}\right) \cdot \tag{15-95}$$

We wish to exploit the fact that as l tends to infinity each term in this sum approaches zero and the distribution of wave numbers k_n becomes continuous. It is therefore expected that the limit of the sum will go over into a definite integral. This transition is obtained by noting that

$$\pi/l = k_{n+1} - k_n = \Delta k_n,$$

and, explicitly writing $\omega_n = k_n c$, we have

$$y(x, t) = \sqrt{\frac{2}{\pi}} \sum_{n=1}^{\infty} \sin k_n x \{\alpha(k_n) \sin k_n ct + \beta(k_n) \cos k_n ct\} \, \Delta k_n. \tag{15-96}$$

As Δk_n tends to zero the limit of this expression* is precisely what is abbreviated by the definite integral

$$y(x, t) = \sqrt{\frac{2}{\pi}} \int_0^\infty \sin kx \{\alpha(k) \sin kct + \beta(k) \cos kct\} \, dk. \tag{15-97}$$

This integral expresses the entire future motion of a semi-infinite string, where $\alpha(k)$ and $\beta(k)$ are determined from the initial conditions by Eq. (15–93). In passing to the limit we have lost explicit reference to n and l. The variable k is now a dummy variable of integration and does not stand for π/l.

---

* Provided that this limit exists. In this brief introductory treatment of Fourier integrals we omit discussion of convergence questions.

The integral can be thought of as the superposition of a number of waves having a *continuous* distribution of frequencies and corresponding wavelengths. An integral of this kind is referred to as a *Fourier integral* as opposed to a Fourier *series* where discrete frequencies are involved. Representation of a given function by means of a Fourier integral is referred to as the process of *Fourier analysis.*

If we set $t = 0$ in Eq. (15–97) and compare with Eq. (15–93), we see that the functions $\beta(k)$ and $y(x, 0)$ bear a symmetrical relation to each other:

$$\begin{aligned} y(x, 0) &= \sqrt{\frac{2}{\pi}} \int_0^\infty \beta(k) \sin kx \, dk, \\ \beta(k) &= \sqrt{\frac{2}{\pi}} \int_0^\infty y(x, 0) \sin kx \, dx. \end{aligned} \tag{15–98}$$

In general, given any function $f(\xi)$, the new function $g(\eta)$ generated by the operation

$$g(\eta) = \sqrt{\frac{2}{\pi}} \int_0^\infty f(\xi) \sin \xi\eta \, d\xi$$

is referred to as the *Fourier sine transform of f.** Thus $y(x, 0)$ and $\beta(k)$ are Fourier sine transforms of each other. The complete symmetry of the relation means that if the Fourier coefficient $\beta(k)$ has been determined for a given pulse shape $y(x, 0) = \phi(x)$, we have actually solved two problems, for we can interpret the function $\beta(x)$ as a pulse shape corresponding to the Fourier coefficient $\phi(k)$. The reader can also show that the functions $\dot{y}(x, 0)$ and $kc\alpha(k)$ are Fourier sine transforms of each other.

The series expression for the total energy

$$\mathcal{E} = \sum_{n=1}^\infty \mathcal{E}_n = \frac{\sigma l}{4} \sum_{n=1}^\infty [\omega_n^2(a_n^2 + b_n^2)]$$

can also be transformed into a definite integral in the limit as l tends to infinity. Substituting for a_n and b_n from Eq. (15–94), we have

$$\mathcal{E} = \frac{\sigma}{2} \sum_{n=1}^\infty \left\{ \omega_n^2[\alpha^2(k_n) + \beta^2(k_n)] \frac{\pi}{l} \right\},$$

and with $\Delta k_n = \pi/l$, the limit of this sum is

$$\mathcal{E} = \frac{\sigma}{2} \int_0^\infty (kc)^2[\alpha^2(k) + \beta^2(k)] \, dk. \tag{15–99}$$

---

* An extensive table of Fourier sine transforms will be found in Volume I of Erdelyi, Magnus, Oberhettinger, Tricomi, *Tables of Integral Transforms* (The Bateman Manuscript Project), New York: McGraw-Hill, 1954.

Thus if we change variables to $\omega = kc$ and define

$$\epsilon(\omega) \equiv \frac{\sigma\omega^2}{2c}\,[\alpha^2(\omega/c) + \beta^2(\omega/c)]\,, \tag{15–100}$$

the total energy can be written

$$\mathcal{E} = \int_0^\infty \epsilon(\omega)\,d\omega.$$

This enables $\epsilon(\omega)$ to be interpreted as the *energy per unit frequency range*; that is, $\epsilon(\omega)\,d\omega$ is the energy contained in the frequency interval from ω to $\omega + d\omega$. In this notation the energy of the nth mode in the case of a discrete distribution can be expressed as

$$\mathcal{E}_n = \frac{\pi c}{l}\,\epsilon(\omega_n). \tag{15–101}$$

The total energy can also be expressed in terms of the initial distributions of kinetic and potential energy:

$$\mathcal{E} = \int_0^\infty \left\{\frac{\sigma}{2}\,\dot{y}^2(x, 0) + \frac{T}{2}\left[\frac{\partial y}{\partial x}\,(x, 0)\right]^2\right\} dx. \tag{15–102}$$

A fundamental identity is obtained by equating the right-hand sides of Eqs. (15–99) and (15–102). In particular, suppose that $y(x, 0) = 0$, so that $\beta(k) = 0$. Then

$$\int_0^\infty \dot{y}^2(x, 0)\,dx = \int_0^\infty [kc\alpha(k)]^2\,dk. \tag{15–103}$$

Since $\dot{y}(x, 0)$ is an arbitrary function and $kc\alpha(k)$ is its Fourier sine transform, we have in fact obtained the general identity

$$\int_0^\infty f^2(\xi)\,d\xi = \int_0^\infty g^2(\eta)\,d\eta, \tag{15–104}$$

where f and g are any pair of Fourier sine transforms.

We shall apply the above formulas to the pulse considered in Section 15–11, namely, the sinusoidal wavetrain of limited extent. Either by calculating directly from Eq. (15–93) or by combining Eqs. (15–94) and (15–60), we find the Fourier coefficients to be

$$\begin{aligned} \beta(k) &= 0, \\ \alpha(k) &= \frac{v}{kc\sqrt{2\pi}}\left\{\frac{\sin\,(k - K)a}{k - K} - \frac{\sin\,(k + K)a}{k + K}\right\}. \end{aligned} \tag{15–105}$$

Since $Ka = M\pi$, the expression for $\alpha(k)$ can be simplified to yield

$$\alpha(k) = (-1)^M\,\frac{v}{kc\sqrt{2\pi}}\,\frac{2K\sin ka}{k^2 - K^2}\,. \tag{15–106}$$

Taking the time derivative of Eq. (15–97) to obtain the velocity-profile function, and substituting from Eq. (15–106), we obtain

$$\dot{y}(x, t) = (-1)^M \frac{2Kv}{\pi} \int_0^\infty \frac{\sin ka}{k^2 - K^2} \sin kx \cos kct \, dk. \quad (15\text{–}107)$$

This expression represents the sinusoidal wavetrain of Eq. (15–59) for all $t \geq 0$, along with its subsequent reflection at the fixed end $x = 0$.

Rather than calculate the energy spectrum for the *exact* representation, we will make an approximation which is suitable for the case $M \gg 1$, that is, the nearly sinusoidal wave. Then Eq. (15–105) can be written approximately

$$\alpha(k) = \frac{v}{kc\sqrt{2\pi}} \frac{\sin (k - K)a}{k - K}. \quad (15\text{–}108)$$

From Eq. (15–100) the energy-distribution function is

$$\epsilon(\omega) = \frac{\sigma c v^2}{4\pi} \frac{\sin^2 [(\omega - \Omega)a/c]}{(\omega - \Omega)^2}. \quad (15\text{–}109)$$

Since this differs only by constant factors from the envelope function $\mathcal{E}(\omega)$ in Eq. (15–63), the curve in Fig. 15–7 can be interpreted as a graph of $\epsilon(\omega)$.

Calculating the area under this graph leads to an interesting problem in integration:

$$\mathcal{E} = \int_0^\infty \epsilon(\omega) \, d\omega = \frac{\sigma c v^2}{4\pi} \int_0^\infty \frac{\sin^2 [(\omega - \Omega)a/c]}{(\omega - \Omega)^2} \, d\omega.$$

Changing variables to $\beta = (\omega - \Omega)a/c$ we obtain

$$\mathcal{E} = \frac{\sigma a v^2}{4\pi} \int_{-\Omega a/c}^\infty \frac{\sin^2 \beta}{\beta^2} \, d\beta.$$

Now $\Omega a/c = Ka = M\pi$, where M is a large integer for the case of the nearly sinusoidal wave. Since the integrand is insignificant for $|\beta| \gg \pi$, the lower limit of integration might as well be replaced by $-\infty$. Then, since the integrand is an even function, we can write

$$\mathcal{E} = \frac{\sigma a v^2}{2\pi} \int_0^\infty \frac{\sin^2 \beta}{\beta^2} \, d\beta \cdot$$

The law of conservation of energy enables us to calculate this quantity from the initial conditions. Using Eq. (15–59) in Eq. (15–102) we obtain

$$\mathcal{E} = \frac{\sigma v^2}{2} \int_0^a \sin^2 Kx \, dx = \frac{\sigma a v^2}{4} \cdot$$

Equating the two expressions for $\mathcal{E}$ implies the integration formula

$$\int_0^\infty \frac{\sin^2 \beta}{\beta^2}\, d\beta = \frac{\pi}{2}. \tag{15-110}$$

This definite integral can be worked out by other means, and is listed in most integral tables.

15-18 FOURIER ANALYSIS OVER THE WHOLE *x*-AXIS

To facilitate comparison with previous solutions obtained for a string with fixed ends, in the last section we considered the *semi-infinite* string obtained by allowing the end at $x = 0$ to remain fixed while the end at $x = l$ moves off to infinity. In this section we shall derive the Fourier integral representation of waves on an unclamped string extending from $x = -\infty$ to $x = +\infty$. The derivation will be based directly on Fourier's theorem, and will otherwise be independent of earlier parts of this chapter.

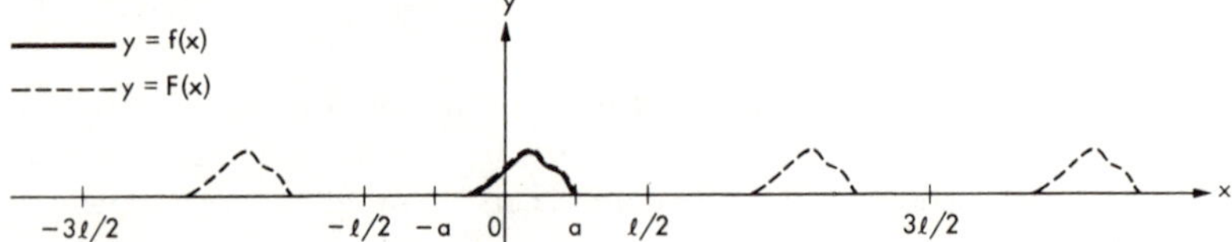

Fig. 15-11. Periodic extension of $f(x)$.

Consider a function $f(x)$ which vanishes outside a finite interval. That is, we assume that there exists some number a such that $f(x) = 0$ for $|x| > a$. (See Fig. 15-11). Now select any value $l > 2a$ and consider a new function $F(x)$ defined by the properties

$$F(x) = f(x) \quad \text{for} \quad |x| \leq l/2 \qquad \text{and} \qquad F(x + l) = F(x).$$

Thus $F(x)$ is defined so that it agrees identically with $f(x)$ in the interval $-l/2 \leq x \leq l/2$, and is extended outside this interval as a periodic function with period l. (See the dotted curve in Fig. 15-11.)

Since ultimately we consider the limit as l approaches infinity, the function $F(x)$ will in fact become identical with $f(x)$ over the entire x-axis. For the moment however, we exploit the fact that $F(x)$ is a periodic function of x. Fourier's theorem can be applied to represent F by means of a trigonometric series, as in Eq. (15-27). Since the independent variable is x and not t, we use different symbols for the parameters in this expansion. Thus from the given space-period l we use $k \equiv 2\pi/l$ to replace ω in Eq. (15-27). Letting

$$k_n = nk = 2\pi n/l$$

we can then write

$$F(x) = A_0 + \sum_{n=1}^{\infty} \{A_n \cos k_n x + B_n \sin k_n x\}. \tag{15–111}$$

The series coefficients A_n and B_n can be calculated from the given function $f(x)$ by a procedure that is based on the fact that the functions $\sin k_n x$ associated with different values of n are orthogonal over the interval $(-l/2, l/2)$. This follows by the same method as the proof of Eq. (15–32). (Note, however, that we now have $k_n = 2\pi n/l$ in place of the former $k_n = n\pi/l$.) It is also a simple matter to prove the analogous identities:

$$\int_{-l/2}^{l/2} (\cos k_m x)(\sin k_n x)\, dx = 0 \quad \text{for all } m, n. \tag{15–112}$$

$$\int_{-l/2}^{l/2} (\cos k_m x)(\cos k_n x)\, dx = 0 \quad \text{if } m \neq n. \tag{15–113}$$

First let Eq. (15–111) be integrated from $x = -l/2$ to $x = l/2$. The right-hand side yields lA_0, since the trigonometric terms integrate to zero over a complete period. The left-hand side yields

$$\int_{-l/2}^{l/2} F(x)\, dx = \int_{-l/2}^{l/2} f(x)\, dx,$$

since the functions F and f are identical throughout the range of integration. Thus we obtain

$$A_0 = \frac{1}{l} \int_{-l/2}^{l/2} f(x)\, dx. \tag{15–114}$$

If Eq. (15–111) is multiplied by $\cos k_m x$ for some m and integrated from $x = -l/2$ to $x = l/2$, the only term on the right-hand side which does not integrate to zero is the term

$$A_m \int_{-l/2}^{l/2} \cos^2 k_m x\, dx = A_m \frac{l}{2}\cdot$$

Hence

$$A_m = \frac{2}{l} \int_{-l/2}^{l/2} f(x) \cos k_m x\, dx, \qquad m = 1, 2, 3, \ldots \tag{15–115}$$

Likewise, upon multiplying by $\sin k_m x$ and integrating we find that

$$B_m = \frac{2}{l} \int_{-l/2}^{l/2} f(x) \sin k_m x\, dx, \qquad m = 1, 2, 3, \ldots \tag{15–116}$$

We further note that since $f(x)$ vanishes outside $(-l/2, l/2)$, the limits of integration in Eqs. (15–114), (15–115), and (15–116) can equally well be written as $-\infty$ and $+\infty$. Now define functions $C(k)$ and $D(k)$ by the integrals

$$C(k) = \frac{1}{\sqrt{2\pi}} \int_{-\infty}^{\infty} f(x) \cos kx \, dx,$$
$$D(k) = \frac{1}{\sqrt{2\pi}} \int_{-\infty}^{\infty} f(x) \sin kx \, dx. \tag{15-117}$$

These functions act as envelope functions for the Fourier series coefficients and enable us to write

$$A_n = (2\sqrt{2\pi}/l)C(k_n)$$

and

$$B_n = (2\sqrt{2\pi}/l)D(k_n).$$

With the view in mind of transforming the sum in Eq. (15–111) into a definite integral, we note that k increases by $2\pi/l$ between successive terms in the sum. Hence we write $2\sqrt{2\pi}/l$ as $\sqrt{2/\pi}(2\pi/l) = \sqrt{2/\pi}\,\Delta k_n$. Making these purely formal changes, we can write Eq. (15–111) as

$$F(x) = A_0 + \sqrt{\frac{2}{\pi}} \sum_{n=1}^{\infty} \{C(k_n) \cos k_n x + D(k_n) \sin k_n x\} \, \Delta k_n.$$

We now consider the limit as l tends to infinity. The function $f(x)$ itself is not changed in this process. We are simply embedding the originally given function into larger and larger intervals on the x-axis. The periodic repetitions of $f(x)$ are ultimately extremely remote from $f(x)$ itself. Suppose for example that we are interested in analyzing a pulse $f(x)$ generated in a laboratory experiment. If l is a distance of a million or so light years, $F(x)$ actually represents not only the laboratory pulse, but also others perhaps propagating through laboratories in distant galaxies. In describing the disturbance which propagates through our own laboratory during the course of the experiment, it is clear that predictions can be based equally well on $f(x)$ or $F(x)$.

We assume that the area of $f(x)$ is finite, and since, according to Eq. (15–114), A_0 is this area divided by l, the limit of A_0 is zero. The remainder of the expression for $F(x)$ has been cast in a form which is directly the kind of sum whose limit is abbreviated by definite integral notation. Thus we have

$$\lim_{l\to\infty} [F(x)] = f(x) = \sqrt{\frac{2}{\pi}} \int_0^{\infty} [C(k) \cos kx + D(k) \sin kx] \, dk. \tag{15-118}$$

This equation expresses the arbitrary function $f(x)$ as the superposition of a continuous distribution of functions of the form

$$[C(k) \cos kx + D(k) \sin kx].$$

That is, the arbitrary waveform $f(x)$ has been represented as a superposition of sinusoidal waveforms.

It is convenient to combine the two real functions $C(k)$ and $D(k)$ into a single complex function:

$$\mathbf{A}(k) \equiv C(k) - iD(k).$$

Then, since

$$C(k)\cos kx + D(k)\sin kx = \mathcal{R}e\,\{\mathbf{A}(k)e^{ikx}\},$$

we have

$$f(x) = \sqrt{\frac{2}{\pi}}\,\mathcal{R}e \int_0^\infty \mathbf{A}(k)e^{ikx}\,dk. \tag{15–119}$$

From Eq. (15–117) we observe that $\mathbf{A}(k)$ is to be obtained from the given $f(x)$ by the prescription

$$\mathbf{A}(k) = \frac{1}{\sqrt{2\pi}} \int_{-\infty}^{\infty} f(x)e^{-ikx}\,dx. \tag{15–120}$$

The function $\mathbf{A}(k)$ is known as the *Fourier integral transform* of $f(x)$. Conversely, $f(x)$ is the *inverse Fourier transform* of $\mathbf{A}(k)$.

EXAMPLE 2. Find the Fourier integral transform of

$$\psi(x) = \begin{cases} \sin Kx, & |x| \le a, \\ 0, & |x| > a, \end{cases} \tag{15–121}$$

where $Ka = M\pi$, M being an integer. Let the Fourier transform of $\psi(x)$ be designated by $\mathbf{B}(k)$. From Eq. (15–120),

$$\mathbf{B}(k) = \frac{1}{\sqrt{2\pi}} \int_{-a}^{a} (\sin Kx)e^{-ikx}\,dx.$$

If we write

$$\sin Kx = (e^{iKx} - e^{-iKx})/2i,$$

the integral is of elementary form. After simplification the result can be written

$$\mathbf{B}(k) = \frac{(-1)^M 2iK \sin ka}{\sqrt{2\pi}\,(K^2 - k^2)}. \tag{15–122}$$

The application of the Fourier integral to waves on the infinite string is mostly a purely formal matter, since the problem of determining the future motion from given initial conditions has already been solved in Section 1–8. Suppose that the general solution is written in the form

$$y(x, t) = f(x - ct) + g(x + ct)$$

and that the given initial conditions are

$$y(x, 0) = \phi(x) \quad \text{and} \quad \dot{y}(x, 0) = \psi(x).$$

Let the Fourier transforms of $f(x)$, $g(x)$, $\phi(x)$, and $\psi(x)$ be respectively $\mathbf{A}_+(k)$, $\mathbf{A}_-(k)$, $\mathbf{A}(k)$, and $\mathbf{B}(k)$. Since $\phi(x)$ and $\psi(x)$ are given, we can assume that $\mathbf{A}(k)$ and $\mathbf{B}(k)$ have been calculated from Eq. (15–120). One aspect of the problem is to use this information to calculate $\mathbf{A}_+(k)$ and $\mathbf{A}_-(k)$, the *coefficients* of the plus and minus waves. Since

$$f(x) = \sqrt{\frac{2}{\pi}}\,\mathcal{R}e \int_0^\infty \mathbf{A}_+(k)e^{ikx}\,dk,$$

then

$$f(x - ct) = \sqrt{\frac{2}{\pi}}\,\mathcal{R}e \int_0^\infty \mathbf{A}_+(k)e^{-i\omega t}e^{ikx}\,dk,$$

where we use the abbreviation $\omega = kc$. Treating $g(x + ct)$ in a similar fashion, we obtain as a representation of the general solution

$$y(x, t) = \sqrt{\frac{2}{\pi}}\,\mathcal{R}e \int_0^\infty [\mathbf{A}_+(k)e^{-i\omega t} + \mathbf{A}_-(k)e^{i\omega t}]e^{ikx}\,dk. \tag{15–123}$$

Setting $t = 0$ yields

$$y(x, 0) = \phi(x) = \sqrt{\frac{2}{\pi}}\,\mathcal{R}e \int_0^\infty [\mathbf{A}_+(k) + \mathbf{A}_-(k)]e^{ikx}\,dk.$$

But, we also have

$$\phi(x) = \sqrt{\frac{2}{\pi}}\,\mathcal{R}e \int_0^\infty \mathbf{A}(k)e^{ikx}\,dk,$$

since $\mathbf{A}(k)$ is the Fourier transform of $\phi(x)$. Subtracting the last equation from the previous one, we have

$$0 = \sqrt{\frac{2}{\pi}}\,\mathcal{R}e \int_0^\infty [\mathbf{A}_+(k) + \mathbf{A}_-(k) - \mathbf{A}(k)]e^{ikx}\,dk.$$

If it is accepted that within a class of sufficiently well-behaved functions the Fourier transform is unique, this last equation shows that

$$\mathbf{A}_+(k) + \mathbf{A}_-(k) - \mathbf{A}(k) = \chi(k)$$

is the Fourier transform of the function $h(x) = 0$. Thus, using the inversion formula, Eq. (15–120), we have

$$\chi(k) = \frac{1}{\sqrt{2\pi}} \int_{-\infty}^\infty h(x)e^{-ikx}\,dx = 0,$$

or

$$\mathbf{A}(k) = \mathbf{A}_+(k) + \mathbf{A}_-(k). \tag{15–124}$$

Likewise, taking the time derivative of Eq. (15–123),

$$\dot{y}(x, t) = \sqrt{\frac{2}{\pi}}\, \mathcal{R}e \int_0^\infty [-i\omega \mathbf{A}_+(k)e^{-i\omega t} + i\omega \mathbf{A}_-(k)e^{i\omega t}]e^{ikx}\, dk \qquad (15\text{–}125)$$

and considering the Fourier transform $\mathbf{B}(k)$ of $\dot{y}(x, 0)$, it can be seen that

$$\mathbf{B}(k) = i\omega[\mathbf{A}_-(k) - \mathbf{A}_+(k)]. \qquad (15\text{–}126)$$

Equations (15–124) and (15–126) can now be solved for the coefficients of the plus and minus waves in terms of the Fourier transforms of the given initial distributions:

$$\mathbf{A}_+(k) = \frac{1}{2}\left[\mathbf{A}(k) - \frac{\mathbf{B}(k)}{ikc}\right],$$
$$\mathbf{A}_-(k) = \frac{1}{2}\left[\mathbf{A}(k) + \frac{\mathbf{B}(k)}{ikc}\right]. \qquad (15\text{–}127)$$

EXAMPLE 3. Consider the initial conditions $y(x, 0) = 0 = \phi(x)$ and $\dot{y}(x, 0) = \psi(x)$, where $\psi(x)$ is the function treated in the previous example, Eq. (15–121). Then $\mathbf{A}(k) = 0$, and $\mathbf{B}(k)$ is given by Eq. (15–122). From Eq. (15–127), we have

$$\mathbf{A}_+(k) = -\frac{\mathbf{B}(k)}{2ikc}$$

and

$$\mathbf{A}_-(k) = \frac{\mathbf{B}(k)}{2ikc}.$$

The velocity profile, Eq. (15–125), then takes the form

$$\dot{y}(x, t) = \sqrt{\frac{2}{\pi}}\, \mathcal{R}e \int_0^\infty \left[\frac{e^{-i\omega t} + e^{i\omega t}}{2}\right]\mathbf{B}(k)e^{ikx}\, dx$$
$$= \sqrt{\frac{2}{\pi}}\, \mathcal{R}e \int_0^\infty (\cos \omega t)\mathbf{B}(k)e^{ikx}\, dk.$$

Substituting from Eq. (15–122), we obtain finally

$$\dot{y}(x, t) = (-1)^M \frac{2K}{\pi} \int_0^\infty \frac{\sin ka}{k^2 - K^2} \sin kx \cos \omega t\, dk. \qquad (15\text{–}128)$$

This result is identical, except for the choice $v = 1$, with Eq. (15–107), which was the answer to an initial-value problem on the semi-infinite string. Comparison of Eqs. (15–59) and (15–121) shows that the specified initial conditions are the same in $x \geq 0$, which is the only region having physical significance in the earlier problem. Since the initial velocity profile for the case of the infinite string is an odd function, the position $x = 0$ is "accidentally" a node, and the problems turn out to be equivalent.

PROBLEMS

15–1. A string of length l is fixed at $x = 0$ but is "free" at $x = l$. (The device of a frictionless slip ring would be required to maintain tension in the string and yet permit no transverse component of the force acting on the free end.) a) Give arguments to show that the general motion is periodic and deduce the period. b) By direct substitution of a function of sinusoidal form determine the frequencies which any sinusoidal solution must have. c) Use the method of separation of variables to deduce the necessary form of a product-form solution satisfying the given boundary conditions.

***15–2.** Consider a string of length l which is fixed at $x = 0$ and is attached at $x = l$ to a slip ring which is free to slide along a transverse post. The mass of the slip ring is m. a) Show that sinusoidal motion cannot take place unless $k = \omega/c$ is one of the roots of the transcendental equation

$$\tan (kl) = \sigma/mk.$$

b) Indicate how you would determine these eigenvalues graphically for the general case. c) Show that the results for $m = 0$ agree with Problem 15–1. d) If the mass of the slip ring is very large in comparison with the total mass of the string, find an approximate value for the lowest eigenfrequency. (This frequency can also be obtained by a simple physical argument in terms of the dynamics of the motion of the slip ring for this case.)

***15–3.** Consider a uniform string extending between rigid supports at $x = -l/2$ and $x = +l/2$ and loaded with a mass point of mass m located at $x = 0$. (Neglect gravity.) Let $y_1(x, t)$ and $y_2(x, t)$ designate the transverse displacement in $x < 0$ and $x > 0$ respectively. a) Show that the boundary conditions at $x = 0$ are

$$y_1(0, t) = y_2(0, t)$$

and

$$\frac{\partial y_2}{\partial x}(0, t) = \frac{\partial y_1}{\partial x}(0, t) + \frac{m}{\sigma c^2}\frac{\partial^2 y_1}{\partial t^2}(0, t).$$

b) Show that the only sinusoidal solutions which will satisfy all four boundary conditions are associated with eigenvalues k_n, $n = 1, 2, 3, \ldots$, where $k_n = n\pi/l$ for n even and $k_{n-1} < k_n < n\pi/l$ for n odd. c) Show that the eigenfunctions are of the form

$$y_n(x, t) = A \sin k_n x \cos \omega_n t \qquad \text{for } n \text{ even}$$

$$y_n(x, t) = A \sin k_n[(l/2) - |x|] \cos \omega_n t \qquad \text{for } n \text{ odd.}$$

d) Show that these solutions reduce to what is expected in the limit as $m \to 0$. e) Find the frequency of the gravest mode (the mode of lowest frequency) for a case in which m is very much greater than the total mass of the string. Check this result by a simple dynamical analysis of the restoring force acting on m under the assumption that the two segments of the string are essentially straight.

15–4. Apply Fourier's theorem to show that a *necessary* form for the general solution to the boundary-value problem of Problem 15–1 is

$$y(x, t) = \sum_{n=0}^{\infty} C_n \sin k_n x \sin (\omega_n t + \varphi_n),$$

where

$$\omega_n = k_n c = (n + \tfrac{1}{2})(\pi c/l) \qquad n = 0, 1, 2, \ldots$$

15–5. The initial conditions for a string with two fixed ends are the following:

$$y(x, 0) = 0, \qquad \dot{y}(x,0) = \sin (2\pi x/l).$$

It is clear that these initial conditions correspond to a particular choice of amplitude and phase of the first harmonic. Show that the formal machinery of the normal-modes expansion leads to the conclusion that this is the only participating mode and exhibit the resulting solution for $y(x, t)$.

15–6. Show explicitly that the condition described by Fig. 15–1(e) is correctly represented by the Fourier series representation given in Eq. (15–48).

15–7. A string with two fixed ends is plucked at the center. Assume that the string is of length l and is at rest at $t = 0$ and that the initial profile is triangular:

$$y(x, 0) = \begin{cases} \dfrac{2h}{l}\, x, & x \leq \dfrac{l}{2}, \\[2ex] \dfrac{2h}{l}\, (l - x), & x > \dfrac{l}{2}. \end{cases}$$

Show that the even harmonics will be missing and that the expansion coefficients are $a_n = 0$ and

$$b_n = (-1)^m \frac{8h}{(2m + 1)^2 \pi^2} \qquad \text{for } n = 2m + 1.$$

15–8. A string of length l is fixed at both ends. If all points on the string are initially at rest and the initial shape of the string is specified by the parabola

$$y(x, 0) = (4h/l^2)(lx - x^2),$$

find the coefficients in the Fourier series representation of $y(x, t)$. Graph the first term in the Fourier series at $t = 0$ and compare with the graph of $y(x, 0)$.

15–9. Consider a string with two fixed ends subject to the initial conditions

$$y(x, 0) = 0, \qquad \dot{y}(x, 0) = \begin{cases} \sin \dfrac{2\pi x}{l}, & x \leq \dfrac{l}{2}, \\[2ex] 0, & x > \dfrac{l}{2}. \end{cases}$$

(Note that this is a special case of the example considered in Section 15–11. Since M is not a large integer, the approximations used in arriving at Eq. (15–61) are no longer justified.) a) Show that the Fourier series expansion of $\dot{y}(x, 0)$ is

$$\dot{y}(x, 0) = \tfrac{1}{2} \sin 2kx + \frac{4}{\pi} [\tfrac{1}{3} \sin kx + \tfrac{1}{5} \sin 3kx - \tfrac{1}{21} \sin 5kx + \tfrac{1}{45} \sin 7kx \ldots],$$

where $k = \pi/l$.

b) By consideration of the propagation and reflection of the plus and minus constituent waves decide what the velocity profile of the string should be at $t = l/2c$. Show that the Fourier series representation of $\dot{y}[x, (l/2c)]$ yields the expected result. c) Show that energy conservation implies the identity

$$\sum_{n \text{ odd}} \frac{1}{(4 - n^2)^2} = \frac{\pi^2}{64}.$$

15–10. A string of length l is fixed at both ends. If all points on the string are initially at rest and the initial shape of the string is specified by $y(x, 0) = x(\sin kx)$, where $k = \pi/l$, a) find the coefficients in the Fourier series representation of this function. b) Draw a bar graph indicating the relative energies of the first few modes. c) Graph the function $y(x, 0)$ and compare this with a graph of the sum of the *two* most prominent modes in the Fourier series at $t = 0$.

15–11. a) Show that Eq. (15–54), referring to the frequency spectrum of a rectangular pulse, implies that not much use is made of modes whose wavelengths are short compared with the width of the pulse. b) Show that the mode number of the highest mode making important contributions is determined by the number of times the pulse width goes into twice the length of the string.

15–12. Verify Eq. (15–58) for the special case $l = 3a$ by juggling with the sum on the left-hand side and making use of the identity

$$\sum_{n \text{ odd}} \frac{1}{n^2} = \frac{\pi^2}{8}.$$

15–13. Let $\Delta\nu$ be the frequency spread in cycles per second of a sinusoidal wavetrain of finite extent. Let Δt be the length of time required for the wavetrain to pass by a fixed point of observation. Show that the relation deduced in Section 15–11 between the spatial and frequency widths is equivalent to the statement

$$(\Delta\nu)(\Delta t) \sim 1.$$

15–14. a) The 6438 Å cadmium red line has a wavelength spread of the order of $\Delta\lambda \sim 0.007$ Å. If this is interpreted as due to the fact that the signal consists of a succession of purely sinusoidal wavetrains of limited duration Δt, calculate Δt and determine the number of cycles of the sinusoidal oscillation contained in each wavetrain. b) A Kerr cell is used to "chop" a monochromatic beam. Consider that the cell is "open" during one half-cycle and "closed" during the other half, so that a succession of limited wavetrains results. If the frequency of the Kerr cell is

10^{-11} sec^{-1}, what line breadth $\Delta\lambda$ should be attributed to the chopped beam? (Assume that $\lambda = 5500$ Å.)

15–15. A string is driven sinusoidally at the two ends $x = \pm l/2$ with frequency ω and amplitude A. Find the steady-state motion of the string and discuss the shape of the string and the conditions of resonance if a) the displacements at the two ends are in phase, and b) the displacements at the two ends are π out of phase.

15–16. A cylinder of gas is driven at one end ($x = 0$) by a piston vibrating sinusoidally with frequency ω. The other end at $x = a$ is a rigid wall. Show that the particle velocity throughout the fluid is given by

$$\dot{\xi}(x, t) = \frac{\sin k(a - x)}{\sin ka}\, \dot{\xi}(0, t)$$

provided that $ka \neq n\pi$.

15–17. A string has a fixed end at $x = 0$, implying that the solution must have the form

$$y(x, t) = f(t - x/c) - f(t + x/c).$$

Suppose that the string is given a specified displacement

$$y(l, t) = A \sin \omega_m t,$$

where $\omega_m = m\pi c/l$ is one of the normal-mode frequencies of the string with two fixed ends. Try a particular solution in which $f(t)$ has the form

$$f(t) = Bt \sin \omega_m t,$$

and find the value of B which enables the boundary condition at $x = l$ to be matched. Show that the resulting particular solution $y(x, t)$ differs from that given in Eq. (15–73) by a solution to the homogeneous boundary value problem.

***15–18.** Find the particular solution to the homogeneous wave equation

$$\frac{\partial^2 y}{\partial x^2} = \frac{1}{c^2}\frac{\partial^2 y}{\partial t^2}$$

which satisfies the following boundary and initial conditions:

$$y(0, t) = A \cos \omega t, \qquad \frac{\partial y}{\partial x}(l, t) = 0;$$

$$y(x, 0) = 0, \qquad \dot{y}(x, 0) = 0.$$

***15–19.** a) Obtain the steady-state solution for a string with two fixed ends subject to a distributed force

$$F(x, t) = \begin{cases} 0, & 0 \le x \le \dfrac{l}{2}, \\[2ex] A \cos \omega t, & \dfrac{l}{2} < x \le l. \end{cases}$$

b) Show that resonance is obtained if $\omega = n\pi c/l$ for all integers n except for the multiples of 4. c) Show that for an appropriately chosen frequency the left half of the string remains motionless and the right half has a motion described by Fig. 15–9(b).

15–20. Discuss the graph of the string profile, Eq. (15–77), for a case in which the distributed force has a frequency close to one of the resonance frequencies; e.g., take $\omega = (3\pi c/l) + \delta$, where $\delta \ll 3\pi c/l$.

15–21. Consider a string which has reached a steady-state after being driven by a distributed force

$$F(x, t) = A \cos \omega t,$$

where ω is one of the *even* eigenfrequencies of the string. At $t = 0$ the distributed force ceases to act. Determine the series expansion which gives $y(x, t)$ for $t > 0$.

15–22. Take $D_2 = -D_1$ and $\gamma_1 = \gamma_2 = 0$ in the combination of eigenfunctions given by Eq. (15–92). Show that this solution possesses nodal lines which divide the membrane into four congruent triangular sectors.

15–23. a) Determine the five lowest eigenfrequencies for a square membrane clamped at all four edges. b) Find at least one quadruply degenerate eigenvalue for the membrane described in (a).

15–24. Let the initial conditions on a semi-infinite string be

$$y(x, 0) = 0$$

$$\dot{y}(x, 0) = \begin{cases} 0, & x < a, \\ v, & a < x < b, \\ 0, & b < x. \end{cases}$$

Find the Fourier integral representation of $y(x, t)$.

***15–25.** Differentiate your result in Problem 15–24 to obtain the Fourier integral representation of $\dot{y}(x, t)$. Given that

$$\int_0^\infty \frac{\sin Ax}{x}\, dx = \begin{cases} -\pi/2, & A < 0, \\ 0, & A = 0, \\ \pi/2, & A > 0, \end{cases}$$

show by explicit integration that $\dot{y}(x, 0)$ reduces in the range $x > 0$ to what was originally given.

***15–26.** Proceed in the same way as in Problem 15–25 to determine $\dot{y}(x, t)$ at $t = (b - a)/4c$. (Compare with Fig. 15–1b.)

***15–27.** Given the integral stated in Eq. (15–110), verify without approximation that

$$\int_0^\infty \epsilon(\omega)\, d\omega = \frac{\sigma a v^2}{4}$$

for the pulse whose Fourier coefficients are given by Eq. (15–105). (Suggestion: use decomposition into partial fractions.)

15–28. Show that if the initial conditions on an infinite string are such that both $y(x, 0)$ and $\dot{y}(x, 0)$ are odd functions, the Fourier integral representation of $y(x, t)$ is identical with that obtained on the semi-infinite string under the same assigned values of $y(x, 0)$ and $\dot{y}(x, 0)$ in $x \geq 0$.

15–29. Find the Fourier integral representation of $y(x, t)$ for an infinite string corresponding to the initial conditions

$$y(x, 0) = 0, \qquad \dot{y}(x, 0) = \begin{cases} v, & |x| < a, \\ 0, & |x| > a. \end{cases}$$

15–30. Let $\mathbf{A}(k)$ be the Fourier transform of a given function $f(x)$ [cf. Eq. (15–120)]. a) Show that the area under the graph of $f(x)$ is given by $\sqrt{2\pi}\, A(0)$. b) Show that the Fourier transform of $f'(x)$ is $ik\mathbf{A}(k)$. c) If f is an even real function of x, show that $\mathbf{A}$ is an even real function of k. Show that under these conditions $f(k)$ can be interpreted as the Fourier transform of the function $A(x)$. [*Note:* Consider for the purposes of this problem that Eq. (15–120) *defines* $\mathbf{A}(k)$ for positive and negative values of k. The general representation of Eq. (15–119) does not make use of negative values of k. However, when the integrand is an even function, the range of integration can be extended over an interval symmetric about the origin if the result is divided by a factor of two.] d) Apply the results of (c) to Problem 15–29. What pulse shape corresponds to a Fourier transform which is constant for $0 \leq k \leq K$ and zero for $k > K$?

15–31. Let $\mathbf{A}(\omega)$ be the Fourier transform of a given function $f(t)$:

$$\mathbf{A}(\omega) = \frac{1}{\sqrt{2\pi}} \int_{-\infty}^{\infty} f(t) e^{-i\omega t}\, dt.$$

Take $f(t)$ to be a function describing damped simple harmonic motion:

$$f(t) = \begin{cases} 0, & t < 0, \\ e^{-t/\tau} \cos \omega_0 t, & t > 0. \end{cases}$$

a) Show that for $\omega \sim \omega_0$ and $\omega_0 \gg 1/\tau$, $\mathbf{A}(\omega)$ is given approximately by

$$\mathbf{A}(\omega) = \frac{1/(2\sqrt{2\pi})}{1/\tau + i(\omega - \omega_0)}.$$

b) Sketch $|\mathbf{A}(\omega)|^2$ versus ω. If the "half-width" is defined as the value of $\Delta\omega = \omega - \omega_0$ at which $|\mathbf{A}(\omega)|^2$ has half its maximum value, show that $\Delta\omega$ is inversely proportional to τ, the time constant of the exponential decay.

15–32. a) If it is given that the disturbance on an infinite string is a + wave, show that a knowledge of *either* $y(x, 0)$ or $\dot{y}(x, 0)$ is sufficient to determine the Fourier integral coefficient. b) Given a + wave for which

$$\dot{y}(x, 0) = \begin{cases} ve^{Kx}, & x < 0, \\ 0, & x > 0, \end{cases}$$

determine the Fourier integral representation of $\dot{y}(x, t)$.

16 Waves in a dispersive medium

16–1 INTRODUCTION

In the preceding chapters we have been able to discuss the propagation of pulses of arbitrary shape only under the assumption that the medium is nondispersive. The analytical properties of a nondispersive medium stem from the wave equation, which contains a unique parameter c, the wave-propagation velocity. Plane waves of arbitrary shape are propagated without distortion in form. The velocity with which the entire form progresses is c.

We have been restricted in the case of a dispersive medium to dealing with pure sinusoidal waves. The wave equation is assumed to hold for a sinusoidal wave provided that an appropriate value is substituted for c. In a dispersive medium this value is not the same for all frequencies. Since no single wave equation governs the general situation, we have yet to consider the analytical procedure to be followed in determining the propagation in a dispersive medium of an arbitrary waveform. Even the case of a sinusoidal wavetrain of *limited* extent is in principle beyond the scope of the framework so far developed.

In the present chapter we will discuss the general procedure to be followed for an arbitrary waveform and will develop specific results for the limited sinusoidal wavetrain. It will be seen that the shape of a wave is not preserved as the disturbance progresses. This poses a difficulty in defining what is meant by "propagation velocity." For example, it turns out that the sinusoidal character of a long wavetrain is preserved to a certain extent, but the boundaries of the disturbance do not remain sharply defined. It will be necessary to make the important distinction between the *group velocity*, the velocity with which the approximate boundaries propagate, and the *phase velocity*, the velocity with which an individual sinusoidal wavecrest propagates within the train. Since these two velocities differ, any individual wavecrest ultimately passes out through one of the boundaries of the region of major disturbance. As it does this, its amplitude diminishes rapidly. The character of the wavetrain as a whole is preserved because of wavecrests which enter at the other end of the region.

16-2 SUPERPOSITION OF TWO SINUSOIDS

The simplest example of a situation exhibiting the distinction between group and phase velocities is provided by the study of the propagation in a dispersive medium of the energy concentrations in the phenomenon of beats. Consider the superposition of two sinusoidal waves of equal amplitude but of different frequency. Let $c(\omega_1)$ be the velocity of a wave of frequency ω_1. The wavelength of this wave is then fixed and can be specified by the wave number

$$k_1 = 2\pi/\lambda_1 = \omega_1/c(\omega_1). \tag{16-1}$$

If the second wave is of frequency ω_2 and wave number k_2, the superposition of the two waves is given by

$$\psi = A \cos(\omega_1 t - k_1 x) + A \cos(\omega_2 t - k_2 x).$$

Applying a trigonometric identity, we can express this in the form

$$\psi = 2A \cos \tfrac{1}{2}[(\Delta\omega)t - (\Delta k)x] \cos(\bar{\omega}t - \bar{k}x), \tag{16-2}$$

where $\bar{\omega} = \frac{1}{2}(\omega_1 + \omega_2)$, $\bar{k} = \frac{1}{2}(k_1 + k_2)$, $\Delta\omega = \omega_2 - \omega_1$, and $\Delta k = k_2 - k_1$. At any fixed time t_1 the graph of ψ as a function of x can be visualized by use of the envelope concept. Let

$$\psi_e = 2A \cos \tfrac{1}{2}[(\Delta\omega)t_1 - (\Delta k)x].$$

Then the curves $\pm\psi_e$ serve as an envelope within which the oscillations of the factor $\cos(\bar{\omega}t_1 - \bar{k}x)$ are inscribed. Since $\bar{k} > \Delta k$, the oscillations of the latter are more rapid than those of ψ_e.

The graph is similar to Fig. 11–8. As drawn, this graph represents the beat phenomenon at a fixed point of observation. For our present purposes it is necessary to examine ψ as a function of x and t to be able to follow the motion. Note that Eq. (16–2) reduces to Eq. (11–9) at $x = 0$.

As time progresses, the envelope curve shifts to a different position along the x-axis. This can be seen from the fact that ψ_e is a function of the form $f[x - (\Delta\omega/\Delta k)t]$. From this form we also see that the envelope advances at the constant velocity $(\Delta\omega/\Delta k)$. This quantity will be referred to as the *group velocity*, u:

$$u = \Delta\omega/\Delta\kappa.$$

On the other hand, the factor

$$\cos(\bar{\omega}t - \bar{k}x) = \cos\bar{k}\left(x - \frac{\bar{\omega}}{\bar{k}}t\right)$$

has the form of a plus wave which advances with velocity $\bar{\omega}/\bar{k}$. This will be referred to as the *phase velocity*, v:

$$v = \bar{\omega}/\bar{k}.$$

Examples below will show that any of the relations $u < v$, $u = v$, and $u > v$ is possible. For purposes of illustration, let us consider $u > v$. We then obtain the following picture of the changes in the graph of Eq. (16–2) as time progresses. The zeros of ψ_e divide the wave into groups which advance with velocity u. The "population" of an individual group does not remain fixed, however, since the crests within the group are moving at the slower velocity v. As the group advances, wavecrests enter at the forward end, grow to a maximum amplitude, diminish, and finally leave at the trailing end, thereupon entering the following group.

This behavior can be exhibited with an oscilloscope and two audio oscillators. The oscilloscope is set on horizontal sweep and the oscillators are connected in parallel to the vertical input. Let the oscillator frequencies be ω_1 and ω_2, with

$$\bar{\omega} = (\omega_1 + \omega_2)/2 \qquad \text{and} \qquad \Omega = (\omega_2 - \omega_1)/2.$$

The signal across the vertical plates is then of the form $\sin\Omega t \sin\bar{\omega}t$. Suppose that the period of the sweep is slightly less than one period of the envelope $\sin\Omega t$:

$$T_s = \frac{2\pi}{\Omega}(1 - \epsilon), \qquad \epsilon \ll 1.$$

Upon successive sweeps, the envelope appears to advance through a distance corresponding to the time difference $2\pi\epsilon/\Omega$. If the velocity of the sweep is V, this distance is $2\pi\epsilon V/\Omega$. The envelope advances this distance every T_s seconds; consequently, the "group velocity" is

$$u = \frac{2\pi\epsilon V}{\Omega T_s} = \frac{\epsilon V}{1 - \epsilon} \doteq \epsilon V.$$

Also suppose that the period of the sweep is very nearly an integral number of cycles of the function $\sin\bar{\omega}t$:

$$T_s = \frac{2\pi n}{\bar{\omega}}(1 - \epsilon'), \qquad \epsilon' \ll \frac{1}{n}\cdot$$

The inscribed curve then advances at the "phase velocity"

$$v \doteq \epsilon' V.$$

If ϵ and ϵ' can be varied independently, different relations between group and phase velocities can be displayed. Let ω_1 and n be selected arbitrarily. Then the question is how to choose ω_2 and the sweep frequency ω_s so as to yield desired values of ϵ and ϵ'. A little approximate algebra shows the requirement to be

$$\omega_2 = \omega_1 + 2\Omega \quad \text{and} \quad \omega_s = \Omega(1 + \epsilon),$$

where

$$\Omega \doteq \frac{\omega_1}{n-1}\left[1 + \frac{n(\epsilon' - \epsilon)}{n-1}\right].$$

For example, choosing $\omega_1/2\pi = 600\ \text{sec}^{-1}$, with $n = 11$ and $\epsilon = \epsilon' = \frac{1}{60}$, we have $\Omega/2\pi = 60\ \text{sec}^{-1}$, $\omega_2/2\pi = 720\ \text{sec}^{-1}$, and $\omega_s/2\pi = 61\ \text{sec}^{-1}$. These conditions will show the nondispersive condition $u = v$. Raising ω_2 slightly creates a situation with $u < v$ and lowering ω_2 produces $u > v$. This oscilloscope arrangement provides an animated graph of Eq. (16–2). It is not properly considered as a demonstration of dispersion itself, since the trace on the oscilloscope is not a propagating wave.

The relation between the group and phase velocities depends on the function $c(\omega)$. If it happens that c is the same for all frequencies, then

$$\Delta k = \frac{\omega_2}{c} - \frac{\omega_1}{c} = \frac{\Delta\omega}{c},$$

and therefore the group velocity is $u = \Delta\omega/\Delta k = c$. Likewise,

$$\bar{k} = \frac{1}{2}\left(\frac{\omega_1}{c} + \frac{\omega_2}{c}\right) = \frac{\bar{\omega}}{c},$$

and the phase velocity is

$$v = \bar{\omega}/\bar{k} = c.$$

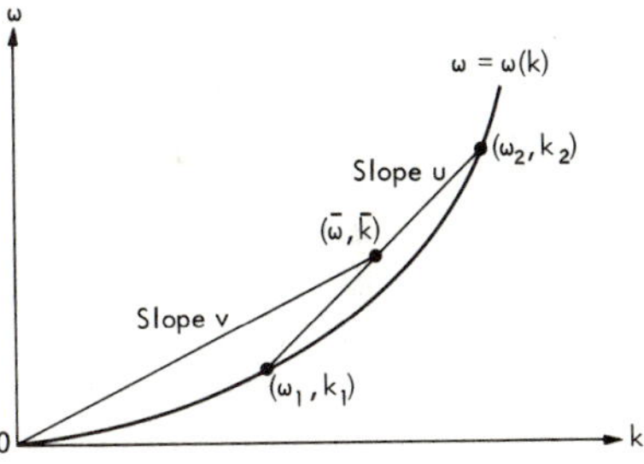

Fig. 16–1. A graph of a dispersion relation.

Thus there is no distinction between group and phase velocities in a nondispersive medium. The term "wave-propagation velocity" is unambiguous in this context.

To study group and phase velocities in a more general case it is convenient to consider that the relation $\omega = kc(\omega)$ defines the frequency ω as a function of the wave number k. That is, although it is not necessary to do so explicitly, we can think of this equation as being solved to yield $\omega = \omega(k)$. The function $\omega(k)$ is referred to as the *dispersion relation* for the medium. In effect, it allows us to choose an arbitrary wavelength and then calculate what frequency is required to produce such a wave in the given medium. As shown in Fig. 16–1, both the group and phase velocities have geometrical interpretations on the graph of the dispersion relation. The group velocity

$u = \Delta\omega/\Delta k$ is the slope of the chord joining the points (ω_1, k_1) and (ω_2, k_2). The phase velocity $v = \bar{\omega}/\bar{k}$ is the slope of the radius vector from the origin to the midpoint of the same chord. If the frequencies ω_1 and ω_2 are sufficiently close to each other, the slope of the chord is very nearly the same as the instantaneous slope of the dispersion curve at $\omega = \omega_1$:

$$\frac{\Delta\omega}{\Delta k} \doteq \left(\frac{d\omega}{dk}\right)_{\omega=\omega_1}.$$

This allows the group velocity to be calculated as a derivative of the dispersion relation

$$u = \left(\frac{d\omega}{dk}\right)_{\omega=\omega_1}. \tag{16–3}$$

Under the same conditions $\bar{\omega}/\bar{k} \doteq \omega_1/k_1$, and the phase velocity can be represented as

$$v = \omega_1/k_1. \tag{16–4}$$

So far the notions of group and phase velocity have been introduced in connection with the example of the sum of two sinusoidal waves of different frequencies. It will be shown in Section 16–5 that the concepts are useful in a wider context. It is in the form of Eqs. (16–3) and (16–4) that the general definitions are made.

The original meaning in Eq. (16–1) of the quantity $c(\omega_1)$ was the velocity of a pure sinusoidal wave of frequency ω_1. On the other hand, the phase velocity v was defined as the velocity of propagation of the individual waves that form a group. Comparison of Eqs. (16–1) and (16–4) shows that these two quantities are identical. They will be used synonymously from now on.

16-3 SURFACE WAVES ON A LIQUID

To illustrate different possible relations between the group and phase velocities defined in the last section we consider waves on the surface of a liquid. Gravity is assumed to be acting and viscosity is neglected. It is also assumed that deformations of the surface are influenced by the surface tension T. If the density of the liquid is ρ and the undisturbed depth is h, the phase velocity is given by*

$$c^2(k) = [(g/k) + (Tk/\rho)] \tanh(kh). \tag{16–5}$$

If the depth is short compared with a wavelength [$kh \ll 1$ and $\tanh(kh) \doteq kh$], the liquid is considered *shallow*. For the opposite extreme [$kh \gg 1$ and $\tanh(kh) \doteq 1$], the liquid is considered *deep*. A further classification results

* C. A. Coulson, *Waves*, New York: Interscience, 1955, p. 83.

from a comparison of the terms (g/k) and (Tk/ρ). Setting the two terms equal defines a critical wavelength

$$\lambda_c = 2\pi\sqrt{T/\rho g}\,.$$

For wavelengths such that $\lambda \gg \lambda_c$ the surface-tension term is of negligible importance and the wave is called a *gravity* wave. For wavelengths such that $\lambda < \lambda_c$ the surface-tension term is dominant and the wave is called a *ripple*. Three limiting cases of Eq. (16–5) are of special interest.

i) Gravity waves in a shallow liquid ($\lambda_c, h \ll \lambda$). In this case $c = \sqrt{gh}$ and the velocity is independent of k. The waves are nondispersive and $u = v = c$. This is the situation considered in Appendix V and referred to as *tidal waves* or *long waves in shallow water.*

ii) Gravity waves in deep liquid ($\lambda_c \ll \lambda \ll h$). In this case $c = \sqrt{g/k}$ and the dispersion relation is $\omega = \sqrt{gk}$. From this the group velocity is

$$u = \frac{d\omega}{dk} = \frac{1}{2}\sqrt{\frac{g}{k}} = \tfrac{1}{2}v.$$

Ship waves fall in this category. Individual wavecrests travel faster than the group and diminish rapidly in amplitude as they pass through the front boundary of the group. A canoeist preparing for the onslaught of motorboat waves has twice as much time as he would estimate from the velocity of an individual wavecrest.

iii) Short ripples in deep liquid ($\lambda \ll \lambda_c, h$). In this case $c = \sqrt{Tk/\rho}$ and the dispersion relation is $\omega = \sqrt{T/\rho}\,k^{3/2}$. The group velocity is

$$u = \tfrac{3}{2}\sqrt{T/\rho}\,k^{1/2} = \tfrac{3}{2}v.$$

Since $u > v$, individual wavecrests fall behind the group, while new crests build up at the forward edge of the group.

The critical wavelength for water is $\lambda_c = 1.7$ cm. In a typical ripple tank experiment $\lambda \sim 1$ cm, and h is in the range of 1 mm to 1 cm. The water cannot be considered to be either shallow or deep, and the full generality of Eq. (16–5) is required. The variation of velocity with depth is slight except when the depth is quite small. When h is much less than a millimeter, viscosity becomes important and there is a high degree of attenuation.

16–4 GENERAL PROCEDURE FOR DETERMINING PROPAGATION IN A DISPERSIVE MEDIUM

In a dispersive medium we can no longer use the wave equation as the basis for a discussion of the behavior of the function $y(x, t)$ which describes a disturbance of arbitrary form. Instead, it is accepted as fundamental that the sinusoidal component waves determined by a Fourier analysis can each be propagated at their own characteristic velocities and that the resultant waveform can be calculated by superimposing these waves at a later time. If the initial waveform is $y(x, 0)$, the composition of wavelengths is obtained by finding the Fourier coefficient [cf. Eq. (15–120)]:

$$\mathbf{A}(k) = \frac{1}{\sqrt{2\pi}} \int_{-\infty}^{\infty} y(x, 0)e^{-ikx}\, dx. \tag{16–6}$$

The initial waveform is expressed as a continuous superposition of sine waves through the inverse Fourier transform [cf. Eq. (15–119)]:

$$y(x, 0) = \sqrt{\frac{2}{\pi}}\, \mathcal{R}e \int_0^{\infty} \mathbf{A}(k)e^{ikx}\, dk. \tag{16–7}$$

The quantity $\sqrt{2/\pi}\,\mathbf{A}(k)$ is the complex amplitude of the contributing sinusoidal wave of wave number k. If each such wave propagates in the positive direction with a velocity $c(k)$, then the contribution $\sqrt{2/\pi}\,\mathbf{A}(k)e^{ikx}$ is given at a later time by $\sqrt{2/\pi}\,\mathbf{A}(k)e^{i(kx-\omega t)}$, where $\omega = \omega(k) = kc(k)$. Thus it is postulated that the net disturbance at a later time is given by the superposition

$$y(x, t) = \sqrt{\frac{2}{\pi}}\, \mathcal{R}e \int_0^{\infty} \mathbf{A}(k)e^{i(kx-\omega t)}\, dk. \tag{16–8}$$

In the event that c is not a function of k, this result reduces to what is expected for a nondispersive medium. For this case we can write

$$kx - \omega t = k(x - ct) = kx',$$

and

$$y(x, t) = \sqrt{\frac{2}{\pi}}\, \mathcal{R}e \int_0^{\infty} \mathbf{A}(k)e^{ikx'}\, dk.$$

Now, since x' is *not a function of* k, the right-hand side of this equation is formally identified from Eq. (16–7) to be $y(x', 0)$. Thus

$$y(x, t) = y(x', 0) = y(x - ct, 0).$$

This is the expected solution in a nondispersive medium, representing a plus wave having the given initial profile.

16–5 GROUP VELOCITY OF A NEARLY SINUSOIDAL WAVE

The concept of group velocity was introduced in Section 16–2 in connection with the superposition of two sinusoids. The resultant waveform consisted of an infinite succession of groups corresponding to the amplitude modulations of the beat phenomenon. In this section we shall show that the same definition of group velocity extends to the discussion of any nearly sinusoidal wave.

By *nearly sinusoidal wave* we understand a wave whose initial shape is specified by a function of the form

$$y(x, 0) = E(x) \cos [Kx + \phi(x)]$$

in which $E(x)$ and $\phi(x)$ are slowly varying functions in comparison with $\cos Kx$. Letting

$$\mathbf{E}(x) = E(x)e^{i\phi(x)},$$

we can write the equation:

$$y(x, 0) = \mathcal{R}e \, \{\mathbf{E}(x)e^{iKx}\}. \tag{16–9}$$

The envelope function $E(x)$ gives *amplitude modulation* to the sinusoidal wave; *phase modulation* is associated with $\phi(x)$. An example is the limited sinusoidal wavetrain of Eq. (15–121). The envelope is the rectangle function

$$E(x) = \begin{cases} 1, & |x| \leq a, \\ 0, & |x| > a, \end{cases}$$

and the phase function is $\phi(x) = -\pi/2$. It was seen that the Fourier transform of Eq. (15–121) is a function whose values are sharply concentrated about $k = K$, provided a large number of cycles are contained in the interval $|x| \leq a$. This will be taken to be an essential feature of a nearly sinusoidal wave. That is, if the Fourier transform of Eq. (16–9) is $\mathbf{A}(k)$, it will be assumed that $\mathbf{A}(k)$ is negligible except when k is very close to K.

Consider then the Fourier integral representation of $y(x, t)$:

$$y(x, t) = \sqrt{\frac{2}{\pi}} \, \mathcal{R}e \int_0^\infty \mathbf{A}(k)e^{i(kx-\omega t)} \, dk.$$

Because of the assumed property of $\mathbf{A}(k)$, it is permissible to replace the remainder of the integrand by an approximation which is good near $k = K$. Such an approximation is provided by the first two terms in the Taylor series expansion about $k = K$ of the phase angle in the exponential. If we let $\omega(K) = \Omega$ and $(d\omega/dk)_{k=K} = u$, the expansion is

$$[kx - \omega(k)t] = [Kx - \Omega t] + (k - K)[x - ut] + \cdots,$$

and the approximation is

$$y(x, t) \doteq \sqrt{\frac{2}{\pi}} \, \mathcal{R}e \left\{ e^{i(Kx-\Omega t)} \int_0^\infty \mathbf{A}(k)e^{i(k-K)(x-ut)} \, dk \right\}. \tag{16–10}$$

Taking $t = 0$ and comparing with Eq. (16–9), we can see that the integral appearing in this equation is to be identified with the complex envelope function $\mathbf{E}(x)$:

$$\mathbf{E}(x) = \sqrt{\frac{2}{\pi}} \int_0^\infty \mathbf{A}(k)e^{i(k-K)x} \, dk.$$

Thus Eq. (16–10) can be written

$$y(x, t) = \mathcal{R}e\,\{\mathbf{E}(x - ut)e^{i(Kx-\Omega t)}\}.$$

This has the same basic form as $y(x, 0)$. At a fixed time t_1 the graph of $y(x, t_1)$ consists of sinusoidal waves of wavelength $\Lambda = 2\pi/K$. These waves are inscribed in an envelope given by $|\mathbf{E}(x - ut_1)|$ and are subject to a phase modulation given by the argument of $\mathbf{E}$. The individual waves propagate with the phase velocity

$$v = \Omega/K = c(\Omega).$$

But the amplitude and phase modulations propagate with the velocity

$$u = \left(\frac{d\omega}{dk}\right)_{k=K}.$$

In particular, if $|\mathbf{E}(x)|$ vanishes outside a limited interval, the disturbance consists of a limited wavetrain, or a *group* of waves. The boundaries of this group advance with velocity u. The significance of the group velocity is more general, however, since it is the velocity with which any modulation of the sinusoidal wave is propagated.

16-6 THE SIGNIFICANCE OF GROUP VELOCITY IN AN OPTICAL CONTEXT

Now that we are aware of the existence of two meanings for "the velocity of light" in a dispersive medium, we should examine different experiments in optics to find whether the group or the phase velocity is involved. In the analysis of the refraction of sinusoidal plane waves at an interface, the waves are treated as being of indefinite extent. It is therefore clear that the ordinary index of refraction $n(\omega)$ is a measure of the *phase* velocity $v = c(\omega)$ through the relation $n(\omega) = c_0/c(\omega)$. In fact, the phenomenological value of $c(\omega)$ which is to be inserted in the wave equation is obtained through this relation (cf. Section 6–4).

On the other hand, most direct measurements of the velocity of light in a material medium involve the chopping of a sinusoidal light beam and the determination of the travel time for the resultant pulses. The velocity obtained is consequently the *group* velocity of a long wavetrain. To examine the expected difference between these two velocities, both measured at the same frequency, it is convenient to express the group velocity in terms of the index of refraction through the relation

$$k = \frac{\omega}{c(\omega)} = \frac{\omega n(\omega)}{c_0}.$$

Thus

$$\frac{1}{u} = \frac{dk}{d\omega} = \frac{n(\omega)}{c_0} + \frac{\omega}{c_0}\left(\frac{dn}{d\omega}\right) = \frac{1}{c(\omega)}\left[1 + \frac{\omega}{n(\omega)}\left(\frac{dn}{d\omega}\right)\right]. \quad (16\text{–}11)$$

In regions of normal dispersion ($dn/d\omega > 0$), the *group index of refraction* $n_g \equiv c_0/u$ is therefore expected to be greater than the regular index of refraction $n = c_0/c(\omega)$ by the factor $[1 + (\omega/n)(dn/d\omega)]$.

Since the Newtonian corpuscular model of light propagation interprets the phenomenon of refraction by assuming that the velocity of light is greater in a material medium than in a vacuum, whereas the wave model predicts the opposite, the direct measurement of the travel time of a light signal in a material medium was once considered an *experimentum crucis.* However, before the experiment was actually performed by Foucault in 1850, the corpuscular theory had already been defeated by the successful analysis of interference experiments by Young and Fresnel on the basis of the wave interpretation. Foucault's experiment merely provided additional confirmation. Michelson repeated the experiment with improved accuracy in 1885. His results are of interest to us because they enable us to compare measured values of velocities with the predictions of Eq. (16–11). Using carbon disulfide, which is a medium of relatively high dispersion, Michelson obtained a value $n_g = 1.758$ for a beam of white light. The regular index of refraction for white light is $n = 1.635$. It is therefore correct that $n_g > n$, and the ratio corresponds to a reasonable "average" value of $(\omega/n)(dn/d\omega)$ to assume for white light.

16–7 THE VELOCITY OF LIGHT IN FREE SPACE AS A LIMITING VALUE

It is fundamental to the theory of special relativity that there are no conditions, whether in free space or in a material medium, under which a signal can be propagated from one point to another with a velocity greater than c_0, the velocity of light in free space. Yet the phase velocity of a sinusoidal wave is $c(\omega) = c_0/n(\omega)$, and the value of $n(\omega)$ measured by refraction experiments is in some circumstances less than one. Figure 6–2 indicates, for example, that the index of refraction of all substances approaches unity through values *less than one* in the limit of high frequencies. It is thus an established fact that the phase velocity can, under certain circumstances, be greater than c_0.

The fictitious character of the infinite sinusoid is at fault in leading us to believe that the preceding is a contradiction of the relativity principle. The pure sinusoidal wave extends uniformly throughout all space for all time. Observers at two points A and B can both detect the sine wave, but the pure wave can hardly be said to be conveying any *information* from A to B. The transmission of a signal requires that the sine wave be interrupted or modulated in some way. From the earlier discussion in this chapter we expect the propagation velocity of such modifications to be the *group* velocity rather than the phase velocity. From Eq. (16–11) we observe that the group velocity is less than the phase velocity in regions of normal dispersion. In most cases this results in a value of u which is less than c_0 and the dilemma is removed.

Nevertheless, a possibility still remains that the value of u computed from Eq. (16–11) will be greater than c_0. This occurs most notably in regions of *anomalous* dispersion where $dn/d\omega < 0$. In these regions it must be recognized, however, that certain approximations were made in arriving at the conclusion that $d\omega/dk$ should be identified with the propagation velocity of the amplitude and phase modulations of a nearly sinusoidal disturbance. These approximations assumed that $\omega(k)$ could be adequately represented by the first two terms in a Taylor series expansion about K, the wave number associated with the sinusoidal portion of the wave. On the other hand, anomalous dispersion occurs in the vicinity of *resonance absorption lines*, where $\omega(k)$ is a rapidly varying function. Furthermore, attenuation due to absorption is not negligible under these circumstances and must be taken into account in a discussion of waveform.

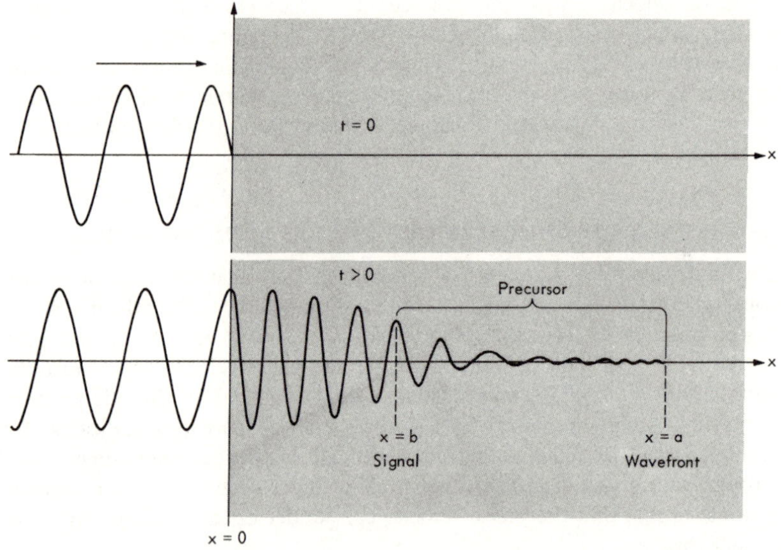

Fig. 16–2. Waveform in a dispersive medium in response to a sinusoidal wave which is incident at $t = 0$.

A complete analysis of the propagation of a limited sinusoidal wavetrain has been carried out by Sommerfeld and Brillouin.* The type of situation considered is indicated in Fig. 16–2. The region $x < 0$ is a vacuum and $x > 0$ is a dispersive medium. A plane wave is normally incident from $x < 0$.

---

* L. Brillouin, *Wave Propagation and Group Velocity*, New York: Academic Press, 1960.

The disturbance at $x = 0$ is taken to be the abruptly initiated sine wave:

$$f(t) = \begin{cases} 0, & t < 0, \\ \sin \Omega t, & t \geq 0. \end{cases}$$

The shape of this wave is not preserved as it propagates through the medium. The following features of the distorted waveform are derived by Sommerfeld and Brillouin.

i) At any time t there is a definite value $x = a$ such that the disturbance vanishes identically in $x > a$, but some form of disturbance, perhaps minor, is present in $x < a$. The location $x = a$ is referred to as the *wavefront. The wavefront velocity is in all cases equal to* c_0, the velocity of light in free space. This is true regardless of the relative values of u, v, and c_0.

The electron theory of matter proposes a sensible interpretation of this analytical result. Interaction between the incident wave and the material medium takes place through an induced oscillation of charge. The mechanical inertia of the charged particles implies that some time lag is required before the charge acquires a significant displacement. To begin with, therefore, the passing wave is "unaware" of the presence of matter and propagates at the same speed that it would in a vacuum.

ii) The portion of the disturbance immediately behind the wavefront is known as the *precursor* wave. The amplitude of the precursor is quite small compared with the steady-state wave which ultimately arrives. The oscillations of the precursor are not of constant frequency. In fact, the characteristics of the initial part of the precursor are not critically dependent on the frequency of the incident wave.

The low amplitude of the precursor is associated with the withdrawal of energy from the incident wave to build up the motion of the charge contained within the molecules of the material medium.

iii) The transition between the precursor wave and the steady-state sinusoidal wave takes place fairly abruptly. We can define the boundary of the *signal* by choosing the location ($x = b$) at which the amplitude reaches half the steady-state value. The velocity at which this locus propagates defines the *signal velocity* $U = b/t$. At frequencies which are not close to an absorption band, the signal velocity turns out to be identical with the quantity $u = d\omega/dk$. Under ordinary circumstances, therefore, the *signal* and *group* velocities are equal. In regions of anomalous dispersion the signal velocity cannot be expressed by such a simple condition. However, it is true in all cases that the signal velocity is less than c_0. In those cases for which $U \neq u$ the "group velocity" is simply a number calculated from the formula $u = d\omega/dk$, but this number has no particular relevance to the situation. Thus the fact that u can be greater than c_0 does not constitute a contradiction to the principle of relativity.

iv) The phase velocity $v = c(\omega)$ serves to determine the *phase* of the steady-state wave, but is not directly connected with the location of the signal boundary. If $v > U$, even in the extreme case $v > c_0$, individual crests move faster than the signal boundary. However, the amplitude decreases rapidly as the crests enter the precursor region at $x = b$. Far inside the precursor region the "inscribed sine wave" loses its identity since the disturbance in this region is not well characterized in terms of the envelope concept. The amplitude of the new disturbance becomes identically zero at $x = a$.

PROBLEMS

16–1. With h as parameter, sketch the family of dispersion curves $\omega = \omega(k)$ for gravity waves [$T = 0$ in Eq. (16–5)].

16–2. a) For waves in deep water, show that the phase velocity is a minimum at the critical wavelength

$$\lambda_c = 2\pi\sqrt{T/\rho g}\,.$$

b) Show that waves of this wavelength are approximately nondispersive (that is, $u = v$ at $\lambda = \lambda_c$). c) Sketch the dispersion curve for deep water waves. What is the geometrical significance on this graph of the feature discussed in (b)?

16–3. The surface tension of water is about 75 dynes/cm. Calculate the minimum phase velocity for waves on deep water and show that this corresponds to the critical wavelength $\lambda_c = 1.7$ cm. Show that there are two different wavelengths corresponding to any other phase velocity, the product of which is equal to λ_c^2.

16–4. Let the Fourier transform of an initial waveform be

$$\mathbf{A}(k) = \begin{cases} 0, & |k - K| > \Delta k, \\ 1, & |k - K| < \Delta k, \end{cases}$$

where $\Delta k \ll K$. Find the complex envelope function $\mathbf{E}(x)$ and sketch the function $y(x, t)$ at a general time t.

16–5. a) If the index of refraction $n = c_0/v$ is given as a function of wavelength, show that the "group index of refraction" can be written as

$$n_g \equiv c_0/u = n - \lambda(dn/d\lambda).$$

b) Given that Δn for glass is the order of -10^{-5} for $\Delta\lambda = 1$ Å, approximately what distance must a sinusoidal group travel to lose one wave at the front of the group?

16–6. The de Broglie relation assigns a matter wave of wavelength $\lambda = h/p$ to a material particle of momentum p, where h is Planck's constant. If we set $\hbar = h/2\pi$ and $k = 2\pi/\lambda$, the relation can be written $p = \hbar k$. Free space is assumed to act as a "dispersive medium," and ω, the angular frequency of the matter waves, is therefore a function of k. The observed "velocity of the particle" is taken to be the group velocity $u = d\omega/dk$ associated with a *wavepacket*, i.e., a nearly sinusoidal wave in the sense of Section 16–5. a) If m is the mass of the particle and the classical relation $p = mu$ is assumed, show that $\omega(k)$ can be determined by integration of the group velocity relation. If the constant of integration is taken to be zero for a free particle (not subject to any forces), show that the result is consistent with the Bohr relation $E = h\nu$, where $E = \frac{1}{2}mu^2$ is the classical expression for the total energy of the particle. Show that the phase velocity is half the group velocity. b) Let m be the rest mass of the particle and c the velocity of light in free space. If the relativistic relation

$$p = \frac{mu}{\sqrt{1 - u^2/c^2}}$$

is assumed, determine $\omega(k)$ and show that this is consistent with the Bohr relation where

$$E = \frac{mc^2}{\sqrt{1 - u^2/c^2}}$$

is the relativistic expression for the total energy. Show that the product of the phase and group velocities is c^2. c) Specialize the dispersion relation $\omega(k)$ obtained in (b) to the case of a particle of zero rest mass (a photon). Deduce the group and phase velocities.

17 The acoustic wave equation in three dimensions

17-1 INTRODUCTION

We have already discussed a number of three-dimensional situations which can be applied to acoustic waves. These problems were approached in previous chapters by a method which falls short of being completely analytical. In every case we have had to make use of auxiliary hypotheses (e.g., the Huygens-Fresnel principle in Section 12-1) or intuitive arguments (e.g., the deduction that $\psi \propto 1/r$ in Section 11-1). In this chapter we shall revert to basic principles and deduce an analytical representation of some three-dimensional acoustic situations. This does not mean that it will not be necessary to make approximations to handle the equations. These approximations will have to be made explicitly, however, and will not be hidden behind tacit assumptions.

For problems of the type considered in this chapter, the acoustic and electromagnetic equations are not sufficiently isomorphic to allow the claim to be made that the results obtained in the acoustic case apply exactly to some electromagnetic situation. The most essential distinction between the two cases is that an acoustic wave, as we shall see below, can be associated with a *scalar* potential function, whereas an electromagnetic wave is a *vector* field and requires a *vector* potential function for its description. The techniques employed in both cases are similar, and the results are in many respects alike. We have chosen to illustrate these techniques through the simpler example of the scalar field, leaving the analogous treatment of the vector field for more advanced courses in electromagnetic theory.

17-2 EXACT HYDRODYNAMICAL EQUATIONS FOR AN INVISCID COMPRESSIBLE FLUID

We adopt the Eulerian formulation in which the numbers x, y, z are the coordinates of a fixed point in space. This is in contrast to the Lagrangian formulation we have used previously in which x, y, z are the labels identifying a particle of the fluid according to its position before the initiation of a disturbance. In the Eulerian notation the functions $P(x, y, z, t)$ and $\rho(x, y, z, t)$ stand for the pressure and density at time t at the point whose coordinates are x, y, z. The velocity and acceleration of the particles of fluid which are passing through this point at time t will be designated by $\mathbf{v}(x, y, z, t)$ and $\mathbf{a}(x, y, z, t)$ respectively.

The particles which are at x, y, z, at time t will be at a different point x', y', z' at time $t' = t + \Delta t$. Their velocity will then be $\mathbf{v}' = \mathbf{v}(x', y', z', t')$. By the differential approximation we can write

$$\begin{aligned} x' &= x + v_x \,\Delta t, \\ y' &= y + v_y \,\Delta t, \\ z' &= z + v_z \,\Delta t. \end{aligned}$$

Again using the differential approximation, the x-component of the velocity of these particles is

$$\begin{aligned} v'_x &= v_x(x', y', z', t') \\ &= v_x(x, y, z, t) + (x' - x)\frac{\partial v_x}{\partial x} + (y' - y)\frac{\partial v_x}{\partial y} \\ &\qquad + (z' - z)\frac{\partial v_x}{\partial z} + (t' - t)\frac{\partial v_x}{\partial t} \\ &= v_x + (v_x \,\Delta t)\frac{\partial v_x}{\partial x} + (v_y \,\Delta t)\frac{\partial v_x}{\partial y} + (v_z \,\Delta t)\frac{\partial v_x}{\partial z} + \Delta t \frac{\partial v_x}{\partial t}\cdot \end{aligned}$$

Introducing the operator symbol

$$(\mathbf{v} \cdot \text{grad}) \equiv v_x \frac{\partial}{\partial x} + v_y \frac{\partial}{\partial y} + v_z \frac{\partial}{\partial z},$$

we obtain

$$v'_x = v_x + \left[(\mathbf{v} \cdot \text{grad})v_x + \frac{\partial v_x}{\partial t}\right]\Delta t.$$

Thus

$$a_x \equiv \lim_{\Delta t \to 0}\left[\frac{v'_x - v_x}{\Delta t}\right] = (\mathbf{v} \cdot \text{grad})v_x + \frac{\partial v_x}{\partial t}\cdot$$

Similar results hold for the other two components of $\mathbf{a}$ and can be summarized by the vector equation

$$\mathbf{a} = \frac{\partial \mathbf{v}}{\partial t} + (\mathbf{v} \cdot \text{grad})\mathbf{v}. \tag{17–1}$$

An observer who records only the velocities of the particles at the fixed point x, y, z will find that these numbers are changing at the rate $\partial \mathbf{v}/\partial t$. This is referred to as the *local derivative* of $\mathbf{v}$. As shown by Eq. (17–1), $\partial \mathbf{v}/\partial t$ alone does not represent the acceleration which the particles of the fluid are undergoing. It is necessary to follow these particles to their position at t' and determine their velocity at the new position.* The additional term $(\mathbf{v} \cdot \text{grad})\, \mathbf{v}$ is known as the *convective derivative* of $\mathbf{v}$. The entire right-hand side of Eq. (17–1) is abbreviated $D\mathbf{v}/Dt$ and is known as the *substantial derivative* of $\mathbf{v}$ or the derivative *following the motion of the substance*. These interpretations apply, in general, to any quantity $\psi(x, y, z, t)$: $\partial\psi/\partial t$ is the local rate of change of ψ;

$$\frac{D\psi}{Dt} \equiv \frac{\partial\psi}{\partial t} + (\mathbf{v} \cdot \text{grad})\psi$$

is the rate of change of ψ as perceived by an observer moving with a particle of the fluid.

Law of conservation of mass. Select an arbitrary fixed volume V within the fluid. Let $d\mathbf{S}$ be a vector directed along the outwardly drawn normal representing an element of area on the surface containing the chosen volume. The mass which flows out of the volume through $d\mathbf{S}$ per unit time is $\rho\mathbf{v} \cdot d\mathbf{S}$. Consequently, the total loss of mass by V per unit of time is

$$\oiint_S \rho\mathbf{v} \cdot d\mathbf{S},$$

where the surface integral is taken over the entire surface enclosing V. The total mass contained by V at time t is

$$\iiint_V \rho\, dV.$$

Thus another expression for the loss of mass per unit time is

$$-\frac{d}{dt}\iiint_V \rho\, dV.$$

Equating these two expressions, we have

$$\oiint_S \rho\mathbf{v} \cdot d\mathbf{S} = -\frac{d}{dt}\iiint_V \rho\, dV.$$

Since we have chosen a fixed volume, the right-hand side can be written as

$$\iiint_V \left(-\frac{\partial\rho}{\partial t}\right) dV.$$

---

* An example in which $\partial\mathbf{v}/\partial t = 0$ but $\mathbf{a} \neq 0$ is provided by the *steady* flow of fluid through a pipe of variable cross section. The assumption of steady flow implies that the velocity observed at any *fixed* point of observation is constant, or, $\partial\mathbf{v}/\partial t = 0$. As a particle moves down the tube, however, its velocity varies according to its position in the tube. (For example, the velocity is greatest where the cross-sectional area is least.)

The left-hand side can be transformed by the divergence theorem (see Appendix VII) to yield

$$\oiint_S \rho\mathbf{v} \cdot d\mathbf{S} = \iiint_V \operatorname{div}(\rho\mathbf{v})\, dV.$$

Thus

$$\iiint_V \left\{\operatorname{div}(\rho\mathbf{v}) + \frac{\partial\rho}{\partial t}\right\} dV = 0.$$

Since V was an *arbitrarily* chosen volume, we conclude that

$$\operatorname{div}(\rho\mathbf{v}) + \frac{\partial\rho}{\partial t} = 0. \tag{17–2}$$

The dynamical equation of motion. The net force exerted by the outside fluid on the fluid within the volume V is $-\oiint_S P\, d\mathbf{S}$. This can be transformed by an analog of the divergence theorem (Eq. VII–14) to $\iiint_V (-\operatorname{grad} P)\, dV$. Since we assume there are no other forces acting on the fluid (gravity and viscosity are neglected), the net force acting on the particles in a small element of volume ΔV is therefore $(-\operatorname{grad} P)\,\Delta V$. The mass of these particles is $\rho\,\Delta V$ and therefore, by Newton's second law

$$(-\operatorname{grad} P)\,\Delta V = (\rho\,\Delta V)\mathbf{a},$$

or

$$-\operatorname{grad} P = \rho\left[\frac{\partial\mathbf{v}}{\partial t} + (\mathbf{v}\cdot\operatorname{grad})\mathbf{v}\right]\cdot \tag{17–3}$$

17–3 LINEARIZATION OF THE EULERIAN EQUATIONS

To obtain tractable differential equations we must now restrict the context to fluid motions in which certain of the variables remain small and the products of small quantities can be neglected. The same definitions as those used in the one-dimensional case apply to the condensation variable $s \equiv (\rho - \rho_0)/\rho_0$ and to the excess acoustic pressure $p \equiv P - P_0$. Likewise, as in Eq. (2–17), the adiabatic equation of state is approximated by the first term in a Taylor-series expansion:

$$p = (\rho - \rho_0)P'_a(\rho_0) = (\rho - \rho_0)c^2 = \rho_0 c^2 s. \tag{17–4}$$

In an acoustic disturbance the variables s, p, and v and their derivatives are considered to be "first-order" small quantities. Products of the type v^2 and sv are "second order" and will be dropped from the equations. Thus $(\mathbf{v}\cdot\operatorname{grad})\mathbf{v}$, the convective contribution to the acceleration, is neglected in comparison with $\partial\mathbf{v}/\partial t$, and ρ is replaced by ρ_0 in $\rho\mathbf{v}$ and $\rho(\partial\mathbf{v}/\partial t)$. Equation (17–2) is thus written

$$\rho_0 \operatorname{div}\mathbf{v} + \frac{\partial\rho}{\partial t} = 0,$$

or, equivalently,

$$\operatorname{div} \mathbf{v} = -\frac{\partial s}{\partial t}, \tag{17–5}$$

and Eq. (17–3) is written

$$\frac{\partial \mathbf{v}}{\partial t} = -\frac{1}{\rho_0} \operatorname{grad} p. \tag{17–6}$$

Elimination of **v** between these last two equations can be carried out by subtracting the divergence of the second from the partial time derivative of the first. The symbol ∇^2 will be used for the *Laplacian operator*, div grad, operating on a scalar function. Using Eq. (17–4) we can express the result as a condition on $p(x, y, z, t)$:

$$\nabla^2 p = \frac{1}{c^2} \frac{\partial^2 p}{\partial t^2}. \tag{17–7}$$

This equation reduces to the one-dimensional wave equation if $\partial p/\partial y$ and $\partial p/\partial z$ are zero. We shall see by consideration of a special example that other solutions of Eq. (17–7) represent a more general kind of wave propagation with propagation velocity c. For this reason an equation of this form is referred to as the *three-dimensional wave equation*. From Eq. (17–4) it can be seen that s and ρ are *linear* functions of p and hence both of these variables also satisfy the three-dimensional wave equation with the same propagation velocity c.

We can obtain an equation involving **v** alone by subtracting $(1/c^2)$ times the partial time derivative of Eq. (17–6) from the gradient of (17–5). The result is

$$\operatorname{grad} \operatorname{div} \mathbf{v} = \frac{1}{c^2} \frac{\partial^2 \mathbf{v}}{\partial t^2}. \tag{17–8}$$

This equation will be written in more suitable form through the following considerations. The Laplacian of a vector such as **v** is defined by the relation

$$\nabla^2 \mathbf{v} \equiv \operatorname{grad} \operatorname{div} \mathbf{v} - \operatorname{curl} \operatorname{curl} \mathbf{v}. \tag{17–9}$$

But if we take the curl of Eq. (17–6), since the curl of the gradient of any quantity is zero, we find

$$\frac{\partial}{\partial t} (\operatorname{curl} \mathbf{v}) = 0.$$

This implies that curl **v** is constant in time. In particular, if curl **v** vanishes at every point in the field at any instant of the time, then the motion of the fluid is such as to preserve the relation curl **v** = 0 at all times. Fluid motions which have the special property curl **v** = 0 are referred to as *irrotational*. Setting curl **v** equal to zero, as we shall do, amounts to restricting the discussion of possible motions of the fluid to the irrotational case. However, it will be found that the solutions which apply to this case are capable of describing the common phenomena of sound propagation in three dimensions. It is therefore not by an argument of *necessity* that we take curl $\mathbf{v} = 0$ at this point, but an argument of *sufficiency*. We are simply inquiring what is possible within the restricted class of irrotational motions of the fluid. It is, incidentally, certain that all *transient* acoustic problems starting with the fluid in a condition of equilibrium must come within this category, for then $\mathbf{v}_0 = \mathbf{v}(x, y, z, 0) = 0$ and curl $\mathbf{v}_0 = 0$.

If we make the assumption of irrotationality, we can combine Eqs. (17–8) and (17–9) and obtain

$$\nabla^2 \mathbf{v} = \frac{1}{c^2} \frac{\partial^2 \mathbf{v}}{\partial t^2}. \tag{17–10}$$

This is the three-dimensional wave equation for a *vector* field.

In a cartesian coordinate system the representation

$$\nabla^2 = \frac{\partial^2}{\partial x^2} + \frac{\partial^2}{\partial y^2} + \frac{\partial^2}{\partial z^2}$$

is valid. This implies, for example, that $v_x(x, y, z, t)$ satisfies an equation of the same form as Eq. (17–7) with v_x in place of p. When working in a curvilinear coordinate system, it is important to realize that $\nabla^2 \mathbf{v}$ is defined by Eq. (17–9). It turns out, for example, that v_r, the radial component of the velocity vector in a spherical polar coordinate system, does not satisfy an equation of precisely the same form as the scalar quantities p and ρ.

17–4 INTRODUCTION OF A VELOCITY POTENTIAL FUNCTION

What is mathematically most significant about the class of irrotational fluid motions is that the criterion curl $\mathbf{v} = 0$ is a necessary and sufficient condition for the existence of a potential function $\phi(x, y, z, t)$ such that

$$\mathbf{v} = -\text{grad}\, \phi. \tag{17–11}$$

A function ϕ having this property is known as a *velocity-potential function.*

In addition to the fact that it is easier to work with a scalar rather than a vector, the velocity-potential function is especially convenient for acoustic fields since it is possible to define it in such a way that we can also derive the pressure and density from it. This can be seen by substituting Eq. (17–11) into Eq. (17–6):

$$\frac{\partial \mathbf{v}}{\partial t} = \frac{\partial}{\partial t}(-\text{grad}\, \phi) = -\frac{1}{\rho_0}\, \text{grad}\, p$$

or,

$$\text{grad}\left[p - \rho_0 \frac{\partial \phi}{\partial t}\right] = 0.$$

The general solution to this equation is*

$$p - \rho_0 \frac{\partial \phi}{\partial t} = C(t),$$

where $C(t)$ is independent of position. To be assured that $C(t)$ can always be chosen equal to zero, let $\phi_0(x, y, z)$ be any function such that $-\text{grad}\, \phi_0 = \mathbf{v}_0$, where $\mathbf{v}_0(x, y, z) = \mathbf{v}(x, y, z, 0)$ is the velocity field at $t = 0$. Then consider the function

$$\phi(x, y, z, t) = \phi_0(x, y, z) + \frac{1}{\rho_0} \int_0^t p(x, y, z, t)\, dt. \tag{17-12}$$

We must show first that this is a velocity-potential function. Taking the gradient of Eq. (17–12) and consulting Eq. (17–6) we find

$$\text{grad}\, \phi = \text{grad}\, \phi_0 - \int_0^t \frac{\partial \mathbf{v}}{\partial t}\, dt = -\mathbf{v}_0 - [\mathbf{v} - \mathbf{v}_0] = -\mathbf{v}.$$

Hence the function defined by Eq. (17–12) does satisfy Eq. (17–11). But it also follows from Eq. (17–12) that

$$\frac{\partial \phi}{\partial t} = \frac{1}{\rho_0}\, p.$$

This means that it is always possible to choose a potential function whose space derivatives yield the components of the velocity and whose time derivative yields the excess acoustic pressure through the relation

$$p = \rho_0 \frac{\partial \phi}{\partial t}. \tag{17-13}$$

It will henceforth be assumed that the term "velocity potential" refers to a function which has been chosen to satisfy Eq. (17–13) as well as (17–11). The equation satisfied by the potential function which has been so defined can be obtained by writing Eq. (17–5) in the form

$$\text{div}\, \mathbf{v} = -\frac{1}{\rho_0 c^2} \frac{\partial p}{\partial t} = -\frac{1}{c^2} \frac{\partial^2 \phi}{\partial t^2}.$$

But also

$$\text{div}\, \mathbf{v} = -\text{div grad}\, \phi = -\nabla^2 \phi;$$

hence ϕ satisfies the scalar wave equation:

$$\nabla^2 \phi = \frac{1}{c^2} \frac{\partial^2 \phi}{\partial t^2}. \tag{17-14}$$

---

* The equation $\text{grad}\, \psi = 0$ is equivalent to the three scalar equations $\partial\psi/\partial x = 0$, $\partial\psi/\partial y = 0$, $\partial\psi/\partial z = 0$. Hence $\text{grad}\, \psi = 0$ implies that ψ is at most a function of t.

17-5 SPHERICALLY SYMMETRIC SOLUTIONS

In a spherical polar coordinate system the velocity potential ϕ will be written as a function of the coordinates r, θ, φ, and of the time t. The Laplacian operator has the form*

$$\nabla^2 = \frac{1}{r^2}\frac{\partial}{\partial r}\left(r^2\frac{\partial}{\partial r}\right) + \frac{1}{r^2\sin\theta}\frac{\partial}{\partial\theta}\left(\sin\theta\frac{\partial}{\partial\theta}\right) + \frac{1}{r^2\sin^2\theta}\frac{\partial^2}{\partial\varphi^2}.$$

We seek an answer to the question, "Does Eq. (17–14) permit any solutions in which ϕ is independent of θ and φ?" With

$$\frac{\partial\phi}{\partial\theta} = 0 = \frac{\partial\phi}{\partial\varphi},$$

the equation for $\phi(r, t)$ is

$$\nabla^2\phi = \frac{1}{r^2}\frac{\partial}{\partial r}\left(r^2\frac{\partial\phi}{\partial r}\right) = \frac{1}{c^2}\frac{\partial^2\phi}{\partial t^2}. \tag{17–15}$$

As a formal identity, the reader can show that for any function ϕ

$$\frac{1}{r^2}\frac{\partial}{\partial r}\left(r^2\frac{\partial\phi}{\partial r}\right) = \frac{1}{r}\frac{\partial^2}{\partial r^2}(r\phi).$$

Thus Eq. (17–15) can be written

$$\frac{\partial^2}{\partial r^2}(r\phi) = \frac{r}{c^2}\frac{\partial^2\phi}{\partial t^2} = \frac{1}{c^2}\frac{\partial^2(r\phi)}{\partial t^2}.$$

(The factor of r can be taken inside $\partial^2/\partial t^2$ since r and t are independent variables.) We observe that with $\psi(r, t) = r\phi(r, t)$, the equation satisfied by $\psi(r, t)$ has the form of the one-dimensional wave equation:

$$\frac{\partial^2\psi}{\partial r^2} = \frac{1}{c^2}\frac{\partial^2\psi}{\partial t^2}. \tag{17–16}$$

Therefore the most general solution is of the form

$$\psi(r, t) = f(t - r/c) + g(t + r/c),$$

where f and g are arbitrary twice-differentiable functions. The implied solution for ϕ is consequently

$$\phi(r, t) = \frac{f(t - r/c)}{r} + \frac{g(t + r/c)}{r}. \tag{17–17}$$

Except under special conditions, this solution has a singularity at $r = 0$. The answer to our original question is therefore that Eq. (17–11) does have

* L. Brand, *Advanced Calculus*, New York: John Wiley & Sons, 1955.

spherically symmetric solutions valid everywhere except possibly at the origin; the general form is given by Eq. (17–17).

So far as the values of $\psi = r\phi$ are concerned, the function $f(t - r/c)$ represents the propagation with velocity c away from the origin, of a wave which does not change form. In contrast, the profile of

$$\phi(r, t) = \frac{f(t - r/c)}{r}$$

drawn at successively later instants of time represents a waveform which moves radially outward, but which does undergo changes in shape. A point of the graph where $\phi = 0$ propagates outward with velocity c. However, a relative maximum on the graph of ϕ cannot be characterized in such a simple fashion. If we follow a given value of f as it propagates outward, the corresponding value of ϕ decreases as $1/r$. At any given time the loci of $\phi =$ *constant* are spheres concentric with the origin. A function of the form $[f(t - r/c)]/r$ is referred to as a *diverging spherical wave*. Similarly, $[g(t + r/c)]/r$ represents a *converging spherical wave*.

We note one especially simple solution in which f and g are taken to be constant. If we set $f + g = B$, the velocity potential is

$$\phi = B/r. \tag{17–18}$$

The pressure and velocity fields associated with this are

$$p = \rho_0 \frac{\partial \phi}{\partial t} = 0, \tag{17–19}$$

$$\mathbf{v} = -\operatorname{grad} \phi = \frac{B}{r^2} \hat{r}. \tag{17–20}$$

The situation represented by this solution is that of a *steady flow* of fluid along radial lines. If we take any sphere concentric with the origin, the mass which flows out of the surface per unit time is

$$\oint\!\!\!\oint_S \rho_0 \mathbf{v} \cdot d\mathbf{S} = \rho_0 \frac{B}{r^2} (4\pi r^2) = 4\pi\rho_0 B, \tag{17–21}$$

and is independent of r. If $B > 0$, the fluid is being fed in at the origin at a constant rate, for example, through the opening of a small pipe. (It must be assumed that the pipe itself is of sufficiently small dimensions so that it does not disturb the spherical symmetry of the exterior flow of fluid.) Alternatively, the source might consist of an expanding sphere whose volume is increasing at a constant rate. Similarly, $B < 0$ corresponds to a *sink* at the origin, through which fluid is removed at a constant rate.

The condition $p = 0$ expressed by Eq. (17–19) means only that there is no *first-order* pressure variation associated with this steady-flow situation.

17-6 BOUNDARY VALUE PROBLEM: A PULSING SPHERE

As an application of the solution exhibited in the last section consider an infinite body of fluid containing a sphere whose radius, a, varies sinusoidally according to the equation

$$a = a_0 + b \sin \omega t, \tag{17–22}$$

where a_0 and b are constants, and $b < a_0$. If we take $r = 0$ to be at the center of the sphere, we expect from symmetry considerations that the disturbance produced in the medium by the motion of the surface of the sphere will be independent of θ and φ and should therefore be described by a spherically symmetric velocity-potential function. Attention will be restricted to the region $r > a$, and the question of a singularity at $r = 0$ in the general solution given by Eq. (17–17) is of no concern.

The boundary conditions to be satisfied are as follows:

i) *No sources of incoming spherical waves.* This condition is met by taking $g(t) = 0$.

ii) *No cavitation.* In general terms this means that the particles of the fluid adjacent to the surface of the sphere must have a velocity whose radial component is the same as that of the surface of the sphere. This boundary condition does not impose any requirements on the other components of $\mathbf{v}$, but these components vanish by the assumption of spherical symmetry. At time t the velocity of the surface of the sphere is $\dot{a} = b\omega \cos \omega t$. From $\mathbf{v} = -\text{grad}\, \phi$ the radial velocity of the fluid is $v_r = -(\partial\phi/\partial r)(r, t)$.

Recalling the Eulerian meaning of the variables, we can evaluate this at the surface of the sphere by taking $r = a$. Thus the second boundary condition is

$$-\frac{\partial \phi}{\partial r}(a, t) = b\omega \cos \omega t. \tag{17–23}$$

Let us try a solution of the form

$$\phi = \frac{f(t - r/c)}{r}, \qquad \text{where} \qquad f(t) = \Re e\,\{\mathbf{A}e^{i\omega t}\}$$

and $\mathbf{A}$ is a complex amplitude to be determined. Then

$$\phi = \Re e \left\{\mathbf{A}\,\frac{e^{i(\omega t - kr)}}{r}\right\}, \tag{17–24}$$

$$-\frac{\partial \phi}{\partial r} = \Re e \left\{\mathbf{A}(1 + ikr)\,\frac{e^{i(\omega t - kr)}}{r^2}\right\},$$

and Eq. (17–23) implies that

$$\Re e \left\{\mathbf{A}\,\frac{e^{-ika}}{a^2}\,(1 + ika)e^{i\omega t}\right\} = \Re e\,\{b\omega e^{i\omega t}\}\,. \tag{17–25}$$

But, realizing that a is a function of the time given by Eq. (17–22), we can see that there is *no* choice for the constant $\mathbf{A}$ which will enable this equation to be satisfied. This implies that our attempt to find a simple solution with $f(t) = \mathcal{R}e\,\{\mathbf{A}e^{i\omega t}\}$ has failed. However, before making further attempts to match the boundary condition Eq. (17–23) as written, it should be questioned whether it makes sense to seek an *exact* solution to an *inexact* problem. That is, the linearized form of the Eulerian equations were obtained after certain approximations. From our experience with sinusoidal plane waves we know that the acoustic approximation requires particle displacements which are small compared with a wavelength. In particular, since the displacement amplitude of the surface of the sphere is b, it is only desirable to seek a solution consistent with the assumption $b \ll \lambda$. It also seems reasonable to demand that $b \ll a_0$, i.e., that the *fractional* change in radius of the sphere be small. In fact, it is the subject of Problem 17–5 to show that it is not proper to drop the term $(\mathbf{v} \cdot \text{grad})\mathbf{v}$ from Eq. (17–3) unless $b \ll a_0$. With these assumptions the left-hand side of Eq. (17–25) can be approximated by replacing a with a_0.† We then obtain

$$\mathbf{A}\,\frac{e^{-ika_0}}{a_0^2}\,(1 + ika_0) = b\omega,$$

or

$$\mathbf{A} = \frac{a_0^2 b\omega e^{ika_0}}{1 + ika_0}\,. \tag{17–26}$$

With this value of $\mathbf{A}$ the velocity potential which satisfies the boundary conditions is given by Eq. (17–24). The associated velocity and pressure fields are

$$v_r = -\frac{\partial\phi}{\partial r} = \mathcal{R}e\,\{\mathbf{v}_r e^{i\omega t}\},$$

where

$$\mathbf{v}_r = \mathbf{A}\left\{\frac{1}{r^2} + \frac{ik}{r}\right\} e^{-ikr}, \tag{17–27}$$

and

$$p = \rho_0\,\frac{\partial\phi}{\partial t} = \mathcal{R}e\,\{\mathbf{p}e^{i\omega t}\},$$

where

$$\mathbf{p} = \frac{i\rho_0\omega\mathbf{A}}{r}\,e^{-ikr}. \tag{17–28}$$

---

† This approximation is tantamount to ignoring the distinction between the Eulerian and Lagrangian interpretations. According to the latter, the boundary surface is rigorously specified by $r = a_0$. In this connection see the remarks at the end of Appendix III.

The function $p(r, t)$ is seen to be of the form $[h(t - r/c)]/r$. This is in accord with the fact that $p(r, t)$ satisfies the scalar wave equation, Eq. (17–7). The proof at the beginning of the preceding section is general and shows that any spherically symmetric solution of the scalar wave equation must have the form of Eq. (17–17).

On the other hand, $v_r(r, t)$ cannot be cast in this form. Indeed, v_r does not satisfy the *scalar* wave equation. By expressing the vector Laplacian in spherical polar coordinates it can be verified that the velocity field given by Eq. (17–27) satisfies Eq. (17–10), the vector wave equation.

A point which is many wavelengths away from the origin of a diverging spherical wave is said to be in the *far zone*. The condition $r \gg \lambda$ is equivalent to $kr \gg 1$. The ratio of the magnitudes of the two terms in parentheses in Eq. (17–27) is kr, and it is the second term which is dominant in the far zone. Thus the complex velocity amplitude can be approximated in the far zone by

$$\mathbf{v}_r = \frac{ik\mathbf{A}}{r} e^{-ikr}.$$

Comparison with Eq. (17–28) shows that

$$\mathbf{p} = \rho_0 c \mathbf{v}_r = Z\mathbf{v}_r.$$

Hence the relation between the pressure and particle velocity is the same in the far zone as for a plane wave. This is to be expected, since a small area on the surface of a spherical wave of large radius would be indistinguishable from a plane wave.

17-7 THE NEAR ZONE OF A DIVERGING SPHERICAL WAVE

The region $kr \ll 1$ is called the *near zone* of the diverging spherical wave specified by Eqs. (17–27) and (17–28). In this region it is the $1/r^2$ term in the velocity which predominates, and Eq. (17–27) can be written

$$\mathbf{v}_r = \mathbf{A}/r^2. \tag{17–29}$$

Since the solution is only meaningful for $r > a_0$, where a_0 is the mean radius of the pulsing sphere, there *is* no near zone unless $ka_0 \ll 1$. Under this condition Eq. (17–26) shows that $\mathbf{A}$ is approximately equal to the real number $a_0^2 b\omega$. Then the velocity function is given by

$$v_r = \frac{a_0^2 b\omega \cos \omega t}{r^2}.$$

This resembles the steady-flow solution of Eq. (17–20) except that the constant B is replaced by $a_0^2 b\omega \cos \omega t$. Substitution of this in Eq. (17–21) yields a mass flux per unit of time given by

$$4\pi\rho_0 a_0^2 b\omega \cos \omega t = 4\pi\rho_0 a_0^2 \dot{a} \doteq \frac{d}{dt}\left[\frac{4\pi}{3}\rho_0 a^3\right].$$

But the latter is just the instantaneous rate at which fluid is being pushed outward by the expanding volume of the sphere. There is thus a consistent interpretation of the velocity in the near zone as a *quasisteady flow* determined by the instantaneous condition of the pulsing sphere. There is no time lag since all points in the region considered are extremely close to the source in comparison with a wavelength.

From Eqs. (17–28) and (17–29) it is seen that pressure and velocity functions are $\pi/2$ out of phase in the near zone. This is similar to the condition in a standing plane wave and should imply zero energy transport when averaged over a cycle. However, this observation clearly points up the need for a less extreme approximation than Eq. (17–29), since all the energy which is radiated must somehow pass through the near zone. The following derivation will be general, and specialization to the near zone will occur only by way of illustration.

Let the ratio of the complex amplitude of the pressure function to the complex amplitude of the velocity function define the *complex impedance* of the field at a given point:

$$\mathbf{Z} = \frac{\mathbf{p}}{\mathbf{v}_r} = R + iX = |\mathbf{Z}|e^{i\delta}.$$

The notations for the rectangular and polar forms of $\mathbf{Z}$ are indicated. The imaginary part, X, will be referred to as the *reactive* part. For the purpose of pure analogy R will be referred to as the *resistive* part, although no *dissipation* of energy is implied. For the divergent spherical wave, $\mathbf{Z}$ is obtained by taking the ratio of Eq. (17–28) to Eq. (17–27); the result is

$$R = \frac{(kr)^2\rho_0 c}{1 + (kr)^2}, \qquad X = \frac{(kr)\rho_0 c}{1 + (kr)^2}, \qquad \tan\delta = \frac{X}{R} = \frac{1}{kr}. \tag{17–30}$$

In the far zone, $kr \gg 1$ and

$$R \doteq \rho_0 c \gg X \doteq \frac{\rho_0 c}{kr}, \qquad \delta \doteq \frac{1}{kr}.$$

In the near zone, $kr \ll 1$ and

$$R \doteq (kr)^2\rho_0 c \ll X \doteq (kr)\rho_0 c, \qquad \delta \doteq \frac{\pi}{2} - kr.$$

The instantaneous intensity vector, $\mathbf{i} = p\mathbf{v}$, has only a radial component, $i_r = pv_r$, for the spherically symmetric case. Let the complex amplitude of the radial velocity be written in polar form $\mathbf{v}_r = v_m e^{i\theta}$, so that

$$v_r = \mathcal{R}e\,\{\mathbf{v}_r e^{i\omega t}\} = v_m \cos(\omega t + \theta).$$

Then
$$p = \mathcal{R}e\,\{\mathbf{p}e^{i\omega t}\} = \mathcal{R}e\,\{\mathbf{Z}\mathbf{v}_r e^{i\omega t}\}$$
$$= v_m\,\mathcal{R}e\,\{(R + iX)e^{i(\omega t+\theta)}\}$$
$$= Rv_m \cos(\omega t + \theta) - Xv_m \sin(\omega t + \theta),$$
and
$$i_r = pv_r = Rv_m^2 \cos^2(\omega t + \theta) - X\frac{v_m^2}{2} \sin 2(\omega t + \theta). \quad (17\text{–}31)$$

The second term, associated with the reactive component of the impedance, is alternately positive and negative and averages to zero. The first term is nonnegative and represents energy which flows outward past the given point of observation and does not return. In the near zone the second term predominates over the first. All but a small part of the energy that flows out during half the cycle flows back during the other half. In the far zone the first term is dominant and the fluctuating term is insignificant.

From Eq. (17–31) the intensity averaged over a cycle is

$$\bar{\imath}_r = \tfrac{1}{2}Rv_m^2.$$

According to Eq. (17–30), R depends on r and hence the average intensity is not simply proportional to the square of the amplitude of the velocity function. On the other hand, after a certain amount of algebraic manipulation it can be shown for the general case that

$$\bar{\imath}_r = \frac{p_m^2}{2\rho_0 c}\cdot \quad (17\text{–}32)$$

This means that the average intensity is proportional to the square of the pressure amplitude. Since the latter is inversely proportional to r (cf. Eq. 17–28), it follows that the net energy flux out of any sphere is independent of its radius. So it has turned out that the phase difference between p and v_r in the near zone is just sufficiently different from $\pi/2$ to allow for the flux of the required amount of radiated energy.

17-8 THE POINT SOURCE

With an arbitrary choice of the complex constant $\mathbf{A}$, the velocity potential function

$$\phi = \mathcal{R}e\left\{\frac{\mathbf{A}}{r}\, e^{i(\omega t - kr)}\right\} \quad (17\text{–}33)$$

is an exact solution to the scalar wave equation except at the point $r = 0$. The associated pressure and velocity amplitudes become indefinitely large as r approaches zero. This means that the small-amplitudes approximation will be violated somewhere in the neighborhood of the origin. That is, although this function is a formal solution to a mathematical problem, the mathematical problem itself ceases to be a correct statement of physical reality. The mathematical fiction represented by Eq. (17–33), without restriction on the value r (provided that $r \neq 0$), is referred to as the field due to a *point source*.

A point source of given frequency is uniquely specified by giving the value of $\mathbf{A}$, the complex amplitude of the velocity-potential function. It can also be characterized, apart from the matter of phase, by stating the *source strength* Q, defined as the maximum value during a cycle, of the volume of fluid per unit time which crosses any sphere of radius $r \ll \lambda$. The relation between Q and $\mathbf{A}$ is as follows: In the near zone $\mathbf{v}_r = \mathbf{A}/r^2$. Hence

$$Q = 4\pi r^2 |\mathbf{v}_r| = 4\pi |\mathbf{A}|.$$

Thus

$$|\mathbf{A}| = Q/4\pi.$$

It is clear that for any given point source it is possible to choose a pulsing sphere of radius $a_0 \ll \lambda$ such that the fields in $r > a_0$ are identical. In part, then, the concept of the point source is a means of avoiding the mention of such details. However, the principal utility of the strength-of-source concept comes from its applicability to small sources which are nonspherical. A sinusoidal source is known as a *simple source* if its maximum dimension is small compared with a wavelength and if all points on the surface oscillate *in phase* with one another. The field at large distances from a simple source is spherically symmetric and is equivalent to the far zone field of a point source. The strength of this equivalent point source is such that the volume flow out of a sphere of small radius is the same as that of the actual source. More explicitly, the field in the near zone of the actual source may not be spherically symmetric. However, under the stated conditions, the volume V of the source is a periodic function of the time. If Q is set equal to the maximum value of dV/dt, then the far zone field of a point source having this strength is the same as the far zone field of the given simple source. A proof of this assertion will not be given here, although it should seem plausible on the basis of the calculations in Chapter 12 relating to particular arrays of point sources.

17-9 THE ACOUSTIC DIPOLE

Consider a point source S located a distance d units from the origin on the polar axis of a spherical polar coordinate system. (See Fig. 17–1.) From symmetry considerations all quantities are expected to be independent of the azimuthal angle φ. The field will be described by a velocity-potential function $\phi = \phi(r, \theta, t)$. In relation to a coordinate system whose origin is coincident with S, the velocity potential is given by

$$\phi = \frac{A}{r'} \cos (\omega t - kr'),$$

where r' is the radial coordinate in this system. If we apply the law of cosines, we obtain

$$r' = \sqrt{r^2 + d^2 - 2rd\cos\theta},$$

and therefore we have

$$\phi(r, \theta, t) = \frac{A\cos\left[\omega t - k\sqrt{r^2 + d^2 - 2rd\cos\theta}\right]}{\sqrt{r^2 + d^2 - 2rd\cos\theta}}. \tag{17–34}$$

It can be verified by direct substitution that this is a solution to the wave equation, Eq. (17–14), except at the singularity $r' = 0$.

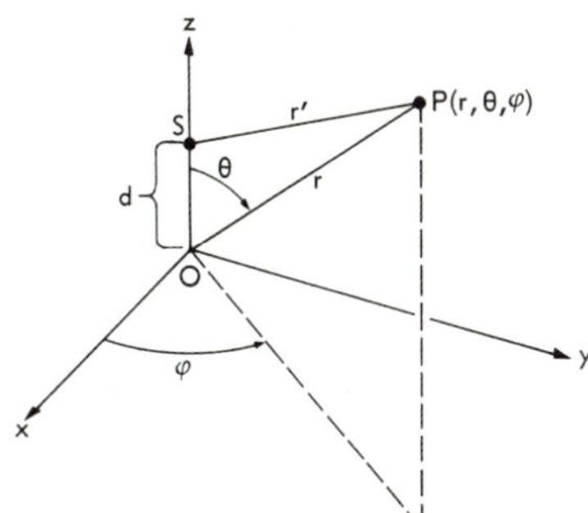

Fig. 17–1.
Point source located at S.

Suppose now that there is a point source at $r' = 0$ and a second point source of equal strength but π out of phase at $r = 0$. The resulting field is described by the velocity-potential function

$$\phi = \frac{A\cos(\omega t - kr')}{r'} - \frac{A\cos(\omega t - kr)}{r}.$$

If d is *sufficiently small*,* we can make use of the differential approximations

$$\frac{\cos(\omega t - kr')}{r'} \doteq \frac{\cos(\omega t - kr)}{r} + (r' - r)\left[\frac{\partial}{\partial r'}\left(\frac{\cos(\omega t - kr')}{r'}\right)\right]_{r'=r}$$

and

$$r' \doteq r - d\cos\theta,$$

from which we obtain

$$\phi = Ad\cos\theta\left[\frac{-k\sin(\omega t - kr)}{r} + \frac{\cos(\omega t - kr)}{r^2}\right]. \tag{17–35}$$

By direct substitution it can be shown that this function satisfies the wave equation everywhere except at $r = 0$.

---

* Examining the neglected terms we find that this approximation requires $d \ll r$ and $kd \ll 1$.

The mathematical fiction described by Eq. (17–35) for all $r > 0$ is referred to as a *dipole field*. In practice it would be used to describe the field at distances $r \gg d$ due to two simple sources of equal strength separated by a distance $d \ll \lambda$.

In Section 11–2 we described a number of features of this field in the far zone ($kr \gg 1$). In this region the second term in Eq. (17–35) can be neglected, and we have

$$\phi = -A(kd\cos\theta)\,\frac{\sin(\omega t - kr)}{r}.$$

The associated pressure field is

$$p = \rho_0\,\frac{\partial\phi}{\partial t} = -\rho_0 A\omega(kd\cos\theta)\,\frac{\cos(\omega t - kr)}{r}.$$

But, the same calculation associated with a *single* source yields

$$p = -\rho_0 A\omega\,\frac{\sin(\omega t - kr)}{r},$$

and from this it can be seen that the average intensities in the two cases differ by a factor of $(kd\cos\theta)^2$. If we take account of the change in significance of the reference axis $\theta = 0$, this result is the same as Eq. (11–7).

Inasmuch as the present discussion is more explicit than the previous one it is now possible to consider the fields in the near zone. In particular, we can obtain an explicit description of the phenomenon of local flow. If we drop all terms except those which are dominant for r small, we obtain, for $kr \ll 1$,

$$\phi = Ad\cos\omega t\,\frac{\cos\theta}{r^2}$$

and

$$\mathbf{v} = -\operatorname{grad}\phi = \frac{Ad\cos\omega t}{r^3}\,[(2\cos\theta)\hat{r} + (\sin\theta)\hat{\theta}]. \qquad (17\text{–}36)$$

We note that in the near zone there is a *transverse* component, v_θ, which is comparable in magnitude to the radial component. Since we are dealing only with distances which are small compared with a wavelength, all particle motions have the same phase.

The feature of interest is the determination of the flow lines of the fluid motion, that is, the paths followed by the particles at different locations within the near zone. Suppose that a given particle moves along a path specified by a function $r = r(\theta)$. The velocity vector for this particle is then

$$\mathbf{V} = \dot{r}\hat{r} + (r\dot{\theta})\hat{\theta} = \dot{\theta}\left[\left(\frac{dr}{d\theta}\right)\hat{r} + r\hat{\theta}\right].$$

But the velocity of this particle when it is at the point (r, θ) is given by

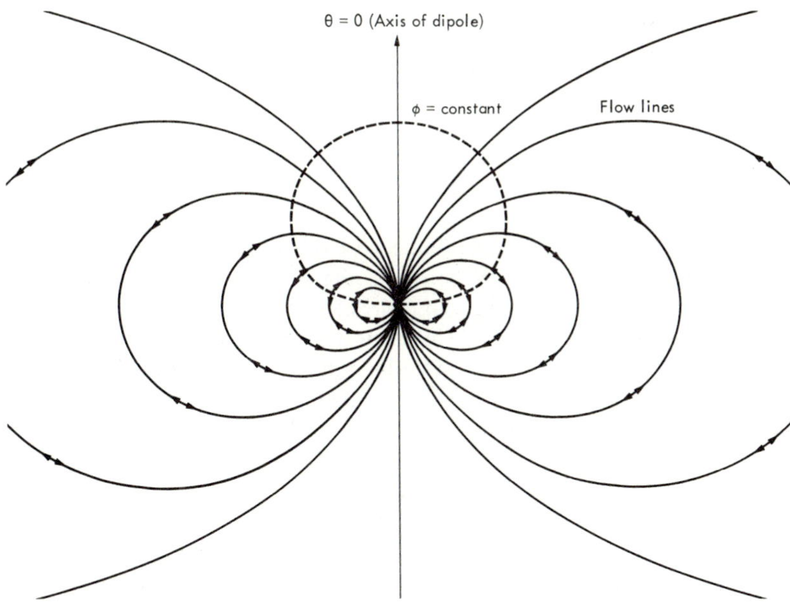

Fig. 17-2. Flow lines in the near zone of an acoustic dipole.

Eq. (17-36). Thus we must have $V_r/V_\theta = v_r/v_\theta$, or

$$\frac{1}{r}\frac{dr}{d\theta} = \frac{2\cos\theta}{\sin\theta}.$$

This equation integrates simply to yield

$$r = \alpha\sin^2\theta, \tag{17-37}$$

where α is the constant of integration. As α is varied, Eq. (17-37) defines a family of surfaces of revolution. An individual particle moves back and forth in a plane φ = constant along a curve determined by Eq. (17-37). A set of these *flow lines* is sketched in Fig. 17-2. The values of α have been chosen so that the volume of fluid which crosses the plane $\theta = \pi/2$ per unit of time is the same between any adjacent pair.

The dotted curve in Fig. 17-2 is the trace of an *equipotential surface*, i.e., a locus along which the value of ϕ is constant at a fixed time t. From analytic geometry we know that in general the normal to a surface $\phi(r, \theta, \varphi)$ = constant has the direction of the vector grad ϕ. But this is just the direction of the vector $-\mathbf{v}$. It is therefore a property of a flow line that it is everywhere perpendicular to the equipotential surface which passes through the same point. Thus the family of equipotential surfaces, $\cos\theta/r^2$ = constant, is orthogonal to the family of surfaces of revolution obtained from Eq. (17-37).

17–10 PLANE WAVE PROPAGATING IN AN ARBITRARY DIRECTION

Consider a cartesian coordinate system S' with coordinates x', y', z'. We know that if $f(t)$ is an arbitrary twice-differentiable function, then $\phi = f(t - x'/c)$ represents a plane wave propagating parallel to the x'-axis. At any given time the values of ϕ are the same at all points along a plane $x' = \text{constant} = x'_1$. At a time Δt later, the same values of ϕ are attained on the plane $x' = x'_1 + c\,\Delta t$.

Now consider another coordinate system S whose origin coincides with that of S'. If we can succeed in expressing ϕ relative to S, we will have a representation of a plane wave whose direction of propagation is not restricted to being parallel to the x-axis. This is most easily accomplished by means of vector notation. Let the unit vector along the x'-axis be designated $\mathbf{n}$. If the position vector of an arbitrary point is $\mathbf{r}$, then the x'-coordinate of this point is $x' = \mathbf{r} \cdot \mathbf{n}$. We can therefore write

$$\phi = f\left(t - \frac{x'}{c}\right) = f\left(t - \frac{\mathbf{r} \cdot \mathbf{n}}{c}\right).$$

In this manner the equation of the plane wave has been given a form which is *invariant* under a rotation of axes. In particular, if the unit vector $\mathbf{n}$ has components (l, m, n) in the system S (l, m, and n are the *direction cosines* of the direction of propagation of the plane wave), we obtain

$$\phi(x, y, z, t) = f\left(t - \frac{\mathbf{r} \cdot \mathbf{n}}{c}\right) = f\left(t - \frac{lx + my + nz}{c}\right). \tag{17–38}$$

It can be checked directly that this is a solution to the equation

$$\nabla^2\phi = \frac{1}{c^2}\frac{\partial^2\phi}{\partial t^2}$$

for arbitrary l, m, and n provided that $l^2 + m^2 + n^2 = 1$. Note that the special cases of plus and minus waves traveling along the x-axis are obtained by taking $m = n = 0$ and $l = +1$ or $l = -1$ respectively.

Equation (17–38) is the general equation for a plane wave of arbitrary waveform propagating in an arbitrary direction. Specializing this to the case of sinusoidal waves, we take

$$f(t) = \mathcal{R}e\,\{\mathbf{A}e^{i\omega t}\}$$

and

$$\phi = f\left(t - \frac{\mathbf{r} \cdot \mathbf{n}}{c}\right) = \mathcal{R}e\,\{\mathbf{A}e^{i(\omega t - k\mathbf{r}\cdot\mathbf{n})}\},$$

where $k \equiv \omega/c$. It is convenient to define a vector

$$\mathbf{k} \equiv k\mathbf{n}$$

called the *propagation vector* of the plane wave.* The direction of **k** is the direction of propagation of the plane wave. Since the magnitude of **k** is $k = \omega/c$, we have the relation

$$|\mathbf{k}| = \sqrt{k_x^2 + k_y^2 + k_z^2} = k = \omega/c.$$

The general sinusoidal plane wave is therefore written

$$\phi = \mathcal{R}e\,\{\mathbf{A}e^{i(\omega t - \mathbf{k}\cdot\mathbf{r})}\}. \tag{17-39}$$

17-11 BOUNDARY VALUE PROBLEM: REFLECTION OF OBLIQUELY INCIDENT PLANE WAVES

Let the plane boundary between two media be designated as $z = 0$ and consider a sinusoidal plane wave incident from $z < 0$ at an oblique angle. The incident wave will be represented by the velocity-potential function

$$\phi_{\text{inc}} = \mathcal{R}e\,\{\mathbf{A}e^{i(\omega t - \mathbf{k}\cdot\mathbf{r})}\}, \qquad z < 0$$

for given values of the propagation vector **k** and the complex amplitude **A**. We shall try to fit all the requirements of the problem by the presupposition that this incident wave gives rise to reflected and transmitted waves which are sinusoidal plane waves of the same frequency as the incident wave.† Prior to the deduction of further specific information about these waves they will be indicated *schematically* as in Fig. 17-3.

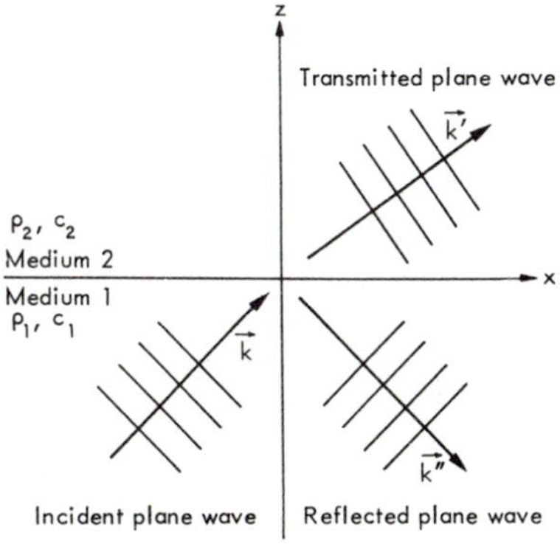

Fig. 17-3. Schematic representation of incident, reflected, and transmitted waves.

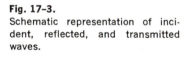

* The symbol **k** is standard for the propagation vector despite the common use of **k** as the unit vector along the z-axis. No case arises in the remainder of this text in which the latter meaning is desired.

† If the fit is successful, we will know definitely only that a possible solution has been found. Physical intuition indicates that the conditions that have been met are sufficient to guarantee that the solution is a unique one. If one does not wish to rely on this presumption of uniqueness, it is possible to proceed on a more general basis. Without presupposing either the *plane wave* or the *sinusoidal* nature of the reflected and transmitted waves, these properties can be shown to be necessary.

The assumed form of the velocity-potential function in medium 1 is $\phi_1 = \phi_{\text{inc}} + \phi_{\text{refl}}$, where

$$\phi_{\text{refl}} = \mathcal{R}e\,\{\mathbf{B}e^{i(\omega t-\mathbf{k}''\cdot\mathbf{r})}\}, \qquad z < 0. \tag{17–40}$$

This satisfies

$$\nabla^2\phi_1 = \frac{1}{c_1^2}\frac{\partial^2\phi_1}{\partial t^2}$$

provided that

$$|\mathbf{k}''| = \omega/c_1 = k_1. \tag{17–41}$$

The wave function in medium 2 is taken to be of the form

$$\phi_2 = \phi_{\text{trans}} = \mathcal{R}e\,\{\mathbf{C}e^{i(\omega t-\mathbf{k}'\cdot\mathbf{r})}\}, \qquad z > 0. \tag{17–42}$$

This satisfies

$$\nabla^2\phi_2 = \frac{1}{c_2^2}\frac{\partial^2\phi_2}{\partial t^2}$$

provided that

$$|\mathbf{k}'| = \omega/c_2 = k_2. \tag{17–43}$$

For ϕ_{refl} to represent a wave propagating *away* from the boundary we require that the propagation vector $\mathbf{k}''$ have a negative component along the z-axis:

$$k_z'' < 0. \tag{17–44}$$

Similarly, the transmitted wave has a normal which points away from the boundary:

$$k_z' > 0. \tag{17–45}$$

One boundary condition to be satisfied is the continuity of the excess acoustic pressure function at the interface:*

$$p_1(x, y, 0, t) = p_2(x, y, 0, t).$$

The necessity for this condition can be seen from the same kind of pillbox argument used in the one-dimensional case (cf. Section 3–4). Expressing this condition in terms of the velocity-potential function, we have

$$\rho_1\frac{\partial\phi_1}{\partial t}(x, y, 0, t) = \rho_2\frac{\partial\phi_2}{\partial t}(x, y, 0, t),$$

or

$$\mathcal{R}e\,\{\rho_1 i\omega e^{i\omega t}[\mathbf{A}e^{-i(k_x x+k_y y)} + \mathbf{B}e^{-i(k_x'' x+k_y'' y)}]\} = \mathcal{R}e\,\{\rho_2 i\omega e^{i\omega t}\mathbf{C}e^{-i(k_x' x+k_y' y)}\}.$$

* In Eulerian notation the interface does not remain at $z = 0$ during the disturbance. However, as indicated in Section 17–6, it is inconsistent to maintain a rigorous distinction between the Eulerian and Lagrangian meanings of the variables after the acoustic approximation has been made.

Since this must be satisfied at all times, we can equate the complex coefficients of $e^{i\omega t}$:

$$\rho_1[\mathbf{A}e^{-i(k_x x+k_y y)} + \mathbf{B}e^{-i(k_x'' x+k_y'' y)}] = \rho_2 \mathbf{C}e^{-i(k_x' x+k_y' y)}. \quad (17\text{–}46)$$

A second boundary condition that must be satisfied is that the velocity component *normal* to the boundary surface must be the same on both sides of the interface. Otherwise the solution would imply that the two media could interpenetrate one another or could leave a gap at the boundary surface. Since the normal component is $v_z = -\partial\phi/\partial z$, the second boundary condition can be written as

$$\frac{\partial \phi_1}{\partial z}(x, y, 0, t) = \frac{\partial \phi_2}{\partial z}(x, y, 0, t).$$

Substituting the assumed solutions, we obtain the equation

$$k_z \mathbf{A}e^{-i(k_x x+k_y y)} + k_z'' \mathbf{B}e^{-i(k_x'' x+k_y'' y)} = k_z' \mathbf{C}e^{-i(k_x' x+k_y' y)}. \quad (17\text{–}47)$$

The pair of equations (17–46) and (17–47) can be solved for **B** and **C** in terms of **A**:

$$\mathbf{B} = \mathbf{A}\frac{\rho_1 k_z' - \rho_2 k_z}{\rho_2 k_z'' - \rho_1 k_z'} e^{i[(k_x''-k_x)x+(k_y''-k_y)y]}, \quad (17\text{–}48)$$

$$\mathbf{C} = \mathbf{A}\frac{\rho_1(k_z'' - k_z)}{\rho_2 k_z'' - \rho_1 k_z'} e^{i[(k_x'-k_x)x+(k_y'-k_y)y]}. \quad (17\text{–}49)$$

These equations must hold for all values of x and y. However, **B** and **C** cannot depend on x and y, for then Eqs. (17–40) and (17–42) would not be functions that satisfy the wave equation. The only way to avoid this* is to have

$$k_x' = k_x = k_x'', \qquad k_y' = k_y = k_y''. \quad (17\text{–}50)$$

Without loss of generality we can assume that the propagation vector of the incident wave lies in the xz-plane, i.e., that $k_y = 0$. It follows then that $k_y' = k_y'' = 0$, which in turn implies that the reflected and transmitted rays also lie in the xz-plane. One outcome of the analytical derivation is thus a proof of the proposition that the planes of reflection and transmission are the same as the plane of incidence. (Each of these planes is defined by the normal to the interface and the respective ray, or propagation vector.) It will be assumed from now on that the y-axis is chosen perpendicular to this plane.

---

* It might be supposed that Eq. (17–50) could have been deduced by inspection of Eq. (17–46). However, this equation alone permits other solutions; for example, $k_x' = k_x \neq k_x''$, but $\mathbf{B} = 0$ and $\rho_1\mathbf{A} = \rho_2\mathbf{C}$.

The value of k_z'' can now be derived from Eq. (17–41):

$$k_1^2 = |\mathbf{k}''|^2 = k_x''^2 + k_z''^2 = k_x^2 + k_z''^2.$$

But also

$$k_1^2 = |\mathbf{k}|^2 = k_x^2 + k_z^2.$$

Therefore $k_z''^2 = k_z^2$, and, consulting Eq. (17–44) for the choice of sign, we see that

$$k_z'' = -k_z.$$

Let an angle of incidence θ and an angle of reflection θ'' be defined by Fig. 17–4. It can be seen that the significance of the result which has just been obtained is that $\theta = \theta''$. This constitutes a proof of the *law of reflection* which states that the angle of reflection is equal to the angle of incidence.

Turning now to the value of k_z', from Eq. (17–43) we have

$$k_2^2 = |\mathbf{k}'|^2 = k_x'^2 + k_z'^2 = k_x^2 + k_z'^2,$$

or

$$k_z'^2 = k_2^2 - k_x^2. \tag{17–51}$$

The possibility arises that this expression for $k_z'^2$ may turn out to be a negative number. Writing $k_x = k_1 \sin\theta$, where θ is the angle of incidence, we have

$$k_z'^2 = k_1^2\left(\frac{k_2^2}{k_1^2} - \sin^2\theta\right) = k_1^2\left(\frac{c_1^2}{c_2^2} - \sin^2\theta\right). \tag{17–52}$$

If $c_1 > c_2$ (the wave is incident from the medium of higher sound-propagation velocity), then $k_z'^2$ is *always* positive. But if $c_1 < c_2$ and θ is greater than the *critical angle* $\theta_c = \sin^{-1}(c_1/c_2)$, then $k_z'^2$ is negative. Setting aside this case until the next section we shall proceed with the analysis valid for $c_1 > c_2$ or for $c_1 < c_2$ provided that $\theta \leq \theta_c$.

With $k_z'^2 > 0$ we can take the square root of Eq. (17–51):

$$k_z' = \sqrt{k_2^2 - k_x^2},$$

where the positive sign has been used to satisfy Eq. (17–45). Since $k_z' < k_2$, we can define an angle of refraction θ' by the relation $k_z' = k_2 \cos\theta'$. From this it follows that $k_x' = k_2 \sin\theta'$. But since $k_x' = k_x = k_1 \sin\theta$, we have $k_2 \sin\theta' = k_1 \sin\theta$, or

$$\frac{\sin\theta}{\sin\theta'} = \frac{k_2}{k_1} = \frac{c_1}{c_2}.$$

Consequently, the analytical process of solving the boundary-value problem has resulted in a proof of *Snell's law* of refraction.

The information concerning the propagation vector of the transmitted wave that has been obtained above is summarized in Fig. 17–5 and in the following set of formulas. We are given a wave of frequency ω incident at an

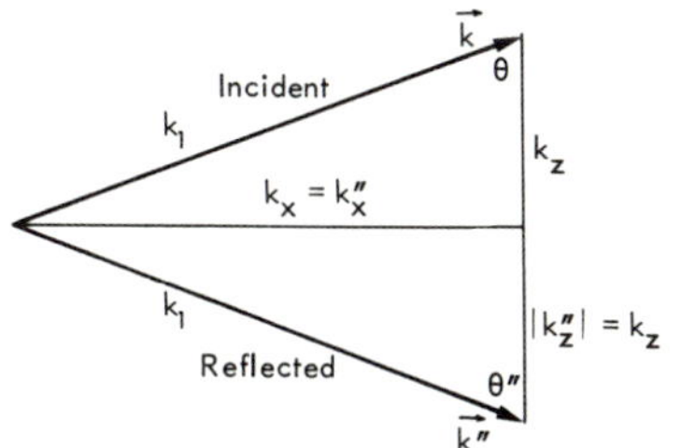

Fig. 17–4. Propagation vectors of incident and reflected waves.

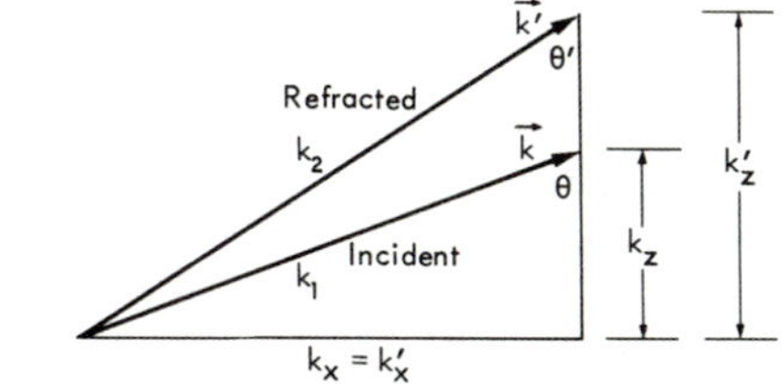

Fig. 17–5. Propagation vectors of incident and refracted waves.

angle θ. From this we can calculate in sequence

$$\begin{aligned} &k_1 = \omega/c_1, \quad k_x = k_1 \sin\theta, \quad k'_x = k_x, \quad k_z = k_1 \cos\theta, \\ &k_2 = \omega/c_2, \quad \sin\theta' = k_x/k_2 \quad (\text{provided } k_x \le k_2), \quad k'_z = k_2 \cos\theta'. \end{aligned} \tag{17–53}$$

The complete solution to the boundary-value problem is now in hand. All that remains is an explicit calculation of the complex amplitudes of the velocity-potential functions in the reflected and transmitted waves. The results are conveniently expressed in terms of the reflection and transmission coefficients for the velocity-potential function, $\mathbf{R} \equiv \mathbf{B}/\mathbf{A}$ and $\mathbf{T} \equiv \mathbf{C}/\mathbf{A}$. We can simplify Eqs. (17–48) and (17–49) considerably. Writing $Z = \rho_0 c$ for the acoustic impedance, we find that the coefficients are real and have the values

$$\begin{aligned} R &= \frac{\rho_2 k_z - \rho_1 k'_z}{\rho_2 k_z + \rho_1 k'_z} = \frac{Z_2 \cos\theta - Z_1 \cos\theta'}{Z_2 \cos\theta + Z_1 \cos\theta'}, \\ T &= \frac{2\rho_1 k_z}{\rho_2 k_z + \rho_1 k'_z} = \left(\frac{\rho_1}{\rho_2}\right)(1 + R). \end{aligned} \tag{17–54}$$

The corresponding coefficients for the pressure and the components of the particle velocity can be obtained through the relations $p = \rho_0(\partial\phi/\partial t)$ and $\mathbf{v} = -\text{grad}\,\phi$. Since we are dealing with progressive plane waves, the magnitude of the radiative intensity is $i = p^2/Z$; the coefficients for this quantity can be calculated as well. The following is a summary of the results:

$$\begin{aligned} R_p &= R, & T_p &= \left(\frac{\rho_2}{\rho_1}\right) T = 1 + R_p; \\ R_{v_x} &= R, & T_{v_x} &= T; \\ R_{v_z} &= -R, & T_{v_z} &= \left(\frac{k'_z}{k_z}\right) T = 1 + R_{v_z}; \\ R_i &= R^2, & T_i &= \left(\frac{Z_1}{Z_2}\right) T_p^2 = \frac{\cos\theta}{\cos\theta'}(1 - R_i). \end{aligned} \tag{17–55}$$

SPECIAL CASES:

a) Normal Incidence

With $\theta = 0$ it follows that $\theta' = 0$ and

$$R_p = \frac{Z_2 - Z_1}{Z_2 + Z_1} = -R_{v_z}.$$

For comparison with Chapter 4, $v_z = \xi$. The sign of R_i is changed since Eq. (4–14) refers to the algebraically signed component i_x, whereas Eq. (17–55) refers to $|\mathbf{i}|$. With these changes all results are seen to be identical.

b) Grazing Incidence

If the angle of incidence is $\theta = \pi/2$, we must clearly be dealing with the case $c_1 > c_2$ if the present formulas are to be applied, since otherwise the critical angle will have been exceeded. The calculated value of θ' is

$$\theta' = \sin^{-1}(c_2/c_1) = \theta_c,$$

which implies that the transmitted wave emerges at the critical angle. However, the transmission coefficients are $T_p = 0 = T_{v_x}$. (The value of T_{v_z} is irrelevant since $v_z = 0$ in the incident wave.) Hence, in fact there are no fields in medium 2 and any statement concerning the "direction" of the transmitted wave is void of meaning. The intensity coefficients are $T_i = 0$, $R_i = 1$, which indicates that 100% of the incident energy is carried by the reflected wave. However, the reflection coefficient for pressure is $R_p = -1$. This means that the pressure functions in the incident and reflected waves are of equal amplitude but π out of phase. Since these two waves are both traveling parallel to the boundary in the same direction, the solution we have before us is actually $p = 0$ identically everywhere. The same analysis applies to the particle velocity, v_x. Hence the trivial case in which all fields vanish identically everywhere is the unique solution to the boundary value problem for a plane wave incident at the angle $\theta = \pi/2$.

c) Incidence at the Critical Angle

With $\theta = \sin^{-1}(c_1/c_2) = \theta_c$ the calculated values are

$$\begin{aligned}
&\theta' = \pi/2; \\
&R_p = 1, \quad R_{v_x} = 1, \quad R_{v_z} = -1, \quad R_i = 1; \\
&T_p = 2, \quad T_{v_x} = 2\rho_1/\rho_2, \quad T_{v_z} = 0, \quad T_i = 4Z_1/Z_2.
\end{aligned} \qquad (17\text{–}56)$$

The intensity of the "transmitted wave" differs from that of the incident wave by the factor $4Z_1/Z_2$, which can be greater or less than unity. At first sight this seems paradoxical. The fact that $R_i = 1$ implies that the intensity of the reflected wave is equal to that of the incident wave. Does it not constitute a violation of the law of conservation of energy to find that there is energy in medium 2? Is it not even worse that the intensity of this wave can be larger than the intensity of the incident wave?

To answer these questions it is necessary to keep in mind exactly what formal conditions have been satisfied by the solution under consideration. What we have proved is that a certain plane wave traveling parallel to the boundary in medium 2 has pressure and velocity fields that match correctly at the boundary surface with the field obtained from the superposition of a wave incident at the critical angle and a reflected wave of equal magnitude. The mathematics is not concerned with whether or not the field in medium 2 is "caused" by the incident wave. When rays of the transmitted wave intersect the boundary surface, this kind of causal interpretation is successful. But in the present example the rays of the plane wave in medium 2 are *parallel* to the boundary. The particle velocity $\mathbf{v}$ is parallel to the boundary surface for the wave in medium 2 and also for the *net* disturbance in medium 1. (Note that $R_{v_z} = -1$.) The intensity vector $\mathbf{i} = p\mathbf{v}$ at no time has a component normal to the surface. The energy present in medium 2 is not flowing in from medium 1. To generate the fields described by this solution there must presumably be sources at $x = -\infty$ within medium 2 as well as the sources of the incident wave within medium 1.

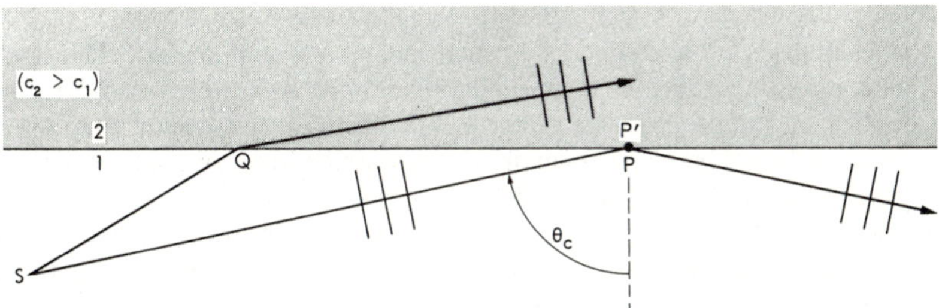

Fig. 17-6. Reflection at the critical angle and how the energy gets into the second medium.

A physical situation *approximately* described by the present equations can be obtained by using a point source S remote from the region in which the observations are made. (See Fig. 17-6.) Near P the spherical wavefronts from S are approximately plane waves. These are totally reflected. The disturbance at P' does not get into medium 2 directly from P. However, there is an energy flux across the boundary at points such as Q where the angle of incidence is less than the critical angle. It is true that the *ray* refracted at Q does not pass through P', and the geometrical ray theory does not account for the magnitude of the disturbance at P'. However, the solution which is being discussed is not restricted by the limitations of the ray-theory approximation. We are dealing with complete solutions to the wave equation and a set of boundary conditions. The disturbance at P' can be interpreted as the result of a lateral spreading of energy along wavefronts in medium 2. In other words, this is a *diffraction* phenomenon. In summary, when S is sufficiently remote, the fields on either side of the interface at P are accurately described by the plane-wave solution being considered. The energy in the wave at P' cannot be thought of as having been transmitted across the boundary at P. But there is no reason to suppose that this mechanism should be the sole energy source for the waves in medium 2; hence there is no contradiction to the law of conservation of energy.

17-12 INHOMOGENEOUS PLANE WAVES

In the previous section we developed a solution which is applicable to all angles of incidence if $c_1 > c_2$, but only to angles of incidence less than or equal to the critical angle if $c_1 < c_2$. The turning point in the argument occurred in Eq. (17–51) where $k_z'^2 = k_2^2 - k_x^2$ turns out to be a negative number if $\theta > \theta_c$. When it was originally postulated in Eq. (17–42) that the transmitted wave would be described by a function of the form

$$\phi_{\text{trans}} = \mathcal{R}e\left\{\mathbf{C}e^{i(\omega t - \mathbf{k}'\cdot\mathbf{r})}\right\},$$

it was implied that $\mathbf{k}'$ should be a vector having *real* components. The associated wave would then be a sinusoidal plane wave traveling in the direction specified by the propagation vector $\mathbf{k}'$. It has now been found that when $\theta > \theta_c$, it is not possible to maintain this assumption and satisfy all the boundary conditions. However, let it be noted that from the purely mathematical point of view Eq. (17–42) defines a perfectly acceptable function even if the value of k_z' happens to be an imaginary number. Thus when $\theta > \theta_c$ we can write $k_z'^2 = -\beta^2 = k_2^2 - k_x^2$ and $k_z' = \pm i\beta$, where β stands for the *positive* square root,

$$\beta = \sqrt{k_x^2 - k_2^2}\,.$$

Then

$$e^{-i\mathbf{k}'\cdot\mathbf{r}} = e^{-ik_x x}e^{-i(\pm i\beta)z} = e^{\pm\beta z}e^{-ik_x x}.$$

The form of the "transmitted" wave function is then

$$\phi_{\text{trans}} = e^{\pm\beta z}\,\mathcal{R}e\left\{\mathbf{C}e^{i(\omega t - k_x x)}\right\} = Ce^{\pm\beta z}\cos(\omega t - k_x x + \gamma),$$

where $\mathbf{C} = Ce^{i\gamma}$.

This function describes a type of disturbance in medium 2 whose characteristics we shall presently examine. We can choose the ambiguous sign in the factor $e^{\pm\beta z}$ by noting that if the plus sign is chosen, ϕ_{trans} becomes indefinitely large for large values of z. Our physical sense tells us that we should rule out this possibility, and the minus sign must be chosen. Tracing back, we see that this choice originates in the choice of the *minus* sign in $k_z' = \pm i\beta$. Hence $k_z' = -i\beta$ is an assumed *boundary condition* which replaces Eq. (17–45).

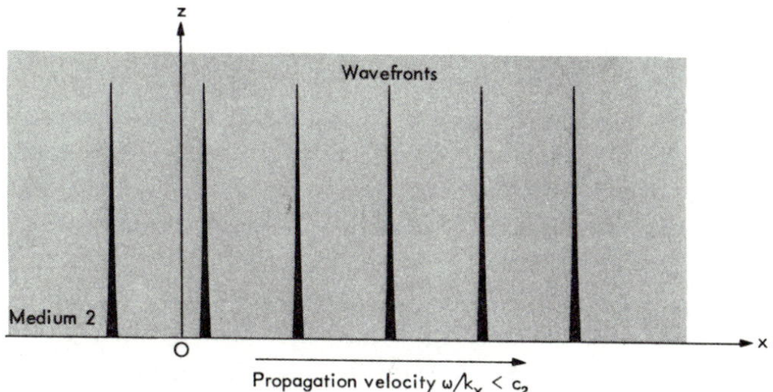

Fig. 17–7. Schematic representation of an inhomogeneous plane wave.

The proposed form of solution for medium 2 is therefore

$$\phi_{\text{trans}} = Ce^{-\beta z}\cos(\omega t - k_x x + \gamma). \tag{17–57}$$

A wavefront is defined as a locus over which the phase maintains a fixed value at a fixed time. It can be seen that the wavefronts of this function are the *planes* $x =$ *constant*. The amplitude of the sinusoidal function is not constant along a given wavefront, but decreases exponentially with distance from the boundary surface. (See Fig. 17–7.) A given wavefront progresses in the positive x-direction with a velocity $\omega/k_x = \omega/(k_1 \sin\theta) = c_1/\sin\theta$. Since the case being examined is for $\sin\theta > c_1/c_2$, this wavefront propagates in medium 2 with a velocity that is *less* than c_2. The field described by the function in Eq. (17–57) is a kind of generalization of a plane wave, and is referred to as an *inhomogeneous plane wave*. The reader should check that Eq. (17–57) defines a solution to $\nabla^2\phi = (1/c_2^2)(\partial^2\phi/\partial t^2)$ provided that $\beta^2 = k_x^2 - \omega^2/c_2^2$.

The pressure and velocity functions associated with this velocity-potential function are

$$\begin{aligned} p &= -\rho_2\omega Ce^{-\beta z}\sin(\omega t - k_x x + \gamma), \\ v_x &= -k_x Ce^{-\beta z}\sin(\omega t - k_x x + \gamma), \\ v_z &= \beta Ce^{-\beta z}\cos(\omega t - k_x x + \gamma). \end{aligned} \tag{17–58}$$

These fields all decrease with distance from the boundary, falling to $1/e$ of their value at the boundary when $z = 1/\beta$. The distance $1/\beta$ is referred to as the *penetration depth.* From Eq. (17–52) we can write $\beta^2 \equiv -k_z'^2 = k_1^2 (\sin^2 \theta - c_1^2/c_2^2)$; and hence

$$\frac{1}{\beta} = \frac{1}{k_1\sqrt{\sin^2 \theta - \sin^2 \theta_c}}. \tag{17-59}$$

The minimum penetration depth occurs at grazing incidence. Then $1/\beta = 1/(k_1 \cos \theta_c)$, which is the order of magnitude of a wavelength. At the opposite extreme, the penetration depth becomes very large as θ approaches θ_c.

The instantaneous intensity vector $\mathbf{i} = p\mathbf{v}$ associated with the functions in Eq. (17–58) has a component parallel to the z-axis. However, since p and v_z are $\pi/2$ out of phase, i_z is a fluctuating quantity which averages to zero. The average energy flow is parallel to the interface and is given by

$$\bar{\imath}_x = \overline{pv_x} = \frac{\rho_2 \omega k_x C^2}{2} e^{-2\beta z}.$$

It is instructive to express this in terms of the pressure amplitude to contrast the behavior of inhomogeneous and ordinary plane waves. From Eq. (17–58) the pressure amplitude at a distance z from the interface is $p_m = \rho_2 \omega C e^{-\beta z}$. This enables us to write the average intensity as

$$\bar{\imath}_x = \left(\frac{p_m^2}{2Z_2}\right)\left(\frac{\sin \theta}{\sin \theta_c}\right). \tag{17-60}$$

Hence the average intensity of an inhomogeneous plane wave is larger by the factor $(\sin \theta/\sin \theta_c)$ than that of an ordinary plane wave having the same pressure amplitude.

Granted that Eq. (17–57) defines a physically meaningful function, we can proceed to try to use this in solving the boundary-value problem for $\theta > \theta_c$. Most of the formulas can be carried over directly from the last section. This is because the same *form* has been assumed for ϕ_{trans} in both cases. Equation (17–57) is simply a more explicit version of Eq. (17–42) where the fact that $k_z' = -i\beta$ is a complex number has been taken into account. It is only necessary to check the steps in the derivation to see whether any operations were carried out which required k_z' to be a real number.

The first modification that is needed comes immediately after Eq. (17–52) where the angle of refraction θ' was defined. This concept is no longer meaningful, and succeeding equations in which θ' appears are to be ignored. For example, Snell's law does not apply. (The closest analog is Eq. (17–59) which is a law determining the penetration depth as a function of the angle of incidence.) All of the equations in (17–53) are correct except the last two which become replaced by

$$\beta = \sqrt{k_x^2 - k_2^2}, \qquad k_z' = -i\beta. \tag{17-61}$$

The only change required in Eq. (17-54) is to designate the reflection and transmission coefficients by **R** and **T** respectively, since they are no longer real. There is now a particular simplification in the form of the reflection coefficient which becomes possible once it is recognized that the numerator and denominator are complex conjugates of each other. Let the numerator be written in polar form

$$\rho_2 k_z - \rho_1 k_z' = \rho_2 k_z + i\rho_1\beta = Ne^{i\chi},$$

where

$$\tan \chi = \frac{\rho_1\beta}{\rho_2 k_z}. \tag{17-62}$$

Then the denominator is

$$\rho_2 k_z - i\rho_1\beta = (\rho_2 k_z + i\rho_1\beta)^* = Ne^{-i\chi},$$

and the reflection coefficient for the velocity-potential function is

$$\mathbf{R} = e^{2i\chi}. \tag{17-63}$$

Let us consider the implications of this result for the physical quantities in the reflected wave. The steps leading from Eq. (17-54) to Eq. (17-55) remain unaltered except for the coefficients relating to the intensity. Because of a relative phase shift it is not convenient to compare *instantaneous* values in the incident and reflected waves. Instead we define the reflection coefficient as the ratio of the *average* intensities†

$$R_i \equiv \bar{I}_{\text{refl}}/\bar{I}_{\text{inc}}.$$

Since both the incident and reflected waves are ordinary plane waves, the relation $\bar{I} = |\mathbf{p}|^2/2Z$ is valid, and we obtain $R_i = |\mathbf{R}_p|^2$. Thus the reflected wave is characterized relative to the incident by the coefficients

$$\begin{aligned} \mathbf{R}_p &= \mathbf{R}_{v_x} = -\mathbf{R}_{v_z} = e^{2i\chi} \\ R_i &= 1. \end{aligned} \tag{17-64}$$

† The definition of R_i adopted in Eq. (4-14) was stated as a ratio of instantaneous values because this allows the concept to be applied to pulses of arbitrary waveform. The equations which are being considered in the present section have been derived under the assumption that the incident wave is sinusoidal. The phase change implied by Eq. (17-63) has interesting implications with respect to the waveform of the reflected wave when the incident wave is a *pulse* of arbitrary shape. [For a specific example see A. B. Arons and D. R. Yennie, *J. Acoust. Soc. Am.* **22,** 231 (1950).]

The *amplitudes* of all quantities in the reflected wave are the same as in the incident, and the intensities are equal. The situation is referred to as *total reflection*. The reflected wave is shifted in phase relative to the incident by the amount

$$2\chi = 2\tan^{-1}\frac{\rho_1\beta}{\rho_2 k_z} = 2\tan^{-1}\left[\frac{\rho_1\sqrt{\sin^2\theta - \sin^2\theta_c}}{\rho_2\cos\theta}\right]. \tag{17–65}$$

This phase shift is independent of frequency, but varies with the angle of incidence, having the value 0 when $\theta = \theta_c$ and π when $\theta = \pi/2$.

The transmission coefficients can also be expressed in terms of χ. From Eq. (17–54) we have

$$\mathbf{T} = \frac{\rho_1}{\rho_2}(1 + \mathbf{R}) = \frac{\rho_1}{\rho_2}(1 + e^{2i\chi}) = \frac{2\rho_1}{\rho_2}(\cos\chi)e^{i\chi},$$

and thus from Eq. (17–55), we have

$$\begin{aligned} \mathbf{T}_p &= 2(\cos\chi)e^{i\chi}, \\ \mathbf{T}_{v_x} &= 2\frac{\rho_1}{\rho_2}(\cos\chi)e^{i\chi}, \\ \mathbf{T}_{v_z} &= 2(\sin\chi)e^{i(\chi-\pi/2)}. \end{aligned} \tag{17–66}$$

As was observed earlier, the functions p and v_z are $\pi/2$ out of phase in the transmitted wave, and hence the average energy flow in a direction perpendicular to the interface is zero. A transmission coefficient for intensity can be defined by taking the ratio of $\bar{I}_x$ from Eq. (17–60), evaluated at $z = 0$, to the average intensity of the incident wave. The result is

$$T_i = \frac{Z_1}{Z_2}|\mathbf{T}_p|^2\left(\frac{\sin\theta}{\sin\theta_c}\right) = \frac{4Z_1}{Z_2}\cos^2\chi\left(\frac{\sin\theta}{\sin\theta_c}\right). \tag{17–67}$$

It is interesting to consider the limiting case $\theta = \theta_c$ and compare it with the same case as obtained in the last section from the formulas valid for $\theta \leq \theta_c$. Setting $\theta = \theta_c$, Eq. (17–65) shows that $\chi = 0$, and Eqs. (17–64), (17–66), and (17–67) become identical with Eq. (17–56).

17–13 FRUSTRATED TOTAL REFLECTION

We can abstract the following purely mathematical results from the boundary value problem treated in the last two sections:

An incident wave of the form

$$\phi_{\text{inc}} = \mathcal{R}e\left\{\mathbf{A}e^{i(\omega t - \mathbf{k}\cdot\mathbf{r})}\right\}, \qquad z < 0$$

corresponds to a transmitted wave of the form

$$\phi_{\text{trans}} = \mathcal{R}e\left\{\mathbf{C}e^{i(\omega t - \mathbf{k}'\cdot\mathbf{r})}\right\}, \qquad z > 0,$$

where the following conditions hold:

$$k'_y = k_y = 0, \qquad k'_x = k_x,$$
$$k_x^2 + k_z^2 = k_1^2 = \omega^2/c_1^2, \qquad k_x'^2 + k_z'^2 = k_2^2 = \omega^2/c_2^2,$$
$$\mathbf{C} = \mathbf{TA}, \qquad \mathbf{T} = 2\rho_1 k_z/(\rho_2 k_z + \rho_1 k'_z).$$

We found that we could apply these equations to a situation of total reflection by taking k'_z to be a pure imaginary number, while k_x, k_z, and k'_x are real numbers. The incident wave in this case was an ordinary plane wave and the "transmitted" wave an inhomogeneous plane wave.

The operations performed in obtaining the above set of equations remain valid if some of the other components of **k** and **k**′ are imaginary quantities. Consider, for example, an *incident* plane wave of inhomogeneous form. Let k_x be a given real number *greater* than k_1. Since this will make k_z imaginary, we set $k_z = -i\beta$, where β is real and positive. We will still have $k'_x = k_x$ and $k_z'^2 = k_2^2 - k_x^2$. The latter quantity can be positive or negative depending on the given value of k_x. The transmitted wave is correspondingly either an ordinary or an inhomogeneous plane wave.

How would a situation arise in which there might be occasion to consider the transmission of an *inhomogeneous* plane wave at an interface? We have seen that inhomogeneous plane waves develop within the medium of higher sound velocity when total reflection is taking place. But suppose that this medium is not infinite and that the inhomogeneous plane waves generated at the first interface have the opportunity to be transmitted through a second interface parallel to the first. Thus consider a given ordinary plane wave incident at an angle greater than the critical angle on the interface designated by $z' = 0$ in Fig. 17–8. The inhomogeneous plane wave which is set up in

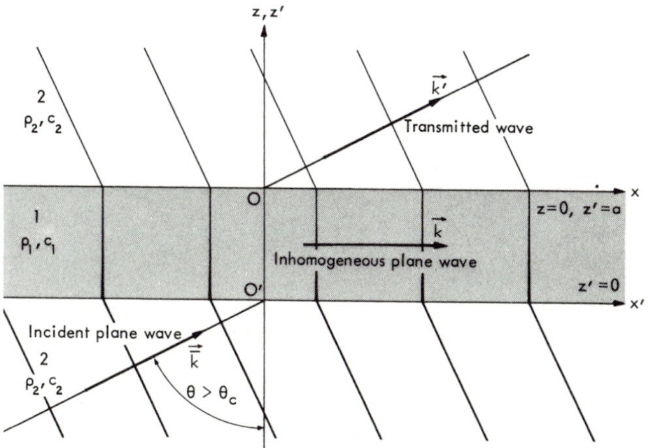

Fig. 17–8. Inhomogeneous plane wave giving rise to a transmitted wave.

$z' > 0$ plays the role of an "incident" wave on the interface $z' = a$. It is the character of the disturbance in $z' > a$ which we wish to determine.

To be specific, let the two outside media be equal. A plane wave of the form

$$\bar{\phi}_{\text{inc}} = \mathcal{R}e\,\{\mathbf{A}_1 e^{i(\omega t - \bar{\mathbf{k}}\cdot\mathbf{r}')}\}$$

is incident from $z' < 0$. The propagation vector $\bar{\mathbf{k}}$ has real components $\bar{k}_x = k_2 \sin\theta$ and $\bar{k}_z = k_2\cos\theta$, and $\mathbf{r}'$ is the position vector measured relative to origin O'. The wave transmitted at the $z' = 0$ interface is

$$\bar{\phi}_{\text{trans}} = \mathcal{R}e\,\{\mathbf{C}_1 e^{i(\omega t - \mathbf{k}\cdot\mathbf{r}')}\}, \qquad z' > 0,$$

where

$$k_x = \bar{k}_x = k_2 \sin\theta, \qquad k_x^2 + k_z^2 = k_1^2,$$

$$\mathbf{C}_1 = \mathbf{T}_1\mathbf{A}_1, \qquad \mathbf{T}_1 = 2\rho_2\bar{k}_z/(\rho_1\bar{k}_z + \rho_2 k_z).$$

Since we are assuming $c_2 < c_1$ and $\theta > \theta_c = \sin^{-1}(c_2/c_1)$, then k_z^2 is negative, and we choose $k_z = -i\beta$, where $\beta = \sqrt{k_x^2 - k_1^2}$. Translating the origin from O' to O we find for the wave incident on the interface $z = 0$

$$\phi_{\text{inc}} = \bar{\phi}_{\text{trans}} = \mathcal{R}e\,\{\mathbf{A}_2 e^{i(\omega t - \mathbf{k}\cdot\mathbf{r})}\},$$

where

$$\mathbf{A}_2 = \mathbf{C}_1 e^{-ik_z a} = \mathbf{C}_1 e^{-\beta a} = \mathbf{T}_1\mathbf{A}_1 e^{-\beta a}.$$

Applying the formulas to transmit this wave through the interface, we find

$$k_x' = k_x = \bar{k}_x = k_2\sin\theta,$$

$$k_x'^2 + k_z'^2 = k_2^2 = \bar{k}_x^2 + \bar{k}_z^2.$$

The latter equation implies that $k_z' = \pm\bar{k}_z = \pm k_2\cos\theta$. We choose the positive sign to obtain a wave traveling *away* from the boundary. Thus the final transmitted wave is an ordinary plane wave traveling *parallel* to the original plane wave. It has the form

$$\phi_{\text{trans}} = \mathcal{R}e\,\{\mathbf{C}_2 e^{i(\omega t - \bar{\mathbf{k}}\cdot\mathbf{r})}\},$$

where

$$\mathbf{C}_2 = \mathbf{T}_2\mathbf{A}_2 = \mathbf{T}_2\mathbf{T}_1\mathbf{A}_1 e^{-\beta a}$$

and

$$\mathbf{T}_2 = \frac{2\rho_1 k_z}{\rho_2 k_z + \rho_1 k_z'} = \frac{-2i\rho_1\beta}{\rho_1\bar{k}_z - i\rho_2\beta}.$$

The overall transmission coefficient for the velocity-potential function is

$$\mathcal{T} = \frac{\mathbf{C}_2}{\mathbf{A}_1} = \mathbf{T}_2\mathbf{T}_1 e^{-\beta a} = \frac{-4i\rho_1\rho_2\beta\bar{k}_z e^{-\beta a}}{(\rho_1\bar{k}_z - i\rho_2\beta)^2}.$$

The simplest way of expressing $\mathfrak{T}$ is in terms of the angle χ associated with the total reflection at $z' = 0$. To review the definitions of the pertinent quantities, we have

$$\begin{aligned} \bar{k}_z &= k_2 \cos\theta, \\ \beta &= k_2\sqrt{\sin^2\theta - \sin^2\theta_c}\,, \\ \tan\chi &= \rho_2\beta/\rho_1\bar{k}_z. \end{aligned} \tag{17-68}$$

A small amount of algebraic manipulation shows that $\mathfrak{T}$ can be written

$$\mathfrak{T} = e^{-\beta a}(-2i\sin 2\chi)e^{2i\chi}. \tag{17-69}$$

Since the final transmitted wave is an *ordinary* plane wave and transports energy in the normal fashion, we can speak of the total reflection which would have occurred at $z' = 0$ as being *frustrated* by the presence of the second interface at $z = 0$. The transmission coefficient $\mathfrak{T}$ depends on a through the factor $e^{-\beta a}$. Consequently, if a is large compared with the penetration depth $1/\beta$, only a small fraction of the energy of the incident wave passes through the layer. The layer can be thought of as a "barrier" which permits only a small fraction of the incident wave to "tunnel" through.

We naturally expect that as $a \to 0$ the effect of the intervening layer of medium 1 will ultimately be small and $\mathfrak{T}$ should approach the value unity. This does not happen with the formula given in Eq. (17–69). (Note that χ does not depend on a.) This calls attention to the fact that the derivation of this formula has been incomplete since it neglected the occurrence of *multiple reflections* between the two boundaries and the subsequent existence of a *series* of waves transmitted into $z > 0$. There are two methods of handling this situation. They correspond to the two approaches which were used to derive the overall transmission characteristics of a layer for the case of normal incidence (cf. Sections 3–7, 8). The procedure of the first method will be outlined. This is followed by an explicit calculation by the second method, which turns out to be relatively simple to perform.

The first method conceives of the problem as a boundary value problem with boundary conditions to be satisfied at the *two* surfaces $z' = 0$ and $z = 0$. The arbitrary separation of the disturbance in medium 1 into a series of multiply reflected waves is avoided. The general solution to the wave equation in medium 1 is a superposition of the form

$$\mathcal{R}e\,\{e^{i(\omega t - k_x x)}[\mathbf{E}e^{-\beta z} + \mathbf{F}e^{\beta z}]\}.$$

The arguments for ruling out the $e^{\beta z}$ solution which were used when the medium extended to $z = +\infty$ can no longer be used, and both solutions are retained. The unknowns **E**, **F**, and two others representing the net reflected wave in $z' < 0$ and the net transmitted wave in $z > 0$ are determined algebraically from the four boundary conditions, two at each surface.

The second method consists in the determination of the contributions from each of the multiply reflected waves and the summation of the resulting infinite series. The reflection coefficient for a wave reflected at $z = 0$ can be obtained from Eq. (17–54) with $k_z = -i\beta$, $k'_z = \bar{k}_z$. It can be seen that $\mathbf{R}$ is of the form

$$\mathbf{R} = -e^{2i\chi},$$

where χ is defined by Eq. (17–68). The complex amplitude before reflection is $\mathbf{A}_2$, and after reflection it is $\mathbf{RA}_2$. The reflected wave then propagates as an inhomogeneous plane wave, and at $z' = 0$ has the amplitude $\mathbf{RA}_2e^{-\beta a}$. Here it is again reflected, with the same value of the reflection coefficient, and arrives back at $z = 0$ with a complex amplitude $\mathbf{R}^2\mathbf{A}_2e^{-2\beta a}$. The complex amplitude of the net transmission through $z = 0$ is therefore given by

$$\mathbf{C}_2 = \mathbf{T}_2\mathbf{A}_2[1 + (\mathbf{R}^2e^{-2\beta a}) + (\mathbf{R}^2e^{-2\beta a})^2 + \cdots]. \tag{17–70}$$

We can see immediately that if $a \gg 1/\beta$, we are justified in neglecting the effects of multiple reflections. Thus Eq. (17–69) is valid with the provision that the layer be sufficiently thick to make the exponential decrement large. Equation (17–70) contains a geometric series of ratio $(\mathbf{R}^2e^{-2\beta a})$ which, when summed, gives

$$\mathbf{C}_2 = \mathbf{T}_2\mathbf{A}_2\left[\frac{1}{1 - \mathbf{R}^2e^{-2\beta a}}\right].$$

The overall transmission coefficient $\mathfrak{T}$ in Eq. (17–69) must therefore be corrected by the factor

$$\left[\frac{1}{1 - e^{4i\chi}e^{-2\beta a}}\right].$$

After simplification the result can be written

$$\mathfrak{T}' = \frac{\mathfrak{T}}{1 - e^{4i\chi}e^{-2\beta a}} = \frac{\sin 2\chi}{\sin 2\chi \cosh(\beta a) + i\cos 2\chi \sinh(\beta a)}. \tag{17–71}$$

In this form it is easy to check that $\mathfrak{T}' \doteq 1$ for $\beta a \ll 1$. That is, a layer whose thickness is small compared with the penetration depth (for the given angle of incidence) fully transmits the incident wave precisely as if there were no intervening medium present.

For an application of the concept of frustrated total reflection to acoustics, consider the transmission of sound through a pane of glass. Although glass is a solid and therefore does not fit immediately into the context of the formulas we have developed for fluid media, it can be shown that isotropic solids propagate compressional waves in a manner similar to fluids. The critical angle for an air-glass interface lies on the air side and is around 4°. Consequently, on a naive basis one might expect that sound waves obliquely incident on a pane of glass would be totally reflected. This is far from the case, however, since an ordinary pane of glass is thin compared with the penetration depth. This means that the factor $e^{-\beta a}$ in Eq. (17–69) does not amount to a drastic reduction in amplitude. Furthermore, when a is sufficiently small, the correction factor coming from the multiply reflected rays compensates for the reduction associated with the contrast in densities. Thus when $\rho_2 \ll \rho_1$, Eq. (17–68) shows that $\chi \ll 1$ and the factor $\sin 2\chi$ in Eq. (17–69) is small. However, when $\beta a \ll 1$, the correction factor is the order of magnitude of $1/\sin 2\chi$ and, as we have seen before, $\mathfrak{T}' \doteq 1$.

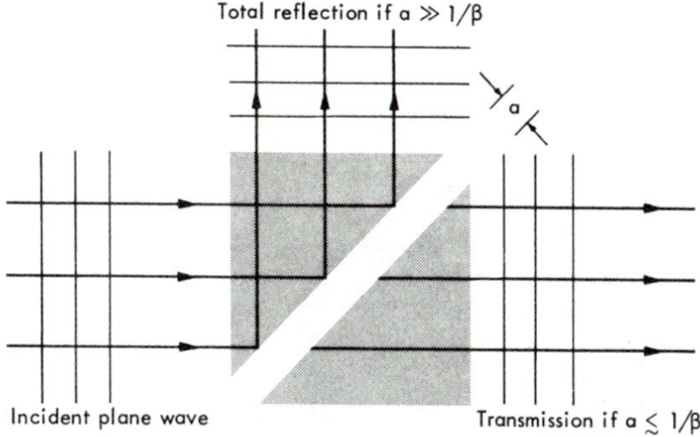

Fig. 17–9. Experimental arrangement for electromagnetic waves to demonstrate the transport of energy by means of inhomogeneous plane waves across a thin layer of higher wave-propagation velocity.

Maxwell's equations lead to an analogous description of the total reflection and frustrated total reflection of electromagnetic waves.* The latter phenomenon can be demonstrated with right-angle prisms as in Fig. 17–9. The experiment is readily performed with microwaves† or infrared radiation.‡ In the optical region the adjustment of the prisms is more delicate. The experiment is usually performed by having a slight curvature to the hypotenuse of the second prism. The surfaces are in contact at a point. Surrounding this there is a circular patch seen as a bright spot in transmission and as a black hole in reflection. When white light is used, the border of the bright spot is tinged red. There is a corresponding bluish tinge in the reflected light. This is evidence that the penetration depth is proportional to the wavelength [cf. Eq. (17–59)].

---

* R. W. Ditchburn, *Light*. New York: Interscience Publishers, Inc., 1953, Sections 14.15 to 14.20.

† A. Sommerfeld, *Optics*. New York: Academic Press Inc., 1954, p. 32.

‡ R. W. Pohl, *Einführung in die Optik*. Berlin: Springer Verlag, 1943, p. 140. See also p. 139 for photographs of total reflection and frustrated total reflection of *surface water waves*.

The concept of frustrated total reflection has an important application in nuclear physics. The escape of α-particles from a radioactive nucleus can be looked upon as a *tunnel effect* which takes place when the wave associated with an α-particle is incident on the potential barrier separating the inside of the nucleus from the outside. A formula analogous to Eq. (17–69) (the asymptotic case $a \gg 1/\beta$) is derived from the Schrödinger equation. The boundary conditions are such that "total reflection" occurs even though the angle of incidence is not oblique.

PROBLEMS

17–1. a) Use vector identities to show that

$$(\mathbf{v} \cdot \text{grad})\mathbf{v} = \tfrac{1}{2} \text{grad} (\mathbf{v} \cdot \mathbf{v}) + (\text{curl}\, \mathbf{v}) \times \mathbf{v}.$$

b) Show then that for *steady irrotational* motion of an *incompressible* fluid (no external forces acting) Eq. (17–3) can be integrated to yield *Bernoulli's equation*,

$$p + \tfrac{1}{2}\rho v^2 = \text{constant}.$$

17–2. As an alternative derivation of Eq. (17–3) use a method paralleling that of Appendix III, beginning with Eq. (III–11). Consider the x-component of the net impulse acting on a small volume element and the flow across all surfaces of momentum in the x-direction. Show that this leads to the equation

$$-\frac{\partial P}{\partial x} = \frac{\partial}{\partial t}(\rho v_x) + \text{div}\,(\rho v_x \mathbf{v})$$

and that by appropriate vector identities this reduces to Eq. (17–3).

17–3. In a field having spherical symmetry let $v_r = v(r, t)$ and $\mathbf{v} = (v/r)\mathbf{r}$. Use identities from Appendix VII to show the following:

a) $\text{curl}\, \mathbf{v} = 0$,

b) $$\text{grad div}\, \mathbf{v} = \frac{\mathbf{r}}{r}\left[\frac{1}{r^2}\frac{\partial}{\partial r}\left(r^2 \frac{\partial v}{\partial r}\right) - \frac{2v}{r^2}\right].$$

c) Show therefore that the wave equation for v_r, Eq. (17–10), takes on a different form from that for p or ϕ. d) Given $\phi = (1/r)f(t - r/c)$ show explicitly that $\mathbf{v} = -\text{grad}\, \phi$ is a solution to Eq. (17–10).

***17–4.** *Spherically symmetric solutions determined by initial conditions in the absence of sources and sinks*

a) The examples of velocity potential functions considered in the text have singularities at $r = 0$; this fact is associated with the presence of point or dipole sources at the origin. In the absence of such sources we require that the pressure remain finite at all times at $r = 0$. Show that this implies that the solution to Eq. (17–16) must satisfy the boundary condition $\psi(0, t) = 0$. b) If the radial velocity $v(r, t)$ is specified at all points at $t = 0$, show that this is tantamount to specifying the initial values of ψ through the relation

$$\psi(r, 0) = -r \int_0^r v(r, 0)\, dr.$$

c) If the pressure distribution is specified at all points at $t = 0$, show that this yields

$$\frac{\partial \psi}{\partial t}(r, 0) = \frac{r}{\rho_0}\, p(r, 0).$$

d) It is clear that the preceding conditions are isomorphic to the problem of determining the future motion of a semi-infinite string from a knowledge of the initial profile and velocity distributions. This problem can be dealt with by the graphical technique illustrated in Fig. 3–1. The techniques of Section 1–8 for an infinite string can also be used if we assume initial conditions which extend $\psi(r, 0)$ and $(\partial\psi/\partial t)(r, 0)$ as odd functions about $r = 0$. Use this method to solve for $\psi(r, t)$ under the initial conditions

$$v(r, 0) = 0, \qquad p(r, 0) = \begin{cases} p_0, & r < a, \\ 0, & r > a. \end{cases}$$

Show that for a point initially outside the disturbed region the pressure and velocity are zero except when t lies in the range

$$\frac{r - a}{c} < t < \frac{r + a}{c}$$

in which case

$$p = \frac{p_0}{2}\left(1 - \frac{ct}{r}\right), \qquad v = \frac{p_0}{4\rho_0 c}\left(1 + \frac{a^2 - c^2t^2}{r^2}\right).$$

17–5. a) Show that for a spherically symmetric situation

$$(\mathbf{v} \cdot \text{grad})\mathbf{v} = \left(v_r \frac{\partial v_r}{\partial r}\right)\hat{r},$$

where v_r is the radial velocity. b) Given the near zone solution for a pulsing sphere

$$v_r = \frac{a_0^2 b\omega \cos \omega t}{r^2},$$

show that the condition

$$(\mathbf{v} \cdot \text{grad})\mathbf{v} \ll \frac{\partial \mathbf{v}}{\partial t}$$

implies $b \ll a_0$, thus showing that the problem of a small pulsing sphere cannot be handled by means of the linearized Eulerian equations unless the amplitude of vibration is small compared with the mean radius.

17–6. With $\mathbf{v}_r$ given by Eq. (17–27) verify explicitly that the vector

$$\mathbf{v} = \hat{r}\,\mathcal{R}e\,\{\mathbf{v}_r e^{i\omega t}\}$$

is a solution to the vector wave equation, Eq. (17–10).

17–7. Verify explicitly that

$$\mathbf{v} = \left[\frac{p_0}{4\rho_0 c}\left(1 + \frac{a^2 - c^2 t^2}{r^2}\right)\right]\hat{r}$$

is a solution to the vector wave equation, Eq. (17–10).

17–8. Show the equivalence of the following expressions for the average intensity at a given point in the field of a diverging spherical acoustic wave:

$$\bar{I}_r = \tfrac{1}{2}Rv_m^2 = \tfrac{1}{2}p_m v_m \cos\delta = p_m^2/2\rho_0 c.$$

17–9. Let $\mathbf{Y} \equiv 1/\mathbf{Z} = G + iB$ be the complex *admittance* at a given point in the field of a divergent spherical acoustic wave. (G and B are referred to as the *conductance* and *susceptance* respectively.) Write

$$p = \mathcal{R}e\,\{\mathbf{p}e^{i\omega t}\} \qquad \text{and} \qquad v_r = \mathcal{R}e\,\{\mathbf{Y}\mathbf{p}e^{i\omega t}\}.$$

Carry out a calculation paralleling that of Section 17–7 to deduce the expression for average intensity

$$\bar{I}_r = \tfrac{1}{2}Gp_m^2 = p_m^2/2\rho_0 c.$$

17–10. a) Evaluate v_m and p_m, the amplitudes of the velocity and pressure functions, at the surface of the pulsing sphere described by Eq. (17–22). b) Use Eq. (17–31) to evaluate the amount of work done by the sphere on the surrounding medium during a quarter-cycle as the velocity goes from 0 to its maximum value. c) What occurs during the following quarter-cycle? In what sense is a sphere of large radius a "more efficient radiator" than a sphere of small radius? d) The contribution in (b) from the reactive term can be associated with the fact that a certain mass of the surrounding fluid must be accelerated from rest to the velocity v_m. Show that for $ka_0 \ll 1$ this is three times the mass of the fluid displaced by the sphere when its radius is at the equilibrium value a_0.

17–11. Verify by direct substitution that the potential functions given by the following equations are solutions to the scalar wave equation, Eq. (17–14): a) Eq. (17–34) (for a point source displaced from the origin); b) Eq. (17–35) (for a dipole source).

17–12. Let α_1 and $\alpha_2 > \alpha_1$ define two flow lines in the near zone of a dipole source [cf. Eq. (17–37)]. Show that the volume rate of flow of fluid across the annular ring in the plane $\theta = \pi/2$ defined by the two values of α is

$$2\pi Ad\left[\frac{1}{\alpha_1} - \frac{1}{\alpha_2}\right]\cos\omega t.$$

17–13. Verify explicitly that $\phi(x, y, z, t)$ given by Eq. (17–38) is a solution to the scalar wave equation, Eq. (17–14), provided that $l^2 + m^2 + n^2 = 1$.

17–14. Find the pressure and velocity fields associated with the scalar potential function for a general sinusoidal plane wave [cf. Eq. (17–39)]. Show that the relation $p = \pm Z\xi$ for $\pm$waves in one dimension generalizes to

$$\mathbf{v} = \left(\frac{p}{Z}\right)\hat{k},$$

where $\hat{k} = \mathbf{k}/k$ is a unit vector in the direction of propagation.

17–15. a) Show from Eq. (17–54) that an obliquely incident plane wave undergoes no reflection if the angle of incidence is

$$\theta_0 = \cos^{-1}\sqrt{\frac{(c_1/c_2)^2 - 1}{(\rho_2/\rho_1)^2 - 1}}.$$

b) Show that this corresponds to a real angle when $c_1 < c_2$ if and only if $Z_1 \geq Z_2$ or when $c_1 > c_2$ if and only if $Z_1 \leq Z_2$. c) Find a numerical value for θ_0 if the plane wave is incident from water upon an interface with castor oil. (See Appendix IV for values of the acoustic parameters.) Is zero reflection possible for a plane wave incident upon the same interface from the castor oil side? d) If $Z_1 = Z_2$, what is θ_0?

17–16. Calculate T_i, the transmission coefficient for intensity, for the case of a wave incident at an angle at which no reflection takes place (cf. Problem 17–15). Show that this result is consistent with the law of conservation of energy.

17–17. a) Show that for two media of equal densities the reflection coefficient for pressure can be written in the same form as Eq. (8–2), the Fresnel reflection coefficient for the component of an electromagnetic plane wave whose E-vector is perpendicular to the plane of incidence. b) Show that for two media of equal bulk moduli the reflection coefficient for pressure can be written in the same form as Eq. (8–3), the Fresnel reflection coefficient for the component of an electromagnetic plane wave whose E-vector is parallel to the plane of incidence.

17–18. Show that in the reflection of an obliquely incident plane wave at an interface the component of the particle velocity parallel to the interface is not continuous, and that

$$\frac{(v_x)_2}{(v_x)_1} = \frac{\rho_1}{\rho_2}.$$

17–19. On which side of an air-water interface does critical reflection of acoustic plane waves occur? What is the numerical value of the critical angle? (See Appendix IV for values of the acoustic parameters.)

17–20. a) Let $\mathbf{A} = 1$ be the complex amplitude of the velocity potential function for a plane wave incident on an interface at an angle greater than the critical angle. Use Eq. (17–63) to show that the net velocity potential function in medium 1 (the sum of the potentials for the incident and reflected waves) is

$$\phi_1 = 2\cos(k_z z + \chi)\cos(\omega t - k_x x + \chi)$$

and that the velocity potential function in medium 2 is

$$\phi_2 = (2\rho_1/\rho_2)(\cos\chi)e^{-\beta z}\cos(\omega t - k_x x + \chi).$$

b) Show that these two functions satisfy the scalar wave equations for their respective media, and that the boundary conditions at the interface are satisfied. c) Find the pressure functions and sketch a graph of $p(z)$ for $z \lesseqgtr 0$ along a plane x = constant at a fixed time t. d) Show that the formal solution for $\theta = \pi/2$ (grazing incidence) is $\phi_1 = \phi_2 = 0$.

***17–21.** Show that when total reflection is taking place the individual fluid particles in either medium describe ellipses whose axes are parallel and perpendicular to the interface (cf. Problem 17–20). *Note*: Unlike the situation of Eq. (17–36), the streamlines of the fluid (the family of curves orthogonal to ϕ = constant) vary with time. The instantaneous streamlines do not give the trajectories of individual particles. To find the latter it is necessary to recall that particle displacements are small compared with a wavelength in acoustic situations. A particle initially at (x_0, z_0) can be assumed to have a velocity given by $\mathbf{v}(x_0, z_0, t)$ throughout the motion.

17–22. For an incident plane wave of given intensity find the instantaneous and average values of the power crossing a unit of area on the interface as a function of the angle of incidence.

17–23. Show that the reflection and transmission coefficients for ϕ, p, v_x and v_z [Eqs. (17–54) and (17–55)] satisfy Stokes' relations, Eqs. (8–6) and (8–7).

***17–24.** Use the method of multiple internal reflections (the second method of Section 17–13) to determine the net reflection coefficient $\mathfrak{R}$ for a wave incident at an angle greater than the critical angle on a layer separating two equal media. Show that $\mathfrak{R} \to 0$ as the thickness of the layer tends to zero. Show also for the general case that

$$|\mathfrak{R}|^2 + |\mathfrak{T}'|^2 = 1.$$

***17–25.** Solve the two-layer boundary value problem by the first method outlined in Section 17–13 to determine the net transmission coefficient $\mathfrak{T}'$ for a layer at which frustrated total reflection is taking place.

APPENDIX I. The representation of sinusoidal functions by complex numbers

An arbitrary complex number $\mathbf{A}$ can be written in either of two forms:

$$\mathbf{A} = a + ib \qquad \text{component (or rectangular) form,}$$
$$\mathbf{A} = Ae^{i\theta} \qquad \text{polar form.}$$

a is referred to as the real part of $\mathbf{A}$ and is written $a = \mathcal{R}e\,\mathbf{A}$,
b is referred to as the imaginary part of $\mathbf{A}$ and is written

$$b = \mathcal{I}m\,\mathbf{A} = \mathcal{R}e\,(\mathbf{A}/i),$$

A is referred to as the amplitude of $\mathbf{A}$ and is written $A = |\mathbf{A}|$,
θ is referred to as the phase (or argument) of $\mathbf{A}$.

Complex numbers will be written in boldface type, real numbers in standard type. Thus, a, b, A, θ are all real numbers. Through the identity

$$e^{i\theta} = \cos\theta + i\sin\theta,$$

we have

$$\mathbf{A} = a + ib = Ae^{i\theta} = A\cos\theta + i(A\sin\theta).$$

But when two complex numbers are equal, their respective real and imaginary parts must be equal. This provides the formulas for converting from the parameters of the polar form to those of the rectangular form:

$$a = A\cos\theta,$$
$$b = A\sin\theta. \tag{I–1}$$

By taking the square root of the sum of the squares of these two equations and by taking the ratio of the second equation to the first, we find the relations for converting from the rectangular form to the polar form:

$$A = \sqrt{a^2 + b^2}, \qquad \tan\theta = b/a. \tag{I–2}$$

[An efficient way to compute these numbers is to find (b/a) first. While looking up $\theta = \tan^{-1}(b/a)$, also record the sine or cosine of this same angle (a parallel entry). Then A can be computed from $A = a/\cos\theta$ or $A = b/\sin\theta$.]

The set of all complex numbers $(a + ib)$ can be put in a one-to-one correspondence with the points in a plane by establishing a cartesian coordinate system and identifying the point $P(a, b)$ whose coordinates are $x = a$, $y = b$ with the complex number $(a + ib)$. Another set of entities which can be put in a one-to-one correspondence with either of the former sets is the set of radius vectors extending from the origin to the point $P(a, b)$. The vector which corresponds to the complex number $\mathbf{A}$ will also be designated by the symbol $\mathbf{A}$. The length of the vector $\mathbf{A}$ is clearly equivalent to the amplitude of the complex number $\mathbf{A}$. The angle between the vector $\mathbf{A}$ and the x-axis is $\tan^{-1}(b/a)$ and hence is equivalent to θ, the phase of the complex number $\mathbf{A}$. Figure I–1 is a convenient mnemonic for the relations discussed so far.

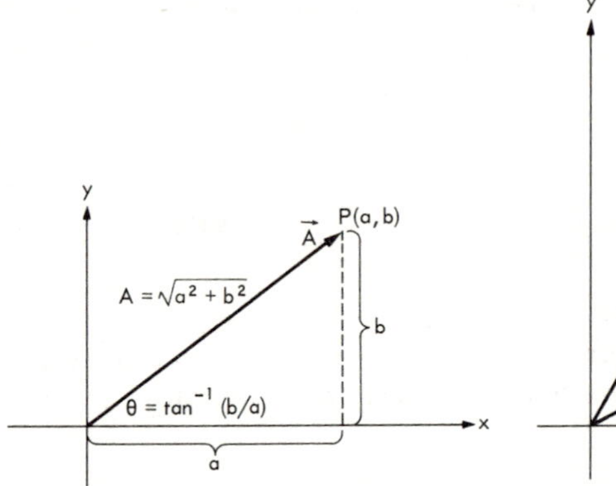

Figure I–1

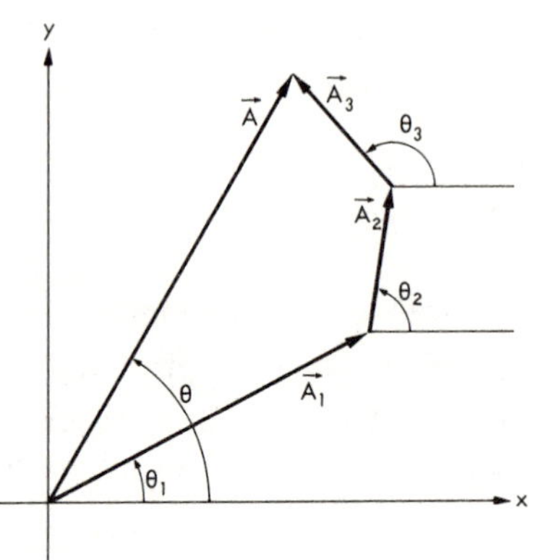

Fig. I–2. Representation of the sum of three complex numbers:
$\mathbf{A} = \mathbf{A}_1 + \mathbf{A}_2 + \mathbf{A}_3$.

The mnemonic can be extended to cover the rules for addition of two or more complex numbers. The sum of the complex numbers $\mathbf{A} = a + ib$ and $\mathbf{A}' = a' + ib'$ is defined as the complex number

$$\mathbf{A}'' = \mathbf{A} + \mathbf{A}' \equiv (a + a') + i(b + b').$$

But the vector sum of the vectors $\mathbf{A}$ and $\mathbf{A}'$ has an x-component equal to $(a + a')$ and a y-component equal to $(b + b')$. We see therefore that the vector associated with $\mathbf{A}''$ is identical with the vector sum of $\mathbf{A}$ and $\mathbf{A}'$. Thus the sum of any number of complex numbers can be obtained from a vector diagram, following the usual rules of vector addition (see Fig. I–2).

PROPOSITION: *The sum of any number of sinusoidal functions all of the same frequency but of differing amplitude and phase is equivalent to a single sinusoidal function of the same frequency.*

Proof: Given the functions $y_k = A_k \cos(\omega t + \theta_k)$ with $k = 1, 2, \ldots, n$, we desire to find $y = \sum_{k=1}^{n} y_k$. Write y_k in the form

$$y_k = A_k \,\mathcal{R}e\, e^{i(\omega t + \theta_k)} = \mathcal{R}e\,\{\mathbf{A}_k e^{i\omega t}\},$$

where $\mathbf{A}_k = A_k e^{i\theta_k}$; $\mathbf{A}_k$ is referred to as the *complex amplitude* of the function y_k. Then

$$\begin{aligned} y = \sum_{k=1}^{n} y_k &= \sum_{k=1}^{n} \mathcal{R}e\,\{\mathbf{A}_k e^{i\omega t}\} = \mathcal{R}e\left\{e^{i\omega t}\left[\sum_{k=1}^{n} \mathbf{A}_k\right]\right\} \\ &= \mathcal{R}e\,\{\mathbf{A}e^{i\omega t}\} = A\cos(\omega t + \theta), \end{aligned}$$

where $\mathbf{A} = Ae^{i\theta} \equiv \sum_{k=1}^{n} \mathbf{A}_k$.

In other words, the complex amplitude of the resultant sinusoidal function is the sum of the complex amplitudes of the individual functions in the sum. The addition of the complex numbers $\mathbf{A}_k$ can be visualized in terms of a vector diagram such as Fig. I–2. A convenient way of computing the resultant is to write each complex amplitude $\mathbf{A}_k$ in component form $\mathbf{A}_k = a_k + ib_k$. Then

$$\theta = \tan^{-1}\left[\frac{b_1 + \cdots + b_n}{a_1 + \cdots + a_n}\right], \qquad \text{and} \qquad A = \frac{(b_1 + \cdots + b_n)}{\sin\theta}.$$

For any complex number $\mathbf{A} = a + ib = Ae^{i\theta}$, the *complex conjugate* $\mathbf{A}^*$ is defined as $\mathbf{A}^* \equiv a - ib = Ae^{-i\theta}$. By means of this we can express

$$\mathcal{R}e\,\mathbf{A} = a = \frac{1}{2}\{(a + ib) + (a - ib)\} = \frac{1}{2}(\mathbf{A} + \mathbf{A}^*)$$

and

$$\mathcal{I}m\,\mathbf{A} = b = \frac{1}{2i}\{(a + ib) - (a - ib)\} = \frac{1}{2i}(\mathbf{A} - \mathbf{A}^*).$$

Thus, for example, we have the useful identities

$$\cos\theta = \mathcal{R}e\, e^{i\theta} = \tfrac{1}{2}(e^{i\theta} + e^{-i\theta}), \qquad \sin\theta = \mathcal{I}m\, e^{i\theta} = \frac{1}{2i}(e^{i\theta} - e^{-i\theta}).$$

EXAMPLES

1) Write $y = \cos x - \sin x$ as a single sinusoidal function.

$$\begin{aligned} y_1 &= \cos x = \mathcal{R}e\, e^{ix}, \\ y_2 &= -\sin x = \cos(x + \pi/2) = \mathcal{R}e\,[e^{i\pi/2}e^{ix}], \\ y &= y_1 + y_2 = \mathcal{R}e\,\{[1 + e^{i\pi/2}]e^{ix}\}. \end{aligned}$$

But $1 + e^{i\pi/2} = 1 + i = \sqrt{2}\,e^{i\theta}$, where $\tan\theta = 1$, that is, $\theta = \pi/4$. Thus, $\cos x - \sin x = \sqrt{2}\cos(x + \pi/4)$.

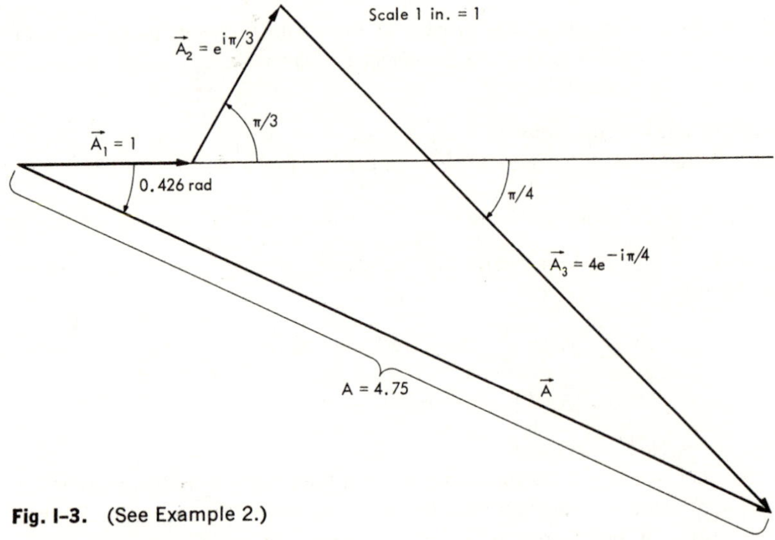

Fig. I–3. (See Example 2.)

2) Write $y = \cos \omega t + \sin (\omega t + 5\pi/6) + 4 \cos (\omega t - \pi/4)$ as a single sinusoidal function.

First note that $\sin (\omega t + 5\pi/6) = \cos (\omega t + \pi/3)$. The complex amplitudes to be added are (see Fig. I–3):

$$\begin{aligned} \mathbf{A}_1 &= 1, \quad \mathbf{A}_2 = e^{i\pi/3} = \tfrac{1}{2} + (\sqrt{3}/2)i, \quad \mathbf{A}_3 = 4e^{-i\pi/4} \\ &= 4[(1/\sqrt{2}) - (i/\sqrt{2})] = 2\sqrt{2} - 2\sqrt{2}\, i; \\ \mathbf{A} &= (1 + 0.5 + 2.828) + (0.866 - 2.828)i = 4.328 - 1.962i; \\ \theta &= -\tan^{-1} (1.962/4.328) = -\tan^{-1} (0.454) = -0.426 \text{ rad}; \\ A &= a/\cos \theta = 4.328/0.911 = 4.75. \end{aligned}$$

Thus $y = (4.75) \cos (\omega t - 0.426)$.

3) Prove the identity $\cos \alpha + \cos \beta = 2 \cos \frac{1}{2}(\alpha + \beta) \cos \frac{1}{2}(\alpha - \beta)$.

$$\begin{aligned} \cos \alpha + \cos \beta &= \Re e \{e^{i\alpha} + e^{i\beta}\} = \Re e \{e^{i(\alpha+\beta)/2}[e^{i(\alpha-\beta)/2} + e^{-i(\alpha-\beta)/2}]\} \\ &= \Re e \{e^{i(\alpha+\beta)/2} 2 \cos \tfrac{1}{2}(\alpha - \beta)\} = 2 \cos \tfrac{1}{2}(\alpha - \beta) \Re e\, e^{i(\alpha+\beta)/2} \\ &= 2 \cos \tfrac{1}{2}(\alpha - \beta) \cos \tfrac{1}{2} (\alpha + \beta). \end{aligned}$$

PROBLEMS

I–1. Show that $\sum_{k=0}^{6} \cos (t + k\pi/6) = -(2 + \sqrt{3}) \sin t$.

I–2. In general, the sum of two sinusoidal functions of equal amplitude but different phase has an amplitude different from that of the original functions. Find the exception. That is, find the value of θ which will allow

$$A \cos x + A \cos (x + \theta) = A \cos (x + \theta').$$

Find also θ', the phase of the resultant.

I–3. When a sinusoidal wave is partially reflected at the end of a string, the resulting shape of the string is given by the function

$$y(x, t) = \cos(kx - \omega t) + R\cos(kx + \omega t),$$

where R is a reflection coefficient. Show that any point at a fixed value of x oscillates sinusoidally with an amplitude which depends on x. In particular, show that if $0 < R < 1$, the largest amplitude occurs at the points $x = n\lambda/2$ and the least at $x = (n + \frac{1}{2})\lambda/2$. Verify that the ratio of the minimum to the maximum amplitude is $(1 - R)/(1 + R)$. (This is the so-called *standing wave ratio.*)

I–4. Prove the identity

$$\int_a^b \cos(\omega t + \beta x)\, dx = A\cos(\omega t + \phi),$$

where ω and β are given constants,

$$\phi = (a + b)\beta/2, \text{ and } A = (2/\beta)[\sin(b - a)\beta/2].$$

Note: This integral is of the basic type required in Fraunhofer diffraction theory, Chapter 12.

I–5. Select several of the identities listed in Appendix II and prove them by the method of Example 3 above.

APPENDIX II. Frequently used trigonometric identities

$$\sin 2\alpha = 2 \sin \alpha \cos \alpha$$

$$\cos 2\alpha = \cos^2 \alpha - \sin^2 \alpha = 1 - 2 \sin^2 \alpha = 2 \cos^2 \alpha - 1$$

$$\sin^2 \alpha = \tfrac{1}{2}(1 - \cos 2\alpha)$$

$$\cos^2 \alpha = \tfrac{1}{2}(1 + \cos 2\alpha)$$

$$\sin (\alpha \pm \beta) = \sin \alpha \cos \beta \pm \cos \alpha \sin \beta$$

$$\cos (\alpha \pm \beta) = \cos \alpha \cos \beta \mp \sin \alpha \sin \beta$$

$$\sin \alpha \pm \sin \beta = 2 \sin \tfrac{1}{2}(\alpha \pm \beta) \cos \tfrac{1}{2}(\alpha \mp \beta)$$

$$\cos \alpha + \cos \beta = 2 \cos \tfrac{1}{2}(\alpha + \beta) \cos \tfrac{1}{2}(\beta - \alpha)$$

$$\cos \alpha - \cos \beta = 2 \sin \tfrac{1}{2}(\alpha + \beta) \sin \tfrac{1}{2}(\beta - \alpha)$$

$$\sin \alpha \sin \beta = \tfrac{1}{2}\{\cos (\alpha - \beta) - \cos (\alpha + \beta)\}$$

$$\cos \alpha \cos \beta = \tfrac{1}{2}\{\cos (\alpha - \beta) + \cos (\alpha + \beta)\}$$

$$\sin \alpha \cos \beta = \tfrac{1}{2}\{\sin (\alpha - \beta) + \sin (\alpha + \beta)\}$$

APPENDIX III. The one-dimensional hydrodynamic equations in Eulerian form

Let X represent the coordinate of a point in space. The pressure and fluid density at this point will be specified by $P_e(X, t)$ and $\rho_e(X, t)$. Since the functional form of the function expressing the pressure at X at time t will be different from the functional form $P(x, t)$ used in the Lagrangian formulation of this problem, we designate the distinction by using a subscript e for the Eulerian functions. [Recall that $P(x, t)$ refers to the pressure in the vicinity of those particles which have position x when the disturbance is not present. At time t these particles have moved to a different location, so that $P(x, t)$ does not represent the pressure *at* x at time t.] The velocity at time t of those particles which are at the point X will be designated by the function $v(X, t)$.

See Chapter 2 for the Lagrangian equations referred to in the following.

Derivation of the Eulerian equations by change of variables from the Lagrangian equations. At time t the physical position in space, X, is occupied by particles whose "home position" is x. The relation between X and x is specified by the equation

$$X = x + \xi(x, t).$$

We can equivalently think of this equation as being solved for x as a function of X and t, thereby determining the home-position coordinate x associated with the particles which are at X at time t. Let this be expressed in the form $x = f(X, t)$. If we are given the functional form $P(x, t)$, the means of finding $P_e(X, t)$ would be to substitute $f(X, t)$ for x in $P(x, t)$. Thus

$$P_e(X, t) = P[f(X, t), t].$$

Since the numerical identity $P(x, t) = P_e(X, t)$ is satisfied, we can express the partial derivatives of P in terms of the partial derivatives of P_e. Thus

$$\frac{\partial P}{\partial x} = \frac{\partial P_e}{\partial X}\frac{\partial X}{\partial x}.$$

But

$$\frac{\partial X}{\partial x} = 1 + \frac{\partial \xi}{\partial x};$$

therefore we have

$$\frac{\partial P}{\partial x} = \left(1 + \frac{\partial \xi}{\partial x}\right)\frac{\partial P_e}{\partial X}. \tag{III–1}$$

Similarly, from the identity $\rho(x, t) = \rho_e(X, t)$ we find

$$\frac{\partial \rho}{\partial x} = \left(1 + \frac{\partial \xi}{\partial x}\right)\frac{\partial \rho_e}{\partial X}, \tag{III–2}$$

and

$$\frac{\partial \rho}{\partial t} = \frac{\partial \rho_e}{\partial X}\frac{\partial X}{\partial t} + \frac{\partial \rho_e}{\partial t} = \frac{\partial \rho_e}{\partial X}\frac{\partial \xi}{\partial t} + \frac{\partial \rho_e}{\partial t} = v\frac{\partial \rho_e}{\partial X} + \frac{\partial \rho_e}{\partial t}. \tag{III–3}$$

Also, from the identity $(\partial\xi/\partial t)(x, t) = v(X, t)$, it follows that

$$\frac{\partial^2 \xi}{\partial x\,\partial t} = \left(1 + \frac{\partial \xi}{\partial x}\right)\frac{\partial v}{\partial X}, \tag{III–4}$$

and

$$\frac{\partial^2 \xi}{\partial t^2} = v\frac{\partial v}{\partial X} + \frac{\partial v}{\partial t}. \tag{III–5}$$

Thus consider the Lagrangian equation expressing Newton's second law:

$$-\frac{\partial P}{\partial x} = \rho_0\frac{\partial^2 \xi}{\partial t^2}. \tag{III–6}$$

By substitution from Eqs. (III–1) and (III–5), Eq. (III–6) becomes

$$-\left(1 + \frac{\partial \xi}{\partial x}\right)\frac{\partial P_e}{\partial X} = \rho_0\left[v\frac{\partial v}{\partial X} + \frac{\partial v}{\partial t}\right],$$

or

$$-\frac{\partial P_e}{\partial X} = \rho_e\left[v\frac{\partial v}{\partial X} + \frac{\partial v}{\partial t}\right]. \tag{III–7}$$

This is Newton's second law expressed entirely in terms of the Eulerian representation.

Consider the Lagrangian equation for the conservation of mass:

$$\rho = \rho_0\left(1 + \frac{\partial \xi}{\partial x}\right)^{-1}. \tag{III–8}$$

By partial differentiation with respect to t this becomes

$$\frac{\partial \rho}{\partial t} = -\rho_0\left(1 + \frac{\partial \xi}{\partial x}\right)^{-2}\frac{\partial^2 \xi}{\partial t\,\partial x}. \tag{III–9}$$

Substituting from Eqs. (III–3) and (III–4) yields

$$v\frac{\partial\rho_e}{\partial X} + \frac{\partial\rho_e}{\partial t} = -\rho_0\left(1 + \frac{\partial\xi}{\partial x}\right)^{-2}\left(1 + \frac{\partial\xi}{\partial x}\right)\frac{\partial v}{\partial X}$$

$$= -\rho_e\frac{\partial v}{\partial X}.$$

This can be rearranged to the form

$$\frac{\partial\rho_e}{\partial t} = -\frac{\partial(\rho_e v)}{\partial X}, \tag{III–10}$$

which is the Eulerian form of the law of conservation of mass.

Direct derivation of the basic Eulerian equations. Consider the volume in space defined by the planes X and $X + \Delta X$ and unit cross-sectional area in the transverse direction. During time Δt the mass which flows into this volume over the surface at X is $\rho_e(X, t)v(X, t)\,\Delta t$. Similarly, the mass which flows in at $X + \Delta X$ is $-\rho_e(X + \Delta X, t)v(X + \Delta X, t)\,\Delta t$. Therefore the net gain in mass during time Δt is

$$-\frac{\partial(\rho_e v)}{\partial X}\,\Delta X\,\Delta t.$$

This may also be expressed as $(\partial M/\partial t)\,\Delta t$, where M is the mass contained within ΔX at time t: $M = \rho_e\,\Delta X$. Equating these two expressions and dividing by $\Delta X\,\Delta t$, we obtain Eq. (III–10).

The net impulse delivered during Δt by external forces on the particles contained in the volume ΔX is

$$-\frac{\partial P_e}{\partial X}\,\Delta X\,\Delta t. \tag{III–11}$$

There is also a direct influx of momentum into the volume due to a transfer of particles over the boundary surfaces. During the time interval Δt the particles which lie within a volume $v(X, t)\,\Delta t$ enter the volume element in question across the end face at X. The momentum per unit volume is $\rho_e v$; consequently, the momentum transported across X during time Δt is $[\rho_e v][v\,\Delta t]$. Taking account of the momentum transported across $X + \Delta X$, we find that the net momentum transported is

$$-\frac{\partial}{\partial X}(\rho_e v^2)\,\Delta X\,\Delta t. \tag{III–12}$$

The total momentum contained within the volume at time t is

$$\mathfrak{M} = (\rho_e v)\,\Delta X.$$

The change in this quantity during time Δt is $[(\partial/\partial t)(\rho_e v)]\,\Delta X\,\Delta t$. Equating this to the sum of Eqs. (III–11) and (III–12) yields

$$-\frac{\partial P_e}{\partial X} = \frac{\partial}{\partial t}(\rho_e v) + \frac{\partial}{\partial X}(\rho_e v^2).$$

Making use of Eq. (III–10) (which has already been proved by the direct method) we find this result to be identical with Eq. (III–7).

Alternative derivation of the dynamical equation of motion. The preceding derivation made use of the impulse-momentum theorem, which is equivalent to Newton's second law. A direct derivation can be given if we pay proper attention to the Eulerian expression for the acceleration of the particles. The quantity $(\partial v/\partial t)(X, t)$ is merely the formal time derivative of the function $v(X, t)$ with respect to t. It represents the rate at which the velocity of the fluid *observed* at X is changing with respect to the time. It does not represent the acceleration which the particles at X are experiencing. The particles which are at X at time t will be at position $X + v(X, t)\,\Delta t$ at time $t + \Delta t$. They will then have velocity

$$v[X + v(X, t)\,\Delta t, t + \Delta t] \doteq v(X, t) + v\,\Delta t\,\frac{\partial v}{\partial X} + \frac{\partial v}{\partial t}\,\Delta t.$$

(The last step makes use of the differential approximation.) Dividing the velocity increment during time Δt by Δt we obtain for the acceleration of the particles which are at X at time t

$$a(X, t) = v\,\frac{\partial v}{\partial X} + \frac{\partial v}{\partial t}\cdot \qquad \text{(III–13)}$$

The quantity appearing on the right-hand side of (III–13) is abbreviated as Dv/Dt and is known as the *substantial derivative* (following the motion of the substance). The term $\partial v/\partial t$ is known as the *local derivative*; $v(\partial v/\partial X)$ is known as the *convective derivative.*

The mass contained in ΔX is $\rho_e\,\Delta X$. When the quantity $(\rho_e\,\Delta X)a(X, t)$ is set equal to the net force $(-\partial P_e/\partial X)\,\Delta X$, we again obtain Eq. (III–7).

Comments on the equations in Eulerian form

$$-\frac{\partial P_e}{\partial X} = \rho_e\left[v\,\frac{\partial v}{\partial X} + \frac{\partial v}{\partial t}\right]\cdot \quad \textit{The force law} \qquad \text{(III–7)}$$

$$\frac{\partial \rho_e}{\partial t} = -\rho_e\,\frac{\partial v}{\partial X} - v\,\frac{\partial \rho_e}{\partial X}\cdot \quad \textit{Conservation of mass} \qquad \text{(III–10)}$$

$$P_e = P_a(\rho_e). \quad \textit{Adiabatic equation of state} \qquad \text{(III–14)}$$

In principle these equations determine the solution to an arbitrary one-dimensional motion of the fluid. Because of the products such as $v(\partial \rho_e/\partial X)$, the equations are nonlinear. There is no convenient elimination which will reduce these equations to a single variable for the general case. (Thus the Eulerian form is not convenient for the discussion of one-dimensional shock waves.) For "small-amplitude" disturbances the following approximations may be made: Assume that $\partial P_e/\partial X$, $\partial \rho_e/\partial X$, v, $\partial \rho_e/\partial t$, $\partial v/\partial X$, and $\partial v/\partial t$ are all "first-order small quantities," as is also the difference between ρ_e and ρ_0. Products of first-order small quantities are referred to as "second-order small quantities," and are to be neglected in comparison with first-order small quantities. In Eq. (III–7), for example, $v(\partial v/\partial X)$ is neglected in comparison with $\partial v/\partial t$. Furthermore

$$\rho_e \frac{\partial v}{\partial t} = \rho_0 \frac{\partial v}{\partial t} + (\rho_e - \rho_0) \frac{\partial v}{\partial t} \doteq \rho_0 \frac{\partial v}{\partial t}.$$

When these approximations are made we obtain

$$-\frac{\partial P_e}{\partial X} = \rho_0 \frac{\partial v}{\partial t}, \tag{III–7$'$}$$

$$-\frac{\partial \rho_e}{\partial t} = \rho_0 \frac{\partial v}{\partial X}. \tag{III–10$'$}$$

The equation of state is approximated in the usual manner by taking the pressure as a linear function of density

$$P_e = P_0 + (\rho_e - \rho_0)P_a'(\rho_0) = P_0 + (\rho_e - \rho_0)c^2, \tag{III–14$'$}$$

where $c^2 = P_a'(\rho_0)$. Substitution of (III–14′) into (III–7′) yields

$$-c^2 \frac{\partial \rho_e}{\partial X} = \rho_0 \frac{\partial v}{\partial t}. \tag{III–15}$$

Taking $\partial/\partial X$ of (III–15) and $\partial/\partial t$ of (III–10′), we can eliminate v:

$$\frac{\partial^2 \rho_e}{\partial t^2} = c^2 \frac{\partial^2 \rho_e}{\partial X^2}. \tag{III–16}$$

Thus $\rho_e(X, t)$ satisfies the same differential equation in X and t that $\rho(x, t)$ satisfies in x and t. Under the conditions of the small-amplitudes approximation, the distinction between x and X is insignificant, and the Lagrangian and Eulerian descriptions are identical. (More precisely, neither method gives a 100% accurate description. The extent to which the two methods differ is of the same order as the error inherent in each.)

See Chapter 17 for a derivation of the Eulerian form of the equations of motion of a fluid in the context of three dimensions.

APPENDIX IV. Table of acoustic parameters of gases and liquids

Table IV–1. Gases*

Substance	ρ_0, kg/m³	c, m/sec	γ	$Z = \rho_0 c$, kg/m²-sec
Air	1.293	331	1.40	429
CO_2	1.977	258	1.30	510
He	0.1785	970	1.66	173
H_2	0.0899	1270	1.41	114
N_2	1.251	337	1.40	421
O_2	1.429	317	1.40	453
Steam (100°C)	0.598	475	1.32	284

* At a pressure of 1 atmosphere (1.013×10^5 N/m²) and a temperature of 0°C.

Table IV–2. Liquids*

Substance	Formula	ρ_0, kg/m³	c, m/sec	$Z = \rho_0 c$, kg/m²-sec	$\mathcal{B}_a = \rho_0 c^2$, N/m²
Benzene	C_6H_6	879	1318	1.16×10^6	1.53×10^9
Castor oil	$C_{11}H_{10}O_{10}$	950	1540	1.46×10^6	2.25×10^9
Ethanol	C_2H_5OH	789	1162	0.92×10^6	1.07×10^9
Mercury	Hg	13,550	1451	19.7×10^6	28.5×10^9
Methanol	CH_3OH	791	1121	0.89×10^6	0.99×10^9
Pure water	H_2O	998	1483	1.48×10^6	2.19×10^9
Seawater	†	1025	1522	1.56×10^6	2.37×10^9
Turpentine	–	870	1330	1.16×10^6	1.54×10^9

* At 20°C and 1 atmosphere pressure.
† Salinity = 35 parts per thousand.

APPENDIX V. Physical situations isomorphic to the one-dimensional acoustic wave

Table V–1 summarizes correspondences which carry the basic equations of the one-dimensional acoustic wave into the basic equations which govern a number of different physical situations. An explanation of variables is given in the numbered sections which follow, and the basic equations are derived from fundamentals appropriate to the given context. In each case it is necessary to check that a correct translation can be obtained of the equations

$$p = -\mathcal{B}_a \frac{\partial \xi}{\partial x}, \qquad \text{and} \qquad -\frac{\partial p}{\partial x} = \rho_0 \ddot{\xi}.$$

From this it follows that a wave equation will hold with a propagation velocity obtained by translating $c = \sqrt{\mathcal{B}_a/\rho_0}$; a conservation law will hold among energy quantities corresponding to

$$W_{\text{kin}} = \tfrac{1}{2}\rho_0 \dot{\xi}^2, \qquad w_{\text{pot}} = p^2/2\mathcal{B}_a, \qquad i = p\dot{\xi};$$

progressive waves will satisfy relations corresponding to $p_\pm = \pm Z\dot{\xi}_\pm$,

Table V–1. Correspondence of variables in situations isomorphic to the one-dimensional acoustic wave

1) Acoustic	ξ	$\dot{\xi}$	p	$\mathcal{B}_a$	ρ_0
2) Transverse on a string	y		F_y	T	σ
3) Electromagnetic		H_z	E_y	$1/\epsilon$	μ
4) Tidal	ξ		$\rho g \eta$	$\rho g h$	ρ
5) Transmission line		i	v	$1/C$	L
6) Longitudinal in a spring	ξ		F	kl	σ
7) Longitudinal in a bar	ξ		F/A	E	ρ
8) Torsion	θ		τ	κl	I
9) Acoustic (alternate)		p	$\dot{\xi}$	$1/\rho_0$	$1/\mathcal{B}_a$

where $Z = \rho_0 c$, and the solution to boundary-value problems can be translated for cases in which the boundary conditions correspond.

For those situations in which there is no variable corresponding to ξ it is necessary to check the equation

$$p = -\mathcal{B}_a \frac{\partial \dot{\xi}}{\partial x}.$$

All other relations then follow as before.

Explanation of variables and derivation of the basic equations

1) ACOUSTIC PLANE WAVES (See Chapters 2 and 4.)
2) TRANSVERSE WAVES ON A STRING (See Chapter 1 and Section 4–10.)
3) LINEARLY POLARIZED ELECTROMAGNETIC PLANE WAVES (See Chapter 7.)
4) TIDAL WAVES (See Fig. V–1.)

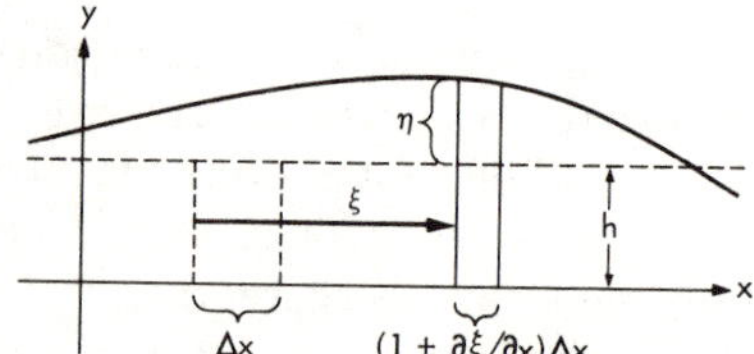

Figure V–1

The situation refers to waves on a liquid surface. In the equilibrium state the liquid is of uniform depth h. The restrictive assumption is made that particles originally lying in a plane perpendicular to the x-axis undergo equal horizontal displacements and hence form a new vertical plane in the disturbed configuration. A more advanced treatment shows that this assumption is a good approximation for sinusoidal waves provided the wavelength is large compared with the depth. For this reason these are sometimes referred to as *long waves in shallow water*. The liquid is assumed to be incompressible, with density ρ. The meaning of the variables is as follows:

h	undisturbed depth of liquid,
ρ	density of liquid,
$\xi(x, t)$	horizontal displacement of particles originally at x,
$\eta(x, t)$	elevation of surface above mean depth h, corresponding to the plane of particles originally at x.

Derivation of the basic relations. Consider unit length perpendicular to the plane of the figure. The Lagrangian method is used to label the particles according to their original positions.

Conservation of mass. The liquid originally in Δx has mass $\rho h \,\Delta x$. In the disturbed configuration this would be written

$$\rho(h + \eta)\left(1 + \frac{\partial \xi}{\partial x}\right)\Delta x.$$

Equating the expressions leads to the relation

$$\left(1 + \frac{\eta}{h}\right)\left(1 + \frac{\partial \xi}{\partial x}\right) = 1.$$

Solving for η this becomes

$$\eta = -\frac{h(\partial \xi/\partial x)}{1 + \partial \xi/\partial x}. \tag{V-1}$$

It will be assumed that $\partial \xi/\partial x \ll 1$; hence η can be expressed to first order as

$$\eta = -h \frac{\partial \xi}{\partial x}.$$

This equation is isomorphic to

$$p = -\mathcal{B}_a \frac{\partial \xi}{\partial x}$$

provided we take $p \rightarrow \rho g \eta$ and $\mathcal{B}_a \rightarrow \rho g h$. Note that $\rho g \eta$ is the hydrostatic pressure difference associated with elevation η of the liquid.

Newton's second law. We ignore atmospheric pressure, since, regardless of the shape of the volume element, the net contribution to the horizontal force is zero. Effects due to surface tension are neglected. The total hydrostatic force exerted on the plane x is equal to the area $(h + \eta) \cdot 1$ multiplied by the mean pressure, i.e. the pressure at a depth $\frac{1}{2}(h + \eta)$ below the surface. (It is again necessary to neglect vertical accelerations in order to treat this as a simple case of hydrostatic pressure.) This pressure is $\frac{1}{2}\rho g(h + \eta)$. Thus the total force acting on the plane of particles originally at x is

$$F(x, t) = \tfrac{1}{2}\rho g(h + \eta)^2 \doteq \tfrac{1}{2}\rho g h^2 + \rho g h \eta.$$

The net force on the volume element is therefore

$$-\frac{\partial F}{\partial x} \Delta x = -\rho g h \frac{\partial \eta}{\partial x} \Delta x.$$

Setting this equal to $(\rho h \, \Delta x)\ddot{\xi}$, the mass times the acceleration, we obtain

$$-g \frac{\partial \eta}{\partial x} = \ddot{\xi}.$$

Note that this shows a simple relation between horizontal acceleration and the *slope* of the surface. This corresponds correctly to the acoustic equation

$$-\frac{\partial p}{\partial x} = \rho_0 \ddot{\xi},$$

with $p \rightarrow \rho g \eta$ as above and with the obvious correspondences $\rho_0 \rightarrow \rho$, $\xi \rightarrow \xi$.

Additional comments

a) The wave propagation velocity follows from $\sqrt{\mathcal{B}_a/\rho_0} \to \sqrt{\rho g h/\rho} = \sqrt{gh}$. It is notable that this does not depend on the nature of the liquid substance.

b) It is assured from the isomorphism that there will be an "energy-conservation law" in which the quantity

$$\frac{p^2}{2\mathcal{B}_a} \to \frac{\rho g}{2h}\eta^2$$

figures as a potential energy density. A direct calculation of the gravitational potential energy of the liquid in Δx is obtained by noting that the center of mass in the disturbed configuration is raised through height $\eta/2$. Thus $(\rho h\,\Delta x)g(\eta/2)$ is the potential energy change. Dividing by the original volume yields the density $\frac{1}{2}\rho g\eta$. This result appears to be in conflict with the above. The difficulty stems from the fact that the actual potential energy density is associated with *convective* and *radiative* terms similar to the acoustic case (cf. Section 4–4). Using Eq. (V–1) we can write

$$\tfrac{1}{2}\rho g\eta = -\tfrac{1}{2}\rho g h\frac{\partial\xi/\partial x}{1 + \partial\xi/\partial x} \doteq -\tfrac{1}{2}\rho g h\frac{\partial\xi}{\partial x} + \tfrac{1}{2}\rho g h\left(\frac{\partial\xi}{\partial x}\right)^2.$$

The first of these terms averages to zero over any interval for which $\Delta\xi = 0$. The second is the radiative part and is the same as that obtained through the isomorphism. Further insight into this situation is gained when it is realized that an *exact* isomorphism exists between the unlinearized tidal wave equations and the equations for a nonlinear sound wave (see Problem V–2).

c) It is questionable how the boundary-value problem should be handled across a step-discontinuity of the bottom surface (a change in the value of h). The assumption that vertical planes move together is certainly not good for particles close to the step. Problems which do translate satisfactorily are:
i) Reflection at a rigid boundary.
ii) Waves produced by the specified motion of a boundary surface (Section 3–3); the corresponding problem is that of waves generated in a canal by the motion of a vertical barrier.
iii) Transmission by a thin plate (Section 3–9); the corresponding problem is that of motion of a freely sliding vertical barrier separating two sections of a canal.

d) Note that $\partial\xi/\partial x \ll 1$ implies, through $\eta = -h(\partial\xi/\partial x)$, that $\eta \ll h$. The linearized equations thus refer to waves of small elevation in comparison with the depth.

e) Ocean tides can be thought of as waves whose periods T are the order of one day. The ratio h/λ can be written in the form $h/\lambda = h/cT = \sqrt{h/g}/T$, where h stands for the depth of the ocean. Now $\sqrt{h/g}$ is the order of the time of free fall over a distance h, and it is clear without further calculation that this is much less than one day. Thus $h/\lambda \ll 1$ and the type of motion considered here is applicable to the theory of tides.

Problem V-1. Consider a progressive sinusoidal tidal wave. Show that a particle at the surface moves in a flat elliptical path. In which direction are the particles moving at the crests and troughs?

Problem V-2. Consider plane tidal waves under the restrictive assumption that vertical planes of particles remain as vertical planes. For this situation it is possible to obtain nonlinear equations if the approximation $\partial\xi/\partial x \ll 1$ is not made. Show that each of the basic equations is isomorphic to the equations for a nonlinear sound wave in an ideal gas with $\gamma = 2$. Use the following correspondences:

$$\xi \rightarrow \xi, \qquad \rho \rightarrow \rho_0(1 + \eta/h), \qquad P \rightarrow F/h.$$

(Assume that vertical accelerations are negligible in order to calculate $F(x, t)$ by hydrostatic formulas.) Examine the quantities which appear in the law of conservation of energy for consistency of the isomorphism.

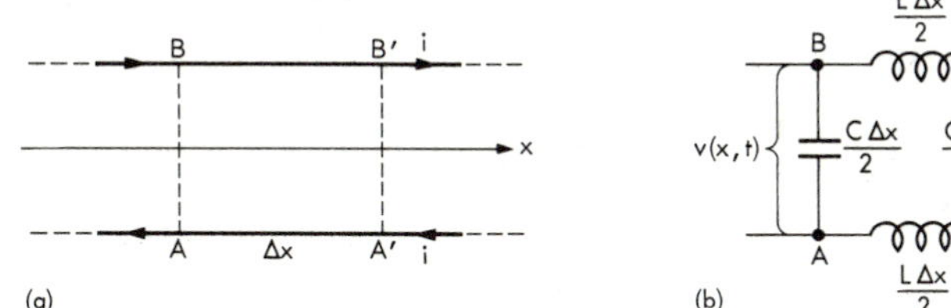

Figure V-2

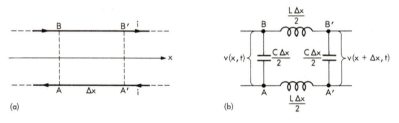

5) WAVES ON A TRANSMISSION LINE

Figure V-2a indicates schematically a portion of a coaxial cable or a two-wire transmission line with current i in one wire and an equal and opposite current in the return wire. Resistance in the wires is neglected and it is assumed that there is no leakage current between the two wires. Due to the finite propagation time of signals along the wire, it cannot be assumed that the instantaneous current is everywhere the same throughout a long transmission line. The notion of distributed circuit parameters is used to analyze the voltage-current relations within a short segment. The meaning of the variables is as follows:

- $i(x, t)$ instantaneous current at position x along the wire,
- $v(x, t)$ instantaneous potential difference between the two wires at position x,
- L inductance per unit length,
- C capacitance per unit length.

Derivation of the basic relations. The quantity $\Delta Q = [i(x, t) - i(x + \Delta x, t)]\,\Delta t$ represents the charge which accumulates in the segment Δx during time interval Δt. The capacitance is $C\,\Delta x$; hence the potential difference between the two wires changes by the amount $\Delta v = \Delta Q/(C\,\Delta x)$. Dividing by Δt and taking the limit as Δx and Δt approach zero, we obtain

$$\frac{\partial v}{\partial t} = -\frac{1}{C}\frac{\partial i}{\partial x}.$$

This is isomorphic to

$$\dot{p} = -\mathcal{B}_a \frac{\partial \xi}{\partial x}$$

if we take $p \to v$, $\mathcal{B}_a \to 1/C$, and $\xi \to i$. Considering the elementary circuit $ABB'A'$ in Fig. (V–2b) we find that the sum of the voltages around the loop is

$$V_{AB} + V_{BB'} + V_{B'A'} + V_{A'A} = -v(x, t) + \frac{L\,\Delta x}{2}\frac{\partial i}{\partial t} + v(x + \Delta x, t) + \frac{L\,\Delta x}{2}\frac{\partial i}{\partial t} = 0,$$

from which we obtain

$$-\frac{\partial v}{\partial x} = L\frac{\partial i}{\partial t}.$$

This is isomorphic to

$$-\frac{\partial p}{\partial x} = \rho_0 \frac{\partial \xi}{\partial t}$$

with the additional correspondence $\rho_0 \to L$.

Additional comments

a) The propagation velocity is obtained from $\sqrt{\mathcal{B}_a/\rho_0} \to 1/\sqrt{LC}$, and the impedance from $\rho_0 c \to \sqrt{L/C}$. The latter is measured in ohms.

b) The boundary conditions at the junction of two lines of different characteristics are $i_1 = i_2$ and $v_1 = v_2$ and are hence isomorphic to the usual acoustic boundary conditions. If the line terminates with a shunt connection between the two wires, the condition is $v = 0$, which corresponds to a free boundary. If the line terminates in an infinite resistance (open circuit), the condition is $i = 0$ which corresponds to reflection at a rigid boundary.

c) The inclusion of a lumped-parameter inductance $\mathcal{L}$ in the lines (Fig. V–3) requires boundary conditions $i_1 = i_2 = I$ and $v_1 = v_2 + \mathcal{L}\dot{I}$, where I is the current through the inductance. This problem is isomorphic to the transmission by a thin plate (Eqs. 3–33 and 3–34) with $\dot{B} \to I$ and $\mu \to \mathcal{L}$.

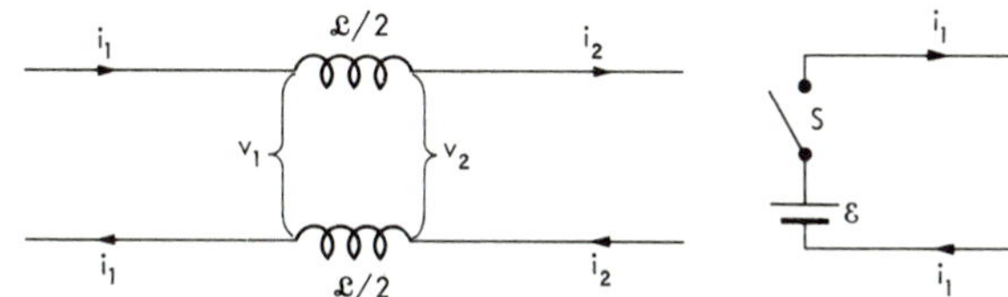

Figure V–3

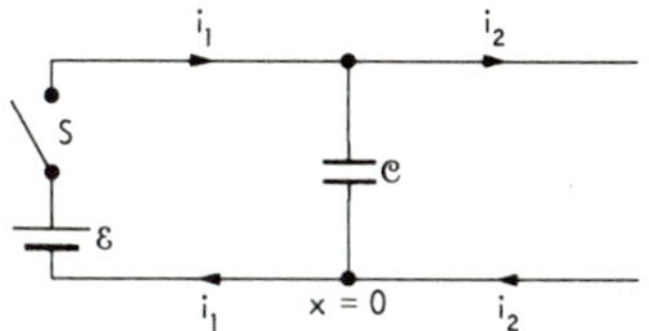

Figure V–4

d) Transmission line theory can be extended to include resistance in the wires and distributed shunt conductance between the wires. Since dissipation is involved, these situations are not isomorphic to ones considered in the text.*

Problem V–3. Consider a lumped capacitance $\mathcal{C}$ connected at $x = 0$ between the two wires of a transmission line (see Fig. V–4). Determine the boundary conditions. Show that this problem is isomorphic to the problem of Section 3–9 provided the correspondence $\xi \to v$, $p \to i$ is adopted. Use the results of Problem 3–18 to determine the potential difference across the capacitor as a function of the time resulting from closing the switch S. (Assume that the resulting incident wave arrives at $x = 0$ when $t = 0$.)

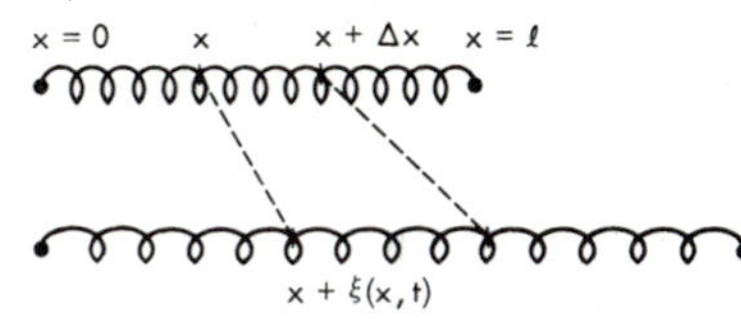

Figure V–5

6) LONGITUDINAL WAVES IN A HELICAL SPRING (See Fig. V–5.)
Consider a helical spring of rest length l. The Lagrangian method will be used to name points on the spring. The meaning of the variables is as follows:

- $F(x, t)$ force exerted by the portion to the left of x upon the portion to the right,
- $\xi(x, t)$ displacement of particle originally at x,
- σ mass per unit length of spring,
- k spring constant of the spring as a whole.

Derivation of the basic relations. We assume that $|F| = A|\partial\xi/\partial x|$. The constant A can be fixed by considering a static situation in which the spring is stretched so that $\xi(0) = 0$ and $\xi(l) = X$. Then the tension throughout the spring is $|F| = kX$. But for this case the displacements ξ are proportional

* A general treatment is given in R. K. Moore, *Wave and Diffusion Analogies*, New York: McGraw-Hill, 1964.

to x, and $\partial\xi/\partial x = X/l$. Hence $AX/l = kX$, or $A = kl$. We can then write

$$F = -kl\,\frac{\partial \xi}{\partial x}.$$

Taking $\xi \to \xi$, $\mathcal{B}_a \to kl$, and $p \to F$, we find that this equation is isomorphic to

$$p = -\mathcal{B}_a\,\frac{\partial \xi}{\partial x}.$$

From Newton's second law applied to the segment Δx we obtain

$$-\frac{\partial F}{\partial x} = \sigma\ddot{\xi},$$

which is isomorphic to

$$-\frac{\partial p}{\partial x} = \rho_0\ddot{\xi}$$

with the additional assignment $\rho_0 \to \sigma$.

Additional comments

a) It is to be noted that kl depends only on the type of spring and not on the actual length l. The wave velocity $\sqrt{kl/\sigma}$ is therefore a characteristic of the type of spring.

Problem V–4. Calculate the length of time required for a pulse to travel from one end of the spring to the other and back. Compare this with the period of oscillation of a mass equal to the total mass of the spring attached to a massless spring of the same constant k.

7) LONGITUDINAL PLANE ELASTIC WAVES IN A BAR (See Fig. V–6.)
The meaning of the variables is as follows:

- $\xi(x, t)$ longitudinal displacement from equilibrium of particles originally at x,
- $F(x, t)$ force exerted by the section to the left of x upon the section to the right,
- E Young's modulus for the material of the bar,
- ρ density of the material of the bar.

Derivation of the basic relations. The proportionality of stress and strain enables us to write

$$\frac{F}{A} = -E\,\frac{\partial \xi}{\partial x},$$

where A is the cross-sectional area of the bar. From Newton's second law,

$$-\frac{\partial F}{\partial x} = \rho A\ddot{\xi}.$$

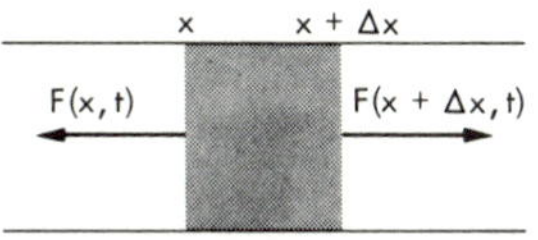

Figure V–6

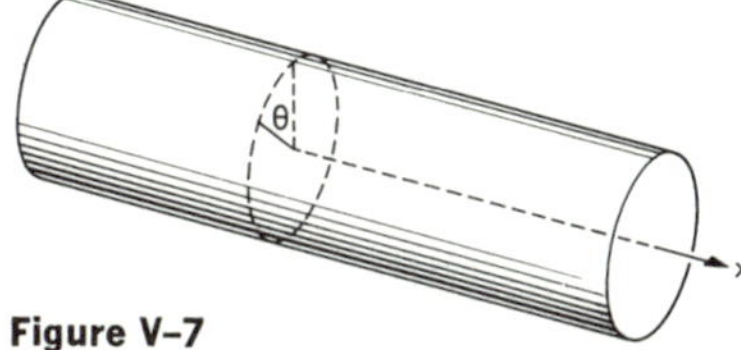

Figure V–7

These equations are easily seen to be isomorphic to the usual acoustic equations using the correspondences listed in the table. The wave velocity $\sqrt{E/\rho_0}$ is independent of the cross-sectional area A.

8) TORSION WAVES (See Fig. V–7.)
The figure illustrates a torsion fiber of circular cross section. The analysis will apply equally well to other shapes provided that buckling does not take place. The meaning of the variables is as follows:

$\theta(x, t)$	angle of twist measured from an equilibrium position,
$\tau(x, t)$	torque exerted by the section to the left of x upon the section to the right,
κ	torsion constant of a total length l of the fiber,
I	moment of inertia per unit length of fiber.

Derivation of the basic relations. By the same method used in (6) above for the helical spring, we obtain the formula

$$\tau = -\kappa l \frac{\partial \theta}{\partial x}.$$

The net torque on the element Δx is $(-\partial \tau/\partial x)\,\Delta x$, which is set equal to the moment of inertia, $I\,\Delta x$, times the angular acceleration:

$$-\frac{\partial \tau}{\partial x} = I\ddot{\theta}.$$

Additional comments

a) The wave velocity $\sqrt{\kappa l/I}$ is independent of the length l.

b) In common demonstration equipment, torsion waves are observed propagating on a wire or a flat steel band which is loaded with equally spaced bars. Since the bars do not have contact with one another, this is a means of increasing I without a corresponding increase in κ. If the separation between bars is small compared with the wavelength of a sinusoidal wave propagating along the system, the average moment of inertia per unit length can be used as if this were a continuous distribution.

c) For a hollow tube or a fiber of circular cross section, it is possible to relate c directly to fundamental parameters of the material. Consider a volume element defined by dx, dr, and $d\phi$ in the equilibrium state (Fig. V–8a). In the disturbed configuration this becomes sheared, as indicated in Fig. V–8b.

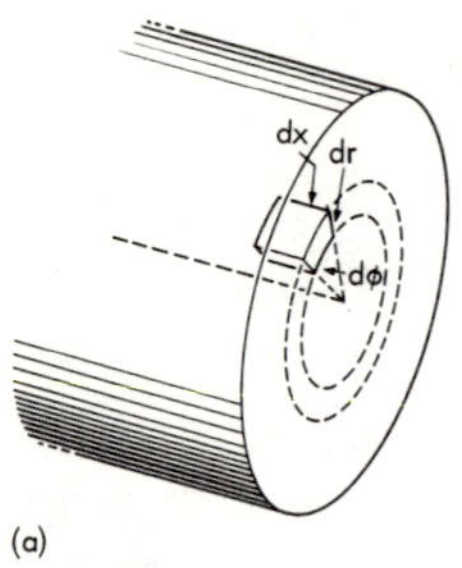

(a)

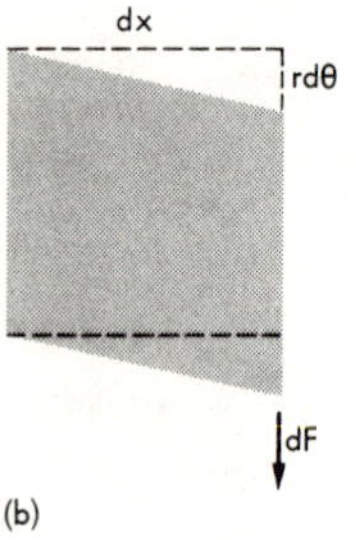

(b)

Figure V–8

The shearing strain $r(\partial\theta/\partial x)$ is associated with a stress

$$\frac{dF}{dA} = G\left(r\,\frac{\partial\theta}{\partial x}\right),$$

where G is the shear modulus. The area dA in question is $dA = r\,dr\,d\phi$, hence the associated torque is

$$d\tau = r\,dF = Gr^3\,\frac{\partial\theta}{\partial x}\,dr\,d\phi.$$

Integrating over the total area from r_1 to r_2, we have

$$\tau = 2\pi G\,\frac{\partial\theta}{\partial x}\int_{r_1}^{r_2} r^3 dr.$$

Hence we identify

$$\kappa l = 2\pi G \int_{r_1}^{r_2} r^3 dr.$$

Considering a unit length along the tube, the moment of inertia per unit length is

$$I = 2\pi\rho \int_{r_1}^{r_2} r^3 dr,$$

where ρ is the density of the material. Hence $c^2 = \kappa l/I = G/\rho$ is independent of the geometry, and is a characteristic of the medium only.

9) SELF-ISOMORPHISM FOR ACOUSTIC PLANE WAVES (See Section 4–11.) Reference to this line of the table provides an alternate method for establishing an isomorphism between any two of the preceding wave situations.

APPENDIX VI. Unabbreviated form of Maxwell's equations

In a cartesian coordinate system with unit vectors **i**, **j**, **k** along the x-, y- and z-axes, Eqs. (6–4) through (6–7) can be considered to be abbreviations for the following equations written out in explicit form:

$$\frac{\partial E_x}{\partial x} + \frac{\partial E_y}{\partial y} + \frac{\partial E_z}{\partial z} = 0, \tag{6–4}$$

$$\frac{\partial H_x}{\partial x} + \frac{\partial H_y}{\partial y} + \frac{\partial H_z}{\partial z} = 0, \tag{6–5}$$

$$\left(\frac{\partial E_z}{\partial y} - \frac{\partial E_y}{\partial z}\right)\mathbf{i} + \left(\frac{\partial E_x}{\partial z} - \frac{\partial E_z}{\partial x}\right)\mathbf{j} + \left(\frac{\partial E_y}{\partial x} - \frac{\partial E_x}{\partial y}\right)\mathbf{k} = -\mu\frac{\partial H_x}{\partial t}\mathbf{i} - \mu\frac{\partial H_y}{\partial t}\mathbf{j} - \mu\frac{\partial H_z}{\partial t}\mathbf{k}, \tag{6–6}$$

$$\left(\frac{\partial H_z}{\partial y} - \frac{\partial H_y}{\partial z}\right)\mathbf{i} + \left(\frac{\partial H_x}{\partial z} - \frac{\partial H_z}{\partial x}\right)\mathbf{j} + \left(\frac{\partial H_y}{\partial x} - \frac{\partial H_x}{\partial y}\right)\mathbf{k} = \epsilon\frac{\partial E_x}{\partial t}\mathbf{i} + \epsilon\frac{\partial E_y}{\partial t}\mathbf{j} + \epsilon\frac{\partial E_z}{\partial t}\mathbf{k}. \tag{6–7}$$

APPENDIX VII. Identities from vector analysis

$$\mathbf{A}\cdot(\mathbf{B}\times\mathbf{C}) = (\mathbf{A}\times\mathbf{B})\cdot\mathbf{C} = \mathbf{B}\cdot(\mathbf{C}\times\mathbf{A}) = (\mathbf{B}\times\mathbf{C})\cdot\mathbf{A} = \mathbf{C}\cdot(\mathbf{A}\times\mathbf{B}) = (\mathbf{C}\times\mathbf{A})\cdot\mathbf{B} \tag{VII–1}$$

$$\mathbf{A}\times(\mathbf{B}\times\mathbf{C}) = (\mathbf{A}\cdot\mathbf{C})\mathbf{B} - (\mathbf{A}\cdot\mathbf{B})\mathbf{C} \tag{VII–2}$$

$$\text{div}\,(\psi\mathbf{A}) = \mathbf{A}\cdot\text{grad}\,\psi + \psi\,\text{div}\,\mathbf{A} \tag{VII–3}$$

$$\text{curl}\,(\psi\mathbf{A}) = (\text{grad}\,\psi)\times\mathbf{A} + \psi\,\text{curl}\,\mathbf{A} \tag{VII–4}$$

$$\begin{aligned}(\mathbf{A}\cdot\nabla)\mathbf{B} = {} & \left(A_x\frac{\partial B_x}{\partial x} + A_y\frac{\partial B_x}{\partial y} + A_z\frac{\partial B_x}{\partial z}\right)\mathbf{i} \\ & + \left(A_x\frac{\partial B_y}{\partial x} + A_y\frac{\partial B_y}{\partial y} + A_z\frac{\partial B_y}{\partial z}\right)\mathbf{j} \\ & + \left(A_x\frac{\partial B_z}{\partial x} + A_y\frac{\partial B_z}{\partial y} + A_z\frac{\partial B_z}{\partial z}\right)\mathbf{k}\end{aligned} \tag{VII–5}$$

$$\text{grad}\,(\mathbf{A}\cdot\mathbf{B}) = (\mathbf{A}\cdot\nabla)\mathbf{B} + (\mathbf{B}\cdot\nabla)\mathbf{A} + \mathbf{A}\times\text{curl}\,\mathbf{B} + \mathbf{B}\times\text{curl}\,\mathbf{A} \tag{VII–6}$$

$$\text{div}\,(\mathbf{A}\times\mathbf{B}) = \mathbf{B}\cdot(\text{curl}\,\mathbf{A}) - \mathbf{A}\cdot(\text{curl}\,\mathbf{B}) \tag{VII–7}$$

$$\text{curl}\,(\mathbf{A}\times\mathbf{B}) = \mathbf{A}\,\text{div}\,\mathbf{B} - \mathbf{B}\,\text{div}\,\mathbf{A} + (\mathbf{B}\cdot\nabla)\mathbf{A} - (\mathbf{A}\cdot\nabla)\mathbf{B} \tag{VII–8}$$

$$\nabla^2\mathbf{A} \equiv \text{grad div}\,\mathbf{A} - \text{curl curl}\,\mathbf{A} \tag{VII–9}$$

$$\text{curl grad}\,\psi = 0 \tag{VII–10}$$

$$\text{div curl}\,\mathbf{A} = 0 \tag{VII–11}$$

The divergence theorem and closely related theorems. Let S be a closed surface containing a volume V; dV is the differential element of volume and $d\mathbf{S}$ a differential area element directed along the outward normal to S. Let $\mathbf{A}$ and ψ be vector and scalar functions of position which are appropriately differentiable throughout volume V. $\oiint_S$ refers to integration over the entire closed surface S, and $\iiint_V$ refers to integration over the enclosed volume V.

The first of the following theorems is referred to as the divergence theorem,

$$\iiint_V \operatorname{div} \mathbf{A}\, dV = \oiint_S \mathbf{A} \cdot d\mathbf{S}, \tag{VII–12}$$

$$\iiint_V \operatorname{curl} \mathbf{A}\, dV = \oiint_S d\mathbf{S} \times \mathbf{A}, \tag{VII–13}$$

$$\iiint_V \operatorname{grad} \psi\, dV = \oiint_S \psi\, d\mathbf{S}. \tag{VII–14}$$

Stokes' theorem and closely related theorems. Let L be a closed curve and S_L any surface having L as perimeter; $d\mathbf{r}$ is a differential element of path length directed along the tangent to L; $d\mathbf{S}$ is a differential element of area directed along the positive normal to S_L. (The positive side of an open surface is defined in relation to the sense in which the perimeter is described by means of a right-hand rule: if the thumb is pointed in the direction of $d\mathbf{r}$ and the fingers lie in the surface, the palm opens outward in the direction of the positive normal.) Let $\mathbf{A}$ and ψ be vector and scalar functions which are appropriately differentiable on L and S_L; $\oint_L$ refers to integration around the closed curve L and $\iint_{S_L}$ to integration over the open surface S_L. The first of the following theorems is known as Stokes' theorem,

$$\oint_L \mathbf{A} \cdot d\mathbf{r} = \iint_{S_L} (\operatorname{curl} \mathbf{A}) \cdot d\mathbf{S}, \tag{VII–15}$$

$$\oint_L \psi\, d\mathbf{r} = \iint_{S_L} d\mathbf{S} \times (\operatorname{grad} \psi). \tag{VII–16}$$

APPENDIX VIII. Introduction to Bessel functions; diffraction by a circular aperture; normal modes of a circular membrane

Definition of $J_0(u)$. If u is a fixed parameter, the function $f(t) = \cos(u \sin t)$ is periodic with period π:

$$f(t + \pi) = \cos[u \sin(t + \pi)] = \cos[-u \sin t] = \cos(u \sin t) = f(t).$$

The graph of $f(t)$ is plotted in Fig. VIII–1 for two values of u. Although the function $f(t)$ itself is simple, the integral which represents the area* under one complete cycle on the graph, $\int_0^\pi \cos(u \sin t)\, dt$, does not reduce by any changes of variable to a form that can be worked out in terms of the so-called elementary functions, that is, rational combinations of polynomials, sine, cosine, logarithmic or exponential functions—in short, any of the functions with which we are familiar. We therefore use this integral, which will be a

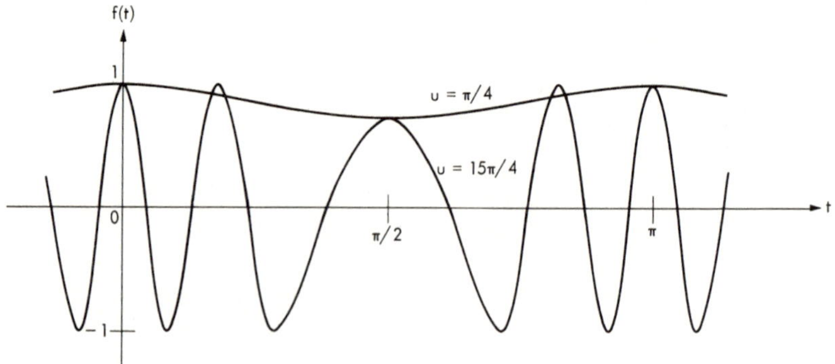

Fig. VIII–1. Graphs of $f(t) = \cos(u \sin t)$ for $u = \pi/4$ and $u = 15\pi/4$.

---

* "Area" is considered negative when below the axis.

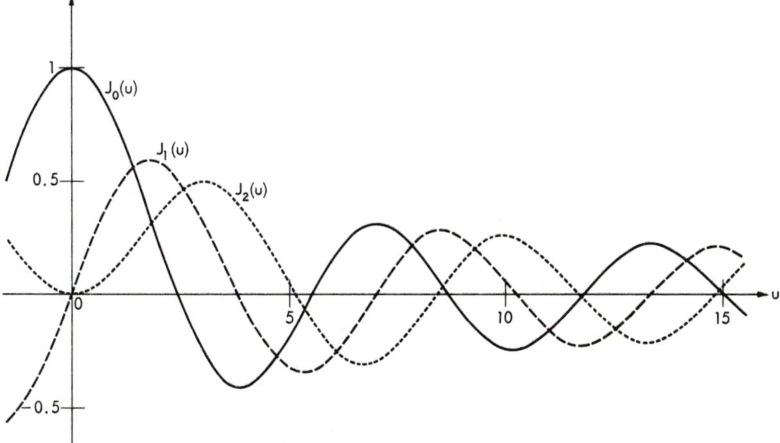

Fig. VIII-2. Graphs of the Bessel functions of order $n = 0, 1, 2$.

function of the parameter u, to define a new function:

$$J_0(u) \equiv \frac{1}{\pi}\int_0^{\pi} \cos\,(u \sin t)\, dt. \tag{VIII-1}$$

The function is referred to as *Bessel's function of zero order.* The factor of $1/\pi$ is included as part of the standard definition; $J_0(u)$ therefore represents the average value of $\cos\,(u \sin t)$ over the interval $(0, \pi)$. Values of the function can be computed from the convergent power series obtained by expanding

$$\cos\,(u \sin t) = \sum_{m=0}^{\infty} \frac{(u \sin t)^{2m}(-1)^m}{(2m)!},$$

and using*

$$\frac{1}{\pi}\int_0^{\pi} \sin^{2m} t\, dt = \frac{(2m)!}{2^{2m}(m!)^2},$$

from which

$$J_0(u) = \sum_{m=0}^{\infty} \frac{(-1)^m}{(m!)^2}\left(\frac{u}{2}\right)^{2m}. \tag{VIII-2}$$

Properties of $J_0(u)$; definition of $J_1(u)$. Most mathematical tables contain a tabulation of $J_0(u)$. [See Fig. VIII-2 for the graph of $J_0(u)$.] The function therefore joins the ranks of the "known" functions, since we can readily look up the value of $J_0(u)$ corresponding to a given value of u. In this sense

---

* *Standard Mathematical Tables* 11th ed., Cleveland: Chemical Rubber Publishing Co., 1957, #405 on p. 303.

$J_0(u)$ is no more recondite a function than $\cos u$, for example, for we must consult tables (computed from a convergent power series) to find the value of $\cos u$ for an arbitrary value of u. However, our feeling of familiarity with $\cos u$ is enhanced by the fact that we know the results of special operations such as $(d/du)(\cos u)$ or $\cos(u + v)$. We know differential equations that are satisfied by $\cos u$. We also know where $\cos u$ equals zero and can find $\cos u$ without using tables, for a number of special values of u. Familiarity with $J_0(u)$ will be gained by a similar study of its special properties. We shall develop only a few which are relevant to the applications which follow.

The derivative of $J_0(u)$ is a new function which cannot be expressed in terms of elementary functions and $J_0(u)$ itself. In analogy with $(d/du)(\cos u) = -\sin u$, we define

$$\frac{d}{du} J_0(u) = J_0'(u) \equiv -J_1(u). \tag{VIII–3}$$

The power series for $J_1(u)$ can be obtained by differentiating Eq. (VIII–2). $J_1(u)$ is referred to as *Bessel's function of first order*, and its values are well tabulated. By differentiation of Eq. (VIII–1) we obtain several integral expressions for $J_1(u)$:

$$J_1(u) = \frac{1}{\pi}\int_0^{\pi} \sin t \sin(u \sin t)\, dt = \frac{u}{\pi}\int_0^{\pi} \cos^2 t \cos(u \sin t)\, dt. \tag{VIII–4}$$

The second form results from integration by parts, the integrated term being zero. Differentiating once again we find

$$\begin{aligned} J_1'(u) &= \frac{1}{\pi}\int_0^{\pi} \sin^2 t \cos(u \sin t)\, dt \\ &= \frac{1}{\pi}\int_0^{\pi} \cos(u \sin t)\, dt - \frac{1}{\pi}\int_0^{\pi} \cos^2 t \cos(u \sin t)\, dt \\ &= J_0(u) - \frac{J_1(u)}{u}. \end{aligned} \tag{VIII–5}$$

Thus the derivative of $J_1(u)$ does not necessitate the definition and tabulation of a new function. The analogy $J_0(u) \leftrightarrow \cos u$ and $J_1(u) \leftrightarrow \sin u$ breaks down in this formula, as of course it must, for any pair of functions f, g such that $g' = -f$ and $f' = g$ are both sinusoidal functions. Setting $u = 0$ in the integral expressions we can see that

$$\begin{aligned} J_0(0) &= 1, \qquad & J_0'(0) &= 0, \\ J_1(0) &= 0, \qquad & J_1'(0) &= \tfrac{1}{2}. \end{aligned}$$

We also observe from Fig. VIII–1 that when u is large, the positive and negative areas will tend to cancel each other and the value of $J_0(u)$ will be small.

More formally, it can be shown that both $J_0(u)$ and $J_1(u)$ tend to zero as u approaches infinity.

Differential equations satisfied by $J_0(u)$ and $J_1(u)$. Substituting $J_1(u) = -J_0'(u)$ in Eq. (VIII–5) we obtain

$$J_0''(u) + (1/u)J_0'(u) + J_0(u) = 0. \qquad \text{(VIII–6)}$$

This is a second-order linear equation with variable coefficients. The equation

$$(d^2y/du^2) + (1/u)(dy/du) + y = 0, \qquad \text{(VIII–7)}$$

known as Bessel's differential equation of zero order, has $y = J_0(u)$ as a particular solution satisfying the initial conditions $J_0(0) = 1$, $J_0'(0) = 0$. The general solution to Eq. (VIII–7) will contain two linearly independent solutions. Thus if $Y_0(u)$ is a solution to Eq. (VIII–7), and $Y_0(u)$ and $J_0(u)$ are linearly independent, the general solution is

$$y(u) = AJ_0(u) + BY_0(u).$$

However, it can be shown that any solution that is linearly independent of $J_0(u)$ must have a singularity at $u = 0$. The applications of Eq. (VIII–7) made below require as a boundary condition that $y(0)$ be defined. This boundary condition therefore always determines $B = 0$; we will not go into a description of a possible choice for $Y_0(u)$.

A second-order linear equation satisfied by $J_1(u)$ is obtained by differentiating Eq. (VIII–5) and setting $J_0'(u) = -J_1(u)$. The general equation which results is Bessel's differential equation of *first* order:

$$(d^2y/du^2) + (d/du)(y/u) + y = 0. \qquad \text{(VIII–8)}$$

As before, solutions to this equation which are linearly independent of $J_1(u)$ are not defined at $u = 0$, and play no role in the applications below.

Higher-order Bessel functions. Since $f(t) = \cos(u \sin t)$ is a periodic function, it can be expanded in a Fourier series. The coefficients A_n and B_n follow from Eqs. (15–114), (15–115), and (15–116) with $l = \pi$ and $k_n = 2\pi n/l = 2n$. It is easily seen that A_0 is just $J_0(u)$. Furthermore, $B_n = 0$, since the integrand is an odd function. For $n \geq 1$ the coefficient A_n is given by $A_n = 2J_{2n}(u)$ where the *Bessel function of order* $2n$ is defined by

$$J_{2n}(u) = \frac{1}{\pi}\int_0^{\pi} (\cos 2nt)\cos(u \sin t)\, dt. \qquad \text{(VIII–9)}$$

Thus the Fourier expansion of $\cos(u \sin t)$ is

$$\cos(u \sin t) = J_0(u) + 2\sum_{n=1}^{\infty} J_{2n}(u)\cos 2nt. \qquad \text{(VIII–10)}$$

In like manner the function sin (u sin t) can be expanded in the form

$$\sin(u \sin t) = 2\sum_{n=0}^{\infty} J_{2n+1}(u)\sin(2n+1)t, \qquad \text{(VIII–11)}$$

where

$$J_{2n+1}(u) = \frac{1}{\pi}\int_0^{\pi} \sin(2n+1)t \sin(u \sin t)\,dt. \qquad \text{(VIII–12)}$$

[Note that this reduces correctly when $n = 0$ to $J_1(u)$ as defined by Eq. (VIII–4).]

Equation (VIII–9) defines the Bessel functions of even order and Eq. (VIII–12) defines the Bessel functions of odd order. A single formula which comprises both definitions is

$$\begin{aligned} J_n(u) &= \frac{1}{\pi}\int_0^{\pi} \cos(nt - u\sin t)\,dt \\ &= \frac{1}{\pi}\int_0^{\pi} \cos nt \cos(u \sin t)\,dt + \frac{1}{\pi}\int_0^{\pi} \sin nt \sin(u\sin t)\,dt, \end{aligned} \qquad \text{(VIII–13)}$$

for, if n is even, the second integral is antisymmetric about $t = \pi/2$ and therefore vanishes. Similarly, the first integral vanishes when n is odd. Using trigonometric identities and integration by parts it can be shown that

$$J_{n+1}(u) = (2n/u)J_n(u) - J_{n-1}(u), \qquad \text{(VIII–14)}$$

which shows how the higher-order Bessel functions can be computed from the tabulated values of $J_0(u)$ and $J_1(u)$. By similar techniques it can be shown that the derivative of the nth order Bessel function can be obtained from

$$J_n'(u) = \tfrac{1}{2}\{J_{n-1}(u) - J_{n+1}(u)\}. \qquad \text{(VIII–15)}$$

Using Eq. (VIII–14), Eq. (VIII–15) can be rewritten:

$$J_n'(u) = J_{n-1}(u) - (n/u)J_n(u) = (n/u)J_n(u) - J_{n+1}(u). \qquad \text{(VIII–16)}$$

[Compare this with Eqs. (VIII–3) and (VIII–5).] From Eq. (VIII–13) it can be seen that $J_n(0) = 0$ for $n \neq 0$ and from Eq. (VIII–15) that $J_n'(0) = 0$ except for $n = \pm 1$.

Differential equation for the *n*th-order Bessel function. From the examples of J_0 and J_1 we suspect that J_n will satisfy a second-order differential equation. Therefore let us compute an expression for J_n'' and attempt to reduce this to terms involving only J_n and J_n'. Differentiating Eq. (VIII–15), we have

$$J_n'' = \tfrac{1}{2}(J_{n-1}' - J_{n+1}').$$

But, from Eq. (VIII–16),

$$J'_{n-1} = \frac{(n-1)}{u} J_{n-1} - J_n \qquad \text{and} \qquad J'_{n+1} = J_n - \frac{(n+1)}{u} J_{n+1}.$$

Subtracting and collecting terms we obtain

$$J''_n = \tfrac{1}{2}\{(n/u)(J_{n-1} + J_{n+1}) - (1/u)(J_{n-1} - J_{n+1}) - 2J_n\}. \qquad \text{(VIII–17)}$$

Now using Eq. (VIII–14), we get $J_{n-1} + J_{n+1} = (2n/u)J_n$, and from Eq. (VIII–15) we have $J_{n-1} - J_{n+1} = 2J'_n$. Therefore

$$J''_n = (n^2/u^2)J_n - (1/u)J'_n - J_n \qquad \text{(VIII–18)}$$

or, $J_n(u)$ is a particular solution to the equation

$$d^2y/du^2 + (1/u)(dy/du) + (1 - n^2/u^2)y = 0. \qquad \text{(VIII–19)}$$

We assert without proof that any solution to this equation (known as Bessel's equation of nth order) which is nonsingular at $u = 0$ is of the form $y(u) = AJ_n(u)$ where A is any constant.

Fraunhofer diffraction by a circular aperture. The integral to be evaluated is of the form

$$\Phi(y', z') = \iint e^{(ik/R)(yy'+zz')}\, dy\, dz,$$

where the limits of integration are determined by the boundaries of a circular aperture (cf. the rectangular aperture, Eq. 12–19). It is clear that the diffraction pattern will have circular symmetry about the x-axis so that for simplicity we can set $z' = 0$, thus determining the pattern at any distance y' from the axis:

$$\Phi(y') = \iint e^{i\kappa y}\, dy\, dz, \qquad \text{(VIII–20)}$$

where $\kappa = ky'/R$. Let the radius of the aperture be a. Integrating first with respect to y, we find that the limits are $y = -\sqrt{a^2 - z^2}$ and $y = +\sqrt{a^2 - z^2}$. The limits on z are then $-a$ and $+a$. Then

$$\Phi = \int_{-a}^{a} \int_{-\sqrt{a^2-z^2}}^{+\sqrt{a^2-z^2}} e^{i\kappa y}\, dy\, dz = \frac{1}{i\kappa} \int_{-a}^{+a} \left\{e^{i\kappa\sqrt{a^2-z^2}} - e^{-i\kappa\sqrt{a^2-z^2}}\right\} dz$$

$$= \frac{2}{\kappa} \int_{-a}^{a} \sin\,(\kappa\sqrt{a^2 - z^2})\, dz.$$

Let $z = a(\cos t)$, and $dz = -a(\sin t)\, dt$; then

$$\Phi = \frac{4a}{\kappa} \int_0^{\pi} \sin t \sin\,(\kappa a \sin t)\, dt = \frac{4\pi a}{\kappa} J_1(\kappa a). \qquad \text{(VIII–21)}$$

Since the intensity distribution on a screen at distance R is proportional to $|\Phi|^2$, it can therefore be written

$$I = 4I_0(J_1(u)/u)^2, \qquad \text{(VIII–22)}$$

where $u = \kappa a = kay'/R = ka\,(\sin\theta)$. The constant of proportionality has been chosen so that I_0 represents the intensity on the axis $\theta = 0$ [cf. Eq. (VIII–4) with $u = 0$]. In general character the graph of I vs. θ is similar to that obtained for a rectangular aperture (Fig. 12–4). However, since the oscillations of $J_1(u)$ decrease in amplitude, whereas $\sin\beta$ does not, the "envelope" in the present case falls off more rapidly than $1/u^2$. Also the zeros of $J_1(u)$ are not equally spaced. The first few zeros of $J_1(u)$ are listed in tables and occur at $u_1 = 3.832$, $u_2 = 7.016$, and $u_3 = 10.173$. The angular radius of the first dark band is therefore given by

$$ka(\sin\theta_1) = u_1 \qquad \text{or} \qquad \sin\theta_1 = 1.22(\lambda/2a). \qquad \text{(VIII–23)}$$

By far the major portion of the energy in the pattern passes through the disk $\theta \leq \theta_1$ (known as Airy's disk). We can calculate the energy flux through a ring of radius y' as follows:

$$\begin{aligned}\mathcal{E}(y') &= \int_0^{y'} I(y')\cdot 2\pi y'\,dy' = 2\pi\left(\frac{R}{ka}\right)^2\int_0^u I(u)u\,du \\ &= 8\pi I_0\left(\frac{R}{ka}\right)^2\int_0^u \frac{J_1^2(u)}{u}\,du. \qquad \text{(VIII–24)}\end{aligned}$$

But from Eq. (VIII–5), we have

$$\frac{J_1^2(u)}{u} = J_0J_1 - J_1J_1' = -J_0J_0' - J_1J_1' = -\frac{d}{du}\left(\frac{J_0^2 + J_1^2}{2}\right).$$

Therefore

$$\int_0^u \frac{J_1^2(u)}{u}\,du = -\tfrac{1}{2}(J_0^2 + J_1^2)]_0^u = \tfrac{1}{2}[1 - J_0^2(u) - J_1^2(u)]. \qquad \text{(VIII–25)}$$

Since both J_0 and J_1 approach zero as u approaches ∞, we can write $\mathcal{E}(\infty) = 4\pi I_0(R/ka)^2$, and

$$\mathcal{E}(u)/\mathcal{E}(\infty) = 1 - J_0^2(u) - J_1^2(u). \qquad \text{(VIII–26)}$$

The relative energy flux through Airy's disk is therefore $\mathcal{E}(u_1)/\mathcal{E}(\infty) = 1 - J_0^2(u_1) = 0.838$. Thus 83.8% of the total energy passes through the central disk. The relative energy flux through the next bright ring is $J_0^2(u_1) - J_0^2(u_2)$ which from the tables is 0.072. Thus the energy flux outside the second black ring accounts for only 10% of the total.

Problem VIII-1. Integrate Eq. (VIII–20) first with respect to z and then with respect to y and show that the result is equivalent to Eq. (VIII–21).

Problem VIII-2. Show that the maxima on the graph of $J_1(u)/u$ occur at the zeros of $J_2(u)$.

The normal modes of a circular membrane. Let the origin be chosen at the center of a membrane of radius a clamped at the outer boundary. Equation (15–78) is converted to polar coordinates through the transformation $x = r\cos\theta$, $y = r\sin\theta$. This is equivalent to expressing the Laplacian in a cylindrical coordinate system and ignoring the z-derivative. The result is

$$\frac{1}{r}\frac{\partial}{\partial r}\left(r\frac{\partial z}{\partial r}\right) + \frac{1}{r^2}\frac{\partial^2 z}{\partial\theta^2} = \frac{1}{c^2}\frac{\partial^2 z}{\partial t^2}. \tag{VIII–27}$$

Seeking an eigensolution, we substitute a function of product form, $z = R(r)\Theta(\theta)T(t)$:

$$\frac{1}{Rr}\frac{d}{dr}(rR') + \frac{1}{r^2}\frac{\Theta''}{\Theta} = \frac{1}{c^2}\frac{T''}{T}. \tag{VIII–28}$$

We see that the equation has been successfully separated, since the portions of the equation referring to r, θ and t respectively can be isolated from one another. In particular, we see that Θ''/Θ must be a constant. Calling this constant b, we obtain

$$\Theta(\theta) = Ae^{\sqrt{b}\theta} + Be^{-\sqrt{b}\theta}, \tag{VIII–29}$$

where A and B are constants. Now if z is to be uniquely determined at a given point on the membrane, there must be no distinction between $\Theta(\theta)$ and $\Theta(\theta + 2\pi)$, or

$$Ae^{\sqrt{b}(\theta+2\pi)} + Be^{-\sqrt{b}(\theta+2\pi)} = Ae^{\sqrt{b}\theta} + Be^{-\sqrt{b}\theta}.$$

This implies that

$$Ae^{\sqrt{b}\theta}(e^{2\pi\sqrt{b}} - 1) = Be^{-\sqrt{b}\theta}(1 - e^{-2\pi\sqrt{b}}) = Be^{-\sqrt{b}\theta}e^{-2\pi\sqrt{b}}(e^{2\pi\sqrt{b}} - 1).$$

Unless $e^{2\pi\sqrt{b}} - 1 = 0$, this equation implies $Ae^{2\sqrt{b}\theta} = Be^{-2\pi\sqrt{b}}$ which cannot hold for all θ except in the trivial case $A = B = 0$. Therefore the only nontrivial solution occurs if

$$e^{2\pi\sqrt{b}} = 1 \qquad \text{or} \qquad \sqrt{b} = in, \tag{VIII–30}$$

where $n = 0, 1, 2, \ldots$ If this is the case we may write Eq. (VIII–29) in the form

$$\Theta(\theta) = C\sin(n\theta + \gamma), \tag{VIII–31}$$

where C and γ are arbitrary constants. The separation constant b has thus been ascribed a set of eigenvalues $b = -n^2$, $n = 0, 1, 2, \ldots$ through the eigenvalue condition Eq. (VIII–30). Substituting $\Theta''/\Theta = b$ in Eq. (VIII–28), the left-hand side refers only to the variable r and allows us to introduce a second constant of separation which we shall call $-\alpha^2$:

$$\frac{1}{Rr}\frac{d}{dr}(rR') - \frac{n^2}{r^2} = -\alpha^2$$

or

$$R'' + \frac{1}{r}R' + \left(\alpha^2 - \frac{n^2}{r^2}\right)R = 0. \tag{VIII-32}$$

This bears close resemblance to the equation for the nth-order Bessel function, Eq. (VIII–19), except that α^2 appears in place of unity. This can be taken care of by a change in the *independent* variable: let $u = \alpha r$, $dR/du = (1/\alpha)(dR/dr)$, $d^2R/du^2 = (1/\alpha^2)(d^2R/dr^2)$. Thus

$$\frac{d^2R}{du^2} + \frac{1}{u}\frac{dR}{du} + \left(1 - \frac{n^2}{u^2}\right)R = 0. \tag{VIII-33}$$

By the context of the problem we cannot have a singularity in the function $R(r)$ at $r = 0$, so that the general solution to Eq. (VIII–33) for our purposes is

$$R(r) = DJ_n(\alpha r). \tag{VIII-34}$$

So far α has been an undetermined separation constant. We now impose the boundary condition at the clamped edge of the membrane: $R(a) = 0$. This requires that

$$J_n(\alpha a) = 0. \tag{VIII-35}$$

This will not occur for arbitrarily chosen α, but restricts α to a set of eigenvalues. If we list the zeros of $J_n(u)$ in order of increasing values of u: $u_{n1}, u_{n2}, \ldots, u_{nm}, \ldots$, the possible eigenvalues of α are

$$\alpha_{nm} = u_{nm}/a, \qquad m = 1, 2, 3, \ldots \tag{VIII-36}$$

Returning to Eq. (VIII–28) once again and writing $\alpha_{nm}c = \omega_{nm}$, we see that the solutions for $T(t)$ are

$$T(t) = E\cos(\omega_{nm}t + \delta).$$

Thus the only solutions of product form are obtained by making an arbitrary choice of constants γ, δ, and $F = C \cdot D \cdot E$ and an arbitrary choice of two integers n and m. The corresponding eigenfunction is

$$z_{nm} = FJ_n(\alpha_{nm}r)\sin(n\theta + \gamma)\sin(\omega_{nm}t + \delta), \tag{VIII-37}$$

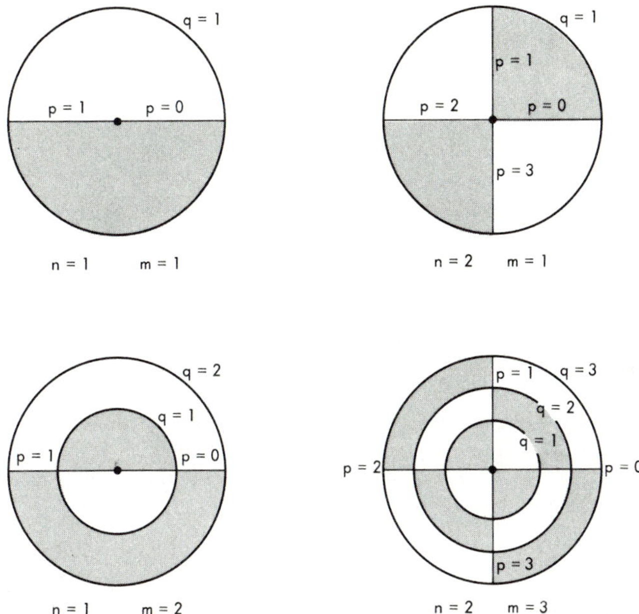

Fig. VIII–3. Sketch of a few normal modes of a vibrating circular membrane. During one half of the cycle the shaded areas are raised and the unshaded areas depressed. In the other half of the cycle the situation is reversed.

where α_{nm} is $(1/a)$ times the mth zero of the nth-order Bessel function. The eigenfrequencies $\omega_{nm} = \alpha_{nm}c$ are clearly not harmonic. The eigenfunction is characterized by two sets of nodal lines, for the value of z is identically zero at all times on the $2n$ radial lines $n\theta + \gamma = p\pi$ with $p = 0, 1, \ldots, (2n - 1)$ and also along the m circles where $J_n(\alpha_{nm}r) = 0$, that is, $\alpha_{nm}r = u_{nq}$ with $q = 1, 2, \ldots, m$. When $n = 1$, for example, these circular nodes have the same spacing as the black bands in the Fraunhofer diffraction pattern for a circular aperture. The nodal lines separate sectors where $z > 0$ from those where $z < 0$. Several normal modes are indicated in Fig. VIII–3 at a particular phase of the motion.

Problem VIII–3. Consult tables and determine the relative frequencies of the normal modes depicted in Fig. VIII–3.

Answers to selected problems

Chapter 1

1–1. a, c, d, e, f

1–3. a) $\sigma = 1.16 \times 10^{-3}$ kg/m   b) 1.61 m/sec

Chapter 2

2–2. $T = T_0 \left(\frac{\rho}{\rho_0}\right)^{\gamma-1}$ **2–4.** nearly two octaves   **2–5.** 421 sec^{-1}

2–9. $\xi_m = 7.74 \times 10^{-7}$ m, $\dot{\xi}_m = 4.85 \times 10^{-3}$ m/sec, $s_m = 1.43 \times 10^{-5}$,
$\Delta T = 1.67 \times 10^{-3}$ °C

2–10. p_m in water 3450 times that in air

2–11. $p(x, t) = -Z\dot{\xi}_m \cos kx \sin \omega t$

2–16. $d/\lambda = 1450$, $d \doteq 500$ m

2–20. $p(x, t) = (Z/2)[\dot{\xi}(x - ct, 0) - \dot{\xi}(x + ct, 0)]$
$+ \frac{1}{2}[p(x - ct, 0) + p(x + ct, 0)]$

Chapter 3

3–2. $$2y(x, t) = \begin{cases} \phi(x - ct) - \phi(-x - ct) - \chi(x - ct) + \chi(-x - ct), & x > -ct \\ \phi(x - ct) + \phi(x + ct) - \chi(x - ct) + \chi(x + ct), & x < -ct \end{cases}$$

where $\chi(x) \equiv \frac{1}{c}\int_{-\infty}^{x} \psi(x)\, dx$

3–6. one-third that of water   **3–7.** $\dot{\xi}, \ddot{\xi}, \partial p/\partial t$

3–11. $Z_2 = \sqrt{Z_1 Z_3}$ **3–16.** a) 2400 cps

Chapter 4

4–1. $W_{\text{pot}} = \dfrac{P_0}{\gamma - 1}\left[\left(\dfrac{\rho}{\rho_0}\right)^{\gamma-1} - 1\right] \doteq P_0\left[-\dfrac{\partial\xi}{\partial x} + \dfrac{1}{2}\gamma s^2\right]$

4–2. $\mathfrak{M}(t) = A\rho_0 \int_{x_1}^{x_2} \dot{\xi}(x, t)\, dx, \qquad \mathcal{I} = A\int_{t_1}^{t_2} [P(x_1, t) - P(x_2, t)]\, dt,$

and $\quad \mathfrak{M}(t_2) - \mathfrak{M}(t_1) = \mathcal{I}$

4–3. $T_i = \frac{3}{4}$, rarefaction, $Z_{12} = 3$

4–4. $p_{\text{rms}} = 2.06 \times 10^{-5}\ \text{n/m}^2$, $\xi_{\text{rms}} = 7.69 \times 10^{-12}$ m,
$\dot{\xi}_{\text{rms}} = 4.83 \times 10^{-8}$ m/sec, $s_{\text{rms}} = 1.46 \times 10^{-10}$

4–6. 3450 to 1, 4 to 1

4–9. 3×10^{-6} joules, 120 db

4–10. 155 db

4–11. -10 db

4–14. 9×10^{-4} joules

4–15. a) 1.62 watts b) 4.86 watts c) -4.86 watts
d) 6.85 watts e) -3.24 watts f) 0

4–16. 23 gm

Chapter 5

5–1. a) 270, 20 b) 8.1 sec, 0.6 sec

Chapter 6

6–4. a) polarized plane waves, **E** parallel to z-axis, propagating in x-direction
b) polarized plane waves, **E** parallel to x-axis, propagating in y-direction

6–5. $a = \dfrac{e^2}{8\pi\epsilon_0 mc^2}$

Chapter 7

7–4. 4% intensity loss at each interface

7–5. $n = 1.225$, $a = 1200$ Å, magenta

7–6. areal velocity $= \frac{1}{2}AB\omega \sin(\phi_2 - \phi_1)$

7–8. b) $\tan\theta = \dfrac{B + B'}{A + A'}$

7–9. a) $\dfrac{2A^2}{\mu c}, \quad \dfrac{2A^2}{\mu c}$

b) $\dfrac{1}{2\mu c}[A^2 + B^2 + A'^2 + B'^2 + 2(AA' + BB')],$

$\dfrac{1}{2\mu c}[A^2 + B^2 + A'^2 + B'^2]$

c) $\dfrac{1}{2\mu c}(A^2 + B^2), \quad \dfrac{1}{2\mu c}[A^2 - 2AB + 3B^2]$

7–10. $H_z = \frac{A}{\mu c} \cos \omega t \cos kx$, $S_x = \frac{A^2}{4\mu c} \sin 2\omega t \sin 2kx$, none except (3)

7–14. 9000 Å

7–16. $5410 \leq \lambda \leq 5590$

Chapter 8

8–1. $I \propto \sin^2 2\theta$

8–8. c) 15

8–12. a) 45°24′ b) 109°5′, 70°55′

Chapter 9

9–1. calcite 0.856 μ, quartz 0.162 mm, mica 0.263 mm

9–2. a) ϕ counterclockwise b) 2ϕ clockwise

9–3. uniform intensity

9–6. a) $I_o/I_e = 1.73$ b) 43°10′

9–9. 4000 Å, 4615 Å, 5447 Å, 6667 Å

9–10. $\Delta n = \frac{\lambda\lambda'}{d(\lambda' - \lambda)}$

9–11. a) 5.52 cm b) 2.87 Å d) $\lambda_0 \pm 190$ Å, approximately 20

9–15. $\alpha = 48°8'$

Chapter 10

10–1. extinguished colors λ = 4330 Å, 5000 Å, 5590 Å, 6125 Å, 6619 Å

10–3. 7.1×10^{-5}

10–4. $\chi \propto 1/\lambda^2$

10–5. b) 5×10^6 amp/m

10–7. four minima 27 cm apart

Chapter 11

11–1. b) $2\nu\left(\frac{v}{c}\right)$

11–11. a) 29 ft b) 270 ft

11–13. 73 cm, 18 m

11–15. a) 2 mm b) 80 m

11–16. 0.16 cm

11–18. a) $V_{\text{rms}} \propto \cos^2 (\pi\alpha d/\lambda)$
c) $d \sim 2$ km for nebula, $d \sim$ diameter of earth would be required for star

11–21. L 63.4 cm from S, A next to L on either side

11–24. 0, 500′, 1000′, etc.

11–25. b) 0.12 mm

11–26. b) 20 m

11–27. 1.5×10^{-2} mm

11–28. b) 1.15 cm

11–30. $\Delta\lambda \sim 2.7$ Å, $\tau \sim 2 \times 10^{-12}$ sec

Chapter 12

12–2. c) 91%

12–5. $I \propto \sin^2 \Omega t, \Omega = \dfrac{kv}{2} \sin \theta$, until such times that the Fraunhofer approximation is no longer satisfied.

12–7. $R \gg a^2/\lambda$ **12–8.** 4

12–9. $\nu \gtrsim 5.5 \times 10^6$ cps **12–11.** b) 1.1 mm

12–12. 2.2×10^{-6} radians, 0.01 mm, 120, 1.5 in

Chapter 13

13–2. a) $u^2 \sim 3.5$, $R \sim 114$ cm b) $R \sim 73$ cm

13–3. $R \sim 33\lambda$(max), 14λ (min), 9λ(max)

13–5. $I \propto a^2$ for $a < 4\lambda$, I uniform for $a > 70\lambda$

Chapter 14

14–3. a) $I(\theta) = 4 I_0 \left(\dfrac{\sin \beta}{\beta}\right)^2 \sin^2 \gamma$ c) $\beta = 1.166$

14–9. a) 982 b) N, the total number of lines
d) $n = 5$, $\theta = 2.32 \times 10^{-2}$ radians

14–11. $a = 5.5 \times 10^{-4}$ cm

Chapter 15

15–1. a) $T = 4l/c$ b) $\omega = (n + \frac{1}{2})(\pi c/l)$, $n = 0, 1, 2 \ldots$
c) $y_n(x, t) = C_n \sin k_n x \cos (\omega_n t + \phi)$

15–2. d) $\omega = \sqrt{\dfrac{T}{ml}}$ **15–3.** e) $\omega = 2\sqrt{\dfrac{T}{ml}}$

15–5. $y(x, t) = \dfrac{1}{2\omega} \sin 2\omega t \sin (2\pi x/l)$

15–8. $y(x, t) = \dfrac{32h}{\pi^3} \left\{\cos \omega t \sin kx + \dfrac{1}{27} \cos 3\omega t \sin 3kx + \cdots\right\}$

15–10. a) $a_n = 0$, $b_1 = l/2$, $b_n = 0$ if n odd and $n > 1$,

$$b_n = \frac{-8nl}{\pi^2(n^2 - 1)^2} \quad \text{if } n \text{ even}$$

b) $\mathcal{E}_1 : \mathcal{E}_2 : \mathcal{E}_4 = 1 : 0.506 : 0.013$

15–14. a) $\Delta t \sim 5 \times 10^{-9}$ sec, 10^6 cycles b) 2 Å

15–15. a) $y(x, t) = A \dfrac{\cos kx}{\cos kl/2} \cos \omega t$ b) $y(x, t) = A \dfrac{\sin kx}{\sin kl/2} \cos \omega t$

15–17. $y(x, t) = (-1)^m \dfrac{A}{l} \{x \sin \omega_m t \cos k_m x + ct \cos \omega_m t \sin k_m x\}$

15–19. $y_s(x, t) = \dfrac{2A}{k^2T} \cos \omega t \left\{\dfrac{\sin kx}{\sin kl} \sin^2 \dfrac{kl}{4} + \beta(x)\right\},$

where $\beta(x) = \begin{cases} 0, & x < l/2 \\ -\sin^2 \dfrac{k}{2}(x - l/2), & x > l/2 \end{cases}$

15–21. $y(x, t) = \displaystyle\sum_{n\,\text{odd}} \frac{4N^2}{n\pi(n^2 - N^2)} \sin k_n x \cos \omega_n t,$ where N is even

15–23. a) With $\omega_0 = \pi c/a$: $\omega_{11} = \sqrt{2}\,\omega_0$, $\omega_{12} = \omega_{21} = \sqrt{5}\,\omega_0$,
$\omega_{22} = \sqrt{8}\,\omega_0$, $\omega_{13} = \omega_{31} = \sqrt{10}\,\omega_0$, $\omega_{23} = \omega_{32} = \sqrt{13}\,\omega_0$

b) $\omega_{17} = \omega_{71} = \omega_{55} = \sqrt{50}\,\omega_0$ (triple),
$\omega_{67} = \omega_{76} = \omega_{29} = \omega_{92} = \sqrt{85}\,\omega_0$ (quadruple)

15–24. $y(x, t) = \sqrt{\dfrac{2}{\pi}} \displaystyle\int_0^\infty \sin kx \left\{\frac{v}{k^2c} \sqrt{\frac{2}{\pi}} [\cos ka - \cos kb]\right\} \sin kct\, dk$

15–29. $y(x, t) = \dfrac{2v}{\pi} \displaystyle\int_0^\infty \frac{\sin ka \cos kx}{k^2c} \sin kct\, dk$

Chapter 16

16–3. 23 cm/sec

16–4. $E(x) = 2\,\dfrac{\sin x\,\Delta k}{x}$

16–5. b) 10^{-3} cm

16–6. a) $\omega = \dfrac{\hbar k^2}{2m}$ b) $\omega = \dfrac{mc^2}{\hbar}\sqrt{1 + \dfrac{\hbar^2k^2}{m^2c^2}}$ c) $\omega = kc,\quad u = v = c$

Chapter 17

17–10. a) $v_m = b\omega,\quad p_m = \dfrac{\rho_0 cka_0}{\sqrt{1 + k^2a_0^2}}\, b\omega$

b) $\bar{\imath} = \dfrac{\rho_0 cka_0 b^2\omega}{2(1 + k^2a_0^2)} [\pi ka_0 + 1]$

17–15. c) $\theta_0 = 44°$ d) $\theta_0 = 0$

17–16. $T_i = Z_1/Z_2$

17–19. $\theta_c = 13°$ at 0°C (on air side)

17–22. For $\theta \le \theta_c$, $i_z = pv_z = \dfrac{4Z_1Z_2 \cos\theta \cos\theta'}{[Z_2 \cos\theta + Z_1 \cos\theta']^2} \cos^2 \omega t;$
For $\theta > \theta_c$, $i_z = \sin 2\chi \sin 2(\omega t + \chi)$, $\bar{\imath}_z = 0$

Appendix I

I–2. $\theta = 2\pi/3,\quad \theta' = \pi/3$

Index

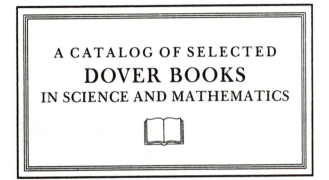

A CATALOG OF SELECTED

DOVER BOOKS

IN SCIENCE AND MATHEMATICS

A CATALOG OF SELECTED
DOVER BOOKS
IN SCIENCE AND MATHEMATICS

THE ELECTROMAGNETIC FIELD, Albert Shadowitz. Comprehensive undergraduate text covers basics of electric and magnetic fields, builds up to electromagnetic theory. Also related topics, including relativity. Over 900 problems. 768pp. 5⅝ x 8¼. 65660-8 Pa. $18.95

FOURIER SERIES, Georgi P. Tolstov. Translated by Richard A. Silverman. A valuable addition to the literature on the subject, moving clearly from subject to subject and theorem to theorem. 107 problems, answers. 336pp. 5⅜ x 8½. 63317-9 Pa. $9.95

THEORY OF ELECTROMAGNETIC WAVE PROPAGATION, Charles Herach Papas. Graduate-level study discusses the Maxwell field equations, radiation from wire antennas, the Doppler effect and more. xiii + 244pp. 5⅜ x 8½. 65678-0 Pa. $6.95

DISTRIBUTION THEORY AND TRANSFORM ANALYSIS: An Introduction to Generalized Functions, with Applications, A.H. Zemanian. Provides basics of distribution theory, describes generalized Fourier and Laplace transformations. Numerous problems. 384pp. 5⅜ x 8½. 65479-6 Pa. $11.95

THE PHYSICS OF WAVES, William C. Elmore and Mark A. Heald. Unique overview of classical wave theory. Acoustics, optics, electromagnetic radiation, more. Ideal as classroom text or for self-study. Problems. 477pp. 5⅜ x 8½. 64926-1 Pa. $13.95

CALCULUS OF VARIATIONS WITH APPLICATIONS, George M. Ewing. Applications-oriented introduction to variational theory develops insight and promotes understanding of specialized books, research papers. Suitable for advanced undergraduate/graduate students as primary, supplementary text. 352pp. 5⅜ x 8½. 64856-7 Pa. $9.95

A TREATISE ON ELECTRICITY AND MAGNETISM, James Clerk Maxwell. Important foundation work of modern physics. Brings to final form Maxwell's theory of electromagnetism and rigorously derives his general equations of field theory. 1,084pp. 5⅜ x 8½. 60636-8, 60637-6 Pa., Two-vol. set $25.90

AN INTRODUCTION TO THE CALCULUS OF VARIATIONS, Charles Fox. Graduate-level text covers variations of an integral, isoperimetrical problems, least action, special relativity, approximations, more. References. 279pp. 5⅜ x 8½. 65499-0 Pa. $8.95

HYDRODYNAMIC AND HYDROMAGNETIC STABILITY, S. Chandrasekhar. Lucid examination of the Rayleigh-Benard problem; clear coverage of the theory of instabilities causing convection. 704pp. 5⅜ x 8¼. 64071-X Pa. $14.95

CALCULUS OF VARIATIONS, Robert Weinstock. Basic introduction covering isoperimetric problems, theory of elasticity, quantum mechanics, electrostatics, etc. Exercises throughout. 326pp. 5⅜ x 8½. 63069-2 Pa. $9.95

DYNAMICS OF FLUIDS IN POROUS MEDIA, Jacob Bear. For advanced students of ground water hydrology, soil mechanics and physics, drainage and irrigation engineering and more. 335 illustrations. Exercises, with answers. 784pp. 6⅛ x 9¼. 65675-6 Pa. $19.95

NUMERICAL METHODS FOR SCIENTISTS AND ENGINEERS, Richard Hamming. Classic text stresses frequency approach in coverage of algorithms, polynomial approximation, Fourier approximation, exponential approximation, other topics. Revised and enlarged 2nd edition. 721pp. 5⅜ x 8½. 65241-6 Pa. $15.95

THEORETICAL SOLID STATE PHYSICS, Vol. 1: Perfect Lattices in Equilibrium; Vol. II: Non-Equilibrium and Disorder, William Jones and Norman H. March. Monumental reference work covers fundamental theory of equilibrium properties of perfect crystalline solids, non-equilibrium properties, defects and disordered systems. Appendices. Problems. Preface. Diagrams. Index. Bibliography. Total of 1,301pp. 5⅜ x 8½. Two volumes. Vol. I: 65015-4 Pa. $16.95
Vol. II: 65016-2 Pa. $16.95

OPTIMIZATION THEORY WITH APPLICATIONS, Donald A. Pierre. Broad spectrum approach to important topic. Classical theory of minima and maxima, calculus of variations, simplex technique and linear programming, more. Many problems, examples. 640pp. 5⅜ x 8½. 65205-X Pa. $16.95

THE CONTINUUM: A Critical Examination of the Foundation of Analysis, Hermann Weyl. Classic of 20th-century foundational research deals with the conceptual problem posed by the continuum. 156pp. 5⅜ x 8½. 67982-9 Pa. $6.95

ESSAYS ON THE THEORY OF NUMBERS, Richard Dedekind. Two classic essays by great German mathematician: on the theory of irrational numbers; and on transfinite numbers and properties of natural numbers. 115pp. 5⅜ x 8½.
21010-3 Pa. $5.95

THE FUNCTIONS OF MATHEMATICAL PHYSICS, Harry Hochstadt. Comprehensive treatment of orthogonal polynomials, hypergeometric functions, Hill's equation, much more. Bibliography. Index. 322pp. 5⅜ x 8½. 65214-9 Pa. $9.95

NUMBER THEORY AND ITS HISTORY, Oystein Ore. Unusually clear, accessible introduction covers counting, properties of numbers, prime numbers, much more. Bibliography. 380pp. 5⅜ x 8½. 65620-9 Pa. $10.95

THE VARIATIONAL PRINCIPLES OF MECHANICS, Cornelius Lanczos. Graduate level coverage of calculus of variations, equations of motion, relativistic mechanics, more. First inexpensive paperbound edition of classic treatise. Index. Bibliography. 418pp. 5⅜ x 8½. 65067-7 Pa. $12.95

MATHEMATICAL TABLES AND FORMULAS, Robert D. Carmichael and Edwin R. Smith. Logarithms, sines, tangents, trig functions, powers, roots, reciprocals, exponential and hyperbolic functions, formulas and theorems. 269pp. 5⅜ x 8½.
60111-0 Pa. $6.95

THEORETICAL PHYSICS, Georg Joos, with Ira M. Freeman. Classic overview covers essential math, mechanics, electromagnetic theory, thermodynamics, quantum mechanics, nuclear physics, other topics. First paperback edition. xxiii + 885pp. 5⅜ x 8½. 65227-0 Pa. $21.95

ORDINARY DIFFERENTIAL EQUATIONS, Morris Tenenbaum and Harry Pollard. Exhaustive survey of ordinary differential equations for undergraduates in mathematics, engineering, science. Thorough analysis of theorems. Diagrams. Bibliography. Index. 818pp. 5⅜ x 8½. 64940-7 Pa. $18.95

STATISTICAL MECHANICS: Principles and Applications, Terrell L. Hill. Standard text covers fundamentals of statistical mechanics, applications to fluctuation theory, imperfect gases, distribution functions, more. 448pp. 5⅜ x 8½. 65390-0 Pa. $11.95

ORDINARY DIFFERENTIAL EQUATIONS AND STABILITY THEORY: An Introduction, David A. Sánchez. Brief, modern treatment. Linear equation, stability theory for autonomous and nonautonomous systems, etc. 164pp. 5⅜ x 8¼. 63828-6 Pa. $6.95

THIRTY YEARS THAT SHOOK PHYSICS: The Story of Quantum Theory, George Gamow. Lucid, accessible introduction to influential theory of energy and matter. Careful explanations of Dirac's anti-particles, Bohr's model of the atom, much more. 12 plates. Numerous drawings. 240pp. 5⅜ x 8½. 24895-X Pa. $7.95

THEORY OF MATRICES, Sam Perlis. Outstanding text covering rank, nonsingularity and inverses in connection with the development of canonical matrices under the relation of equivalence, and without the intervention of determinants. Includes exercises. 237pp. 5⅜ x 8½. 66810-X Pa. $8.95

GREAT EXPERIMENTS IN PHYSICS: Firsthand Accounts from Galileo to Einstein, edited by Morris H. Shamos. 25 crucial discoveries: Newton's laws of motion, Chadwick's study of the neutron, Hertz on electromagnetic waves, more. Original accounts clearly annotated. 370pp. 5⅜ x 8½. 25346-5 Pa. $10.95

INTRODUCTION TO PARTIAL DIFFERENTIAL EQUATIONS WITH APPLICATIONS, E.C. Zachmanoglou and Dale W. Thoe. Essentials of partial differential equations applied to common problems in engineering and the physical sciences. Problems and answers. 416pp. 5⅜ x 8½. 65251-3 Pa. $11.95

BURNHAM'S CELESTIAL HANDBOOK, Robert Burnham, Jr. Thorough guide to the stars beyond our solar system. Exhaustive treatment. Alphabetical by constellation: Andromeda to Cetus in Vol. 1; Chamaeleon to Orion in Vol. 2; and Pavo to Vulpecula in Vol. 3. Hundreds of illustrations. Index in Vol. 3. 2,000pp. 6⅛ x 9¼. 23567-X, 23568-8, 23673-0 Pa., Three-vol. set $44.85

CHEMICAL MAGIC, Leonard A. Ford. Second Edition, Revised by E. Winston Grundmeier. Over 100 unusual stunts demonstrating cold fire, dust explosions, much more. Text explains scientific principles and stresses safety precautions. 128pp. 5⅜ x 8½. 67628-5 Pa. $5.95

AMATEUR ASTRONOMER'S HANDBOOK, J.B. Sidgwick. Timeless, comprehensive coverage of telescopes, mirrors, lenses, mountings, telescope drives, micrometers, spectroscopes, more. 189 illustrations. 576pp. 5⅜ x 8¼. (Available in U.S. only) 24034-7 Pa. $11.95